AF529863

Food Crops:
Evolution, Diversity and Advances

Related Books

- Agriculture and Food Technology in Human Life — *Prem Nath*
- Applied Microbiology (Agriculture, Environmental, Food and Industrial Microbiology) — *S.M. Reddy*
- Arid Fruit Research — *A. Chandra*
- Beneficial Fungi and their Utilization *2nd Ed* — *M.C. Nair*
- Beverages: Processing and Technology — *Deepak Mudgil*
- Biopesticides in Environment and Food Security: Issues and Strategies — *Opender Koul*
- Food Preservation and Safety: Principles and Practice — *S.J. Vangarde*
- Fruit and Vegetable Processing — *FAO*
- Functional Foods: Sources and Health Benefits — *Deepak Mudgil*
- Glossary of Dairy Technology — *Deepak Mudgil*
- Grah Vigyan Evam Prasar Shiksha (Home Science and Extension Education) — *Renu Arya*
- Herbaceous Plants as Natural Protective Food — *M.K. Rana*
- Horticulture and Livelihood Security — *Prem Nath*
- Human Development: Capability Poverty — *R.K. Jain*
- Leafy and Edible Plants of North-East India — *Sudhir Kumar*
- MCQs for Food Technology — *P.M. Kotecha*
- Objective Food Science & Technology *3rd Ed* — *Deepak Mudgil*
- Objective Home Science at a Glance — *Shanti Balda*
- Phenolic Antioxidants and Health Benefits — *U.D. Chavan*
- Physico-Chemical Constituents & Engineering-Properties of Food Crops — *R.P. Kachru*
- Quinoa *(Chenopodium Quinoa Willd)* — *Dharm Singh*
- Rural Growers Manual — *Don Zasada*
- Sustainable Horticulture Development and Nutrition Security *(Vol. 3)* — *Prem Nath*
- Tamarind Science and Technology — *A. Balasubramanian*
- The Basics of Human Civilization Food, Agricutlure and Humanity *(Vol. 3)*: Agricutlure — *Prem Nath*
- Vegetables and their Allied as Protective Food — *M.K. Rana*

Food Crops:
Evolution, Diversity and Advances

By

Dr. D.A. Patil
Former Professor and Principal,
S.S.V.P. Sanstha's L.K. Dr. P.R. Ghogrey
Science College, Dhule-42005 (M.S.), India.

Published by

SCIENTIFIC PUBLISHERS (INDIA)
5 A, New Pali Road, P.O. Box 91
Jodhpur 342 001 (India)
E-mail: info@scientificpub.com
Website: www.scientificpub.com

ISBN : 978-93-88449-37-3
eISBN : 978-93-88449-39-7

Printed in India

GRATEFULLY DEDICATED

TO

THE INDIAN FARMERS

FOR SHOULDERING SILENTLY

THE WHEEL OF NATIONAL PROGRESS,

PRESTIGE AND FUTURE.

About Author

Dr. D.A. Patil did his M.Sc. and Ph.D. in Botany from Dr. B.A. Marathwada University (Aurangabad, M.S.) with Angiosperms as a subject of specialization in 1978 and 1983 respectively. He has about 34 years of research and teaching experience at graduate and post-graduate classes. His major research interest includes Taxonomy of Angiosperms, Anatomy, Floristics, Ethnobotany, Medicinal Plants and Etymology. He has authored (i) Flora of Dhule and Nandurbar Districts (Maharashtra) (2003), (ii) Origins of Plant Names (2007), (iii) Useful Plants : Origin, History & Civilization (2008), (iv) Herbal Cures: Traditional Approach (2008), (v) Medicinal Plants : History, Culture and Usage (2010), (vi) A Key To The Angiospermic Families of North Maharashtra (1992), (vii) Flora of Jalgaon District (Maharashtra) (2008), (viii) Ethnobotany of Nasik District (Maharashtra) (2006), (ix) Ethnobotany of Jalgaon District (Maharashtra) (2008), (x) Ethnobotany of Buldhana District (Maharashtra) (2010), (xi) Spices And Condiments: Origin, History & Applications (2013), (xii) Herbal Materia Medica of Maharashtra (2013). (xiii) Ethnobotany of Badwani District (Madhya Pradesh) (2016) and (xiv) Maharashtratil Aushadhi Vanaspati Sangrah (In Marathi) (2016). He has also edited proceedings (i) Plant Diversity and Biotechnology (2005), (ii) Biodiversity, Sustainable Development and Human Welfare (2010), (iii) Biodiversity and Environmental Crisis: Past, Present & Future (2013). He has 335 research papers to his credit published in national, international journals and journals of CSIR, India. Apart from his participation in several national and international conferences and symposia in India and abroad, many students in botany received his guidance for Ph.D. and M.Phil. degrees. He is member of Editorial Boards of six national journals and acted as a referee for Ph.D. and M.Phil. degrees for other universities. He has been awarded (i) Dr. M.S.Swaminathan Award (1996) by Indian Botanical Society, (ii) Scientist of The Year 2004 by National Environmental Science Academy,

(iii) Jivaka Ayurved Vishwa Sanman (2006) by International Academy of Indian Medicine, (iv) Lifetime Achievement Award-2007 for Health and Education by Health & Education Development Association. (v) Best Teacher Award-2010 by S.S.V.P.Sanstha, Dhule (MS) India. (vi) Talented Scientist Award-2012, Awarded at 4th International Conference on Medicinal Plants and Herbal Products, Johns Hopkins University and The George Washington University, (USA). (vii) Mahatma Jyotirao Phule National Best Teacher Award-2013, awarded by Dr.Panjabrao Deshmukh Shikshak Parishad (M.S.), India. (viii) Shri Gurudeo Mahalingdas Ahirrao Award-2012 by Shri Gurudeo Mahalingdas Ahirrao Dyanprabodhini, Wade (M.S., India), (ix) Award for excellence in Research-2013 by Education Expo (TV), Research & Branding Company at Bangalore, and (x) Eminent Botanist Award-2013 by organizing committee for National Conference on "Recent Emerging Trends in Plant Sciences-2013 at PSGVP Mandal's Arts, Science and Commerce College, Shahada (M.S., India). Recently, he is awarded a prestigious Prof. V. Puri Award-2014 by Indian Botanical Society in recognition of his significant contribution in the field of Taxonomy. He is Ex-Dean, Faculty of Science, North Maharashtra University (Jalgaon, M.S.) during 2000 to 2005 and also Ex-Chairman, Board of Studies in Botany of the same university during 2000 to 2005 and 2010 to 2015. He is a member of many prestigious academic and scientific bodies and societies. He also contributes for Marathi daily newspapers and creates interest in plants by delivering talks especially for younger generation. He also headed his college as Professor and Principal.

Preface

On account of superior intellect, an average man thinks himself not a part of the ambient organic world. He is ignorant that he is absolutely dependent on other organisms, whether plants or animals, for his very existence on this blue planet. Nay, he is a guest of green plants which utilize sunlight through the process of photosynthesis. This process and the plants are still to be realized and appreciated widely. Lord Buddha had gone ahead of this aspect and stated that plants afford protection to all beings, offering shade even to the axeman who destroys it. We all are surrounded by the plant world. Some of them, particularly very useful for our life, are cultivated by us for immediate help, but we do not extend any effort to understand them in right perspective. Naturally some questions strike to human mind. What is the native of cultivated plants? What are their ancestors and from where they brought in our vicinity? How and when they reach to us? Some intellectuals have endeavored to attend these questions. It was Alphonse De Candolle (1883) who attempted to decide place and period of cultivated plants based on conclusions on a variety of evidence such as earlier works of pioneers (Theophrastus, Dioscorides, etc.), historians, ancient writings, archaeological remains, ethnological and philological data (from ancient languages Hebrew, Sanskrit, etc.), ancient monuments, etc. Another milestone work on this line was done by N.I. Vavilov (1992). He based his conclusions on anatomy, genetics, cytology, distribution and plant diseases, etc. According to him, some crop plants had a diversified rather than a single origin. Interestingly, he explained origin of some cultivated species from weeds *e.g.* rye, oats. He concluded that the great centres of distribution of cultivated plants were always in mountainous regions, and the greatest amount of diversity occurred in such areas.

The most essential need of mankind is food which primarily comes from plants in the form of cereals, millets, pseudocereals, pulses (legumes), fruits, vegetables, sugar, starch and oil yielding plants. To meet an ever-increasing demand, an application of man's know-how about plants is of great importance for better utilization of plant products. To meet his requirements, man has endeavored to tap plants as sources for his varied uses and comfort by exploiting his scientific knowledge. He has been successful on this path to a greater extent. But he has to work more because of ever growing human population and a great variety of demands.

Plants are studied for utilitarian (applied) and philosophical (basic) point of view. Both aspects are essential for the progress of plant science and to ease the sufferings of man. Plants are considered, for convenience, in various compartments like agriculture, horticulture, medicine, economic botany, ethnobotany, etc. Plants are thus learned or taught touching usually a single compartment. Inter-connections between these are usually ignored while framing syllabi of universities and academic bodies. Study of plants incorporating various fields or disciplines of study and research is not usually the aim of such syllabi. Hardly, there are books or treatises based on different footings. This book will familiarize the history of plant introductions. The movement of plants about the world by man is the most practically important results of botanical study and has had economic, cultural and political results on a world scale. These facts must be revealed to kindle the readers. The present author has made a headway in this direction by presenting this book to render the readers broad-based. This will not allow one to think plants in isolation. The present account has mainly focused food crops. A careful attempt has been also made to include the underutilized crop plants of the world belonging to the aforesaid group of crop plants. This will help reveal the complete history and development of the world crops.

The idea of writing a book on crop plants struck to my mind while surveying flora and agrobiodiversity especially in northern part of Maharashtra State (India). The tribal farmers and rurals revealed me not only the diversity of crops but also crop species usually unknown to urban elites and even teachers in botany. I was fully impressed above all else with the wealth of plant materials of great interest. Some crop species were exotic, while others were indigenous but underutilized. I also feared that these underutilized crop species will be, with time, forgotten because of the fast rate of acculturation in Indian tribal and rural societies. I, therefore, extended my study particularly selecting to crops especially focusing food yielding plants. This gave a chance to know and reveal the state of knowledge of these important groups of crops particularly from the standpoint of agro-biodiversity, their evolution and emphasis vis-à-vis negligence on them by the present human societies.

This treatise is designed principally for tertiary education in Botany, Agriculture and Food Science keeping in view all worldwide universities. Major part of it is scientific while other aspects such as archaeology, history, warfare, rulers, travelogues, ethnology, ancient treatises or writing, religious treatises, study of various languages, geography, etc. are also dilated wherever found necessary. In these aspects, technical language is obviously kept minimum. Research development for each crop species has been conveniently divided into history and further developments. An attempt has been particularly made to provide relevant bibliography separately for each crop so that it can be useful to the researchers in these fields of study and render this treatise useful as a reference book. It includes ancient as well as recent developments in the said sciences and disciplines. Nomenclature is updated, except those referred originally in research works to avoid further nomenclatural chaos.

I wish to extend my humble thanks to Shri Shivaji Vidya Prasarak Sanstha for library facilities, and Agricultural Universities and colleges in the state of Maharashtra for the same. I am thankful to my most revered teacher Professor R.M. Pai (Dr. B.A. Marathwada University, Aurangabad, Maharashtra, India) for his ever guiding force to continue study even after my retirement from active service. All members of my family endured the many occasions that I have been forced to ignore during the gestation of this treatise.

Place : **Dhule** (Maharashtra)

Date : 13/09/2018 **Dr. D.A. Patil**

About The Book

Food Crops: Evolution, Diveristy and Advances

The treatise presents a wide range of information on different facets of world food crops. It justifies the present state of knowledge with reference to crop history, initial and latest developments with evolutionary approach based on karyotaxonomic and geographical evidence. The author has endeavored to dilate attempts for crop domestication and cultivation by worldwide human societies. These are supported by various considerations like ethnology, philology, ancient and religious treatises, archaeological evidence, works of pioneers, various human affairs, etc. He further based his writing on botanical conclusions accrued from distribution, number of varieties, occurrence of wild types, agro-botanical characteristics, time and place of origin, and worldwide usage of each crop species.

The treatise is a highly-acclaimed source of information of crop plants useful to degree and post-graduate students of botany, agriculture and food science as a main text and reference work for researchers. It is recommendable for use in any country especially in the English speaking world. It will also appeal to the general reader, being rich in contents.

In a nutshell, this treatise is a history of development of food crops. And therefore, a humble attempt has been made, first, to ascertain what was actually done or said, and second, to emphasize the causes of success or failure. Truly, it is the study of causes that makes history more fruitful and a basis for evaluating the future events with confidence.

Contents

1

History of Food Crops

The most noteworthy fact about the food plants in vogue in the world today is their great antiquity. Majority of them were domesticated from wild ancestors long before the beginning of the historical period. All available records warrant that they were as well-known to the peoples of the ancient world as they are to us. Few new crop plants have been developed in the past, although the older ones have been greatly altered and improved in response to the increasing dependency and complexity of man's existence. The history of our useful plants and their influence on human culture has always been of interest to botanists and ethnologists. Many investigations have been extended out to determine their age and place of origin, as well as their cultural history. In the following, brief history of their development is provided.

During earlier attempt to domesticate plants, man was unable to form settlements. His entire time was devoted in search of food. Presumably, prehistoric man depended on berries, succulent herbage and underground plant parts. These people lived in small groups and had a fierce struggle for existence against the carnivorous animals of that time and the vagaries of nature. Certainly, they led life of hardships. During the longest palaeolithic period or 'old stone age', which began one and three quarter million years ago, there was no farming as such. The people of this age were able to use fire and chipped pieces of hard stones into rough implements which were used to remove flesh from animal hides. During this period human population was confined to Africa, with a density of probably only 0.00425 person per square kilometer and a total population only about 125000. Agriculture presumably began in the Mesolithic (middle stone age) (from 12000 to 6000 BC.), when man lived by the spear, the bow and the fishing net. This change from food gathering to food producing started independently at different times in different parts of the world. The Neolithic (new stone age) began about 6000 BC. when ground and polished stone tools became the necessity and development of agriculture continued. The cultivation of plants was an evolutionary accomplishment. It relieved some members of a tribe from food gathering and hunting. This made them possible to devote time and taught to invent, discover and develop other facets of life.

Man's oldest occupation is cultivation of plants. Probably it began when he discovered that certain seeds spilled on disturbed ground grew in some mysterious way into new plants. Early domestications were made more or less concurrently and independently on the lower slopes of the Zagros mountains and the 'fertile crescent' of Tigris and Euphrates valleys in northern Iraq, and in the Tehuacan Valley of Mexico. The earliest evidence of genuinely cultivated forms so far discovered, date about 7000 BC. in both hemispheres. Our prehistoric ancestors who, long before the dawn of recorded history, discovered the virtues of certain plants, selected the most useful wild species and tremendously altered them. Many of them have been changed so much that their wild ancestors cannot be decided with certainty. In fact, early man was a plant breeder, without any scientific knowledge. He displayed a noteworthy wisdom in domesticating plants.

The foundation of the scientific study of plants was laid by Greek and Roman physician-botanists like Theophrastus (370-285 BC.) (the Father of Botany), Dioscorides, Pliny the Elder (77 AD), Galen and others. Dioscorides' De Materia Medica (60 AD) is probably the most significant of these ancient writings. No significant progress was made in the subsequent period upto 14th century. The earlier works of Aristotle and Theophrastus were considered so complete and authoritative that there was little else to learn. The revival of learning in the 'Renaissance' had a great impact on agriculture.

One of the old beliefs regarding to the origin of cultivated plants was that they were thought 'Creations of Almighty'. Also, there was a belief that the process of cultivation itself improved the heredity of plants. People started questioning these suggestions by the end of 18th century.

Darwin (1868) considered that the cultivated plants arose by profound modifications in the wild plant under cultivation. Mendel's work, first published in 1865, remained obscure until 1900. He formulated the laws of inheritance and attributed the origin of cultivated plants to natural selection and hybridization.

Alphonse de Candolle (1886) in his *Origin of Cultivated Plants*, studied 247 species of cultivated plants. He attempted to solve the mystery about the ancestral form, region of domestication and history of most of our important cultivated plants. He based his conclusions on the works of ancient historians like Theophrastus and Dioscroides, etc., Chinese writings, traveller's accounts, archeological and ethnological findings, such as Egyptian monuments, the remains of Pompeii, the relics of Swiss lake dwellers and the ruins of the Inca civilization in South America, philological evidence from the names of crops in ancient languages such as Hebrew and Sanskrit and botanical conclusions based on geographical distribution and variation, etc. De Candolle attached great importance to the presence of wild relatives while deciding the place of origin of cultivated plants.

De Candolle (1886) classified the economic plants into six classes as given in the following:

Plants of Old World Origin:

(i) Plants cultivated for at least 4000 years and possible longer include: almond, apple, apricot, banana, barley, cabbage, date, eggplant, fig, flax, grape, hemp, mango, millet, olive, onion, peach, pear, rice, sorghum, soybean, tea, watermelon, wheat, etc.

(ii) Plants cultivated for at least 2000 years and probably longer: alfalfa, asparagus, beet, oats, breadfruit, carrot, radish, cherry, chestnut, cotton (some types), celery, citrus fruits, lettuce, mustard, nutmeg, pea, pepper (black), plum, poppy, rye, sugarcane, walnut and yam.

(iii) Plants cultivated for less than 2000 years: artichoke. buckwheat, coffee, currant, gooseberry, muskmelon, okra, parsley, parsnip. raspberry, rhubarb and strawberry (some types).

Plants of New World Origin:

(i) Plants cultivated certainly over 2000 years and probably more than 4000 years: cacoa, kidney bean, maize, mate, sweet potato and tobacco.

(ii) Plants cultivated before the time of Columbus, but whose antiquity is not known: avocado, cotton (some species), groundnut, guava, Jerusalem artichoke, pineapple, potato, pumpkin, quinoa, pepper (red), squash, tomato and vanilla.

(iii) Plants cultivated since the time of Columbus: allspice, blackberry, black walnut, blueberry, cinchona, cranberry, dewberry, gooseberry, pecan, plum, rubber and strawberry.

The deductions about the domestication of cultivated plants by De Candolle (1886) are still largely accepted. He opined that cultivated plants originated at some time in the remote past from wild ancestors in rather restricted areas of the world with no communication whatsoever with each other, particularly China, southwest Asia including Egypt and intertropical America. He believed that each of the various crops commonly had a single region of origin. It also thought that all the cultivated races of Brassica were still to be seen wild in Europe and Siberia. Species, in his opinion, change directly under cultivation.

Investigations in crop geography and genetics, extensive collections of cultivated plants and their wild relatives were made by sending expeditions all over the world under the direction of Vavilov. He concluded on the uses of a variety of facts gathered from sources different from those earlier thinkers *e.g.* morphology, anatomy, cytology, genetics, distribution and reaction to diseases. Vavilov invented the diverse forms of our most important cultivated plants and their distribution over various parts of the globe. He noticed that the distribution of plant species is not uniform. A wide range of genetic variability is encountered in some restricted areas *e.g.* Ethiopia for wheat and South America for potato. More than 50% of the diversity found in the world was present in these regions.

World Centres of Origin of Cultivated Plants (After Vavilov)

I. Chinese Centre: It is the earliest and largest independent centre for the origin of cultivated plants, and includes the mountainous regions of central and western China and the adjacent lowlands. Total 136 endemic species are listed, some of which are species of buckwheat, soybean, crucifers, cucurbits, pear, cherry, quince, citrus, persimmon, many legumes, bamboo, onion, lettuce, eggplant, sugarcane, cinnamon, tea, etc.

II. Indian Centre:

A. The main centre: It includes Assam and Burma (Myanmar). Total 117 species are endemic. These are species of rice, many legumes, mango, orange, and tangerine, jute, sugarcane, coconut palm, eggplant, oriental cotton, yam, black pepper, cinnamon tree, etc.

B. The Indo Malayan Centre: It includes Indo-china and the Malay Archipelago. Total 55 species are listed belonging to banana, coconut, mangosteen, clove, nutmeg, black pepper, sugarcane, manila hemp, etc.

III. Central Asiatic Centre: It includes northwest India (Punjab, Northwest Frontier Provinces and Kashmir, Afghanistan, Tadjikistan and Uzbekistan (USSR) and western Tian-shan (China). Total 43 species are listed belonging to beans, lentil, wheat, peal, hemp, cotton, carrot, pistachio, apricot, almond, apple, pear, garlic, spinach, etc.

IV. Near Eastern Centre: It includes the interior of Asia Minor, all of Transcaucasia, Iran and the highlands of Turkmenistan (USSR). Total 83 species are included in this region. At least nine species of wheat as well as rye are indigenous to this centre. Many of our subtropical and temperate fruits *e.g.* walnut, cherry, pomegranate, quince, almond and fig and several forage crops *e.g.* alfalfa, Persian clover and vetch are also native to this region.

V. Mediterranean Centre: It includes the borders of the Mediterranean Sea. Total 84 species are thought originated including olives and many cultivated vegetables *e.g.* garden beet, cabbage, tunips, rhubarb, asparagus; forage plants *e.g.* Egyptian clover, white clover, crimson clover; oil yielding plants *e.g.* rape, black mustard; wheats (durum and emmer) and ethereal oil and spice plants *e.g.* caraway, anise, sage, hops, thyme, peppermint, etc.

VI. Abyssinian Centre: It comprises Abyssinia (Ethiopia), Eritrea and parts of Somalia. Total 38 species belong to this region. Wheat and barley are especially rich in diversity, besides species belonging to sesame, castor bean, coffee and okra.

VII. South-Mexican and Central American Centre: It includes the southern parts of Mexico, Guatemala, Honduras and Costa Rica. Species native to this region are extremely varied *e.g.* squash, chayote, maize, bean, sweet potato, red pepper, upland cotton, sisal, papaya, guava, tobacco, cacao, etc.

VIII. South American Centre:

A. **The Pervuvian-Ecuadorean Bolivian Centre:** It consists mainly of high mountainous areas representing the centre of pre-Inca civilization. Plants native to the Puna and Sierra uplands are also included. This centre is known to be the original home of lima bean, pumpkins, red pepper, coca, Egyptian cotton, potato, tomato, tobacco, quinine tree, etc.

B. **The Chiloe Centre:** It is an island near the coast of southern Chile. It is thought to be region of origin of the common potato.

C. **The Brazilian-Paraguayan Centre:** It is thought to be the region of origin of groundnut, cassava, pineapple, cashewnut, rubber tree, etc.

These areas of diversity constitute only a small proportion of the total land area of the earth. They are geographically distinct, being isolated by deserts or mountain ranges. Eighty-five per cent of the 640 species listed by Vavilov originated in the Old World (Asia, Europe and Africa) and the remainder from the New World (the Americas).

In the opinion of Vavilov, many cultivated plants had a multiple rather than a single origin. This was also suggested by De Candolle (1886). These primary crops were domesticated directly from wild plants, *e.g.* wheat, barley, rice, soybean, flax, cotton, maize and potato. Vavilov further opined that there were, at least, two distinct centres of distribution for wheat: (i) soft wheats originating in southwest Asia and (ii) hard wheats came from the Mediterranean region. Similarly, barley was domesticated in southwest Asia, north Africa and south-east Asia. Secondary crops *e.g.* rye, oats, etc. originated as weeds growing with the primary crops. However, as the primary crops were taken to areas with harsher climatic conditions and poor soils, the percentage of weeds increased as they were better adapted to such climates, until finally the weeds became the crop and the crop the weed.

The basic idea of the Vavilovian concept of centres of diversity remains unchanged, some of the generalisations have been questioned in the light of new knowledge. In fact, he did work on his concept of gene centres, modifying it until his death in 1943, at the age of 56. He began with five centres in 1926, but raised them to eight with three sub-centres (Vavilov, 1935, 1951). Darlington and Janaki Ammal (1945) have increased the number of centres to 12; in 1956, Darlington brought it to 15 and subsequently in 1973 to 16; whilst Zhukovsky (1968, 1975), one of Vavilov's colleagues, proposed a series of twelve 'macrocentres' or 'megagene' centres, covering almost the whole world. The thought-provoking alternative schemes of critics of the Vavilovian concept of centres of diversity such as the of Zhukovsky, Harlan and Hawkes are included to help to give a clear idea of the gene centre theory.

Zhukovsky (1968) put forward his idea about 'megagene centres' where species were domesticated. These enlarged centres cover much of the world's land surface, excepting Canada, Brazil, southern Argentina, northern Siberia and small nations such as Norway and Britain. He divided the world into regions and implied that there were no centres. This hardly seems to agree with the concept of 'centres' at all. Also, he gave 'microgene centres' of wild species genetically related to cultivated plants. The megagene centres are: China, Indochina-Indonesia, Australia, New Zealand, Indian subcontinent, Central Asia, West Asia, Mediterranean coastal and adjacent regions, Africa, Europe-Siberia, Central America, Bolivia-Peru-Chile, and North America.

The history of plant domestication is more intricated than had been supposed. A pattern that stems rather consistently from geobotanical and genetic studies is one which Harlan (1956, 1961, 1966) has called 'diffuse origin'. Crop origins can be diffuse in both time and space. Each crop may have been repeatedly domesticated at different times in different locations or may have been brought into cultivation in several regions simultaneously (Harlan, *loc. cit.*). At least, we cannot pinpoint with any precision a single centre of origin for these particular crop species. Cultivated species were not domesticated as the 'crops' we know today. They were, at first, something like their wild ancestors but as they spread out of their area of origin they picked up additional germplasm from their wild relatives on the way in a number of crops *e.g.* wheat, maize, sugarcane, cassava, sorghum and potato. *Helianthus annuus* L. is also a good example of a crop with a vast polytypic complex and an extremely large gene reservoir It has considerably expanded its geographical distribution and ecological range by absorbing genes, through interspecific hybridization, from it least half a dozen other *Helianthus* species.

Perhaps the most serious criticism of the gene centre theory has come from Harlan (1971) who developed the idea of centres and noncentres. In fact, he was prompted in his thinking by evidence that seems to demonstrate that plant domestication occurred almost everywhere south of the Sahara and north of the equator from the Atlantic to the Indian Ocean. Such a vast region could hardly be termed a 'centre' without distorting the very meaning of the word, so he called it a non-centre.

Thus according to Harlan, three independent systems are:

	Centre		*Non-Centre*
A_1	Near East	A_2	African
B_1	North Chinese	B_2	South-east Asian and South Pacific

	Centre		*Non-Centre*
C_1	Central American or Meso-American	C_3	South American

Recently, Hawkes (1983) postulated an alternative scheme to clarify the concept of nuclear centres of agricultural origins and regions of diversity. The nuclear centres of agricultural origins, in his opinion, are the places where agriculture first began. The regions of diversity are the areas into which the crops spread, from the nuclear centres, and in which additional cultivars arose, by both unconscious and conscious selection. These regions correlate fairly closely with Vavilov's centres but it is necessary to redesignate some of the names (Table-1).

Table 1: Nuclear centres and regions of diversity of domesticated plants.

Nuclear Centres	Regions of Diversity	Outlying Minor Centres
i) Northern China	I. China II. India III. Southeast Asia	1. Japan 2. New Guinea 3. Solomon Islands, Fiji and South Pacific
ii) The Near East	IV. Central Asia V. The Near East VI. The Mediterranean VII. Ethiopia VIII. West Africa	4. Northwestern Europe
iii) Southern Mexico	IX. Meso-America	5. United States, Canada 6. The Caribbean
iv) Central to Southern Peru	X. Northern Andes (Venezuela to Bolivia)	7. Southern Chile 8. Brazil

Hawke's concept includes a number of outlying minor centres probably of more recent origin, from which only a few crops, sometimes not more than one or two, have originated. These include Japan [origin of *Vigna angularis* (Willd.) Ohwi & Ohashi], New Guinea (origin of sugarcane), and the Solomon Islands to Fiji (origin of Fei Bananas) (Hawkes, 1983). In the opinion of author, this scheme is tentative and much more archaeological data from Africa, India, and Southeast Asia are necessary before the nuclear centres for such areas can be pinpointed precisely. Presently, we have a fairly good idea of the problem of how, where and when agriculture originated but the concerted efforts of botanists, archaeologists, social anthropologists and geographers would help us to get nearer to the truth. Harlan's and Hawke's schemes would provide stimulation and feedback in terms of ides so as to enable us to gain a fuller understanding of the sequence of events in agricultural origins and evolution.

It will not be inappropriate to emphasize that today the centres of production of nearly all important agricultural crops are far removed from the centres of origin (Table-2).

Table 2: The centres of world production and centre of origin of major economic crops

Crop	Centre of Origin	Centre of Production
Citrus spp.	S.E. Asia	United States
Clove	Moluccas	Zanzibar
Cocoa	S. America	Ghana
Coffee	Ethiopia	C. and S. America
Groundnut	S. America	India, China
Maize	Tropical America	Middlewest US
Nutmeg	Moluccas	Grenada, W. Indies
Pineapple	Brazil	Hawaii
Potato	Peru	Eastern Europe
Rubber	Brazil	Southeast Asia
Soybean	N.E.Asia	United States
Wheat	Central Asia	North-Central North America

An adequate food supply is, and always has been, man's most outstanding necessity. Major share of it was obtained obviously from plants and nearly same share he is enjoying even in modern time from them. In modern period, just 150 plant species are significant to meet our food requirement. Only about 20-30 plant species are mostly emphasized (*cf.* Harlan, 1975; Bermejo and Leon, 1954). Over 70% of the world's food is hailed from only nine crops *viz.,* rice, wheat, maize, barley, potato, cassava, soybean, oats and sugarcane (*cf.* Ewal *et al.,* 1999). Out of these crops, 50% of the global requirement for proteins and carbohydrates are provided by only three crops *viz.,* rice, wheat and maize (Jaenicke and Hoschle-Zeledon, 2006). It is also worth to note that in the name of 'Green Evolution' only few crops were targeted in which these three crops found place. It is how again other crop species remained underutilized and underdeveloped. The overall world scenario of gramineous, leguminous and few other crops is presented.

References:

Darlington, C.D. (1956) Chromosome Botany And The Origins of Cultivated Plants (Rev. Ed 2) George Allen and Unwin Ltd., London, UK.

Darlington, C.D. and E.K. Janaki Ammal (1945) Chromosome Atlas of Cultivated Plants. G.Allen and Unwin Ltd. London, UK.

Darwin (1868) The Variation of Animals and Plants Under Domestication. John Murray, London, U.K. (Ist Ed.).

De Candolle A. (1886) Origin of Cultivated Plants (Translated from 2nd Ed. *In:* French, 1959), Hafner, New York, USA.

Harlan, J.R. (1956) Distribution and utilization of natural variability in cultivated plant. *In:* Genetics In Plant Breeding. Brookhaven Symp. Biol. 9:191-206.

Harlan, J.R. (1961) Geographic origin of plants useful to agriculture. *In:* Germplasm Resources Publs. *Am. Ass. Advmt. Sci.* 66:3-19.

Harlan, J.R. (1966) Plant introduction and biosystematics. *In:* Plant Breeding. Iowa State University Press. Ames.

Harlan, J.R. (1971) Agricultural Origins: Centres and Non-Centres. *Science* 174:468-474.

Hawkes, J.G. (1983) The Diversity of Crop Plants. Harward University Press, Cambridge, Massachusetts.

Vavilov, N.I. (1926) Studies on The Origin of Cultivated Plants. Trudy Byuro Prikl. Bot. 16:139-248. (Bull. Appl. Bot. Pl. Breeding: 161-245).

Vavilov, N.I. (1935) Theoretical Basis of Plant Breeding. Moscow (Russian).

Vavilov, N.I. (1951) Phytogeographc basis of plant breeding: The origin, variation, immunity and breeding of cultivated plants. *Chronica* Bot. 13:1-366.

Zhukovsky, P.M. (1968) New centre of origin and new gene centres of cultivated plants including specifically endemic microcentres of species closely allied to cultivated species. *Bot. Zh.* 53:430-460.

2

Cereals

INTRODUCTION

Even before the Christian era, the ancient Romans held festivals at seed time and harvest in honour of the their goddess Ceres. They worshipped her as the giver of grain. They offered wheat and barley to her as the 'cerealia munera' or 'Gifts of Ceres'. Since then, these came to be known as 'cereals'. The Greeks had similar religious observations. In the New World, the Mexican natives worshipped an agricultural deity to whom they brought the first fruits of their harvest. Nearly every culture or aboriginal societies worship some form of deity who presided over its crops.

Even long before the beginning of recorded history, cereals were among the first plants to be domesticated. No civilization worthy of the name has ever been founded on way agricultural basis other than the cereals (Mangelsdorf, 1953). All important civilizations were based on one or another of the cereal grains. The ancient cultures of Mesopotamia (Iraq) – Sumerian, Babylonian, Assyrian, and Chaldean depended on wheat and barley. The Egyptian, Greek and Roman civilization also were based on wheat and barley. The ancient people of India, China and Japan depended on rice for their daily food. In the Western Hemisphere, the pre-Columbian civilizations of the Mayans, Aztecs and Incas were built on maize. Cereals were first used for food by merely parching or popping the grain rendered it suitable for consumption. The scorching and parching of grain is practised even today in parts of the Near East. Later, parched grains were ground into a coarse meal that on soaking in water made a kind of gruel or porridge. This primitive recipe is consumed presently by the toothless, both old and young. Heavy, baked, unleavened cakes or flat bread made from coarsely ground meal appeared later. Modern bread making or leavened bread is a rather recent innovation.

The cereals are unquestionably the most important sources of plant food for man and the lower animals. This holds true today and even in the earliest time. During their long period of cultivation their original wild ancestors have been deleted and several new species, varieties and races have been evolved. Much of this evolution occurred prior to the historical period for the older civilizations were already familiar with several kinds of wheat, barley, and other grains. Also their actual home had been so long forgotten. They were given supernatural powers and integrated with the religious ceremonies of the various regions of antiquity.

The cereals are important as food plants for various reasons. Many of these grasses are available for each kind of climate. The northern regions have barley and rye, the temperate regions wheat, and the tropics and warmer temperate areas maize and rice. Cereals enjoy a wide range of soil and moisture requirements. Their cultivation is rather simple and involves less of labour but more yield. The grains are easy to handle, store low water content, with a very high food value. They contain a higher percentage of carbohydrates than any other food plants, together with a fair amount of proteins, fats and vitamins.

As food plants, cereals are prominent. This is so because of: (a) great adaptability to every type of ecological habitat, (b) more ease in cultivation, (c) tillering habit, and higher yield per unit area, (d) the grains being compact, dry and easy to handle, transport and stockpile without undergoing damage and (e) the grain being nutritive with a higher percentage of carbohydrates than any other food plant, sufficient protein (7-10 per cent), plus fats, vitamins and minerals. They can be really regarded the staff of life.

The cereals belongs to the grass family, Gramineae (Poaceae). They produce characteristic fruit of the family, the caryopsis. In this fruit the wall of the seed becomes fused with the ripening ovary wall to form the husk. Caryopsis is an indehiscent, one-seeded fruit. The term 'grain' is applied either to this type of fruit or to the plant that produces it. Out of a total of about 9000 species of this family, hardly about 35 taxa have been cultivated as cereals. Only about a dozen of them are important today. Truly barley, maize, oats, rice, rye, and wheat, are designed as cereals. Of these, wheat, maize and rice are the most important, and each has played a significant role in the development of human civilization.

The cereals are mostly herbaceous annuals. The culm is often erect, generally cylindrical, and mostly hollow, except at the nodes, with characteristic intercalary meristem. In addition, secondary and tertiary or even quaternary tillers may rise from the subterranean nodes of the primary, secondary and tertiary or even quaternary tillers respectively. Because of tillering habit, the plants are rendered tufted appearance. Leaves are alternate, distichous with parallel venation. Leaves are divisible into distinct parts *viz.*, the leaf sheath and the lamina. At the junction of the two, a thin membranous outgrowth, the ligule, generally occurs. The inflorescence may be a spike of spikelets (*e.g.* wheat, barley, rye) or panicle of spikelets (*e.g.* oats, rice). The axis of the main inflorescence is called the rachis, while that of a spikelet a rachilla. Each spikelet consists of a series of one to seven florets, usually invested by a pair of glumes. Each floret, in turn, is enclosed by a lemma and palea. They are best adapted for wind pollination. The cereal grain is a one-seeded indehiscent fruit, a caryopsis.

A major segment of the world's population subsists mainly on either wheat, rice or maize, each accounting for nearly a quarter of the world's total cereal supply. Sorghum, barley, rye, oats and millets together make up the rest.

2.1 BARLEY

Hordeum vulgare Linn. (Poaceae)

Resume: *It is one of the earliest crops domesticated in the Near East. Archaeological remains are recovered from Mureybat on the Euphrates in Syria dating perhaps to 8000 BC. They are also discovered at Beidha in the southern Jordan Highlands dating about 6800 BC. and at Jarmo in the Zagras foothills in Iraq nearly at the same age. The domesticated forms are diploid with chromosome number n=7. Barley is the world's 5th important cereal. Its grains are early maturing with high yield potential. It is useful as staple food for mankind and has considerable nutritive value. It is also used in various industrial products.*

1. Introduction: Ceres, the Roman goddess of agriculture, is often shown with ears of barley plaited in her hair. The ears are also pictured on ancient Greek and Roman coins. It is an important cereal, domesticated from wild races found today in southwestern Asia. It is a short-season, early maturing grain with high yield potential. It extends far into the arctic, reaches the upper limits of cultivation in high mountains and may be grown in desert oases, where it is more salt tolerant than other cereals. It is a cool-season crop. It can withstand high temperatures if the humidity is low, but is not suited to warm-humid climates. It is little grown in the tropics except in cool highlands, as in Mexico, the Andes and East Africa. Major production areas are: most of Europe, the Mediterranean fringe of North Africa, Ethiopia, the Near East, erstwhile USSR, Canada, USA, China and India. It is mentioned as 'Yava' in vedic literature *viz.,* Rigveda, Yajurveda, Atharveda and also in other ancient Sanskrit treatises *viz.,* Brahmana, Upnisada, Kalpsutra, Paninis Astadhyayi and Vartik and Patanjal's Mahabhasya. This is indicative of its domenstication and cultivation even before vedic period in India. It was probably the principal food during the vedic period (1500-600 BC.) (Singh, 2008; Bakshi and Rana, 1974).

2. Plant Characteristics: An annual, tufted plant growing to a height of about 0. 6-1. 2 m, three 1-flowered spikelets borne on each node, arranged alternately on opposite sides, heads thus having 6-rowed spikelets, grains (caryopsis) more of less similar to wheat, invested by lemma and palea (hull), naked forms do exist.

3. Useful Parts: Grains, straw.

4. Chemical Constituents: Carbohydrate (78-83%), starch (63.65%), protein (10-12%), minerals (Calcium, magnesium, phosphorus, sodium, potassium), thiamine, riboflavin, niacin, pantothenic acid, folates and pyridoxena..

5. Uses And Applications: Barley grains are useful for human consumption including medicine. Until 16th century, barley was the principal source of bread flour and remained a staple food. Its nutritive value is considerable, although it is gluten deficient. It is used in the form of unleavened cakes, soups, breakfast food, children's food or baby gruels, as parched grains, as 'chapatis' after mixing with wheat flour, summer drink, etc. Medicinally, it is also significant. Decoction of barley prepared in water is called 'Barley Water'. It benefits in treating inflammation of membranes of the chest and feverish disorders. It is used as a diuretic and demulcent drink. Powdered grains in the form of gruel are employed in cases of painful and atomic dyspepsia.

It has also wide industrial uses. Barley is the principal grain used in malting for brewing beer, alcohol and liquors like whisky, brandy, etc. Barley malt is used in pharmaceutical preparations, malted milk concentrates and candies. Barley is used for hay and pasturage. The straw is used for stock feed and bedding. Young plants provide nutritious green fodder for cattle. The surplus grain is used as a livestock feed.

6. Karyotaxonomic Evidence: Barley is the world's fifth important cereal. It is diploid with 2n = 2× = 14 (Anonymous, 1968; Nilan, 1971; Takahashi, 1955). Tetraploids appeared spontaneously but are a negligible part of the crop. The wild and weed races are although designated as *Hordeum spontaneum*, biologically they belong to the same species as the cultivated races. Hybrids between wild and cultivated forms occur naturally where the two are found together. Hybrid plants are completely fertile. The chromosomes pair and segregate normally. In the spontaneous forms, the spike breaks up at maturity and the grains fall down. In case of domesticated races, the spike is tough with the persistent grains. The difference is governed either one of two tightly linked brittle (Bt and Bt_1) genes. Wild-type is dominant.

Two-rowed *Hordeum* are all the truly wild forms. They bear three spikelets at each node of the ear, the two lateral ones being female-sterile and only the central one develops a grain. Under domestication, six-rowed races appeared in which all three spikelets produce grains. There are two genes involved, both with multiple allelic series, but a single recessive mutation (vv) is adequate to cause a two-rowed barley to become six-rowed. Six-rowed genotypes with fragile ears are probably derived from six-rowed cultivars.

The husks do not adhere to the grain in case of naked barley which fall free on threshing. The naked grain is much more favourable as human food and such cultivars are preferred where barley is a major part of the human diet. The character is controlled by a single recessive gene (n). Barley when crossed with several other species of *Hordeum*, the hybrids were highly sterile or anomalous. No other species appeared to be involved in barley evolution.

7. History: It is one of the earliest crops domesticated in the Near East (Harlan, 1968; Harlan and Zohary, 1966). Archaeological remains of what appear to be wild forms with fragile ears were found in considerable quantities at Mureybat on the Euphrates in Syria dating, perhaps, to 8000 BC. Similar remains were found at Beidha in the southern Jordan Highlands dated about 6800 BC. and at Jarmo in the Zagros foothills in Iraq at about the same age. It is not possible to determine if these remains were from wild harvests or from early attempts to grow the crop.

Cultivated barley was recovered from Ali Kosh (Iran), Ramad (Syria), Jericho (Palestine) and at Catal Huyuk and Hacilar (both in Turkey) at time ranges between 6000 and 7000 BC. (Renfrew, 1969). Barley usually appears with remains of other primary crops of the region such as emmer wheat, einkorn wheat, flax, pea and lentil, etc. All the earliest barley finds were of two-rowed, covered (*i.e.* non-naked) sorts but both six-rowed and naked types had appeared by 600 BC. (Helbaek, 1966; Renfrew, 1969; Murray, 1970).

In the Near East, agriculture was first started on dry land and at elevations sufficient for adequate rainfall. The arts of irrigation have been identified at which irrigation was evidently practised by 6000 BC. Crops may have been watered considerably earlier at the oasis of Jericho. Barley was a basic crop for the early irrigated agriculture in Mesopotamia and ancient Egypt. Our first evidences of agriculture in Egypt are from the Fayum and date to about 4500 BC. Cultivation probably began much earlier than this but the evidence has been buried under silts deposited by the annual floods of the Nile. Early barley remains from the city states of Mesopotamia and dynastic Egypt are more common than those of wheat. The earliest literatures in cuneiform, linear B and Egyptian hieroglyphic all indicated that barley was more important than wheat for human food. It is hence Sumerians had a god for barley but not one for wheat. Toward the end of the third millennium BC., the irrigated lands of southern Mesopotamia began to salt up and wheat production there declined quickly. A near monoculture of barley was established by about 1800 BC., since it was the only crop that could withstand the high salt content (Jacobsen and Adams, 1958).

The most cheapest abundant grain of the ancient Near East war barley. It was the standard fare of the poor, the ration of the soldier, serf and slave, and the staff of life for the Greek peasantry. It had the esteem of being a strong food and was awarded to victors in the Eleusinian games. Ancient gladiators were called 'hordearii' or 'barley-men' because they were trained on barley. The shift to wheat as human food came in classical times.

Barley bread was much used in earlier days, but has been condemned by experience, and barley is now mostly fed to animals, although beer was highly tried to persuade people for its healthful properties (Pliny, 1st AD). Barley reached Spain in the fifth millennium BC. and the lower Rhine a little later. It spread eastward to the Indus where it was grown in the third millennium BC., and reached China late in the second. At some time, it reached the highlands of Ethiopia where it became one of the dominant crops. This is evidenced by the archaeological investigations.

Thus barley originated in a limited area from a two-rowed ancestor; that three recessive mutants were of key importance; and that, as a corollary and contrary to earlier views, six-rowed 'wild' barleys (such as Agriocrithon types) reflect introgression from cultivars. The situation is taxonomically much simpler than was once believed. It is not necessity to guess multiple domestication. Harlan *et al.,* (1940) showed that, inexplicably, the yielding ability of 2R × 6R and 6R × 2R combinations was inferior to that of crosses within groups. Inter-group crosses often generate morphologically peculiar or otherwise uninterpretable segregates. Here one recalls Takahashi's studies of the distribution of *bt* alleles and other genes which are at least suggestive of geographical differentiation related to row number (Takahashi, 1955). It appears that acceptable expression of the six-row character would have demanded so much adjustment of the genetic background as to generate an incipient differentiation between the groups.

8. Further Developments: Columbus, on his second voyage, took barley to the New World. It did not succeed in Hispaniola but, early in the 16th century, it was introduced to the highlands of Mexico where it continues today as a minor crop. In Peru, it had success among the Indians because of its short season and cold tolerance which adapted it to the cold highlands. It satisfied special need as fodder for guinea pigs. By the beginning of the 18th century, barley was being grown in the Spanish colonies from California to Chile at appropriate elevations.

Wheat was more in demand for early subsistence farming in North America and barley was relatively little grown until the cities developed an appreciable demand for meat and dairy products. Because of its early maturity, barley was particularly suited to the western plains of Canada and production increased enormously as these were settled. The Canada and USA are the only countries with really significant productions.

All barleys must have existed as highly heterogeneous 'land races', mixtures of more or less inbred lines with a significant admixture of hybrid segregates, the products of a low level of random crossing in earlier generation up to the late 15th century. Many barleys exist in this state today but, over the past century or so in advanced agricultures, the land races have been virtually wholly displaced by pure-line cultivars. These were, at first, selections from land races. Later, they have come from successive cycles of crosses between established pure lines, sometimes of diverse geographical origins. The result has surely been a considerable narrowing of the genetic base in many, probably all, advanced agricultures, partially concealed by a great diversity of cultivar names. There has been a recent morphological change. Barleys have mostly become shorter in stature, tolerant of high-N fertilizer regimes and also more efficient biologically.

Ellis *et al.* (2000) investigated wild barley as a source of genes for crop improvement in the 21st century. The development of new barleys tolerant of abiotic and biotic stresses is an essential part of the continued development of the crop. A range of assasy are designed to explore abiotic stress tolerance in barley. The assays included sodium chloride uptake in wild barley and mapping of population, effects for $\delta^{13}C$ and plant dry weight in wheat aneuploids, effects of photoperiod and vernalisation in wild barley and measurements of root length in wild barley given drought and nitrogen starvation treatments in hydroponic culture. The authors concluded that wild barley thus offers the prospect of a 'goldmine' of untapped genetic reserves.

Goets *et al.* (2008) experimented for the highly efficient generation of transgenic barley. They achieved advancements through establishment of a protocol for *Agrobacerium*-medicated gene transfer to immature embryos. The analysis of representative number of transgenic lines revealed that the obtained T-DNA copy numbers are typically low, the generative transmission of the recombinant DNA is in accordance with the Mendelian rules and in vast majority of the primary transgenics produce progeny that expresses the respective transgene product. This newly established protocol turned out to be useful to transform not only the highly amenable cultivar (CV.) 'Golden Promise' but also other spring and winter barley genotypes, albeit with lower efficiency. As a major result of this study, a very useful tool is now available for future functional gene analyses as well as genetic engineering approaches. Numerous transgenic plants have been generated using diverse expression cassettes. These plants represent an example of how transformation technology may contribute to further our understanding of important biological processes.

9. Epilogue: Cultivars have been developed suitable for such altered conditions *e.g.* short-strawed or semi dwarf types that are less vegetative, more resistant to lodging, that are fertilizer-responsive and have greater photosynthetic efficiency. There is a need for yield genes and heterotic combinations that will increase yield, improve test weight and provide a wider range of adaptation. The trend in this direction is already evident. The digestibility of protein and efficiency of conversion from grain to animal gain is critical. Simple increases in grain protein do not always result in a better feeding efficiency. Animals can be fattened on a full barley ration but a number of nutritional problems need to be attended to develop high quality feed grain cultivars. The trend is in the direction of much improved feed barleys for livestock nutrition.

Considerable emphasis is placed on genetic resistance to the most serious diseases of barley. Among the most common and damaging are powdery mildew (*Erysiphe*), rusts (*Puccinia*), smuts (*Ustilago*), net blotch (*Pyrenophora*); spot blotch (*Helminthosporium*), leaf blotch (*Septoria*), scald (*Rhynchosporium*), barley yellow dwarf virus (BYDV) and barley stripe mosaic virus (BSMV) etc. are common and damaging. General resistance and tolerance are sought as well as genes conferring specific resistance to individual races of pathogen.

Ethiopian barleys have been particularly useful in supplying resistance to leaf diseases. Emphasis is being given to problems of adaptation, *e.g.* greater hardiness for winter barleys; special cultivars for cool, wet regions; early short-season drought resistant cultivars for more difficult environments; and wide-range adaptation for general use. If the barley acreage is to be expanded, then it is necessary to exploit the crop's unique characteristics of earliness, drought resistance, salt tolerance, vigorous root system, high tillering capacity, etc.

Although trends in malting barleys are somewhat along the same lines, but tempered by special demands for quality and uniformity. The specifications for good malting quality are very different from those for good feed barley and selection. Two-rowed cultivars are preferred in Europe and in some parts of the USA, inspite of highly acceptable six-rowed barleys have been produced. Uniformity is important for malting barleys.

Ethiopian barley cultivars have been an immensely valuable source of genetic resistance to diseases and of improved nutritional quality. High resistance to powdery mildew, loose smut, leaf rust, net blotch, scald yellow dwarf virus and stripe mosaic virus has been noted in them. Not only is resistance common in these materials but many lines have multiple resistances and some lines have resistance to all races of a disease tested. Protein as high as 18% and lysine levels of 4. 4% of protein have been found in Ethiopian barleys.

2.2 MAIZE (CORN)

Zea mays L. (Poaceae)

Resume: *Its origin in wild state is the lowlands of tropical South America. During prehistoric time, it was cultivated in the Andes by the American Indians before the arrival of Columbus and prior to the period of the Inca civilization and played important role in the civilization of the Mayas and Aztecs. Chromosome number of Zea mays and its close relative viz., Teosinte share the same number n=10. Maize plant is monoecious and diclinous. It bears characteristic ear separately. Maize is used mainly as a staple human food, a seed for livestock and as a raw material for many industrial products. Maize grain is very nutritious, with a high percentage of easily digestible carbohydrates, proteins and fats.*

1. Introduction: Maize is a great gift from American continent to mankind. It is said to be originated in a wild state in the lowlands of tropical South America. Its cultivation goes back to the Andes in pre-historic time. It was cultivated there by the American Indians long before the arrival of Columbus. It must have grown for many centuries prior to the period of the Inca Civilization. It was grown in new Mexico as early as 2000 BC. It also played a prominent part in the civilization of the Mayas and Aztecs.

There is no evidence of any kind-archaeological, linguistic, ideographic, pictorial or historical–for the existence of maize in any part of the Old World before 1492. The antiquity of maize in India is believed that the Portuguese introduced it to India from Europe during the early part of the 16th century. No evidence for the existence of maize on the Indian plains in pre-Columbian times is available from the literature including Vedas. Earliest unambiguous reference to maize in India is found in the 'Ain-i-Akbari' given by Abul Fuzl only in connection with the description of another plant in the administration report of Emperor Akbar (1590 AD.) (Watt, 1889-1893).

Pre-Columbian introduction of maize in India is sometimes advocated. The Arabs, and not the Portuguese, introduced it into India. This is supported by some Indian names *viz.,* Mekkajola (Meccasorghum), Makkai (grain of Mecca), Mekka Cholam (Mecca sorghum) (Vishnu-Mittre, 1974).

Anderson (1943, 1945), Stonor and Anderson (1949) studied maize varieties from Assam and Burma Hills and provided evidence from ethnological, linguistic, distributional and morphological point of view. They concluded that there has been at least two major movements of maize in Asia, of which one was pre-Columbian and the other early post-Columbian. Complex of characters in Assam maize varieties resembled to some extent to those of Bolivia and the eastern lowlands but unknown in Mexican and Central American material. 'Waxy' maize was discovered in Western China and described by Collins (1909) but could not prove its existence in China before 16th century. In Europe, maize was introduced obviously by Columbus. It was unknown in Europe at the time of Roman Empire as stated by De Candolle (1886).

2. Plant Characteristics: An erect annual, attaining a height of 3-15 ft; stem jointed; roots seminal and adventitious; leaves distichous, large, narrow; male and female inflorescence borne separately, male inflorescence branched panicle (tassel), spikelets paired in rows at the top of stem; the female inflorescence (cobs) with large leafy bracts covering pistillate flowers and long silky styles (silk); grain (caryopsis) large, depending upon structure and constituents, maize is called pod maize, pop maize, flint maize, dent maize, soft (flour) maize, sweet maize and waxy maize.

3. Useful Parts: Grains, stalks, foliage, cobs.

4. Chemical Constituents: Protein (5-10%), starch (72%), carbohydrates (66-75%), fats (3-18%), minerals (P, K, Ca, Mg, Cu, Mn, Zn).

5. Uses And Applications: Grains are chiefly used as food for man and seed for livestock. In food industry, it is used in the preparation of puddings, pop corns, pies, salad dressings and confections. Maize starch is used in the production of dextrose and corn syrup, corn oil in cooking. employed also as a diluent for pharmaceutical preparations, ingredients of oil-well drilling muds. Maize starch is used for toilet powder. Various maize products such as maize grits, brewer's grits, maize flakes and maize oil are used in various food preparations, breweries, distilleries, etc. The list is still exhaustive.

The fibres in the stalks are used in making of paper and paper-boards, inner husks for cigarette papers, cobs for fuel, besides as green fodder for cattle.

6. Karyotaxonomic Evidence: Maize (*Zea mays*), teosinte (*Euchlaena mexicana or Zea mexicana or even Zea mays pp. mexicana*), and tripsacum (*Tripsacum species*) are the three New World members of the tribe Maydeae. Of these, Teosinte is a close relative of maize (Wilkes, 1967) and both share the same chromosome number 2n=20. They have rather similar chromosome morphology. These, however, differ most in the structure of their female inflorescences and in their chromosome knob patterns.

These knobs are inherited in Mendelian fashion, the knobs of maize being much less often at terminal and subterminal positions. Teosinte plants often resemble maize morphologically. The stems are less robust and individual plants tend to have more tillers. The terminal (male) inflorescences are similar but the teosinte tassel does not possess the prominent central spike which is notable in maize. The female inflorescences of maize usually are one to three or rarely ten, short lateral branches terminating in ears which have 8 to 24, or rarely more, row of perhaps fifty kernels each. The female inflorescences of teosinte are also lateral branches, each of which may terminate in a two-rowed spikes of perhaps a dozen hardened fruit-cases. Each fruit-case contains a single seed enclosed by a indurated glume. The larger lateral branches may bear terminal male spikes. The ear of maize is enclosed by numerous husks (modified leaves) of the lateral branch and the kernels (caryopses) adhere to the tough cob or rachis. The teosinte lateral spike is loosely enclosed by a few husks, the rachis of the spike rendered fragile upon maturity, and the fruit-cases disseminating easily. Maize is wholly dependent upon man for its propagation.

Tripsacum species are perennials with chromosome numbers in multiples of x=18 (Randolph, 1970). *Tripsacum* seems to be more closely related cytologically and morphologically to the genus *Manisuris* (tribe Andropogoneae) than to maize or teosinte. The *Tripsacum* species are quite variable vegetatively. They are variable in plant size. Their inflorescences differ from those of maize and teosinte in that male and female flowers are borne separately, but in tandem, in terminal spikes. The female flowers are born on the lower parts of the inflorescence with the male flowers present above them. The seeds are embedded in cylindrical, indurated rachis segments. These break apart at maturity. *Tripsacum* also has chromosomal knobs. Its general genetic and chromosomal structure is different from that of maize (Galinat, 1973). It has alleles of maize genes, although not always on the same chromosomes. Chromosome shapes and knob positions but differ greatly from those of maize.

Maize and Teosinte cross readily and the descendants are fertile. These two species are cross- (wind-) pollinated and grow sympatrically in Mexico and Guatemala and reciprocal introgression between them is possible and doubtless occurs. *Tripsacum* can be crossed with maize but the offspring show varying degrees of sterility. Through back-crossing, small portions of the *Tripsacum* genome can be incorporated into that of maize.

The Oriental Maydeae includes *Coix, Sclerachne, Polytoca* and *Chionachne* and are thought distantly related to maize (Mangelsdorf, 1974). Of these, *Coix* has knobbed chromosomes in multiple of ×=5 species and is more closely related to maize than are the other Oriental genera. It is suggested that the ×=10. Maydeae arose as a result of amphidiplody between two ×=5 species, such as the ×=5 species of *Coix* and *Sorghum*. The Maydeae seem heterogenous and since the separation between the Andropogoneae and the Maydeae rests solely upon the single character monoecy (vs. dioecy), it may well be that the Oriental Maydeae and the American Maydeae each evolved from more than one Andropogonoid ancestor.

There are three hypotheses about the origin of the American members of the Maydeae. First is that of maize, teosinte and *Tripsacum* which are descendants of a common ancestor (Weatherwax, 1954), the last-named having differentiated earlier than maize and teosinte. A second one is that of maize derived from teosinte. This suggestion has arisen sporadically over the years. Galinat (1971) and Mangelsdorf (1974) remarked for no archaeological evidence to support the idea that teosinte is the ancestor of maize. Almost all the earliest known maize is both polystichous and soft-glumed, as well as having paired spikelets. Mangelsdorf (1974) and Weatherwax (1954) have emphasized that these traits appear to be evolutionarily more specialized in teosinte than in maize. For these reasons, among others, Mangelsdorf (1974) suggested a third hypothesis, namely that teosinte originated from maize.

Studies initiated by Galinat and Mangelsdorf to test the hypothesis that Teosnte arose as a hybrid between maize and *Tripacum* (Galinat, 1971, 1973). The arrangements of loci in maize and teosinte are quite similar, with linkage between several of the distinguishing loci. In *Tripsacum*, in contrast, the same loci are often found on different chromosomes. These findings do not support the hypothesis. This formed a part of the Mangelsdorf and Reeves' 'tripartite hypothesis' about the origin and evolution of maize.

7. History: Pollen samples belonging to maize, teosinte or their common ancestor have been dated at 60,000 to 80,000 years ago in drill cores collected in Mexico city in the mid-1950s. At the earliest Tehuacan levels, dated about 5000 BC., very small cobs, little larger than an ordinary pencil eraser, some of which are thought 'wild maize' (Mangelsdrof, 1974). Some of these cobs have long, soft glumes, lending support to some extent the second part of the 'tripartite hypothesis' that modern maize is descended from pod corn, the most common form of which is determined by a series of alleles at the *Tu* (tunicate) locus on chromosome 4. In the later levels, larger cobs with firmer glumes were disclosed. It was hypothesized that these cobs represented maize which had introgresed with teosinte, from experimental evidence that teosinte introgression resulted in greater induration of the rachises and lower glumes of the cobs. This evidence is thought to support the third part of the 'tripartite hypothesis' that teosinte contributed to the evolution of modern maize by wider spread introgression.

Archaeological evidence on domestication is scanty and mostly availed from the dry coastal areas of Peru. The earliest materials date to about 1000 BC. Complete ears dated at about 500 BC. are similar to Andean races still found in Peru and Bolivia. They are quite distinct from current or archeological Mexican maize.

Kernel size depends mainly upon the size of the endosperm. The smallest popcorns have kernels measuring only several mm in length, width and thickness. The largest Cuzco Gigante kernels are about two cm in length and width and perhaps one cm in thickness, white, yellow or orange. Colouration is noted in the endosperm. The aleurone layer may be colouless, purple, red, lemon-yellow or brown and purple. The aleurone and or pericarp may be patterned (stippled, speckled, dotted, stiped, streaked, etc.). Most commercial maize is either yellow or white but flint or dent, much of the maize cultivated by American Indians, is highly coloured and also floury in texture.

Mexico and/or lowland Central America is the centre of variability for commercially important dent types (Brandolini, 1970; Mangelsdorf, 1974). These forms have spread around the tropics since AD. 1500. Derivatives of the Mexican dents apparently spread into the southern United region. These dents were crossed in the mid 1800s with flints indigenous to the northern USA by mid-western farmers, who often replanted poor stands of late-maturing southern dents with the early-maturing flints. The Corn Belt dents were among the descendants upon which most of the world's corn production is now based. The Corn Belt dents are well adapted to southern Europe.

The Cuban and Argentine Flints, or Catetos, were apparently scattered along the Atlantic coast from the Caribbean to Argentina after 1600. These yellow to orange flints have had several different suggested origins. But these lack archaeological or historical evidence for their occurrence anywhere within the Atlantic costal area until the 1800s. Descriptions of other types of maize for the same region date back to about 1500. The true flints which are used throughout the tropics and subtropics trace to this source of germplasm. However, the flints of the temperate zones trace their origin back to the northern flint and flour corns of the USA. Their precise origins are obscure. These do not resemble closely either the 8-rowed Mexican or Guatemalan races which are sometimes thought ancestral to them.

The Caribean and the northern edge of South America encompass the centre of diversity for the Coastal Tropical Flints, a group of semi-flints, which, together with their close relatives, the Tusons, a group of cylindrical flinty-dents from the same region, appear to have arisen from crosses between dents of Mexican origin and indigenous flings. Exact place of origin of this type of maize is but unknown.

In Latin America, several other centres of variability exist, but materials from them have been of less world-wide importance than those dilated above. In northwestern Mexico, there are several long-eared, flinty and floury races of maize. These have low row numbers and ears tapering at both ends. In central Mexico at high elevations (ca 2000 m), a series of maize varieties with conically shaped ears and narrow kernels are predominant. The plants are heavily pubescent, have few tassel branches and tend to root lodge badly. The 'shoepeg' types of maize, once popular in the southern USA, are thought to trace their origin to these corns.

Highland Guatemala has been a centre of variability for a series of long-eared flint corns. These appeared extending to south and the northern Andes. The greatest single source of variability in maize in the form of the greatest arrays of kernel, cob and plant colours and kernel sizes is found at mid to high elevations in the central Andes. While floury endosperm predominates, flints are also present. Dents are scarce and apparently introduced. Ears are generally grenade-shaped, with size reduced with increasing altitude. East of the northern and central Andes, in the Amazon basin and surrounding lowlands, a single type of maize predominates. It has typically long, narrow ears with bronze-coloured floury kernels. Its most distinctive feature, however, is the modification of the usual paired arrangement of kernel rows so that kernels from what would ordinarily be adjacent rows are positioned alternately in the same row. This results in a lower row number and an increase in length of ear.

The centres of development of variability in the New World parallels the development of American Indian civilizations and the spread of intensive agriculture. The Mexican and lowland Central American dents are thought associated with the Mayan civilization, while the conical corns from higher elevations in central Mexico appear to have been associated with the Aztecs and their predecessors. The variability of maize in the central Andes correlates with the extensive agricultural development of the Incas and their predecessors.

Prior to 1500, the major portion of the variability now found in maize has developed. Most widely grown races notably Corn Bent Dent have developed later. The large differences between the types of maize found in the various centres of variability have led to suggest the hypothesis that the domestication of maize occurred independently in different regions from different types of wild maize. Mangelesdorf (1974) further postulated that six still-living races of maize represent products of such independent domestications. After 1500, secondary centres of variability have developed in regions in which maize has undergone widespread use, *e.g.* the north central USA and southeastern Europe. Maize from the Caribbean region and the northern coast of South America also represents secondary variability as a result of the trading activities of the early colonists. The agricultural activities of American Indians at the time of Columbus were described by Weatherwax (1954). The most common form was the milpa system. Forest was cleared by the 'slash and burn' method still practiced in many tropical areas. Crops, which usually included maize, were grown on such land for above three years. As yields decreased and weed problems increased, such plots were then abandoned, for about a decade, before being cleared again.

Genetic and archaeological data indicate that maize was domesticated in Mexico-8700 calibrated years before the present (Piperno *et al.,* 2009; Ranere *et al.,* 2009). It then spread into other tropical regions in and south of Mexico in the following millennia (Pohl *et al.,* 2007; Piperno *et al.*; 2000). In South America, preceramic-aged microfossils, like starch grains and phytoliths, are reported from southwest Ecuador by 7000-5500 cal BP (Piperno, 2006; Peersall *et al.,* 2004) in the southern Andean highlands, by 4000 cal BP (Perry *et al.,* 2006) and south-eastern South America, by 4500 cal BP (Iriate *et al.,* 2004).

Lane *et al.,* (2008) provided the earliest evidence of Ostionoid maize agriculture from island of Hispaniola. The two sedimentary records contain abundant maize pollen dating back to around AD. 1060. These prehistoric pollen grains constitute the earliest evidence of maize from the interior of Hispaniola. This represents, in the opinion of these authors, some of the earliest and most securely dated evidence of maize agriculture on the island of Hispaniola and from the Caribbean as a whole.

8. Further Developments: Since the time of the colonization of the Americas upto the mid-1800th century there was little formal breeding of maize. The European settlers accepted local American Indian varieties or planted similar varieties from neighbouring settlements. In Europe, tropical varieties, apparently introduced from the Caribbean, were widely used in the south and southeast.

In the USA, in the 1800s, maize growers began to show their product at various fairs and exhibitions. This was the beginning of the corn show era. This era ended early in the 1900s. In the corn shows, emphasis was placed upon uniformity of the sample of ears centered and conformity to an 'ideal' type of Corn Belt dent ear. The southern dents were tall, late maturing, soft-kernelled, non-tillering, many-rowed and frequently with tapering ears. The northern flints were largely eight-rowed, early maturing, short in stature, tillering, prolific (many-eared) and usually with cylindrical ears. As a result of the influence of the corn shows, the Corn Belt dents were selected toward singly- and cylindrically-eared, tillerless plants. The emphasis of the shows on selecting uniformly large ears resulted in widescale elimination of prolific and tillered plant types which tended to produce more but smaller ears. The publicity which accompanied such shows ensured wide-spread use of specific open-pollinated varieties of corn across the US Corn Belt and abroad. The first of these were the observations, by Charles Darwin in England and by Beal at Michigan State, of hybrid vigour in varietal crosses of maize. This vigour or heterosis resulted in increases of yields. Later, the concept of yield-testing during the various corn show entries to determine whether the judges rankings corresponded to the relative yields of the offspring seems to have been developed, first in Iowa. As might be expected, the judges' choice were often poor yielders, as were seedstocks supplied by commercial seedsmen. At about the time that yield-testing began, interest developed in inbreeding corn. The resulting inbred lines were weak and poor yielding, but East at Connecticut and Harvard, and Shull at Cold Spring Harbour New York, found that crosses between two inbred lines often produced uniform, F_1 (single cross) hybrids which were superior in yield to any open pollinated varieties of the time. The early inbred lines were unproductive. Their use for seed production seemed impracticable. It remained for D. F. Jones at Connecticut in 1918 to report a cross of two different single-crosses [*e.g.* (A × B) × (C × D)] which resulted in an economically feasible, reasonably uniform, and highly productive 'double-cross' hybrid. Within fifteen years, double-cross hybrids became economically important and, by the early 1950s, essentially all the maize grown in the US Corn Belt was from double cross hybrid seed.

The leading three sources of inbred lines for double cross hybrids were three open-pollinate varieties. The first, Reid's Yellow Dent was developed by the Reid family of Illinois in the late 1800s. It spread to Iowa and Indiana later. It was used widely in subsequent corn breeding studies. A second source of useful inbreds was the variety Krug, developed by George Krug of Illinois, from a combination of a strain of Reid's and another variety. The superiority of Krug to other open pollinated varieties was discovered as a result of yield tests conducted in the early 1920s. The variety was then incorporated into the breeding programme of Lester Pfister, one of the early commercial producers of hybrid corn seed. The third source of useful inbred lines was the variety Lancaster Sure Crop, developed by Isaac Hersey in Lancaster County,

Pennsylvania. High yields of this variety impressed F.D Richey of the US Department of Agriculture. This was incorporated in early inbreeding programmes. Inbreds from Krug and Lancaster were frequently crossed with inbreds of Reid origin in the production of commercial hybrids (Wallace and Brown, 1956).

US Corn Belt farmers after the late 1950s, planted single-cross rather than double cross hybrid seed. Because the single-cross seed must be produced on an inbred line, the kernels are less uniform and more expensive than double-cross seed. This type of selection programme resulted in a new generation of inbreds. These were not only themselves higher yielding than were their predecessors. These also yield better in hybrid combinations. It has also resulted in a considerable loss of variability in breeding materials, which may increase the vulnerability of our maize crop.

A stable type of cytoplasmic male sterility in the late 1940s was discovered in Texas material obtained from the variety Mexican June. Thus, by backcrossing inbred lines to this source of sterility, it succeeded to obtain male-sterile inbred lines for use seed production and avoid at least a portion of the detasseling effort by blending seed produced by such lines with normal seed. Soon dominant restorer genes were discovered. These genes restored male fertility to plants carrying the Texas source of cytoplasmic male sterility. In 1956, a US patent on the use of such restorer genes was issued (Mangelsdorf, 1974). By incorporating restorers into inbred lines used as males, the seed produced would be heterozygous and plants developing from them would shed normal pollen. These discoveries came at a time when labour costs were increasing and labour supplies decreasing. Therefore, industry quickly adopted them.

During maize breeding in the past, there has been the emphasis, chiefly by researchers trained in the area of quantitative genetics, on development of improved maize populations rather than on the immediate production of improved inbred lines. Various breeding methods such as mass, recurrent and reciprocal recurrent selection procedures have been used to improve population performance before extraction of inbred lines is begun.

Khan (2016) concluded experiments to investigate the impact of different N-fertilizer sources (urea, calcium ammonium nitrate and ammonium sulphate) and their levels (50, 100, 150 and 200 kg ha-1) and phenology, seed weight and composition (seed protein, oil and starch content) of maize (*Zea mays* L.) genotypes "Local cultivars (Azam and Jalal) vs. hybrid maize (Pioneer, 3025)". The results revealed that (all the experimental plots applied with N) had delayed phenological development and had produced heavy grains with higher protein oil and starch contents over control (N not applied). In both years, the maturity was delayed, seed weight and seed protein content increased with the application of two higher N rate (150 and 200 kg ha-1) as compared with the two lower N rates (50 and 100 kg ha-1). Applications of N in the form of ammonium sulphate enhanced maturity and decreased seed weight but increased seed oil content over urea and calcium ammonium nitrate. The maize hybrid (P-3025) with delayed maturity produced heavy seeds and higher seed protein and oil contents as compared to the two local cultivars. The hybrid (P-3025) was informed to be the more efficient maize genotype that responded very well to different sources and rates of N than the local maize cultivars.

Bhusal *et al.* (2017) estimated the extent of genetic variability and traits association in maize. Total 55 genotyes were tested under irrigated condition in randomized block design with free replications. High genotypic and phenotypic coefficient variation was observed for grain yield/plant, biological yield/plant and cob weight coupled with high heritability, ear height, leaf area index cobs/plant, cob weight, cob length, cob girth, grains/row and biological yield/plant both at genotypic and phenotypic levels. They inferred that traits showing variability and strong positive correlation both at genotypic and phenotypic levels need to be paid attention while formulating breeding strategies for improvement of grain yield of maize.

9. Epilogue: Rapid breeding systems may become increasingly important, not simply because of the speed and economy with which can be implemented, but because of necessity. Use of sources of tropical germplasm in temperate (long day) environments to increase available genetic diversity has been dramatically hindered by the photoperiodic response of such material. It requires a day length of 12-14 hr to produce tassels and ear shoots. Tropical material, therefore, does not flower in temperate regions until late fall, at plant heights of about 6 m. Those tropical materials which do appear to mature normally in temperate regions are usually extremely early under conditions to which they are adapted, with the usual limitations on yield and plant strength that extreme earliness implies.

In a nutshell, the area of plant breeding, it is apparent that development of breeding materials needs to be coordinated more closely with agronomic studies of the best methods of growing maize. While results for maize will likely not be as spectacular as the Green Revolution with rice and wheat, much of the yield increase which has occurred since the introduction of hybrid corn has been due not only to genetic improvement of yield.

2.3 OATS

Avena sativa L. (Poaceae)

Resume: *It is a temperate cereal and important crop of Mediterranean climates. Its cultivation began much later than wheat and barley. Oats were grown by the prehistoric peoples of Central Europe where archaeological remains dating the 1st millennium BC. have been recovered. Species of*

Avena occur at three ploidy levels viz., diploids, tetraploids and hexaploids, the haploid chromosome number being n=7. Oats are the most nutritious of all cereals for human consumption in various forms. They are also useful as feed for livestock.

1. Introduction: Oats are not found in wild. It is difficult to determine their native homes. Probably, they have a multiple origin *e.g.* Ethiopia or Mediterranean region, Abyssinia. Traditionally, it has been the dominant cereal in temperate latitudes with moist maritime climates and, on the other hand, an important farm crop in Mediterranean climates. During the last decades, world production has remained fairly constant and yields have risen steadily but areas sown to the crop have declined sharply in several countries, mostly due to a decline in the use of oats for stock feeding on the farm and also because of the high husk content which reduces the attractiveness of oats as a raw material.

2. Plant Characteristics: Plants growing 2-5 ft, leaves bluish and devoid of auricles at base of leaf blade, leaf sheath firmly enveloping the internode; inflorescence branched panicles, the branches either equally distributed on all sides of falling to one side, erect or drooping, ultimate branches terminating in large pendulous and long pedicelled spikelets, each spikelet 3-flowered, often one being sterile, lemmas awned or awnless, grain caryopsis, greyish-yellow, more slender than wheat, invested by a hull (lemma and palea), however, naked in *A. nuda*.

3. Useful Parts: Grain (Caryopsis), husk and straw.

4. Chemical Constituents: Carbohydrates (66.3%), protein (16.9%), fat (6.9%), minerals (calcium, iron, magnesium, manganese, phosphorus, potassium, sodium, zinc), thiamine (B1), riboflavin (B2), niacin (B3), pantothenic acid (B5), vitamin (B6), folate (B9).

5. Uses And Applications: Oat grains are most nutritious of all cereals for human consumption being rich in nutrients as stated above. Oatmeal is prepared by grinding the grains rather coarsely between stones. It is popular and good food for muscle building and particularly well adapted for use in cold weather by people leading an active outdoor life. It is mainly used in biscuits, cakes and breakfast food but not as bread flour since it lacks gluten. Oatmeal is popular breakfast in USA, UK and many European countries. Rolled oat is produced by running groats between rollers in a steam chamber and then used as porridge and oaten bread. Oat grains constitute the chief grain food for horses.

Oat hulls are used as a raw material for further industrial solvents which is widely applied in oil refining, nylon, synthetic rubber production and manufacture of antiseptics. Oat hull is useful at packing material and fuel. Oat straw is used as feed for domestic animals and also as animal bedding. Oats are also grown for hay and frequently in crop rotation (Kochhar, 1998).

6. Karyotaxonomic Evidence: *Avena* species exhibit three ploidy levels *viz.*, diploids (2n=2× =14), tetraploids (2n=4× =28) and hexaploids (2n=6× = 42), all being regular bivalent formers. Cultivated forms with non-shedding grains occur in each group: *strigosa-brevis* diploids (fodder oats of negligible agricultural significance); tetraploid *abyssinica* (confined to Ethiopia); and, most important, as hexaploids of the *sativa-byzantine-nuda* group. Either *magna* or *murphyi* can be regarded as the tetraploid ancestor of *sterilis* and cytogenetic studies implicate *strigosa* as a genome donor to the hexaploids. The occurrence of at least one other diploid species can be conjectured common ancestor to *magna, murphyi* and *sterilis*; contributing the characteristic spikelet morphology of these three species, which is lacking in all diploids known to date. The diploid donors of the third genome of *sterilis* and the second genomes of *magna* and *murphyi* are still unknown.

The *barbata* group of tetraploids is autopolyploid in origin and derived directly from the *strigosa* diploids (Holden, 1966). It does not appear, from either cytogenetic or morphological evidence, to have contributed to the evolution of *sterilis* but rather to have derived separately as a distinct group of wild, weedy and cultivated forms.

Independent evolution of at least two tetraploid groups is indicated. The two major cultivated species, *sattua* and *byzantina*, are diversed taxonomically by the position at which the rachilla fractures during threshing (a difference governed by one or two major genes). *Nuda* forms appear free-threshing variants derived from *sativa* or *byzantina* having characteristic distribution range, eastwards from the centre of origin into China. All three types lack a dispersal mechanism and can exist only under cultivation. The two candidates for the position of hexaploid progenitors are *viz., sterilis* and *fatua*. Both have seed dispersal mechanisms; in *sterilis* the unit is the spikelet (minus glumes); in fatua, it is the single grain and both have effective seed dormancy systems. The former occurs as a major component of the herbaceous flora around the shores of the Mediterranean and through Asia Minor to the slopes of the Zagros mountains in the east. It forms dense stands in the (often relic) oak park-forest belt in relatively stable primary habitats. Geographically, it overlaps its putative tetraploid and diploid ancestors. It is also an aggressive pioneer of disturbed land and abandoned agriculture and a troublesome weed of crops. The latter is virtually confined to crops and disturbed land on the margins of cultivation. Self supporting populations in primary habitats are unknown and *fatua* should be regarded as a specialized weedy type, derived either directly from weedy ecotypes of sterilis or, perhaps, from *sativa or byzantine, sterilis* may, therefore, be regarded as the basic hexaploid type from which the others evolved. (Lardizinsky, 1971; Rajhathy and Sadasivaih, 1969; Rajathy and Thomas, 1994).

7. History: The first oats appear about 1000 BC. and then in central Europe (Helbaek, 1959). Oats are generally regarded as a secondary crop that evolved in western and northern Europe from the weed oat components of the primary grain crops wheat and barley. Oats grow best in cool, moist and maritime climates. As cultivation of 'einkorn' and 'emmer' moved westwards and northwards from the gene centre, they became progressively less; well adapted and were superseded by the weed oat component. Recent researches demonstrate the aggressiveness and adaptability of *A. sterilis* and support this view of the origin of the cultivated hexaploids. The critical genetic changes (a solid, non-shedding grain base and loss of dormancy) that occurred at this time would have been at a strong selective advantage under cultivation. Towards end of the first century AD. hexaploid *sativa-byzantina* oats appear to have become established as a major crop in Europe.

It appears likely that cultivated diploid forms of the *strigosa* group evolved, possibly in Iberia. Relics of this culture are not found on the poorest soils of the northern and western fringes of the European land mass. Although diploid oats were probably widely cultivated as dual purpose grain-fodder varieties in early times, they were replaced by the early hexaploid forms carried northwards and westwards by the migrants.

A literary reference of the growing of Pilcorn (a naked oat) in the British Isles in mediaeval times is available by John Gerard in Herball (Coffman, 1961). It is considered as a form of naked diploid *nudibrevis*, since lost to cultivation, or of hexaploid *A. nuda*. The culture of naked oats (*A. nuda*) has been practised since early times in China and the mountainous parts of southwestern Asia. It seems to have been restricted to this region. The unpopularity of *nuda* in Europe is traditionally because of poor storage characteristics and poor germination of seed grain. This is probably due to exposed embryo and its liability to damage during threshing.

The cultivated tetraploid, *abyssinica*, is opined by Zohary (1971) as a 'tolerated man-dependent weed' restricted to Ethiopia. It probably originated from weedy *barbara* types introduced into Ethiopia with barley from the Middle East. It is sown, and consumed as an integral part of the barley crop. The various diploid and tetraploid cultivated types are of little agricultural importance presently and played little part in the development of oats as a world wide crop. The large grained *satiua-byzantina* hexaploids, on the other hand, had become well established in Europe in time for them to play their part in the colonization of North America, Argentina and Australia. They have, of late, returned to the Middle East as crops from Europe.

During 17th to 19th centuries, the European 'landraces' were almost certainly heterogeneous mixtures of homozygotes with the capacity for some adaptive response to the conditions of their new habitats. The common names of the early oat varieties of North America indicate their origins in Russia,

Poland, Sweden, Finland, France, England, Scotland and Ireland (Coffman, 1961). Also, colonization of temperate regions of the southern hemisphere was coupled with the introduction of the landraces characteristic of the country of origin of the colonists. The genetic variability of these landraces permitted an adaptive response to environmental change. It was an important factor undoubtedly in the establishment of a successful culture of old world crops in new places.

8. Further Developments: There were two basic types on which conscious selection was first practised in the latter half of the 19th century.

Attempts to improve oats were first made, probably more or less contemporaneously in several European countries at the end of the 18th and the beginning of the 19th centuries. Similar successful attempts were followed in America by the middle of 19th century. In all cases, the material for selection was the variation, often considerable, in the landraces grown at the time. The principles of selection differed from broadly based mass selection to rigid single-plant pedigree schemes. The most effective improvement followed single-plant selection and landraces were supplanted by the first pure line varieties. As a result, genetic variation declined and further progress by simple selection came to a halt. This resulted in the widespread use of higher yielding, more uniform spring and winter varieties and a reduction in the *sativa-byzantina* gene-pool.

The next phase appears started with the use of controlled hybridization to generate new variability. This approach was rapidly adopted by many breeders in Europe and North America. Parents were selected on phenotype with the object of combining desirable complementary characters in the progeny. Many noteworthy varieties were yielded which existed in cultivation for several years.

Breeding has been based subsequentially on inter-crosses between the original pure line varieties, maintained in breeders' collections, and their derivative. Sources of new genetic variation have been obtained through exchange of cultivars between different countries and continents. Major-gene resistances to fungal diseases (rusts, smuts, mildew) generally lacking in *sativa* varieties, have been especially transferred from *byzantina* types. Exceptionally, and more recently, breeders have outcrossed to ludoviciana (a weedy *sterilis* type) for resistance to eelworm and powdery mildew. There are no attempts to exploit the gene pool of the related wild and weedy forms of *sterilis* and *fatua* and none at all to explore the resources of putative tetraploid and diploid ancestral stock.

Mushtaq Ahmad *et al.* (2013) assessed genotype environment interaction and determined stable oat (*Avena sativa* L.) cultivar in Kashmir division over three locations for grain yield and its components in 10 genetically diverse genotypes using randomized block design. They observed considerable variation in grain yield within and across environments. They conducted stability analysis for grain yield to check the response to genotype × environment (G×E) interactions. The mean squares due to G×E (linear) were significant. Mean squares due to pooled deviations were highly significant, reflecting considerable differences among genotypes for non-linear response. Out of 10 genotypes, only two oat lines *viz.*, Sabzaar and SKO-208 showed non-significant deviation from regression and their regression coefficient value of 0.011, the smallest deviations from regression (S^2di) value of the highest grain yield could be considered the most widely adopted cultivar. The other test cultivars were sensitive to production-limiting factors, their wider adaptability, stability and general performance to the fluctuating growing conditions within and cross environments being lowered.

Mut *et al.* (2015) investigated one hundred varieties of worldwide origin to study hay yield and quality of oat genotypes harvested at the late milk stage. They recorded significant differences between the tested oat varieties varied in respect of plant height (76.2 to 141.2 cm), hey yield (6.03 to 11.83 t ha^{-1}), crude protein ()58.8 to 136.49 kg^{-1} dry matter), acid detergent fiber (333.2 to 424.89 kg^{-1} dry matter) and neutral detergent fiber (522.5 to 652.49 kg^{-1} dry matter). Total digestible nutrients raged from 465.1 to 583.39 kg^{-1} and relative feed value (80.9 to 112%). Cluster analysis grouped the 100 genotype within seven clusters, each of which having 17, 21, 13, 12, 2012 and 5 genotypes. Of these genotypes *viz.*, Sisko. Akiyutaka, Longchamp. Sanova, Matre. Flamingslord and Reviser were identified as the high hay yield potential genotypes. However, quality traits of these genotypes were lower than some other genotypes. Furthermore, some macro minerals were insufficient and others were in excess regarding healthy feeding. These authors concluded that to meet animal needs in oat-based forage systems, crossing high yielding genotypes with genotypes having high quality should be proposed for future breeding programs.

9. Epilogue: The competitive disadvantage of oats, as a cash crop, compared with wheat and barley, is principally due to the high percentage of husk. Differences in gross yields of grain/hectare for the three cereals are not sufficient to account for the decline. Stray attempts are on record in the past to breed improved free threshing varieties of *nuda* but with little success. Yield increases were achieved, but the varieties so produced were all *nuda* type, with large, pendulous, multiflorous spikelets and rather weak straw, serious lodging hazards in wet and windy climates. The breeding of modern sativa types with free threshing grain would doubtless reverse the declining area under the crop. While some modest projects with this aim have been started. It is clear that a much greater effort is needed if the considerable problems involved are to be solved. The natural advantages of oats over wheat and barley, in terms of animal nutrition, then become exploitable attributes. The potential of wild populations remains to be determined. Isolated researches into variation in protein and fat content in Avena and examples of the successful manipulation of chemical composition of other crops all point to the possibility of reviving the culture of oats as a high energy crop of high nutritional value. Resistance to the major fungal pathogens is certain to remain a major breeding objective but new approaches and new genetic resources are required. Again, sterilis with its wide range of wild and weedy ecotypes would appear to be the most likely source of new variation.

2.4 RICE

Oryza sativa L. (Poaceae)

Resume: *The Indian subcontinent is the ancestral home of Oryza sativa. Archaeological remains are on record dating as India 2500 BC., China 2750-3280 BC. and Thailand 3500 BC. The recorded history of its cultivation in China goes back to the 3rd millennium BC. It is mentioned in Vedic texts viz., Yajurveda and Atharveda as 'Vrihi'. O. sativa is diploid, however, wild rices are tetraploids. Rice varieties are either glutinous mealy endosperm contains dextrin instead of starch or non-glutinous (possessing starchy endosperm). Rice equals wheat in importance as a staple food for mankind. A closely related annual wild form viz., O. nivara is found in the Deccan Plateau (India) and parts of southeast Asia and Oceania.*

1. Introduction: Rice is an indispensable food of over world's half population. Densely populated countries *viz.,* China, India, Indonesia and Japan are rice-eaters. Major share, about 95%, of rice production, is met with these countries. Thus rice replaces all other cereals as the staff of life. It dominates the economic and social existence particularly in tropical countries.

In China, a ceremony was established about 2800 BC. by the Emperor Chin-nuna. The reigning Emperor has himself to sow the rice. It is used to pelt newly wed couples in order to bring them good luck and assure them of many children. The Chinese were the first to cultivate rice (about 4000 BC.). In classical Chinese works, the words 'agriculture' and 'rice' are used synonymously. Remains of rice were found in the Yung Shao excavations in China, dating as far back as 2600 BC. Some assume origin of rice in China, but it appears untenable as wild relatives of rice are rare in this country.

In India, the rice entered extensively into the ordinances of Hinduism. Rice is used as votive offerings at many religious ceremonies associated with birth, marriage and death. It is a symbol of fertility and wealth. In the Ordinance of Manu called 'Manusmriti', ancient legal text of Hindus (2nd century BCE to 3rd century CE), the word 'food' and 'rice' are used synonymously. Sanskrit names in ancient literature *e.g.* 'dhanya', 'vrihi', 'syali', 'jiva-sadhana', 'tanonu' and 'shashtika' are traceable to roots which mean or denote life, existence and subsistence. The word 'vrihi' is met with in the Atharaveda (1500-1000 BC.). The most general local names of rice in southern India and Bengal are suggestive of rice cultivation. Then rice has spread under cultivation from these Indian habitats to all over and ultimately crossing the Himalaya. The factors such as marshy expenses, periodic inundations and low intervening mountains are ideally suited to the rice cultivation. Carbonised rice has been recovered during archaeological excavations of Lothal (Gujarat) in India, dated to 2300 BC. Lothal is thought to be a southward extension of the Harappa and Mohenjo-daro culture of the Indus Valley Civilization (Sindhu Culture). It is similarly recovered from Rangpur (2000-1800 BC.), Navdatoli (1550-1400 BC.) and Hastinapur (1000-800 BC) in India. Thus its origin, cultivation and spread in India can be strongly supported by the evidences revealed by habitat, wild relatives, archaeological excavations, social and religious culture, philological studies of local names for rice and ancient Sanskrit scriptures. It is mentioned in Vedic literature as 'Vrihi' especially in Yajurveda and Atharveda and also in other ancient Sanskrit literature *viz.,* Brahmana, Kalpasutra, Panini's Astadhyayi and Vartik, and Patanjali's Mahabasya. It is named also as 'Raktashali' in Kalpsutra, as 'Sastik' in Panini's Astadhyayi and Vartik and simply as 'Shali' in Kalpsutra, Panini's Astadhyayi and Vartik, and Patnjali's Mahabhasya (Singh, 2008).

2. Plant Characteristics: Semi-aquatic, free tillerinig height of 2 to 4 ft; culms noded; leaves linear, alternate; panicle composed of many branches by a husk; each spike compressed, 1-flowered, short-pedicelled, lemma and palea make the hull or husk attached to the grains (caryopsis).

3. Useful Parts: Grains.

4. Chemical Constituents: Carbohydrates (80%), protein (7.1%) fat (0.66%), minerals (calcium, iron, magnesium, phosphorus, potassium, sodium, zinc, copper, manganese, selenium), thiamine (B1), riboflavin (B2), niacin (B3), pantothenic acid (B5), vitamin (B6), folate (B9), vitamin-E, vitamin (K9).

5. Uses And Applications: In South-East Asia, rice is the most important cereal or diet. It is consumed in the form of cooked plain boiled rice often along with cooked pulses, curd, vegetables, etc. In India, it is used as food articles like kheer, firni (boiled rice in milk and sugar) and pulao. Particularly in South India, fermented preparations *e.g.* idli, dosa and uppama are prepared. Rice is also eaten in the form of parched rice and parched paddy. Its flour is used in ice-creams, puddings, pastry and confectionery. Rice starch is used in laundry, in cosmetic industry as a thickener, in calico printing, textile finishing and for making adhesives, glucose and dextrins. Alcoholic drinks *e.g.* 'sake' in Japan and 'Wang-tsin' in China are prepared from rice after fermentation of fungal agents. Rice bran oil is edible.

Hulls (husk) are useful as a fuel, bedding for poultry, for insulation and packing. Furfural is made sometimes from hulls. Ash obtained from burnt hulls is used as a filler in concrete and bricks and also as a source of sodium silicate in making soaps, polishes and some cleaning agents. Bran and rice polish are useful as stock-feed. Straw is useful for cattle feeding, manufacture of straw-boards, ropes, baskets, mats, sacks and paper making.

6. Karyotaxonomic Evidence: Chang (1970) has listed 20 species and their distributions. *Oryza sativa, O. glaberrima* and their wild relatives are diploids (2n = 24), while more than half the wild rices are tetraploids (2n = 48).

The 'sativa complex' includes *O. sativa* and its wild relatives (*O. rufipogon* and *O. nivara*) as well as *O. glaberrima* and its relatives (*O. barthii* and *O. longistaminata*). Members of this complex have genome A which is subdivided into A, Ab, Acu, and Ag to indicate partial sterility and minor pairing aberrations in hybrids between members of different subgenomes. The A genome shows partial homology with genome B (*O. minuta, O. eichingeri* and *O. punctata* are BBCC) and with genome C of *O. officinalis* (International Rice Research Institute, 1964).

The putative ancestor of *O. sativa* include: (i) *O. officinalis*, (ii) *O. fatua* (referred to here as spontanea forms of *O. sativa*), (iii) *O. perennis* (an ambiguous name often used to include *O. rufipogon, O. nivara* and *O. fatua* of Asia, or *O. longistaminata* and *O. barthii* of Africa, or even both Asian and African wild relatives of the two cultivated species and (iv) *O. perennis subsp. balunga* (The Asian race of *O. rufipogon*). Studies of the origin of *O. glaberrima* have pointed to either the rhizomatous *O. longistaminata* (formerly called *O. barthii*) or the annual wild race, *O. barthii* widely known as *O. breviligulata*) as the putative ancestor (Chang, 1964; Nayar, 1973).

7. History: Many workers have thought that the Indian subcontinent is the ancestral home of *O. sativa*. The earliest archaeological evidence from India goes back only to 2500 BC. On the other hand, rice glumes belonging to the neolithic period in China have been dated to 2750-3280 BC. and the recorded history of rice cultivation in that country goes back to the third millennium BC. Some sonsidered south-east Asia as the centre of origin. Recently excavated pottery from Thailand dated to 3500 BC., bears imprints of rice glumes. In general, there is no archaeological evidence about time and place.

There are two cultivated species of rice, the Asian and the African. The cosmopolitan O. perennis complex has been proposed as the probable common progenitor (Oka, 1974). The two cultigens represent two independent and parallel domestications. The pan-tropical and subtropical distribution of the wild relatives of the two cultivated species in Africa, south and southeast Asia, Oceania and Australia, and Central and South America strongly warrant a common progenitor which existed in the humid zone of the Gondwanaland continents prior to their drifting.

Among the wild relatives of *O. sativa*, the perennial and weakly rhizomatous, *O. rufipogon* is widely distributed over southern and southeast Asia, southern China, Oceania and South America, usually in deep water swamps. A closely related annual wild form, *O. nivara,* occurs in the Deccan plateau (India), many parts of southeast Asia and Oceania. The habitats of *O. nivara* are ditches, water holes and edges of ponds. Morphologically similar to *O. nivara* are the very widely distributed spontanea forms of *O. sativa* (*O. fatua*), which represent numerous intergrading hybrids between *O. sativa* and its two wild relatives. Throughout southern and southeast Asia, the spontanea rices occur in canals and ponds adjacent to rice fields and in the rice fields themselves.

The wild and the above weed races are long awned (often pigmented), with a high frequency of natural cross-pollination, light and shattering spikelets and strong seed dormancy. These traits enabled them to persist as weeds in spite of the hot, dry season alternating with the monsoon. Because of extensive hybridization, typical specimens of *O. rufipogon* and *O. nivara* are now rarely found. Also rice agriculture has become more intensive and hence many wild populations have depleted from their known homes.

The ultimate progenitor of *O. sativa* was a perennial form. Domestication most likely started from an annual which may have resembled *O. nivara.* Similarly, *O. glaberrima* is most likely originated from the annual *O. barthii.* Although annual wild forms appear to be the immediate progenitors of both the Asian and African cultivated rices, the associated weed races (the spontaneas in Asia and *O. barthii* in Africa) have played a significant role in the development of the more recent cultivars (IRRI 1964; Harlan, 1969). Recent samples of the weed races suggested that introgressive hybridization took place and that gene flow from the cultivated rices to the wild forms. The principal barriers to gene flow between cultivated and wild forms are hybrid sterility, non-viability or weakness. Some are governed by complementary or duplicate genes (Oka, 1974).

Several morphological and physiological changes occurred in the evolution of *O. sativa.* Larger leaves, longer and thicker culms and longer panicles resulted in a larger plant size. There were also increases in the number of leaves and in their rate of development, in the number of secondary panicle branches, grain weight, fate of seedling growth, tillering capacity, synchronization of tiller development and panicle formation. There was a slight increment in the net photosynthetic rate of individual leaves and the period of grain filling lengthened. Hand in hand, there were decreases in pigmentation, rhizome formation, ability to float in deep water, awning, shattering, duration of grain dormancy, photoperiod response and sensitivity to low temperatures. A decline in the frequency of cross-pollination resulted into inbred than its wild ancestors. These changes were called forth by semi-natural selection imposed by the facts of cultivation in diverse climates, soils and seasons and under diverse cultural practices (Oka and Morishima, 1970).

The continuous distribution of *O. rufipogon, O. nivara* and the spontanea forms of *O. sativa* in a belt 2,000 miles long from the foothills of the Himalayas to the Mekong region (Chang, 1975) suggests a diffuse origin. The domestication of a crop is not necessarily restricted to the centre of diversity of its wild relatives. The area of greatest diversity of cultivated forms may also provide a more useful clue to the centre of domestication. Based on evidences from philology, palaeo-climatology and ethnology, the area including northeastern India, northern Bangladesh and the triangle adjoining Burma, Thailand, Laos, Vietnam and southern China appears to be the primary centre of domestication. From this region, rice was introduced into the Yellow River valley of China where the temperate race (Keng or Japonica) evolved. From China, rice was introduced to Korea and, later, to Japan (300 BC.). The tropical race (Indica or Sen) was introduced into the Yangtze River valley about AD. 200. The tall, large and bold grained bulu varieties (javanica) of Indonesia appear to be more recent derivatives from the tropical continental forms. The bulu varieties spread from Indonesia to the Philippines, Taiwan and Japan.

Oryza glaberrima exhibits less diversity and its distribution is limited to tropical West Africa (Porteres, 1956; Chang, 1975). *O. glaberrima* probably originated about 1500 BC. or later. The swampy area of the upper Niger is its primary centre of diversity. It has two secondary centres which lie to the southwest, on the Guinea coast (Porteres, 1956). In the field, *O. glaberrima* is often grown mixed with its annual weed race. Africans gathered both species, as well as the rhizomatous. *O. longistaminata*, for food. *Oryza glaberrima* is more closely related to *O. barthii* than it is to *O. longistaminata*.

8. Further Developments: Man made full use of natural variability in the crop, spontaneous mutations, natural hybrids and introductions from foreign lands. The introduction of the early maturing Champa varieties from central Vietnam into China in the 11th century was largely responsible for the practice of double cropping in south China. The significant change in Thailand from predominantly bold-grained rices between the 14th and 18th centuries to long-grained varieties 200 years later, illustrates the relative ease with which rice varieties could be replaced.

Trials of local varieties were started in Japan about 1893 and the earliest report of rice breeding by hybridization came from the same country in 1906. The hybrid variety Ominishika was developed by 1913. Hybridization of rice in the United States began about 1922 when breeders crossed varieties introduced from Japan, China, India and the Philippines. These programmes emphasized nitrogen responsiveness, smooth hulls and stiff straw. Varieties selected or bred in the USA later spread to South America and Australia and there attained local dominance.

The first variety of deliberate hybrid origin was bred in Indonesia about 1913. After the late war, tropical and Japanese varieties were crossed extensively under a region-wide Indica-Japonica hybridization project. Hybrid sterility and extreme differences in ecological adaptation between the parents were obstacles. The modest progress attained in improving rice yields during the periods 1934-1938 and 1956-1960 was achieved largely through pure-line selection and, to a more limited extent, by hybridization but was not sufficient to cope with the increase in population in tropical Asia (Parthasarathy, 1972).

Crosses among Japanese material led to the development of the ponlai varieties. These were non-seasonal, high yielding and adaptable over a wide area inclusive of the tropics, subtropics and even several temperate areas. Later, in the mid 1950s, Indica type of rice was developed, because of its short straw and profuse tillering, and outperformed the Japonica type in nitrogen-responsiveness and ability. The success of Taichung Native-1 promoted wide adoption of semi-dwarfs in India during 965-66 (Huang *et al.,* 1972).

A cross between the tall tropical variety Peta (from Indonesia) and the subtropical semi-dwarf Dee-geo-woo-gen (from Taiwan) produced the semidwarf IR8. The IR8 selection established records of grain yield. IR8 and many other improved semi-dwarfs from several national breeding programmes or from IRRI combined most of the desired features in the improved plant type that was sought: plant stature of about 100 cm; erect, relatively short, dark green leaves; high tillering; stiff culms; early maturity and photoperiod insensitivity; nitrogen responsiveness and a high ratio of grain to straw (Shastry, 1974).

Since pre-historic times, rice has featured in the agriculture of South and East Asia. New sights have been offered by recent improved methods of archaeobotanical research. Archaeobotanical evidence revealed gradual evolutionary process of domestication through rice spikelet bases and grain size change. Separate trends in grain size change are identified in India and China. The earliest centre of rice domestication was in the Yangtze basin of China, but a largely separate trajectory into rice cultivation is traced in the Ganges plains of India (Fuller *et al.,* 2010). It is revealed that early rice cultivation in the Yangtze basin and southern China was based on wet and paddy-field systems from early before 4000 BC. Early rice in northern India and Thailand was predomently dry rice at 2000 BC., with a transition to flooded rice for India at C. 1000 BC. (Fuller *et al.,* 2011). Evidence from ecology, modern genetics and archaeology together inform about early domestication of rice. Early rice cultivation followed two paths towards domestication in India and China, with selection for domestication traits in early Yangtze 'japonica' and a non-domestication feedback system inferred for 'proto-indica'. The protected domestication process finishd about 6500-6000 years ago in China and around two millennia later in India, when hybridization with Chinese rice took place. Later on, farming populations grew and expanded by migration and incorporation of pre-existing populations. These dispersals took place from China southwards by the Sino-Tibetan and Austronesian groups. In South Asia, dispersal of rice took place after Indo-Aryan and Dravidian speakers adopted rice from speakers of lost languages of northern India (Fuller, 2012).

Pragnya *et al.* (2018) evaluated 75 specialty rices, called soft rice genotypes, collected from various parts of North-East India to study genetic diversity, to estimate genetic parameters among the genotypes of yield and component traits. The authors grouped the genotypes into seven clusters based on the relative magnitude of D^2 values. They found cluster I as the largest comprising of 62 genotypes, cluster II with 08 genotypes and cluster III, IV, V, VI and VII with one genotype each. Based on the inter-cluster distances, they suggested crossing between the genotypes as: IV & VII, III & VII, I & VII and III & VI. The data on characters means for seven clusters indicated that the cluster III is having highest mean value for per cent filled grains per cereals panicle, cluster V for seed yield per plant and cluster I for test weight (1000-grain weight) and cluster II for number of productive tillers per plant and cluster IV for panicle length. The characters per cent filled grains per panicle, plant height, seed yield per plant contributed maximum to the genetic divergence. These characters should be given priority, according to these authors, for selection of genotypes for further crop improvement programmes.

Muhammad Sabar and Muhammad Arif (2004) conducted study using eight genotypes of diverse origin to explore their response against variable drought stress. Two water stress treatments for one week and two weeks were given sixty days after seeding. Significant differences for genotypes and water stress levels were observed from phenotypic performance. Correlation studies indicated a positive and significant association of paddy yield with spikelet fertility and 1000 grain weight. Rice genotype IR55419-04 showed least effect of water stress treatments for 1000 grain weight i.e. 6.28% and 40.01% reduction and paddy yield per plant of 24.97% and 51.35% under one week and two weeks water stress respectively. Basmati 2000 and Super Basmati were found to be the most sensitive to drought for paddy yield with 92.8% and 91.5% reduction under severe water stress given for two weeks respectively.

Rice strain IR55419-04 showed the potential for drought stress tolerance amongst all the tested genotypes and need attention of the breeders to explore the genetic tolerance through modern mapping techniques and then incorporating it through advance biotechnological approaches.

Gangashetty *et al.* (2013) carried out to know the association among yield and yield attributing traits in local non-basmati aromatic genotypes of rice. Totally 42 genetically diverse genotypes during Kharif 2010 were analyzed. The authors recorded the significant and positive association for number of tillers per plant, number of productive tillers per plant, panicle weight, grain breadth and test weight with grain yield per plant at both genotypic and phenotypic levels. The negative significant correlation noticed for plant height and days to 50 per cent flowering at both genotypic and phenotypic levels. These associated characters, in their opinion, will help for selection process during handling of segregating material for various traits. The high direct effect on grain yield per plant was observed for number of tillers per plant, number of productive tillers per plant, panicle weight and test weight at both genotypic and phenotypic levels. The authors concluded that these characters must be taken into consideration while selecting genotypes for high grain yield.

9. Epilogue: Majority of the semi-dwarf varieties emerged from primary crosses. Wider crosses and random mating of hybrids, along with a judicious use of induced mutations or protoplast fusions, have promise for obtaining unusual recombinations or breakup of tight linkages.

The semi-dwarfs are mostly grown in areas of the tropics where water control is effective, where the soil is free of toxicity or deficiency problems and where there are no climatic constraints. To assist farmers in less favoured areas, varietal adaptation requires tolerance of deep water and photoperiod sensitivity in ill-drained areas, drought resistance in upland or rain-fed lowland fields, tolerance of adverse factors in problem soils, and tolerance of low temperatures at high elevations. Incorporation of the desired features into improved varieties for specific environments requires both multi-disciplinary evaluation and the utilization of diverse germplasm.

2.5 RYE

Secale cereale L. (Poaceae)

Resume: *It is of more recent origin. It resembles to barley in habit and to wheat in grains. The region between the Austrian Alps and the north of the Casparian seas is thought to be the original areas of rye. Some consider Secale montanum and S. antolicum to be wild ancestors of rye. It was domesticated in the 1st millennium BC. It is a classic example of a secondary crop having arisen as a weed of wheat and barley. It is diploid with chromosome number n=7. It is primarily a grain crop but also used as fodder. It is chiefly used for bread in European countries, besides as a source of some alcoholic drinks.*

1. Introduction: Amongst the cereals, rye is of more recent origin. De Candolle (1886) thought the region between the Austrian Alps and the north of the Casparian seas original area of rye. Aitchinson (1888) considered *Secale montanum*, a wild species of Afghanistan and Turkistan, to be the wild ancestor. Some considered *S. antolicum* of Asia minor as the progenitor. No traces of rye have been found in the Swiss Lake Dwellings or ruins of Egypt. It was known to Greeks and Romans. It was probably domesticated in the 1st millennium BC. It resembles to barley in habit and to wheat in grains.

2. Plant Characteristics: Annual reaching upto six ft., with slender and tough tillers or stalks; leaves bluish; a large number of spikelets clustered into heads, each spikelet with two fertile florets and borne singly at the joints of axis; grains resembling wheat grains.

3. Useful Parts: Grains, straw.

4. Chemical Constituents: Grains contain protein (13. 4%) (rich in gluten), flat (1. 8%), carbohydrates (80. 2%) and fibres (2-3%).

5. Uses And Applications: It is chiefly used in European countries for bread (known as 'black bread' because of dark colour). Grains are useful as stock feed. Grains are valuable as a source of Dutch gin, rye whisky and alcohol. Rye is used for hay and pasturage, as a winter cover to prevent erosion and leaching, as a sand binder, and in crop rotation. The straw has demand in making of hats, manufacture of paper, packing and bedding purposes.

6. Karyotaxonomic Evidence: All Secale species are diploid ($2n = 2x = 14$) including from artificial polyploids. There are numerous reports of B-chromosomes in both the wild and cultivated outbreeders *S. montanum* and *S. cereale*. The taxonomy of *Secale* is rather dubious. Roshevitz (1947) recognised 14 species but it is uncertain whether, in fact, all these should be accorded specific ranks. Some taxonomists recognize fever, even as few as five species. Two distinct groups of species (or subspecies) can readily be distinguished as being important in the evolution of the cultivated forms. The first group is a group of annual weeds *e.g. S. ancestrale, S. dighoricum, S. segatale* and *S. afghanicum* which cytologically similar to each other and cultivated rye. Actually these should be included as subspecies of *S. cereale*. This group is virtually restricted to agricultural lands, the weedy types being widespread in cereal crops in northeastern Iran, Afghanistan and Transcaspia (Zohary, 1971). Another group is an aggregate of wild perennial races widely distributed from Morocco eastwards through the Mediterranean countries and the plateau region of central and eastern Turkey to northern Iraq and Iran. These have sometimes been accorded distinct species status but are most probably best considered as variants of a single species, *S. montanum*. Also members of this group resemble cytologically to each other and completely interfertile. They differ from the *S. cereale* complex by two major interchanges (reciprocal translocations) involving three pairs of chromosomes.

Time and again, there are two other species which have been implicated in the evolution of cultivated rye. These are: (i) *S. vavilovii*, an annual self-pollinating species of limited geographical distribution, and (ii) *S. sylvestre*, an annual inbreeder but widely distributed rom central Hungary to the steppes of southern Russia. There is general concensus that *S. sylvestre*, although morphologically distinct from *S. montanum*, has a similar karyotype. The status of *S. vavilovii* is still doubtful. It is not clear whether the chromosome arrangement is the same as that of *S. cereale* and therefore different from *S.*

montanum by two reciprocal translocations involving three pairs of chromosomes (Stutz, 1972) or whether it differs from *S. montanum* by one reciprocal translocation and from *S. cereale* by two (Khush, 1962).

7. History: Rye is a classical example of a secondary crop. The evidence confirms the original hypothesis of Vavilov (1971) in general. Having first arisen as a weed of wheat and barley, it would have been introduced as a crop in its own right in later period. It is likely that the immediate progenitors of the cultivated form would have emerged from the aforesaid weedy races. They are chromosomally identical to it. It is plausible that these weedy races would themselves only have evolved with the development of agriculture in western Asia. It is generally agreed that the original ancestor of these weedy races (and also of cultivated rye itself) was *S. montanum*. It is perennial, outbreeding and widely distributed.

The change from truly wild *S. montanum* to the agriculturally dependent *S. cereale* complex is still not clear. It is thought that no other distinct intermediate species were involved (Khush and Stebbins, 1961; Zohary, 1971). The question as to how a major structural rearrangement, involving a double translocation, could have become established in the first place, bearing in mind the initial handicap of reduced fertility which would certainly have been incurred. Adaptive superiority of the original structural heterozygote was thought likely (Riley, 1955; Khush and Stebbins, 1961). Eventually, structural homozygosity for both translocations became established. The partial fertility barrier between the new chromosome arrangement and the old would serve as an isolating mechanism thus preventing the swamping' of the evolving weedy races of *S. cereale* by their progenitor *viz., S. mont*anum

A stepwise evolution of *S. cereale* from *S. montanum* was proposed (Stutz, 1972). The annual weedy races were thought derived from the introgression of *S. montanum* into *S. vavilovii*. This annual inbreeder was in turn derived from the annual *S. sylvestre* as a consequence of chromosome translocation. *S. sylvestre*, itself capable of a degree of cross pollination, would have had the perennial outbreeder *S. montanum* as its progenitor.

This hypothesis implies the evolution from the cross-fertilized *S. montanum* of self-fertilization (*S. sylvestre* and *S. vavilovii*) followed by cross-fertilization once more in *S. sereale*. This is appears a drawback. An attractive feature, however, is that the fixation of the double translocation is more easily conceived because it is assumed to have occurred in a self-pollinating species (*S. vavilovii*).

The probable place of origin of the weedy races of rye is near about the central and eastern Turkey, north-western Iran and Armenia. This area of maximum genetic diversity coincides in part, at least, with a considerable percentage of variability of the perennial *S. montanum*. The northward spread of wheat and barley cultivation from the more fertile areas of the Jordan valley (now Israel to this plateau region) would have penetrated one of the natural habitats of *S. montanum* and some of it allies. The cold, harsh climate of this area is not ideal for the two main cereals. It is reasonable to consider that under such conditions rye could have become established as a cornfield weed. It is doubtless that the weedy annual races (with brittle or semi-brittle rachises) were and still are, successful colonizers. A second probable area showing considerable genetic diversity of these annual types exists further east centred on Turkestan, Afghanistan and north eastern Iran.

After establishment as a weed of the larger grained cereals, it is so far to imagine the kind of selection, that would have been exerted. An erect culm would result from selection for competitive ability. There would also have been unconscious selection for genes controlling non-shattering of the ears. Cultivations sowing would have favoured to destroy seedlings originating from seeds dropped by shattering spikes of the previous year. Seeds of non-shattering genotypes would have been harvested with the main cereal crop and subjected to the same cultural procedures during the subsequent season. Also, there would have been selection for larger grain, since even the most primitive winnowing procedure favours the retention of grains approaching the size of those of barley and wheat.

The actual region of domestication of the crop may be doubtful. It was probably domesticated at several locations independently and presumably within the general area marked out above. The timing of the event (or events) is also difficult to determine with any accuracy since archaeological information is lacking. A reasonable estimate for domestication is about 3000 BC. and, for introduction into Europe, about 2000-2500 BC. (Khush, 1963). The latter might have occurred by two separate routes, first, northwards and westwards through the Caucasus from the primary centre and secondly through central Asia from the secondary centre further east.

8. Further Developments: The information on spread of rye cultivation especially in Europe during Roman and post-Roman times is scarce as well as archaeological evidence. Some account is made available by Helbaek (1971). Rye was probably grown originally in the same areas as the other temperate cereals. Its toleranc of low rainfall, cold winters and poor light soils rendered it suitable to northern and eastern Europe and parts of European Russia. It had become the major cereal of the region. The 'black bread' was popular in eastern Europe. In addition, the elongate straw was valued for thatching. Rye bread was the main cereal food of a third of the population of Europe upto the early 20th century. Since then, however, it has been gradually replaced by wheat.

Conscious plant breeding began, in the later part of the 19th century, several locally adapted 'land races' arose and the early plant breeders relied heavily on them as their immediate source of variability. Breeding methods have, naturally, been much influenced by the outbreeding nature of the crop. Rye has a gametophytic two-locus incompatibility system. (Lundquist, 1956). Early breeding techniques were in the form of simple recurrent selection. With improved knowledge of the genetic structure of outbreeding populations, more sophisticated methods then appeared. Rye breeding have not been dominated by aspects of disease resistance. Improvement of grain yield, protein content and quality together with cold tolerance and shorter straw have been the objective breeding. The poisonous sclerotia (*Claviceps purpurea*) which totally replace the grain and occasionally get into flour was reported as causing hallucinations in people and abortion in cattle. The utilization of rye as a forage has led to the breeding varieties bred solely for the purpose. Emphasis was then placed on characteristics other than grain production, total dry matter yield, digestibility of the herbage being important growth in winter and early spring.

Targonska *et al.* (2016) evaluated genetic diversity in *Secale cereale* L. based on SSR markers. The plant used 367 rye accessions comprising historical and contemporary cultivars, cultivated materials, landraces and breeding strains from the Polish breeding company Danka. They analysed genetic diversity among these accessions and determined population structure, principal coordinate analysis (PCoA), neighbor-joining (NJ) and Bayesian clustering. They also defined a core collection of 25 rye accessions representing over 93% of SSR alleles. The results showed that accessions from the rye gene bank are clearly divergent in comparison with materials received directly from European breeding companies. They revealed a relatively limited genetic diversity within contemporary rye cultivars and breeding strains, indicating the common genetic background and germplasm exchange. They also found a lack of correlation between clustering of rye accessions, including improved cultivars and their geographic origin. The results indicated a need for broadening the genetic diversity in breeding programme.

Hansen *et al.* (2004) studied grain characteristics, chemical composition and functional properties of rye as influenced by genotype and harvest year in 19 different cultivars in one location in up to three years. The cultivars included eight adopted hybrid, seven adapted population cultivars and four non-adapted population cultivars. The results showed a significant influence on both harvest year and genotype on features stated above. They recorded variations in respect of plant height, harvest yield and protein content mainly due to genotype differences and to a lesser extent to differences among harvest years. The kernel weight, hardness index and content of dietary fiber components, however, were more strongly influenced by the harvest year than the genotype. Differences in starch properties, amylograph peak viscosity and temperature at peak viscosity were strongly influenced by the harvest year. The water absorption was strongly influenced by genotype effects, compared to yearly differences.

9. Epilogue: Some other developments are of greater potential evolutionary importance. First, there are autotetraploids (4×=28) (Kranz, 1973). These have been rather neglected, even for grain production, for some years and this use is likely to develop. In common with other synthetic autopolyploids, there is a decrease in fertility which, however, in rye is compensated by increased grain size and protein content. Response to selection for fertility has been slow but some progress is made. Secondly, there is the potential of interspecific and intergeneric hybrids. Attempts to breed perennial rye for forage production are common. Hybrids of *S. cereale* and the perennial *S. montanum* have shown considerable promise in terms of dry matter production. High yield can be fixed in genotypes which are both perennial and sufficiently fertile.

2.6 TRITICALE

Triticosecale sp. (Poaceae)

Resume: *It is the result of first attempt to synthesize a new crop species from intergeneric hybridization between Triticum (wheat) and Secale (rye). Thus it is a made-made amphidiploid cereal with tetraploid, hexaploid and octaploid levels of ploidy. They belong to an artificial genus Triticosecale. The very first variety of Triticale was released in Canada. Triticale consistently exhibited higher contents of protein and essential aminoacids than wheat.*

1. Introduction: Triticale is a small-grain man-made cereal. It is the result of first successful attempt to synthesize a new crop species from intergeneric hybridization. As implied by common name, Triticale is the outcome of the combination of the two genera *viz.*, wheat (*Triticum*) and rye (*Secale*). Breeders succeeded in the synthesis of hybrids (amphidiploids) between wheat and rye as a means of combining the desirable winter hardiness of rye with the agronomic and commercial properties of wheat. Triticale resembles morphologically to wheat in plant-type and kernel characteristics. The main difference lies in its greater vigour relative to wheat and its larger spike and kernel size. Depending upon the varieties (or species) of wheat and rye used in its synthesis, triticale can be of either spring or winter-habit and can be used as either a grain or forage.

2. Plant Characteristics: A bunch type grass with upright like wheat and rye; stem hollow, smooth; roots fibrous; leaves rolled in the whorl; spike with one spikelet at each node containing many florets; seeds wheat-like, of same size and colour.

3. Useful Parts: Grains, straw.

4. Chemical Constituents: Protein (26%), Carbohydrate (46%), fats (3%), minerals (Ca,Cu, Mg, P, K, Zn, Fe, Mn), Vitamins A, C, D, E, K, B_6, B_{12}.

5. Uses And Applications: It is mostly grown as fodder. Triticale-based foods are available at health food stores. Flour is used for biscuits and breads. It is used as protein supplement.

6. Karyotaxonomic Evidence: Hexaploid (2n=6×=42) and octaploid (2n=8x=56) both forms of triticale are developed. The 6x forms are obtained from the hybridization of tetraploid wheats (2n=4×=28) with rye (2n=2×=14), with subsequent doubling. The 8× triticales arise similarly from the combination of hexaploid wheats (2n=×=42) with rye. Octaploid triticales are more easily made than hexaploids, hence received more attention during the early years. With the advent and improvement of embryo culture and chromosome doubling techniques, however, the hexaploid form has proved more useful as it is more seed fertile. Tetraploid forms (2n=4×=28) have also been experimentally developed (Krolow, 1973) but, in their present state, are agronomically inferior to forms of higher ploidy.

Usually, triticale has exhibited varying degrees of aneuploidy, depending upon the wheat-rye parentage and also upon the level of ploidy. Hexaploid triticale, generally exhibits greater cytological stability than other forms.

The development of hexaploid triticale, of late, as a new crop has centred on the production of 'secondary' hexaploids. They are superior to primary hexaploids and the improvement is attributable to modified cytoplasm carried by the octaploid parent (Larter and Hsam, 1973; Hsam, 1974). Moreover, the combination of the two forms (the octaploid, genomically AABBDDRR, and the hexaploid, genomically AABBRR), results in viable hexaploids in which R-chromosomes are replaced by D-chromosomes of common wheat. The 'pivotal' genome AB from the tetraploid wheat remain unaltered while the D-R genomes become reconstructed through substitution. Secondary hexaploids selected from these recombinants have been found to possess expected agronomic traits from both the D and R genomes (Gustafson and Zillinsky, 1973). Recently, the identification of each of the rye chromosomes by heterochromatic banding enabled to decipher the rye chromosomes that have been replaced by D-genome wheat chromosomes in a number of hexaploid lines (Darvey and Gustafson, 1975).

7. History: Natural hybrids between rye and wheat were frequently observed. Most of them were sterile and consequently lost. Rimpau (1891) noted partially fertile tillers on an otherwise sterile plant. It was the first spontaneous report. The frequency of such hybrids, however, too low to be of practical value. The improvement of embryo culture and chromosome doubling techniques in the 1930s encouraged further work *e.g.* the first reports that of Muntzing (1939). He began work in 1932, using an amphidiploid (8×) which arose as a restitution product from a wheat (6×) X- rye cross. He crossed this hybrid with the early Rimpau strain from Germany. He concluded that superior triticales could be developed through traditional methods of plant breeding. By 1950, 18 years after the initiation of his programme, Muntzing's best triticale lines were already yielding 90 per cent of the bread wheats grown at that time. Their fertility and seed quality were steadily improved. Later, Pissarev (1966) who worked in Russia in the early 1940s also reported progress in octaploid material.

With primary hexaploids meanwhile suggested that they were superior in meiotic stability and fertility relative to octaploid forms (Kiss 1966; Sanchez-Monge, 1958). There was thus renewed interest in the crop but at the hexaploid instead of the octapoid level. The first large-scale programme in North America was initiated at the University of Manitoba, Winnipeg, Canada in 1954 under the direction of Drs. L. H. Shebeski and B. C Jenkins. The early stages of the programme were stared with the objective of the introduction, synthesis and evaluation of primary hexaploids under Canadian conditions. Intercrossing of promising lines started 1958 and, by 1967, the grain yields of at least a few lines were equal to those of the then standard Canadian bread wheat varieties. Feeding trials meanwhile indicated that triticale grain was nutritionally equal to wheat and barley and preliminary commercial-scale milling tests showed that the grain could be used for human consumption. Various forms of experimental breakfast cereal were made having a characteristic 'nutty' taste. Good quality bread was made from the flour by using modified mixing and fermentation times (Lorenz *et al.,* 1972).

8. Further Developments: The very first triticale variety was released for commercial production in Canada. It was the variety Rosner, a product of a double cross involving four hexaploids: three from *T. durum* (4x) and one from *T. persicum* (4×) (Larter *et al.,* 1970). Rosner represented a considerable improvement in straw strength, fertility and earliness as compared to earlier triticales. Its adaptability, however, was limited because of partial sensitivity to day-length. There was also a need for performance of improved yield. On an international basis, the initiation of a triticale programme by the Centro Initernational de Majoramiento de Maizy Trigo (CIMMYT) in 1965 provided the necessary stimulus to develop triticale. Using the Canadian material as the initial breeding stock, lines were quickly developed that were both day-length insensitive and highly fertile (Zillinsky and Borlaug, 1971). These lines were called 'Armadillo'. They represented a major advance. In 1971, the Canadian International Development Agency (CIDA) approved major financial support for an international triticale developmental programme involving CIMMYT and the University of Manitoba. The main objective was the development of triticale for human consumption. Results from advanced trials in Canada in 1973 showed that 60 per cent of all lines tested out yielded Rosner and 2 per cent out yielded the highest yielding Canadian wheat variety. Also, Zillinsky and Borlaug (1971) reported CIMMYT results which showed triticale as outyielding wheat for the first time since the initiation of the programme in Mexico in 1965. Gustafson *et al.,* (1972) reported that yields of some triticales exceeded 6700 kg/ha in California, USA, a return which closely approached that of the highest yielding wheat varieties. Similar reports from other workers indicated that progenies derived from the 'Armadillo' type material were widely adapted than the later-maturing, light-sensitive strains previously used.

Dominik Losert *et al.* (2017) analysed phenotypic and genotypic diversity and breeding progress in European triticale (X Triticosecale Wittmack). Such study is central for the optimum design of breeding programmes. They evaluated a collection of 885 diverse European triticale lines in multilocation field trials in two years. Based on a subset of the population of 121 varieties registered in Europe between 1983 and 2014, they observed a substantial breeding progress for grain yield with a significant rate of increase of 53 kg/ha or 0.67% per year. All lines were genotyped by a genotyping-by-sequencing approach yielding 58888 polymorphic markers. These analyses revealed the absence of major population structure but a certain grouping of lines according to their origin. Breeding history and results on triticle germplasm will provide important information for breeding programs and selection of gain in this crop for future.

McGoverin *et al.* (2011) reviewed uses and the effect of growth environment on grain quality of triticale. The potential of triticale has remained largely unrealized. Since its first cross, triticale has mostly been used as animal feed. Growing demand for food resources has led to an increased interest in triticale development. Triticale, in past, has not become a routine replacement for wheat in the baking industry. Triticale has been surveyed for use in biofuel production and nitrogen capture in double-cropping system. Many triticle quality parameters, and therefore cultivar suitability for biofuels, animal feed or baking, are significantly influenced by both environment and genotype. These authors concluded that triticale could satisfy many of the hopes originally placed upon it, and may be useful in foodstuffs and fuel, but only when growth environment is carefully considered.

9. Epilogue: Triticale has consistently showed higher contents of protein and essential amino acid than wheat. Villegas *et al.,* (1970) reported extensive analyses and found a wide range upon which selection could be based. Feeding trials in Canada proved that certain lines of triticale support higher growth rates in rats, mice and voles than other cereal grains, except the oats. The quality of triticale as a forage and grazing crop has been paid less attention than its potential as a grain. However, in certain regions of the United States fall-sown triticales have been used successfully for grazing and have exhibited excellent recuperative virtues.

2.7 WHEAT

Triticum Spp. (Poaceae)

Resume: *The wheats have had a multiple origin. The first record of grains of domesticated wheat date about to 7500-6500 BC. in the 'Fertile Crescent' (Mesopotamia). Its cultivation is prehistoric in the Old World, particularly* *in the temperate parts of Europe, Asia and Africa. It is mentioned as 'Godhum' in Vedic treatises viz., Yajurveda and Atharveda. De Candolle strongly favoured India as the home of some forms of wheat. The various species of Triticum form a polyploidy series based on ×=7 and consisting of three different ploidy levels viz., diploids, tetraploids and hexaploids. It is the staple food of at least one-third of world's population. It is an important source of carbohydrates in human diet. It is also rich in proteins. Its pre-eminence over most other cereals is largely due to its baking quality.*

1. Introduction: The history of cultivated wheat and that of human civilization have been closely interwoven. In the course of domestication, the wheat plant has lost the ability to disseminate its seeds effectively and is now completely dependent on man for dispersal. But man has fostered this cereal to such an extent that it is now the world's foremost crop plant. This, in turn, has led to community settlement, population increase and rapid cultural evolution. The plant is high yielding in a wide array of environments from 67°N in Norway, Finland and Russia to 45°S in Argentina but in the subtropics and tropics. Its cultivation is confined to higher elevations. The main wheat-producing regions are southern Russia, the central plains of the United States and adjacent areas in Canada, the Mediterranean basin, Argentina and south-western Australia, north-central China and India.

The long period of cultivation of wheat has resulted in the production of various species and varieties as a result of selection. Generally different eight types are recognized: (i) Einkorn (*T. monococcum*), (ii) Emmer (*T. dicoccum*), (iii) Spelt (*T. spelta*), (iv) Polish Wheat (*T. polonicum*), (v) Poulard Wheat (*T. turgidum*), (vi) Club Wheat (*T. compactum*), (vii) Durum Wheat (*T. durum*) and (viii) Common Wheat (*T. aestivum*).

The cultivation of wheat is prehistoric in Old World. Very ancient Egyptian monuments show the establishment of this cultivation. However, when Egyptians or Greeks speak of its origin, they associate it to mythical deties *viz.*, Isis, Ceres, Triptolemus, etc. A small-grained wheat has been found at the earliest lake dwellings of Western Switzerland, the inhabitants of which were at least contemporary with the Trojan war, and perhaps earlier. The same wheat, *Triticum vulgare antiquorum* Heer, was found by Unger in a brick of the pyrmid of Dashur in Egypt dating 3359 BC. Another small-grained wheat, *T. vulgare compactum muticum* Heer, was less common in Switzerland in the earliest stone age. A third intermediate form was cultivated in Hungary at the same period. However, De Condolle (1886) believes that the wheat cultivation in the temperate parts of Europe, Asia and Africa is probably older than the most ancient known languages. The Chinese certainly grew it in 2700 BC. They thought it a gift direct from heaven. In the annual ceremony in China of sowing five kinds of seeds, instituted by the Emperor Shen-nung or Chin-nung, wheat was one of the crop species.

De Candolle (1886) mentioned in his 'Origin of Cultivated Plants' that the original home of wheat in very early prehistoric times was in Mesopotamia. According to Berosus the earliest of all Western Historians, and a Chaldean priest who wrote some twenty-three centuries ago, it occurred wild there. According to Strabo, (born 50 BC.) a grain very similar to wheat grew wild upon the banks of the Indus river.

De Candolle (1886) opined that there is strong evidence in favour of India being the home of some of the forms of wheat as can be shown for any other part of the globe. India possesses perhaps as comprehensive a series of time honoured forms of wheat as can be shown for any other country. India has its hard wheats and soft wheats; its starch wheats and spelt wheats; its beared and beardless wheats. It has been established that most of these have been grown for countless ages on the same fields as they are to be found at the present day. The wheat grains are discovered during Indus valley excavations at Mohenjo-Daro. Accordingly, north-western India was one of the ancestral lands of this cereal (Pal, 1966). The carbonized grains resemble those of *Triticum sphaerococcum* Percival, an endemic species which is still found in a few places in India. Vavilov opined that the wheats have had a multiple origin, the soft wheats flourished in the mountains of Afghanistan and the southerwestern Himalaya; the durum wheats came from Abyssinia, Algeria and Greece, whereas einkorn is from Asia Minor The archaeological evidences establish that the wheat has been cultivated for at least 6000 years. It is mentioned as 'Godhum' in Vedic texts *viz.*, Yajurveda and Atharveda and also in other ancient treatises *viz.*, Brahmana, Upnishada and Kalpasutra (Singh, 2008). This is indicative of its domestication even before vedic period in India.

2. Plant Characteristics: Erect annuals, culms cylindrical, nodes solid, internodes hollow, roots seminal as well as adventitious, tillering; leaves linear-lanceolate, 20-37 cm long, lingulate; inflorescence terminal, 2-stichous spike, each spikelet consisting 2 to 5 florets, each floret with lemma and palea, grain dry, indehiscent (caryopsis), deeply furrowed ventrally.

3. Useful Parts: Grains.

4. Chemical Constituents: Protein (8-15%), fats (1. 5-2. 0%), carbohydrates (60-80%), vitamin-E, thiamine, miacin, riboflavin, foliac acid, pantothemic acid, biotin, minerals *e.g.* calcium, phosphorous, iron, magnesium, potassium, sodium, magnesium, zinc.

5. Uses And Applications: Wheat is used for human consumption in the form of leavened bread, roti, chapatti, nan, paratha, cakes, biscuits, pastries, crackers, etc. Thus it is a staple food in most part of the world. Industrial uses include the manufacture of starch, gluten, distilled spirits, malt, pasta, etc. Wheat products include peeled wheat, wurid wheat, Bulgur (paraboiled product), wheat flakes, puffed wheat, shredded wheat, etc.

Wheat straw is used as feedstuff for livestock, animal bedding and compost, for corrugated paper, insulated building chairs, stuffing mattresses, straw carpets, baskets, hats, packing and thatching, etc.

6. Karyotaxonomic Evidence: The various species of *Triticum* form a polyploid series based on ×=7 and consisting of three different ploidy levels: diploids (2n=2×=14), tetraploids (2n=4×=28), and hexaploids (2n=6×=42) (Kihara, 1954; Morris and Sears, 1967; Riky, 1965). The diploid species are in wild presumably monophyletic in origin. They have diverged considerably from each other. This divergence is especially visible in the morphologically well defined seed dispersal units of the species and their specific ecological requirements, besides geographical distributions. Each diploid contains a distinct genome (Kihara, 1954). The related chromosomes of the different genomes show no affinity with each other and do not pair regularly in interspecific hybrids. They lead to sterility and isolation of the diploid species from each other.

The polyploid species evolved through amphidiploidy. They behave like typical genomic allopolyploids. Their chromosomes pair in a diploid-like fashion and the mode of inheritance is disomic. The allopolyploid nature of the *Triticum* polyploids is tested from cytogenetic analysis of hybrids between species of different ploidy levels. Each polyploid species is marked out as a product of hybridization followed by chromosome doubling. The different genomes are intimately related (Morris and Sears, 1967) and hence polyploid wheats are segmental rather than typical genomic allopolyploids. The diploid-like behaviour of polyploid wheats is due to suppression of pairing of homologous chromosomes by a specific gene. In case hexaploid *T. aestivum*, this gene is located on the long arm of chromosome 5 of genome B and known as the 5BL gene (Riley, 1965). Plants devoid of this gene behave like segmental allopolyploids. Their homologous chromosomes pair and form multivalents.

To date, three groups of polyploids are identified (Zohary and Feldman, 1962). Species of each group possess one genome in common and differ in their other genomes. Polyploids of group A share the genome of diploid wheat, *T. monococcum*; those of group D share the genome of *T. tauschii* (= *Ae. squarrosa*); and those of group Cu genome of *T. umbellulatum* (= *Ae. umbellulata*). The polyploids of each group resemble the diploid donor of the common pivotal genome in basic morphology and the structure of the seed dispersal unit. The cultivated polyploid wheats belong to group A and the arrowhead shaped dispersal unit of wild diploid wheat, *T. monococcum* var. *boeoticum*, can be identified in all the wild polyploids, whereas the non-brittle ear of cultivated *T. monococcum* var. *monococcum* expressed in all the cultivated polyploids. Group A polyploids comprise the tetraploids, *T. turgidum* (AABB) and *T. timopheevii* (AAGG), and the hexaploid *T. aestivum* (AABBDD).

Hybrids obtained from the two tetraploid species divulge partial asynapsis and high sterility. In other words, the B genome of *T. turgidum* is non-homologous with the G. genome of *T. timopheevii*. The two tetraploid taxa could arose independently from crosses between wild forms or *T. monococcum* and two distinct diploid species or could have a monophyletic origin. Hexaploid *T. aestivum* (AABBDD) contains two genomes homologous with the A and B genomes of *T. turgidum*. Hence *T. aestivum* arose from hybridization between *T. turgidum* and a diploid species having genome D. The donor of the D genome of hexaploid wheat is recognized as *T. tauschii* (Morris and Sears, 1967). Synthetic hexaploids are produced from crosses of different varieties of *T. turgidum* with *T. tauschii* and simulate certain established hexaploids. Hybrids between synthetic and natural hexaploids are generally completely fertile.

The AB genomes of hexaploid wheat could have been donated by either wild or by cultivated *T. turgidum*. While distribution of the wild tetraploid *T. turgidum* var. *dicoccoides*, overlaps with that of *T. tauschii* to only small area if any. Cultivated *T. turgidum*, especially var. *dicoccum*, has been grown throughout the distribution area of *T. tauschii* at one time or another in western Iran and eastern Turkey. Because of the different mode of seed dissemination of wild tetraploid wheat and of *T. tauschii*, the amphidiploid which could evolve from these two species would have been unable to disseminate its seeds effectively and would therefore have been depleted early. It is likely that the hexaploid originated in wheat fields after the spread of cultivated *T. turgidum* into the distribution area of *T. tauschii*. Such an origin could clarifies why hexaploid wheat has no wild relatives in nature.

Hexaploid wheats were probably formed recurrently by numerous hybridizations involving different genotypes of tetraploid varieties and various races of *T. tauschii*. The donor of the D genome grows today within and at the edges of wheat fields in Iran and Armenia. Natural hybrids between tetraploids and *T. tauschii* can be found there and they usually set some seeds with 2n = 42 chromosomes due to the formation of unreduced gametes. *T. tauschii* also has ample geographical contacts with both wild and cultivated *T. timopheevii* (AAGG), however natural hexaploid derived from these two species is yet unknown.

The B genome has so far defied conclusive identification. Morphological, geographical and cytological evidence has been used to implicate *T. speltoides* (genome S) or a closely related species (Riley, 1965). However, cytogenetic data (Kimber, 1974) suggests that *T. speltoides* may have donated the G genome of *T. timopheeii* rather than the B of *T. turgidum*. An incomplete homology between S and G is accounted for by the supposition that G has undergone some modification at the tetraploid level.

If it is assumed that the B and G genomes are monophyletic, B could have differentiated from the G of wild *T. timopheevii* or from an earlier tetraploid of constitution AASS. Accordingly, both B and G would have to be thought as modified S genomes. The occurrence of a modified genome or genomes side by side with a stable one is characteristic of polyploids of group A but also of all polyploids of the genus *Triticum*. Such a constitution is conceived (Zohary and Feldman, 1962) to have originated from hybridization between initial polyploids sharing one genome and differing in one or two additional ones. In these hybrids, the two differential genomes, brought together from different parents, can exchange genetic material and become assimilated to each other. Accordingly, in the polyploids of group A, the initial amphidiploid (AASS) could have exchanged chromosome segments with other amphidiploids or with diploids such as *T. longissimum* (genome S1), *T. bicorne* (genome Sb) or others. The S genome could have become modified resulting into G and B.

T. speltoids (genome S) is in contact with wild *T. monococcum* (*i.e.* var. *boeoticumi)* in eastern Turkey, northwestern Iraq and western Iran. There are numerous mixed populations in that area of these two species, which contain also wild *T. timopheevii* and, more sporadically, wild *T. turgidum*. The main distribution of wild *T. turgidum* var. *dicoccoides* is, however, southwest of this speltoides-monococcum area. It occurs in northeastern Israel, northwestern Jordan and southern Syria, wherein it does not form mixed populations with *T. spleltoids* but flourish there with *T. longissimum*.

7. History: The first record of grains of domesticated wheat date to about 7500-6500 BC. found in prehistoric sites of mountain chains flanking the plains of Mesopotamia, called 'Fertile Crescent', and the Syrian desert and also in Anatolia and the Balkans. The domestication of wheat should have taken place in the fertile crescent being the centre of their wild progenitors (Harlan and Zohary, 1966). Wild diploid wheat ('einkorn', *T. monococcum* var. *boeoticum*) occurs naturally in open herbaceous oak park-forests and steppe-like formations in southwestern Iran, northwestern Iraq and southeastern Turkey. Wild einkorn is also very common in secondary habitats in this area and beyond it in Transcaucasia, Syria, central and western Turkey and Greece. Wild tetraploid or 'emmer' wheat, *T. timopheevii* var. *araticum* and *T. turgidum* var. *dicoccoides*, occur in primary habitats as the wild diploid but are less continuous in their distribution. *Triticum timopheevii* grows in western Iran, north-western Iraq, eastern Turkey, Armenia and Transcaucasia, whereas *T. turgidum* is distributed in southern Syria, nothwestern Jordan and northeastern Israel. *T. turgidum* has also been recorded in certain sites in the southernmost area of wild *timopheevii*.

One of the characters which was of no pre-adaptive advantage for cultivation is the wild mode of seed dissemination. Wild wheats are characterized by a brittle spike. This disarticulates into arrowhead-shaped spikelets when mature. These seed dispersal units facilitate self-burial in the soil and hence protection during the long dry summer and successful germination after the first rain. But they must have proved a nuisance to the ancient farmer as he had to collect most of the spikelets from the ground or cut the culms before the grains were mature. Types with non-brittle heads were uncounsciously selected from the earliest stages of domestication later by ancient farmers. The loss of self-propagation was followed by the loss of self-protection. The wild forms have tightly closed glumes resulting in a 'hulled' grain after threshing. Only a few of the modern cultivated wheats retain this feature. Mutations which spread under domestication led to loosely closed glumes. These release the naked 'free-threshing' grain.

The wild, brittle diploid wheat, *T. monococcum* var. *boeoticum* and the cultivated, non-brittle var. *monococcum* are hulled types. The earliest known carbonized grains of brittle diploid wheat exist in the prehistoric settlement of Tell Mureybit, northern Syria (7th millenium BC.) where the cereal was apparently collected rather than farmed. (Renfrew, 1973; Helbaek, 1966.) At somewhat later sites (*e.g.* Ali Kosh in Iranian Khuzistan, 7500-6750 BC.; Jarmo in Iraqi Kurdistan, ca. 6750 BC.; Cayonii Tepesi in southeastern Turkey, ca. 7000 RC.; and Hacilar in west-central Anatolia, ca. 7000 BC.) brittle diploid wheat found side by side with a non-brittle type and is gradually replaced by it. Towards the end of the 7th millennium, there are also records of nonbrittle *T. monococcum* in the southern Balkans. It is significant that no remains of either wild or cultivated *T. monococcum* have been found in irrigated settlements in the lowlands of the Mesopotamian plains or in the Nile valley, to which wheat farming spread in the 5th millennium. Cultivated einkorn (*T. monococcum*) spread to central and western Europe through the Danube and Rhine valleys during this millennium. In subsequent eras *viz.*, Bronze and early Iron Ages, einkorn wheat attained a wide distribution in Europe and the Near East. In modern times, it occurs only as a relic crop in mountainous parts of Yugoslavia and Turkey. *Monococcum* wheat is resistant to cold and to rust diseases.

The wild tetraploids *viz., T. timopheevii* var. *araraticum* and *T. turgidum* var. *dicoccoides* cannot be marked out morphologically. Present day geographical distribution of the two species, the carbonized grains, spikelets, and clay impressions found at Jarmo (ca. 6750 BC.) and at Cayonu Tepesi (ca. 7000 BC.) could belong to both of them. The brittle form of *T. timopheevii* gave rise to only a restricted number of non-brittle cultivars, all of which are cultivated in Armenia and Transcaucasia, that of *T. turgidum* is the predecessor of most cultivated tetraploid and hexaploid wheat. The non-brittle, hulled tetraploid, wheat *viz., T. turgidum* var. *dicoccum* (cultivated emmer) existed in prehistoric villages of the Near East as early as 7500 BC. Together with cultivated einkorn the non-brittle emmer spread rapidly to all 7th millenium farming areas of the Near East, including valleys not penetrated by

T. monococcum cultures. Brittle forms of both diploids and tetraploids were harvested in the wild state throughout the fertile crescent long before actual farming.

Emmer was the most principal cereal in the early farming in the Near East. From the mountainous areas of the fertile crescent, it reached to the lowlands of Mesopotamia in the 6th millennium BC. and, during the 5th and 4th millenia, to Egypt, the Mediterranean basin, Europe, central Asia and India. It reached Ethiopia some 5,000 years ago by the Hamites. Presently, emmer is grown on a small scale in Ethiopia, Iran, Transcaucasia, eastern Turkey and the Balkan.

The *durum* wheat evolved probably from cultivated emmer by an accumulation of mutations that reduced the toughness of the glumes culminating into free-threshing. Most other naked grain tetraploid wheats as probably of relatively recent origin, deviated from var. *durum* in only single characters and share the genetic system and expressed the free-threshing habit. *T. turgidum* var. *carthlicum* is the tetraploid known exclusively to have a different gene complex determining free threshing. This complex, known as the Q factor, also occurs in hexaploid wheats (Morris and Sears, 1967). The narrow distribution of var. *carthlicum* in Transcausia may suggest that this variety derived recently by hybridization of an unknown tetraploid with a hexaploid of the *aestivum* group.

Hexaploid *T. aestivum*, probably evolved and entered cultivation only after the simultaneous domestication of diploid and tetraploid forms. Archaeological research suggests it from the 7th millenium BC. The earliest finds, attributed to club wheat, *T. aestivum* var. *compactum*, are at Tell Ramad, Syria (about 7000 BC.). Finds from the 6th milleniurn, identified as ancestral forms of free-threshing hexaploids, have been unearthed at Tepe Sabz in Iranian Khuzistan, at Tell es-Sawwan in Iraq, at Catal Hiiyuk in central and Hacilar in west-central Anatolia, and at Knossos on Crete. Between 6000 and 5000 BC., *T. aestivum* penetrated along with cultivated emmer, into the irrigated agriculture of the plains of Mesopotamia and western Iran and, in the 5th millenium, into the Nile basin. In the 5th millenium, *T. aestivum* also occurs from the central and western Mediterranean basin. Compact forms of hexaploid wheat were under cultivation in central and western Europe at the end of the 4th milleniurn. They are associated, together with einkorn and emmer, with the first traces of agricultural operations.

While tetraploid wheats, in keeping with their Near Eastern origin, are adapted to mild winters and rainless summers, the addition of the central Asiatic *tauschii* (D) genome must have contributed to the adaptation of hexaploid wheats to more continental climates. This could have promoted the spread of bread wheat into central Asia and, by way of the highlands of Iran, to the Indus valley, where it occurred at the beginning of the millennium BC.

The earliest recorded evidence in China is from the middle of the 3rd millennium.

The hulled hexaploid wheats are more primitive than the free-threshing forms. The genetic data are not in accord with archaeological chronology. While *aestivum* was abundant in the prehistoric Near East from the 6th millenium onwards, there are no indications of the cultivation of *spelta* or other hulled forms before 2000 BC. Moreover, in central Europe, *spelta* appears about 1,000 years later than compact forms of free-threshing wheat.

The economically more advantageous free-threshing hexaploids may have competed with or replaced hulled tetraploid emmer at a time when free-threshing tetraploids had not yet evolved. Var. *aestivum* is by far the most important hexaploid and is now grown on a world-wide scale. It has given rise to *compactum* and *sphaerococcum* through mutations. Var. *compactum* is grown today in certain restricted areas of Europe, the Near East, and the northwestern United States, whereas *sphaerococcum* is grown to some extent in India and central Asia and is known from India as early as the 3rd millenium BC.

8. Further Developments: Modern wheat cultivars have evolved through three main phases of selection: (i) subconscious selection by the earliest food grower simply by the processes of harvesting and planting, (ii) mainly deliberate selection in the polymorphic field of the primitive farmer; and (iii) scientifically planned modern breeding. The main attainments of the first phase were non-brittle spikes, simultaneous ripening of grains, rapid and synchronous germination, perhaps also erect versus prostrate culms and, to some extent, the free-threshing trait. Through the expansion of wheat culture into new areas during that phase, an additional achievement was wider adaptation to different environments. Primitive farmers selected and planted the grains most desirable for their specific needs. The efforts of selections resulted in increased yield, larger seed size, better flour quality and adaptation to a wider range of climatic and farming regimes. Since numerous genotypes were grown in mixtures in the field, the genetic material at the disposal of the primitive farmer would have been enhanced by the production of occasional recombinants. Under modern breeding, the wheat field has become genetically uniform and is no longer conducive to spontaneous gene exchange. On the other hand, gene migration on a large scale has been promoted by world-wide introduction services. Today's selection techniques can achieve the objectives of the primitive farmer with much greater certainty. One of the modern breeding aims is the production of early maturing varieties which can, by virtue of a short life cycle, escape adverse climatic conditions and certain diseases. In breeding for grain quality, the main achievements have been improvements of milling and baking characteristics.

Of late, more light has been shed on wheat origin and domestication. About 25000 different cultivars of wheat crop have been reported in the world.

Most of these are varieties of two groups *viz.,* (i) common wheat, and (ii) durum wheat. Common or bread wheat (*Triticum aestivum*) accounts for 95% of all the consumed wheat in the world today, the other 5% is durum or hard wheat (*Triticum turgidum* ssp. *durum*) used in pasta and semolina products. Bread and durum wheat are both domesticated forms of emmer wheat (reported variously at *T. araraticum, T. turgidum* ssp. *dicoccoides* or *T. dicoccoides*). Another early form of wheat called einkorn (*T. monococcum*) was domesticated as about the same time, but has limited distribution today.

The origins of modern wheat are found in the Karacadag mountain region (Southeastern Turkey). Both emmer and einkorn wheats were domesticated some 12000 years ago or so. The earliest collected wheat was wild emmer, at the Ohalo II site, about 23000 years ago. Emmer was first cultivated in the southern Leavant (Netiv Hugdud, Tell Aswad, Mureybet, PPNA Sites), whereas Einkorn is found in the north PPNA sites (Abu Hureyra, Mureybet, Jerf el Ahmar, Gobekli Tepe, other southern PPNA). Splelt (*T. spelta*) and Timopheev's wheat (*T. timopheerii*) were ancient forms of emmer wheat developed by the late Neolithic, but neither has much of a market today.

The earliest evidence for domesticated emmer and einkorn wheats found to date was at the Syrian site of Abu Hureyra dated to the late Epipaleolithic period, the beginning of the Younger Dryas, ca 13000-1200 cal BP. It is also suggested that one-grained einkorn was domesticated in southeastern Turkey after the end of the Younger Dryas (~11600-10400 cal BP). It has been recovered from the earliest layers of the pre-pottery Neolithic B (PPNB) Sites of Nevali Cori and Cafer Hoyuk about 11600 cal BP. (Kris Hirst, 2017).

Shafeeq *et al.* (2006) studied genetic variability of different wheat (*T.aestivum* L.) genotypes/cultivars under induced water stress. They screened a large collection of wheat gene pool using cell membrane stability (CMS) as a screen tool. They screened 50 genotypes/cultivars for CMS by exposing the flag leaf discs with instant drought using PEG (6000). They recorded significant differences among the wheat genotypes for CMS, number of tillers and 10 seed weight. They noted significant interaction between water regimes and genotypes which showed the genetic variation among the 50 cultivars/genotypes. They further noted significant correlation between with 100 seed weight. They also noted the association between the number of tillers and 100 seed weight non-significant. CMS, in their opinion, was the most reliable screening parameter for characterization of cultivars/genotypes for drought.

Bahrani and Joo (2012) studied response of some wheat (*T.aestivum* L.) genotypes to salinity at germination and early seedling growth stages. They used 15 cultivars to investigate the effect of three salinity concentrations including 4, 8 and 16 dSm^{-1} Nacl solution and distilled water as control on their germination percent, radicle and hypocotyle length, seedling fresh and dry weight, radicle and hypocotyle dry weight. They recorded increasing NaCl concentrations reduced germination percentage, radicle and hypocotyle length, seeding fresh and dry weight, radical and hypocotyle dry weight. In all cultivars generation percentage was decreased with increasing salinity up to 8 dSm^{-1} NaCl. Decrease in radicle length was more pronounced than the root in all NaCl salt treatments. This decrease was more prominent in Dez cultivars *viz.*, Inia, Kavir and Chamran appeared the highest salt tolerant, whereas cultivars *viz.*, Dez, Vinak, Cross Adl and Star showed the lowers tolerance. The authors concluded that to select cultivars for better salt stress tolerance and seedling stage, coleoptile and root elongation may be used as breeding criteria. The cultivars *viz.*, Inia and Kavir are useful breeders for future development of salt tolerant cultivars.

Kosner and Pankova (2002) studied vernalization response of winter wheat cultivars. They detected different vernalisation and photoperiod responses of 17 cultivars of wheat (*T.aestivum* L.). They noted the effect of photoperiod sensitivity not sifnificantly changed by vernalization. They further observed that different vernalization responses were probably due to the presence of multiple alleles at *Vrn* loci. The delay in their opinion, depended on the vernalization deficit exponentially: y=parameter (1) + [y_0-parameter (1)] x EXP [parameter (2) x (X-X_0)]. The dependence was shown to be general and significant for the given model in all the studied cultivars. Cluster analysis according to the characterization obtained full vernalization treatment, minimum vernalization requirement, insufficient vernalization and parameters of the dependence showed the relationships between cultivars and enabled their grouping by similar profiles of vernalization, and possibly, of photoperiod response.

Gazi (2012) extended research to study phenological development-yield relationship in Durum Wheat Cultivars under late-season high-temperature stress in a semiarid environment in Maru (Northern Jordan). He evaluated the phenological variation using heat-accumulated system and its relation with yield in 16 durum wheat genotypes. He concluded that variation in length of the pre-heading stage influences grain yield more than variation in length of grain filling. Lengthening the pre-heading period increased grain yield, but lengthening the grain fill stage did not affect the yield. Simultaneously, selection for early maturity and long pre-heating is recommended in the development of early maturing and high yielding genotypes for the Mediterranean semiarid areas. The genotypes *viz.*, Masara-1, Waha-1, Omrabi-5 were the best performing genotypes at late stage drought and high temperature stress conditions as they gave the highest grain yields among genotypes studied. Plants of Omrabi-5 genotype also showed to stay green for longer period after physiological maturity.

9. Epilogue: Our incomplete understanding of the genetic architecture of the wheat is an impediment to further developments. The large number of chromosomes and the occurrence of duplicate loci make polyploid wheat unsuitable for analysis and breeding by conventional genetic methods. A technique which promises to overcome these difficulties makes use of the aneuploid lines produced in the bread wheat variety Chinese Spring. With the aid of these lines and of substitution lines, genes can be assigned to specific chromosomes, linkage maps can be constructed and genetic differences between varieties or even species can be established (Morris and Sears, 1967; Sears, 1969; Law, 1968). Moreover, by using these lines, entire chromosomes or chromosome segments can be precisely transferred from one variety to another. The utilization of wild wheats is another area for research. A simple way to use them is to add their entire genomes to tetraploids or hexaploids. The new crop, triticale, produced from different cultivars of *T. turgidum* (2n=28) and rye, Secale cereale (2n=14), is one promising result of such experiments. Attempts are also made to raise yields through heterosis (Wilson, 1968). The basic tools for producing F_1 hybrid wheat, *viz.,* male-sterile lines and fertility restoration genes, are available. High seed set under field conditions has already been reported for several hybrid combinations. Recent trends are the adding emphasis on gene conservation. Attempts are made to collect and preserve genetic resources for eventual future use, not only from wild relatives but also from primitive or underutilized cultivated varieties.

References:

Aitchinson, Y.E.T. (1888) The botany of the Afghanistan delimitation commission. *Trans Linn. Soc.Bot.London* Ser.2 (Bot.) 3.

Anderson, E. (1943) A variety of maize from Rio Loa. *Ann. Mo. Bot. Gard.* 30:469-476.

Anderson, E. (1945) What is Zea mays? A report of Progress. *Chronica Botanica* 9:88-92.

Anonymous (1968) Barley: origin, botany, culture, winter-hardiness, genetics, utilization, pests. USDA Agriculture Handb., 338, pp. 127.

Bahrani, A. and M. Hagh Joo (2012) Response of some wheat (*T.aestium* L.) genotypes to salinity at germination and early seedling growth stages. *World Applied Sciences Journal* 16(4):599-609.

Bakshi, J.S. and R.S.Rana (1974) Barley. *In:* Evolutionary Studies In World Crops. (Ed.Joseph Hutchinson). Cambridge University Press, London, U.K. pp.47-52.

Bhusal, T.N., Lal, G.M., Marker, S. and G.J. Synrem (2017) Genetic variability and traits association in maize (Zea mays L.) genotypes. *Annals of Plant And Soil Research* 19(11):59-65.

Brandolini, A. (1970) Maize. *In:* Frankel, O. H. and E. Bennett (Ed.), Genetic resources in plants-their exploration and conservation. *Philadelphia* 273-309.

Chang, T.T. (1964) Present knowledge of rice genetics and cytogenetics. Intnl Rice Res. Inst. Tech. Bull., 1, pp. 96.

Chang, T.T. (1970) Rice. *In:* Genetic Resources in Plants. (Ed. Frankel, O. H. and E. Bennett) Oxford, U.K. and Edinburgh, Scotland pp.267-272.

Chang, T.T. (1975) Exploration and survey in rice. *In:* Frankel, O. H. and J. G. Hawkes (eds), Crop Genetic Resources For Today And Tomorrow, Cambridge, UK, pp.159-65.

Coffman, F.A. (Ed.) (1961) Oats and oat improvement. American Society, Madison, Agronomy, USA.

Collins, G.N. (1909) A new type of Indian corn from China. U.S.D.A. *Bur. Plant Industry Bull.* 161:7-30.

Committee on Genetic Vulnerability of Major Crops (1972) Genetic vulnerability of major crops. National Academy of Sciences, Washington D.C., pp. 307.

Darvey, N.L. and P. Gustafson (1975) Identification of rye chromosomes in wheat-rye addition lines and triticale by heterochromatic bands. *Crop. Sci.* 15:239-243.

De Candolle A.P. (1886) Origin of Cultivated Plants (Translated from 2nd Ed. *In:* French, 1959) Hafner, New York, USA.

Dominik Lsert, Hans Reter Maurer, J.J. Marulands and T. Wurschum (2017) Phenotypic and genotypic analyses of diversity and breeding progress in European triticle (X Triticosecale Wittmack) *Plant Breeding* 136(1):18-27.

Eadizinsky, G. (1971) Chromosome relationships between tetraploid (2n = 28) Avena murphyi and some diploid, tetraploid and hexaploid species of oats. *Can. J. Genet. Cytol.* 13:203-209.

Ellis, R.P., Forster, B.P., Robinson, D., Hadley, L.L., Dordon, D.C., Rusell, J.R. and W.Powell (2000) Wild barley: a source of genes for crop improvement in the 21st century. *Journal of Experimental Botany* 51(342):9-17.

Feldman, M. (1968) Regulation of somatic association and meiotic pairing in common wheat. Proc, 3rd intnl Wheal Genet. Symp., Canberra, pp.169-178.

Fuller D.Q., Van Etten J. Manning K., Castillow C, Kingewell-Banham, E., Weisskopf A., Qin L., Sato V. and R.Hijmens (2011) The contribution of rice agriculture and livestock pastoralism to prehistoric methane levels: An archaeological assessment. *The Holocene* 21:743-759.

Fuller, D.Q. (2012) Pathways to Asian Civilizations: Tracing The Original And Spread of Rice And Rice Cultures. Rice 4:78-92.

Fuller, D.Q., Yo-Ichiro Sato, Christina Castillow, Linig Qin, Alison R. Weisskopf, Eleanor J. Kingwell-Banham, Jixiang Song, Sung-Mo Ahn and Jacob van Etten (2010) Consilence of genetics and archaeobotany in the entangled history of rice. *Archaeological And Anthropological Sciences* 2(2):115-131.

Galinat, W. C. (1971) The origin of maize. *Ann. Rev. Genet.* 5,447-478.

Galinat, W. C. (1973) Intergenomic mapping of maize, teosinte, and Tripsacum. *Evolution* 27:644-655.

Gangashetty, P.I. Salimath, P.M. and N.G. Hanamaratti (2013) Association analysis in genetically diverse non-basmati local aromatic genotypes of rice (Oryza sativa L.) *Molecular Plant Breeding* 4(4):31-37.

Gazi N. Al-Karaki (2012) Phenological development – yield relationships in Durum wheat cultivars under late-season high-temperature stress in a semi-arid environment. *ISRN Agronomy* Vol. 2012 : 7 pages.

Goetz Honsel, Vladimir Valkov, Jill Middlefell-Williams and Jochin Kumlehn (2008) Efficient generation of transgenic barley: The way forward to modulate plant-microbe interactions. *Journal of Plant Physiology* 165(1):71-82.

Gustafson, J.P. and F.J. Zillinsky (1973) Identification of D-genome chromosomes from hexaploid wheat in a 42-chromosome triticale. Proc. 4th Intl. Wheat Genet. Symp., Columbia, USA, pp.225-31.

Gustafson, J.P., Qualset, C.O., Prato, J. D., Puri, Y.P., Isom, W.H. and W.F. Lehman (1972) Triticale in California. *Calif. Agr.* 26:3-5.

Hansan, H.B., Miller, B., Anderson, S.B., Jersensen, J.R. and Ase Hansen (2004) Grain characteristics, chemical composition and functional properties of rye (*Secale cereale* L.) as influences by genotype and harvest year. *J.Agric.Food.Chem.* 52(8):2282-2291.

Harlan, J.R. (1968) On the origin of barley. USDA Agriculture Handb. 338:9-31.

Harlan, J.R. (1969) Evolutionary dynamics of plant domestication. *Jap. J. Genet.* 44 (Suppl. 1): 337-343.

Harlan, J.R. and D. Zohary (1966) Distribution of wild wheats and barley. *Science*, N. Y, 153:1074-1080.

Harlan, J.R. and J.M.J. De Wet (1971) Toward a rational classification of cultivated plants. *Taxon*, 20, 509-517.

Harlan, J.V., Martini, M. L. and H. Stevens (1940) A study of methods in barley breeding. USDA tech. Bull., 720, pp. 25.

Helbaek, H. (1959) Domestication of food plants in the Old World. *Science* 130:365-372.

Helbaek, H. (1966) Commentary on the phylogenesis of Triticum and Hordeum. *Econ. Bot.* 20:350-360.

Helbaek, H. (1971) The origin and migration of rye, Secale cereale: A palaeo-ethnoboranical study. In : P. H. Davis *et al.* (Ed), Plant Life of South-West Asia. Edinburgh, pp. 265.

Holden, J.H.W. (1966) Species relationships in the Avenae. *Chromosoma* 20:75-124.

Hsam, S.L.K. (1974) A study of the application of nucleocytoplasmic relationships to the improvement of hexaploid triticale. Ph.D. Thesis, University of, Manitoba, Winnipeg, Canada, pp. 194.

http://dx.doi.org/10.5402/2012/456856.

Huang, C.H., Chang, W.L. and Chang, T.T. (1972) Ponlai varieties and Taichung Native I. In Rice Breeding, IRRI, Los Banos, Philippines, pp.31-46.

International Rice Research Institute (1972). Rice breeding, Los Banos, Philippines, pp. 738.

Iriorte *et al.* (2004) Evidence of cultivar adoption and emerging complexity during the mid-Holocene in the La Plata basin. *Nature* 432:614-617.

Jacobsen, T. and R.M. Adams (1958) Salt and silt in ancient Mesopotamian agriculture. *Science*, N.Y., 128:1251-1258.

Khan, A. (2016) Maize (Zea mays L.) genotypes differ in phenology seed weight and quality (protein and oil contents) when applied with variable rates and source nitrogen. *Plant Biochem. Physiol.* 4:164. Doi 10.4172/2329-9029.1000164.

Khush, G.S. (1962) Cytogenetic and evolutionary studies in Secale-II. Interrelationships of the wild species. *Evolution* 16: 484-496.

Khush, G.S. (1963) Cytogenetic and evolutionary studies in Secale. III. Cytogenetics of weedy ryes and 'origin of cultivated rye. *Econ. Bot.* 17:60-71.

Khush, G.S. and G.L. Stebbins (1961) Cytogenetic and evolutionary studies in Secale-1. Some new data on the ancestry of S. cereale. *Amer.J. Bot.* 48:721-730.

Kihara, H. (1954) Considerations on the evolution and distribution of Aegilops species based on the analyser-method. *Cytologia* 19:336-357.

Kimber, G. (1974) The relationships of the S-genome diploids to polyploid wheats. Proc. 4th Intnl Wheat Genet. Symp. *Columbia*, pp.81-85.

Kiss, A. (1966) Neue Richtung in der Triticale-Zuchtung. *Z. Pflanzenzucht.* 55:309-329.

Kochhar, S.L. (1998) Economic Botany In The Tropics (2nd Ed.) Macmillan India Ltd. Delhi, India.

Kosner, J. and K.Pankova (2002) Vernalization response of some winter wheat cultivars (*Triticum aestivum* L.) Crech. *J. Genet. Plant Breed.* 38(3-4):97-103.

Kranz, A.R. (1973) Wildarten und primitivformen des Roggens (Secale). *Fortschr. PflZucht.* 3:1-60.

Kris Hirst (2017) Wheat Domestication: The History And Origins of Floury Grains. *https://www.thoughtsco.com/wheat-domestication_the_history_170669.*

Krolow, K.D. (1973) 4x-Triticale production and use in triticale breeding. Proc. 4th intnl. Wheat Genet. Symp., Columbia, USA, pp.237-241

Lane Chad S., horn Sally P., Orris Kenneth H., and Claudia I. Mora (2008) The earliest evidence of Ostonoid maize agriculture from the interior Hispaniola. *Caribbean J. Science* 44(1):43-52.

Larter, E. N. (1974) A review of the historical development of triticale. In Tsen (cd.) Triticale first mall-made cereal, American Association of Cereal Chemists, Minnesota, USA, pp.35-52.

Larter, E.N. and S.L.K. Hsam (1973) Performance of hexaploid triticale as influenced by source of wheat cytoplasm. Proc. 4th intnl. Wheat Gene). Symp., Columbia, USA, pp.245-251.

Larter, E.N., Shebeski, L.H., McGinnis, R.C., Evans, L.E. and P.J. Kaltsikes (1970) Rosner, A hexaploid Triticale cultivar. *Can.J. Plant Sci.* 50:122-124.

Law, C.N. (1968) Genetic analysis using inter-varietal chromosome substitutions. Proc. 3rd intnl Wheat Genet, Symp., Canberra, pp.331-342.

Lorenz, K., Welsh, J., Norman, R. and J. Maga (1972) Comparative mixing and baking properties of wheat and triticale flours. *Cereal Chem.* 49 187-193.

Lundquist, A. (1956) Self-incompatibility in rye. I. Genetic control in the diploid. *Hereditas* 42: 293-348.

Mangelsdorf, P.C. (1953) Wheat. *Sci. Amer.* 189:5059.

Mangelsdorf, P.C. (1974) Corn. Its Origin, Evolution And Improvement. Cambridge, Mass, UK.

McGoverin, C., Snyders, F., Muller, N. and M.Manley (2011) A review of triticale uses and the effect of growth environment of grain quality. *J.Sci.Food Afric.* 91:1155-1165.

Morris, R. and E.R. Sears (1967) The cytogenetics of wheat and its relatives. In Quisenberry and Reitz (Eds), Wheat And Wheat improvement, Madison, USA, pp.19-87.

Muhammad Sabar and Muhammad Arif (2014) Phenotypic response of rice (Oryza sativa genotypes to variable moisture stress regime. *Int.J.Aric.Biol.* 16:32-40.

Muntzing, A. (1939). Studies on the properties and the ways of production of rye-wheat amphiploids. *Hereditas* 25:387-430.

Murray, J. (1970) The first European agriculture. Edinburgh University Press, Edinburgh, Scotland.

Mustaq Ahmad, Gul Zaffer, S.M. Razvi, S.D.Mir and M.A. Rather (2013) Stability properties of certain oats (Avena sativa L.) genotypes for major grain yielding characteristics. *International Journal of Plant Breeding And Genetics* 7:182-187.

Mut, Z., Akay, H. and O.D. Erabas (2015) Hay yield and quality of oat (Arena sativa L.) genotypes of worldwide origin. *International Journal of Plant Production* 9(4):507-522.

Nayar, N.M. (1973) Origin and cytogenetics of rice. *Adv. Genet.* 17:153-292.

Nilan, R. A. (Ed.) (1971) Barley Genetics II. Proc. Znd internatl Barley Genet. Symp., Pullman, Washington, USA, pp. 622.

Oka, H.I. (1974) Experimental studies on the origin of cultivated rice. Proc. 13th Intnl. Congr. Cenet. 1:475-486.

Oka, H.I. and H. Morishima (1970) The dynamics of plant domestication: cultivation experiments with wild / Oryza populations. *Evolution* 25:356-364.

Pal, B.P. (1966) Wheat. ICAR Monograph, New Delhi, India.

Parthasarathy, N. (1972) Rice breeding in tropical Asia up to 1960. *In:* Rice breeding, IRR I, Los Banos, Philippines, pp.5-29.

Pearsall, D.M., Chandler-Ezell, K. and A.Chandler-Ezell (2004) Maize in ancient Euador: Results of residue analysis of stone tools from the Real Alto Site. *J. Archaeol.Sci.* 31:423-442.

Perry *et al.* (2006) Early maize agriculture and interzonal interaction in southern Peru. *Nature* 440 (7080):76-79.

Piperno, D.R. (2006) Phytolith Analysis: A Comprehensive Guide for Archaeologists And Paleoecologists. Alta Mira Press, Lanham, MD.

Piperno, D.R., Ranere, A.J., Holst, I., and P.Hansell (2000) Starch grains reveal early root crop horticulture in the Panamanian tropical Brest. *Nature* 407:894-897.

Piperno, D.R., Ranere, A.J., Holst, I., Iriarte, J. and R.Dickau (2009) Starch grain and phytolith evidence for ear ninth millennium B.P. maize from the Central Balsas River Valley, Mexico. *Proc.Natl.Acad.Sci.USA* 106:5019-5024.

Pissarev, V. (1966). Different approaches in Triticale breeding. Proc. 2nd intnl Wheat Genet. Symp., Lund. *Hereditas* Suppl. 2:279-290.

Pohl MED, Piperno D.R., Pope K.O. and J.G.Jones (2007) Microfossil evidence for Pre-Columbian maize dispersals in the neotropics from San Andres, Tabasco, Mexico. *Proc.Natl.Acad.Sci.USA* 105:6870-6875.

Porteres, R. (1956) Taxonomie agrobotanique des riz cultives, O. saliva *et al.,* O. glaberrima. *J. Agr. trop. BOI.* Appl. 3:343-384, 541-580,627-700, 821-856.

Pragnya, K. Radha Krishna, K.V. Subba Rao, L.V. and K. Suneetha (2018) Studies on genetic divergence analysis of soft rice (Oryza sativa L.) genotypes. *Int.J.Pure App.Biosci.* 6(2):968-975.

Rajhathy, T. and H. Thomas (1974) Cytogenetics of oats (Avena). Genet. 50. Canada, misc. Publ., 2, pp. 90.

Rajhathy, T. and R.S. Sadasivaiah (1969) The cytogenetic status of A. magna. *Can. J. Genet. Cytol.* 11:77-85.

Randolph, L.F. (1970) Variation among Tripsacum populations in Mexico and Guatemala. *Brittonia.* 22:305-337.

Ranere, A.S., Piperno. D.R., Holst, I., Dickau, R. and J. Iriarte (2009) Preceramic human occupation of the Central Balsas River Valley, Mexico: Cultural context of early domesticated maize and squash. *Proc.Natl.Acad.Sci.USA* 106:5014-5018.

Renfrew, J. M. (1969) The archaeological evidence for the domestication of plants: methods and problems. *In:* P.J. Ucko and G.W. Dimbleby (Ed). The domestication and exploitation of plants and animals. Chicago, 149-172.

Renfrew, J.M. (1973) Palaeoethnobotany - The prehistoric food plants of the Near East and Europe. Methuen and Co. Ltd., London, UK.

Riley, R. (1955) The cytogenetics of the differences between some Secale species. *J. Agric. Sci.* 46:277-283.

Riley, R. (1965) Cytogenetics and evolution of wheat. *In:* J.B. Hutchinson (Ed.) Essays On Crop Plants: Evolution, Cambridge, pp.103-122.

Rimpau, W., (1891) Kreuzungsprodukte landwirtschaftlicher Kulturpflanzen. *Land. Fbuch.* 20:335-371.

Roshevitz, R.J. (1947) A monograph of the wild, weedy and cultivated species of rye. *Acta Inst, Bot. Acad. Sci.* U.S.S.R., ser. 1(6): 49-58.

Sanchez-Monge, E. (1958) Hexaploid triticale. Proc. 1st intnl Wheat Genet. Symp., Winnipeg, pp.181-194,

Sears, E.R. (1969) Wheat cytogenetics. *Atm. Rev. Genet.* 3:451-468.

Shafeeq, S., Mahboob-Ur-Rehman and Yusuf Zafar (2006) Genetic variability of different wheat (*Triticum aestivum* L.) genotypes/cultivars under induced water stress. *Pak.J.Bot.* 38(5):1671-1678.

Shastry, S.V.S. (1974) Rice. *In:* Evolutionary Studies In World Crops. (Ed. Joseph Hutchinson, Cambridge University Press, London, UK pp.55-61.

Singh Ajay (2008) Plants In Ancient Indian Civilizations. Agam Kala Prakashan, Delhi, India.

Stonor, C.R. and E.Anderson (1949) Maize among the hill peoples of Assam. *Ann. Mo. Bot. Gard.* 36:355-404.

Stutz, H.C. (1972) The origin of cultivated rye. *Amer. J. Bot.* 59:59-70.

Takahashi, R. (1955) The origin and evolution of cultivated barley. *Advance Genet.* 7: 227-266.

Torgonska, M., Boliboc-Brogoszewska, H. and M.Rako (27-Trojanoska (2016) Assessment of genetic diversity in scale cereal based on SSR markers. *Plant Mol. Biol. Report.* 34:37-51.

Vavilov, N.I. (1971) On the origin of cultivated rye. *Bull. Appl. Bot.* 10:561-590.

Villegas, E., McDonald, C.E. and K.A. Gilles (1970) Variability in the lysine content of wheat, rye and triticale proteins. *Cereal Chem.* 47:746-757.

Vishnu-Mittre (1974) The beginnings of agriculture: Paleobotanical evidence in India. *In:* Evolutionary Studies in World Crops. (Ed. Joseph Hutchinson) Cambridge University (1889-1893), UK., pp.3-30.

Wallace, H.A. and W.L. Brown (1956) Corn and its early lathers. East Lansing, Michigan, USA.

Watt, G. (1889-1893) A Dictionary of Economic Products of India. Vol.1-6. Periodical Expert, Shahadero, Delhi, India.

Weatherwax, P. (1954) Indian corn in old America. Macmillan, New York. USA.

Wilkes, H.G. (1967) Teosinte: The closest relative of maize, Bussey Institution, Harvard University, Cambridge, Mass, UK.

Wilson, J.A. (1968) Problems in hybrid wheat breeding. *Euphytica* 17 (suppl. 1):13-33.

Zillinsky, F.J. and N.E. Borlaug (1971) Progress in developing triticale as an economic crop. CIMMYT Res. Bull. pp.17.

Zohary, D. (197l) Origin of south-west Asiatic cereals: wheat, barley, oats and rye. *In:* P. Davis *et al.* (Ed.) Plant Life in Southwest Asia, Edinburgh, pp.235-263.

Zohary, D. and M. Feldman (1962) Hybridization between amphidiploids and the evolution of polypioids in the wheat (Aegilops-Triticum) group. *Evolution* 16:44-61.

3

Millets

INTRODUCTION

Grasses yielding small-seeded edible grains are generally termed as 'millets'. They are grown predominantly in the dry areas of the Far East and in Africa. They are a staple food in the poorer regions and so have become known as 'poor main's cereal'. They contain low protein content but are richer in fat than wheat. They are thus comparable with rice in nutritional status. Millets may be boiled and consumed whole. In southeastern Europe, the grains are ground into meal which is used for making a flat, heavy bread or for porridge. Millets may be fermented to make alcoholic beverages. Millet species are generally non-glutinous and hence they are unsuitable for making bread or leavened loaves. In the West, millets are used principally as animal feed, in particular for poultry and other birds.

Millets have long been under cultivation. None of them has been able to compete with cereals in world commerce. This appears due to their small size or to assume any place of importance in feeding the world population. They are typically grown under conditions of intense heat, limited rainfall and relatively infertile soils. Cultivation practices are invariably primitive and millets are often grown with legumes. They attain maturity within 3 to 4 months. They are chiefly sown in Africa, China and India.

Millets were originally cultivated by the ancient Egyptians, Greeks and Romans. They are still part of the diet of the people of China, Japan, African countries and India, although in western countries they are used mainly as birdseed. The more important millets are *viz.,* finger millet, [*Eleusine coracana* (L.) Gaertn.], known as ragi in India; foxtail millet [*Setaria italica* (L.) Beauv.], Japanese barnyand millet, [*Echinochloa frumentacea* (Roxb.) Link] and pearl or bulrush millet [*Pennisetum americanum* (L.) K. Schum.; syn. *P. typhoides* (Burm. *f.*) Stapf *et* Hubb.]. In some regions of Africa and India, pearl millet (known as bajra in India) and finger millet are important staple crops, apart from few other minor millets.

Selective utilization of crops and their varieties, of late, have threatened world's agro-biodiversity. This trend has, therefore, also led to rapid erosion of food sources and consequently affected the nutritional security of people. Certainly, the possible pathway for conservation of the neglected agrobiodiversity resources is to bring them into use thereby making them viable crops within the contemporary economic and social context. There is a general complaint worldwide in the last few decades that this agro-biodiversity has been declining due to some factors *e.g.* expansion of commercial agriculture, market links, unsustainable use and modification of the landscape, land alienation and land use changes, cultural erosion, etc. All these result in the shrinkage of area of the neglected or underutilized crops and their varieties. One such group is millets. The introduction of cash crops, drudgery involved in the processing of millets, lack of market linkages for millets, expanding transport facilities, availability of other food grains at reasonable cost through the Public Distribution System, consumption of some crop species and varieties of social status (*e.g.* Basmati rice in India) and mobility are the major factors responsible for the decline in the cultivation and consumption of millets, especially minor millets. The interest and altitude of some communities towards commercial horticulture is also one such factor that shaped the mind of people to neglect minor milets. Poor financial returns from millets do not attract farmers or cultivators.

On account of fast rate of acculturation, erosion of traditional knowledge especially about millets is operating worldwide *e.g.* knowledge on processing, cooking, seed collection, seed sowing, crop management, harvesting and storage, etc. To overcome this problem, strategies for millet conservation is a dire necessity in the present times. The strategies for conservation of millets should involve seed collection, multiplication, seed distribution and farmer-to-farmer exchange mediated though the traditional seed storage system. It is high time to remember that a 'Green Revolution Programme' focusing on a small number of major food crops carries the danger of marginalization of other 'useful' and 'utilized' biodiversity.

3.1 COMMON MILLET (PROSO MILLET)

Panicum miliaceum Linn. (Poaceae)

Resume: *Its native is said to be India or eastern Mediterranean region. It is ancient crop antedating wheat in cultivation. The chromosome number n=18 is reported for it. The genus Panicum has several species with diploid and tetraploid levels. It is cultivated as a food crop for man, livestock and used in the same way as the rice. It is nutritious and rich in starch and protein.*

1. Introduction: It is also popular as Hog Millet, Russian Millet, Indian Millet and Broomcorn Millet. This is the true millet, the 'milium' of the Romans. It is probably most ancient, perhaps antedating wheat in cultivation. Its native home was probably India or the eastern Mediterranean region. The grains have been used by the ancient Chinese as well as the Swiss lake dwellers. It is mentioned as 'Anu' in ancient Indian Sanskrit literature *viz.*, Upnisada, Kalpasutra and Paninis Astadhyaya and Vartik (Singh, 2008).

2. Plant Characteristics: A shallow rooted annual growing to a height of 2 to 3 ft; with an open branching inflorescence, compact or one-sided drooping; unawned, panicle; grains surrounded the scales of spikelet, variously coloured, hard, invested by hard, shining lemma and palea; grain white, ovate, shedding easily when ripe.

3. Useful Parts: Grains, straw.

4. Chemical Constituents: Carbohydrates (68. 9%), protein (10-18%), fat, fibre, minerals (calcium, iron, zinc, potassium, magnesium, copper, phosphate), vitamin-D, vitamin B12, and carotene, folate, thiamine, riboflavin, niacin, lysine, etc. (Deshaprabhu, 1966; http:www.hort.purdue.edu/ newcropplafcm/ millet html.)

5. Uses And Applications: Grains are used for human consumption, eaten whole, boiled and cooked like rice, also in the form of 'Chapati', palatable bread, and parched into 'marha' or 'mard', besides porridge, fermented beverage, etc. It is useful as hay and forage grain being very nutritious. Green plants or straw make excellent fodder for cattle and horses. Starch is useful as sizing agent in the textile industry.

6. Karyotaxonomic Evidence: *Panicum miliaceum* (2n=2n=36) may have been domesticated first in central or eastern Asia. Particularly, it was domesticated in Manchuria and introduced in Europe about 3000 years ago, followed by Near East and India. There are several diploid and tetraploid Asian species of the genus *Panicum* among of which an origin might me sought (Seetharam *et al.,* 2007). *Panicum miliaceum* is a tetraploid. Hunt *et al.,* (2014) attempted to identify likely ancestors within the genus. They investigated nuclear and chloroplast DNA sequences from diploid and tetraploid relatives. *In situ* hybridization showed that half the chromosomes of *P. miliaccum* hybridized more strongly with labeled genomic DNA from *P. capillare* and half the labeled DNA from *P. repens*. Their data supported the allotetraploid origin of *P. miliaceum* with the meternal ancestor being *P. capillare* and other generas shared with *P. repens*.

7. History: It is a native of central and eastern Asia. Probably, it is the most ancient antedating wheat in cultivation. It is cultivated in China, Japan, India, southern Europe and former USSR. It is introduced into the United States many years ago (Purseglove, 1988; Seetharam *et al.,* 2007). Presently, it is important mainly in eastern and southern Asia. In India, it is cultivated in the states *viz.,* Tamil Nadu, Andhra Pradesh, Karnataka, Uttar Pradesh and Maharashtra.

8. Further Developments: India has released six varieties after selecting features like high yield, earliness and suitability for late sowing. Seetharam *et al.,* (2007) recognized four landraces. About 842 germplasm accessions from 30 countries are conserved at the ICRISAT gene bank. The origin of common millet from Neolithic China has generally been accepted. Phytoliths and biomolecular components are identified solely as common millet from newly excavated storage pits at the Neolithic Cishan site, China, dated between ca. 10,300 to ca. 8700 calibrated years before present. It is indicated that the earliest significant common millet cultivation system was established in the semiarid regions of China by 10000 cal yr. BP. It was used as staple food in northern China. It is also suggested that common millet might have been domesticated independently in this area and later spread to India, Russia, the Middle East and Europe (Lu *et al.*, 2009).

Salini *et al.* (2010) evaluated 364 proso millet (*Panicum miliaceum* L.) germplasm collections to study genetic variability, diversity, correlation and path analysis of yield and its components. Highest variability was observed for grain yield per plant and total numbers of productive tillers showed highest heritability. High genetic advance as per cent of mean was recorded for all characters except days to 50 per cent flowering and plant height. High heritability coupled with high genetic advance was observed for grain yield, number of productive tillers and total number of tillers indicated the scope of simple selection for improvement of these characters. All characters exhibited highly significant positive correlations with grain yield. Positive direct effect of plant height, number of productive tillers and 100 grain weight indicated direct selection for these characters. These would improve the grain yield in proso millet. It is indicated that great yield advantage can be achieved by selecting accessions with more productive tillers, tall stature and medium duration. Seventeen distinct clusters were observed by cluster analysis. Principal component analysis suggested that panicle length, days to 50 per cent flowering, grain yield per plant, plant height, total number of tillers and 100 grain weight could be used as characters to distinguish the germplasm entries.

Graybosch and Bultensperger (2009) evaluated for the presence of accessions with waxy (amylose-free) endosperm starch of the entire USDA-ARS maintained collections of 650 accessions of proso millet (*P. miliaceum* L.). Segregation ratios for wax endosperm were evaluated in F2 and F3 populations derived from crosses between two waxy accessions, PI43662 (Lung Shu 1) and PI436626 (Lung Shu 18) and several wild-type accessions. The waxy trait was found to be under the control of duplicate recessive alleles at two loci herein designated Wx-1 and Wx-2. Wildtype alleles at these loci were designed Wx-1 and Wx-2. Iodine binding revealed a mean grain-starch amylose concentration of 3.5% in waxy lines and 25.3% in wild-type proso. Expression of the granule-bound starch synthase (waxy protein) in waxy lines was reduced to approximately one-tenth that of wild-type accessions. The waxy accessions identified now are available for the introgression of this trait into breeding lines adapted to the Great Plains of North America.

9. Epilogue: The crop is not subject to any serious pastes or diseases. It is adapted for short rainy season and thin or poor soil. It is a short duration and quick growing crop. It is valued for areas where other crops would not succeed (Anonymous, 1952). Further study from the Old World collections is required. It is important to identify why the water-efficient *P. miliaceum* is now of minimal importance in agriculture, and it may be valuable to exploit the diversity within species and its ancestors.

3.2 FINGER MILLET

Eleusine coracana Gaertn. (Poaceae)

Resume: *It is of African origin and Eleusine is predominantly an African genus with about nine species confined to Africa. Finger millet probably reached India at an early date over 3000 years ago. In India, it has Sanskrit name 'Rajik' or 'Ragi'. India is a secondary centre of evolution of the crop. Carbonised grains of E. coracana are recovered from the Neolithic site Hallur dating about 1800 BC. The cultivated E. coracana is a tetraploid with n=18. Its wild relatives are diploids with n=9. It is now staple food in parts of eastern and central Africa and southern India. Its grains can be stored for ten years.*

1. Introduction: Finger millet is a warm country crop which have peculiar merits. It is grown under rainfed conditions and most common in drier parts. The grains of *Eleusine* are capable of prolonged storage. *Eleusine coracana*, finger millet (also known as African millet, Koracan and, in India as Ragi), extends in Africa from Nigeria eastwards to Eritrea and southwards to southwest Africa and Natal. It is a staple food in parts of eastern and central Africa, particularly in northern and parts of western Uganda, north-eastern Zambia and in India, besides Nepal. Since ancient times, it has deep religious and ceremonial ties with tribal societies of Africa. Finger millet is said to be of African origin. De Candolle (1886) opined India to be its native. Finger millet was domesticated in Western Uganda and the Ethiopion highlands at least 5000 years ago before introduction to India approximately 3000 years (Dida *et al.*, 2008).

2. Plant Characteristics: Tall annual, stems tufted, leaves linear, sheathed; each with 4 to 6 spikes, spikes digitate, alternately arranged on rachis; seeds brown or white, globose, smoother and rugose.

3. Useful Parts: Grains, straw.

4. Chemical Constituents: Protein (7.3%), fat (1.3%), carbohydrates (76.3%) minerals (calcium, manganese, potassium, copper, zinc, phosphorous, iron, magnesium), carotene (vitamin-A), vitamin-B_1, nicotinic acid.

5. Uses And Applications: Seeds ground into flour and made into cake, sweetmeats, a thick porridge, and used for puddings. After fermentation, they produce beer. Enzymatic extract of germinated grains is used for unhairing hides and skins. The grains are tonic and astringent, useful in biliousness. They are particularly advised for diabetic patients as a wholesome food. Grains can be stored for ten or more years, and hence can be consumed as feminine food. Moreover, it is free from insects. In eastern and central Africa, it is useful as a staple food. The straw constitutes nutritious feed for the milk animals.

6. Karyotaxonomic Evidence: *Eleusine* is predominantly an African genus. It consists of six of its total nine species confined to tropical Africa. One species *E. tristachya*, is but South American. Also another species, the highly successful annual weed, *E. indica*, is pantropical. The genus has a narrow distribution, mainly in eastern and north-eastern tropical Africa. The species can be divided into annuals and perennials. Within each group, the differences are often small and, among the annuals in particular, introgression is frequent, with intermediates occurring (Phillips, 1992).

The cultivated *E. coracana* is a tetraploid with 2n=4x=36. There are diploids (2n=18) among its wild relatives. It was considered for many years to have originated from *E. indica* either in Africa or India. The African tetraploid form (2n=36), if separated from the pantropical diploid *E. indica* (2n=18) as a distinct species which was named *E. africana* (Kennedy O'Byrne, 1957). Phillips (1972) later made this a subspecies of *E. indica*, subsp. *africana*. *Eleusine indica* subsp. *indica* is widespread throughout the tropics and subtropics. In Africa, it occurs mainly in West Africa, along the coast of east and south east Africa and in Madagascar. *Eleusine indica* subsp. *africana* is restricted to Africa and is distributed mainly in the uplands of eastern and southern Africa. It occurs commonly as a weed in the fields of *E. coracana* with which it hybridizes freely and forms many intermediates. Its seeds are about half the size of those of the cultivated species. They are shed about three weeks earlier than the latter.

7. History: Two groups of cultivars of *E. coracana* exist: (a) African highland types, bearing a resemblance to *E. indica* subsp. *africana*, with long spikelets, long glumes, long lemmas and grains enclosed within the florets, (b) Afro-Asian types, with a close resemblance to *E. indica* subsp. *indica*, with short spikelets, short glumes, short lemmas and the mature grains exposed distally. Both groups are tetraploids with 2n=36. Probably both types originated from subsp. *africana* and that the African highland forms were selected from large-grained mutants (Kennedy-O'Byrne, 1957; Mehra 1963a,b). Mehra (*loc. cit.)* believes that Afro-Asian cultivars then evolved from African highland forms by mutations for short glumes and exposed grains. Emphasizing this hypothesis, Purseglove (1972) thought Uganda or a neighbouring territory as the centre of origin. The crop is certainly of very ancient cultivation in the area. This is associated with religious and tribal ceremonies and even by the many customs connected with its cultivation.

E. coracana was probably taken at an early date to India, over 3,000 years ago (Mehra, 1963 a,b). *Sorghum* and bulrush miller could have been taken about the same time. The crop was unknown in ancient Egypt, but it has a Sanskrit name 'Rajik' or 'Ragi' in India. It seems likely that it was taken up the seaward edge of the Arabian peninsula along the Sabaean Lane. India became a secondary centre of evolution of the crop.

Carbonised grains of *E. coracana* are known from Mysore state (presently Karnataka State of India) from the Neolithic of Hallur dated about 1800 BC. (Vishnu-Mittre, 1974). Discovery together with its relative *E. indica* is suggestive of an early stage in the history of the crop. These Indian records are the oldest known in the world (Hutchinson, 1974).

Finger millet is mainly self-pollinated. It is reported that about one per cent cross-pollination takes place by wind, which accounts for contamination by *E. indica* subsp. *africana* when the two grow together. Numerous cultivars are recognized in Africa and India. Line breeding and selection from single plants produced few worth-while results in Uganda. A number of improved cultivars have been obtained by breeding in India, including 'E. C. 593', a general purpose type evolved in Madras (Anonymous, 1952).

8. Further Developments: Several cultivars are identified in Africa and India emphasizing height (drawrf or tall), colour of vegetative parts (green or purple), tillering (poor or profuse), inflorescence (straight or incurred, length), number of spikelets per spike, length of spikelets, length and colour of glumes, grain colour (orange red, deep brown, purple, black or white), grains (exposed or covered), yielding ability, etc. In India, nearly 23 varieties have been released during 1986-2001. Also, 29 landraces simply from Eastern Ghat region in peninsular India are reported (Seetharama *et al.*, 2007).

Several improved strains have been evolved in India by breeding and selection *e.g.* EC593 (Ragi Co. 1), AR525, AR325, H22, R. 6009, R. 0871, R. 0883, R. 0870, E31, B11, Jagudan 44, etc. These have improved in view of dry conditions, higher yield, grain colour (*e.g.* white), medium duration, drought resistant, irrigated conditions, etc.

Ramkrishnan *et al.,* (2016) evaluated the genetic variation and population structure in Indian and non-Indian genotypes of finger millet using 87 genomic SSR primers. They isolated genomic DNA from 128 genotypes. All noted genotypes are genetically diverse and grouped into three subpopulations based on their geographic regions. The authors opined that the data generated may be utilized for mapping of any important agronomical trait for breeding programme to improve the finger millet.

Kalyana Babu *et al.,* (2007) evaluated genetic diversity of 32 finger millet genotypes. They fingerprinted them using DNA (RAPD) markers. Out of the total 529 loci generated using the 50 RAPD primers, 479 loci were polymorphic and informative to differentiate accessions. Two accessions *viz.,* GEC 182 and CO 12 are found distantly related. These two accessions with higher diversity at molecular level, phenology and grain weight, in the opinion of these authors, will be ideal as parents in hybridization programme to develop improved finger millet varieties suitable for peninsular region of India.

Ramchandra *et al.* (1997) analyzed 32 varieties of finger millet (*Eleusine coracana*) comprising of both brown and white seed coat material from Indian and African source. They reported that the white grain variety contained lower levels of polyphenols (0.04-0.09%) than the brown grain varieties (0.08-3.47%). The total polyphenol content of the Indian varieties ranged from 0.08-0.96%, reported at chlorogenic acid equivalents and tannins in the range of 0.04-1.05%, reported as catechin equivalents respectively. On the contrary, in the African varieties, the corresponding value were 0.54-2.44% and 0.5-3.47% respectively.

Malting of finger millet is a traditional process largely practiced for preparation of local beer in Africa and also at the Himalayan belt. But n India, the millet malt is mainly used to prepare 58 specialty foods. During the process as malting, the bioavailability of proteins, carbohydrates, vitamins and minerals are enhanced and the concentrations of anti-nutritional factors are considerably reduced (Malleshi and Klopfenstein, 1998). The malted millet is nutritionally superior to that of native millet and is a good source of amylases. When the malt flour mixed with water or milk is heated to boiling, the amylases hydrolyse the starch to simple sugars thereby reduce its water holding capacity and as a result from low bulk and nutrient-dense foods. This natural process of enhancing the nutritional density has been used for developing various specialty and health foods such as infant food, weaning food, enteral food and milk-based beverages and also in confectionary (Malleshi, 2002). The millet malt, therefore, could be a new ingredient for the food industry.

9. Epilogue: Grains are insect free and can be stored for about 10 or more years, hence of great use particularly in famine period. As a crop, it is suited to dry farming and can withstand severe drought. Thus, it is more beneficial to famine-prone areas. Finger millet evolved by local selection for large grains and non-shattering heads from wild forms which persist in the area. It existed in variable populations capable of continued genetic exchange with their wild relatives. It migrated early to India where secondary centres of diversity became established. De Candolle thought India as its native, while others think its origin in Africa.

3.3 FOXTAIL MILLET, ITALIAN MILLET

Setaria italica (L.) P. Beauv. (Poaceae)

Resume: *It is probably the first crop to be domesticated in Central and Eastern Asia. China is its native place and also the likeliest place of domestication. It was one of grain or crop to be sown by princes of the family of emperor Chin-nung (2800 BC.) in China. Setaria italica is not reported in wild. Its progenitor is said to be S. viridis both having n=9 chromosomes. The grains are useful as food and for brewing beer, besides as bird seed.*

1. Introduction: It is also known as Italian millet, Hungerian millet, Kangu or Priyangu in Sanskrit in India. In India, it is called 'Rala' or 'Kand'. A ceremony in China was established by the Emperor Chin-nung (2800 BC.) in which the sowing of five kinds of grain is the chief observance. The emperor himself has to sow the rice, but he may delegate the sowing of the other four kinds of grain to the princess of his family. Foxtail millet was one of these four kinds of grains. Perhaps, China is the likeliest place of domestication. The grain is reported widely used by the Lake Dwellers of Swiss during a time when it was unknown in Greece and Syria. Now, it is a important crop in Asia, North Africa and south-eastern Europe. Seeds have been recovered from archaeological sites.

2. Plant Characteristics: Erect annuals, 2.5 to 3 ft in height; leaves 0.5 to 1 ft long, linear-lanceolate; stem slender, leafy; panicles compact, 3 to 5 in long, spikelets ellipsoid, bristly, lower glumes ovate, 1-nerved, upper glumes ovate, grain small (±2 mm), encased in thin hull, oval, convex, colours varied.

3. Useful Parts: Grains, straw.

4. Chemical Constituents: Protein (11.7%), carbohydrate (59.1%), fat (3.9%), minerals (Ca, P, Fe, Mg, Mn, Zn, K, Na) (Purseglove, 1988), thiamine, riboflavin, niacin.

5. Uses And Applications: Grains are used for food, for brewing beer particularly in Russia. Grains are consumed in the form of cakes, porridge, parched or boiled. It is commonly used as bird seed. In USA, it is cultivated as fodder crop. Grain is popular in medicine for alleviating pains after parturition and as a sedative to the gravid uterus. It also benefits in rheumatism and used as diuretic and astringent. In India, it is a favourite food, called 'Sawank' and is eaten by Hindus on fast days. It is commonly used for bird feed, livestock forage and hay production.

6. Karyotaxonomic Evidence: *Setaria italica* (2n=2×=18) is not reported in wild but has frequently naturalized as a weed, an escape from cultivation, in moist warm parts of the world. Its progenitor is said to be *S. viridis* (2n=2×=18), a common Old World weed. Two other species *viz., S. geniculata* (Lam.) Beauv. And *S. macrostachya* HBK. were among the earliest domesticated crops in the Tehuacan valley of Mexico, dating back 7000 years (Smith, 1976; Kochhar, 2006).

7. History: Foxtain millet is probably the first crop to be domesticated in central or eastern Asia (Purseglove, 1988). Presently, several forms and cultivars are recognized based on (i) plant height, (ii) period of maturity, (iii) length, breadth and compactness of inflorescence, (iv) number, colour and length of bristle, (v) colour of grain, etc. About 19 varieties or cultivars are recognized in Eastern Ghats of India (Seetharam *et al.,* 2007).

8. Further Developments: About twelve hybrid varieties are released in India emphasizing the features such as high yield, resistancy to stem borer, green colour, bold seeds, etc. (Seetharam *et al.,* 2007). At the ICRISAT genebank, 1535 germplasm accessions from 26 countries are conserved for use in research and development.

The history of foxtail is more complex. Sequence diversity of 250 Chinese genotypes was found to be quite high, averaging 20. 9 alleles per locus when examined with 77SSRs (Wang *et al.,* 2012). Alleles clustered into two main geographic diversity centres, indicating the possibility of two domestication events within China. Work on this line is necessary. It is also suggested that foxtail millet was independently domesticated in Europe based on archaeological evidence (Jusuf and Pernes, 1985; Hunt *et al.,* 2008). It is closely related to *Setaria viridis* which is supposed to be its progenitor. Interspecific hybridization between these species is common (Li *et al.,* 1942; Jusuf and Pernas, *loc. cit.).* These crosses have resulted in resistance to a variety of herbicides (Darmency and Pernes, 1985; Wans *et al.,* 1996). After its domestication in China, foxtail millet spread throughout Asia, Europe and eventually to North America (Jusuf and Penes, *loc. cit.).* Its large range has resulted in three different races, each with multiple subraces as: (i) Moharia in Europe, Russia and the Middle East, (ii) Maxima in Eastern China, Georgia, Japan, Korea,Nepal, northern India and USA, and (iii) Indica in southern India and Sri Lanka (Jusuf and Pernes, *loc. cit.)*

Trivedi *et al.* (2018) studied foxtail millet [*Setaria italica* (L.) P.Beauv.] from Himalayan region for genetic variability and nutritional quality. A total of 30 accessions having unique traits of agronomic importance were collected and molecular profiling was performed. A total of 63 alleles were generated with an average of 2.52 alleles per locus and average expected heterozyosity of 0.37 ± 0.231. Significant genetic variability was revealed through the genetic differentiation (Fst) and gene flow (Nm) values. Structure-based analysis divided whole germplasm into three sub-groups. Rich variability was found in nutritional traits such as thiamine content. The collected germplasm may be useful for developing nutritionally rich and agronomically beneficial varieties of foxtail millet and also designing strategies for utilization of unexploited genetic diversity for food and nutrition security in this and other similar agro-ecological regions.

Sintho *et al.* (2017) analyzed the genetic diversity of foxtail millet [*S.italica* (L.) P.Beauv.] genotypes using RAPD markers. Genomic DNA of ten foxtail millet genotypes was amplified using 26 random primers through RAPD analysis. Of these primers, 22 produced reproducible amplicons and were polymorphic among the 10 foxtail millet genotypes. The number of polymorphic markers for each primer varied from 1 (primer E15) to 14 (primer M17). The amplified product size ranged from 120 to 2500 base pairs (bp). Diverse genotypes identified in this study can be used as potential parents in an efficient crop improvement programmes.

9. Epilogue: Foxtail millet ranks second in the total millet production of millets. The crop requires only 6 to 10 weeks to attain maturity. Looking to this rapidity, it is better to use this crop as 'Catch crop' after some other crop has failed. It will help cultivating for both, food and fodder or feed for his cattle or birds. In United States, it is mainly used as forage crop. It is considered there very desirable for it produced as many as eight crops in a year and has large leaf area. In Japan, it is cultivated mainly in areas where rice will not grow. This can be also adopted in other countries. Its place and importance has been replaced by other higher yielding varieties of rice, wheat, maize, etc. The total area of cultivation is thus thereby reduced. Moreover, traditional food habit is being changed by some modern food articles in market. This trend will deplete the traditional methods of cooking food, preparing beer, etc. Knowledge on processing, cooking and consumption will rapidly erode with modern wave. The crop, therefore, needs measures for conservation and perpetuation of traditional knowledge.

3.4 INDIAN BARNYARD MILLET

Echinochloa frumantacea Link (Poaceae)

Resume: *Species of Echinochloa are native to tropical, subtropical and temperate areas of the Old World. Generally two species of the genus Echinochloa are found under cultivation viz., E.frumantacea and E.crus-galli. The latter was cultivated in Iron age as evidenced by impressions on pottery. The former is mentioned as 'Syamak' in Vedic treatises viz., Yajurveda and Atharveda. E.frumantacea has 2n=36, and 54 chromosomes, whereas E.crus-galli possess 2n=36, 48 and 54 chromosomes. When the paddy fails, then as a substitute of rice, it is cultivated especially in China and Japan. It is used as a food and forage crop.*

1. Introduction: Species of *Echinochloa* have not been cultivated in Europe for many years. But *E. crus-galli* may once have had some importance as a crop plant. It was certainly cultivated in Iron age as evidenced by impressions on pottery (Smith, 1976). *E. frumantacea* is called 'sewan', 'shamule' or 'Barti' in India. It is mentioned as 'Shyamak' in Vedic texts *viz.*, Yajurveda and Atharveda, besides other ancient Sanskrit treatises *viz.*, Brahamana, Kalpasutra and Paninis Astadhyayi and Vartik (Singh, 2008).

2. Plant Characteristics: Plant erect, 2. 5 to 3 ft in height, robust, with sparse tillers; leaves linear-lanceolate, 1 to 1.5 ft; spikelets, ovoid; seeds turgid and purplish, shining, rounded with pointed apex (Purseglove, 1988; Anonymous, 1952).

3. Useful Parts: Grains, straw.

4. Chemical Constituents: Protein (6.2%) fat, carbohydrate (65.5%), fibre, minerals (calcium, phosphorous, iron, magnesium, potassium, copper, manganese, zinc), thiamin, niacin (Purseglove, 1988), catotene traces, vitamin, B_1. thiamine, riboflavin, niacin

5. Uses And Applications: In southeast Asia, grains are consumed as porridge or mixed with rice. In India, a very potent beer is also made from it. It is particularly used as a forage and food crop in India. In Asia, it is cooked as rice or parched. It is boiled with milk and sugar for human consumption. Straw is used as fodder in India.

6. Karyotaxonomic Evidence: *Echinochloa frumantacea* (2n=4×=36; 2n=56) is very similar to the widespread weedy, grass *E.crus-galli* (2m=4×, 6×=36, 54; also 48) which is also cultivated (called Bharti). It is a native over a large area of the Old World, in tropical subtropical and temperate zones. Sino-Japanese and Indian hexaploid strains (6×=54) have different genome constitutions and hence must have partially different origins. The Sino-Japanese plants, presently recognized as *E. utilis* are thought to have a far Eastern hexloid strain of *E. crus-galli* as their immediate ancestor. The Indian strain of *E. frumantocea* is identical genomically to the wild species *viz., E. colonum*, native to Malaysia and Java, and may have been selected from that (Yabuno, 1962).

It is thought originated either from *E. colonum* or *E. crus-galli* and possesses characters intermediate between these two species. It is treated as a variety of these two species (Blatter and McCann, 1934; Anonymous, 1948).

7. History: It is used in the United States exclusively as a forage crop. It produces eight crops in a year. It has large leaf area. In Japan and China, it is grown as a substitute crop when paddy fails. In Egypt, it is grown as a reclamation crop on lands which are too saline for other crops (Anonymous, 1952). Indian barnyard millet contains antifeedants which are present at concentrations higher than in rice. It display resistance to the feeding activity of brown planthopper (Kim *et al.,* 2008). Significant phenotypic variation is observed in it. Four morphological races *viz.,* laxa, robusta, intermedia and stolonifera were recognized by de Wet (1983) based on the lengths of flag leaves, peduncles, racemes, plant height and tiller number.

8. Further Developments: Nearly six varieties have been released in India for all states with special emphasis on high yield. In USA, it attracted attention as a fodder crop. It is useful as a late season feed, taking the place of maize. In Japan, it is used for seeding dairy cows (Anonymous, 1948). More recently, a variety of morphological parameters were examined and PCA indicated three morphotypes corresponding to races *viz.*, robusta, intermedia and stolonifera. Laxa was absent suggesting that efforts must be made to collect more of this race (Gupta *et al.*, 2009). These authors observed high variability in grain yield, straw yield and number of productive tillers. They opined that the number of racemes, flag leaf width and internode length showing high correlation with grain yield should be considered by breeders when performing selections, and promising donor genotypes of these and other traits reported earlier. Variation across genotypes in photosynthesis and related traits such as transpiration and stomatal conductance have also been observed (Subrahmanyam and Rathore, 1999). A study by Hilu (1994) using RAPD markers suggested that the sequence diversity of Indian barnyard millet is significantly higher than the Japanese species (*Echinochloa esculenta*), perhaps because of multiple domestication events in different locations across India. In his opinion, DNA sequence analyses are lacking in Indian barnyards millet.

Barnyard millet (*Echinochloa frumantacea* Link.) is one of the underutilized, nutritious minor millet. It is a source of nutraceutical components. It provides total dietary fiber to an extent of 13.02 per cent including 8.18 per cent insoluble and 4.66 per cent soluble fractions and antioxidants besides other macro-and micronutrients. This millet exhibited relatively low carbohydrate content (68.76%) with lower starch digestibility (25.88%) and higher protein digestibility (81.13%). It has, therefore, proved to be an ideal natural food for obese and weight watchers (Veena *et al.*, 2005). To add to the potential benefits, it was reported that glycemic index of thermally processed barnyard millet was low as 41.71%. The benefits of its consumption were expressed in terms of lowered blood sugar levels and improvement in lipid profile in normal and diabetic volunteers after dietary intervention period of one month (Ugre, 2008).

9. Epilogue: *Echinochloa frumantacea* is the quickest growing of all the millets. It is short duration crop. It is quick growing and attains maturity as little as six weeks. In Japan, it is cultivated in areas where rice will not grow. It needs no manuring and special attention. In China and Japan, it is cultivated as a substitute for rice, when the paddy fails (Purseglove, 1988). This is also advisable for other countries. It contains principal protein prolamine rich in lysine, cysteine and histidine, with a nutritive value markedly superior to that of polished rice.

3.5 KODO MILLET

Paspalum scrobiculatum L. (Poaceae)

Resume: *It is a native to India and is domesticated nearly 3000 years ago. It is mentioned in ancient Sanskrit literature and entered in food economy after*

Satavhana period (250-400 AD.). Other wild species related to it grow in India. It is presently a minor food crop and being neglected from cultivation. Its grain flour is useful for bread.

1. Introduction: It is also called 'Kodra' or 'Harik' in India. It is minor millet domesticated in India 3000 years ago and grown throughout in poor soils especially in tribal dominated drier areas. It is grown only once in rainy (kharif) season. It is a short duration crop. It is mentioned as 'Kodrav' and 'Uddalak' in ancient Sanskrit literature *viz.,* Kalpasutra (Singh, 2008). It entered the food economy in the post-satavannah period (250-400 AD.) (Vishnu-Mittre, 1974).

2. Plant Characteristics: Annuals, ascending, growing to a height of 2. 5 to 03 ft., culms noded; leaves two-ranked, 18-20 inch long, linear, hairy; ears compact, spikelets ovoid, lower glume absent, upper glume ovate-oblong, oval; grains whitish, tinged red and rounded. Inflorescence of 2 or more sessile alternate spikes.

3. Useful Parts: Grains, straw.

4. Chemical Constituents: Protein (10.6%), carbohydrate (59.2%), fat (4.2%), minerals (calcium, magnesium, potassium, manganese, zinc, phosphorous, iron), thiamine and niacin, etc.

5. Uses And Applications: In India, grain flour is used for bread. They are also boiled and cooked like rice and consumed with other vegetables or pulses. Straw is used as fodder for cattle. Grains are also used as cattle feed.

6. Karyotaxonomic Evidence: *Paspalum* is a large genus containing about 400 species and the chromosome numbers range from 2n=20 to 160 (Darlington and Wylie, 1955). *Paspalum scrobiculatum* is tetraploid with 2n=40. Meiosis in it is quite normal. Cytological evidence indicates that cultivated species *P. scrobiculatum* is closer morphologically to wild *P. commersonii* but are distinct species (Clayton, 1975).

7. History: It is grown throughout India in Deccan and south India. Other species *P. distichum* Linn. is wild in India and useful as soil binder and pasture grass. *Paspalum notatum Fluegge*, native of America is introduced in India and useful as *P. distichum*. This is also the of *P. plicatulum* Mich. It is important fodder grass (Anonymous, 1952).

Two varieties are recognized *viz.,* (i) wholesome and named a Pechadi, Pakod or Harkin in Konkan and Goa (India). It is small seeded and paler in colour, (ii) unwholesome named as Dhone or Majari-harik in Konkan and as Mana Kodra in Gujarat (Watt, 1889-1893). Kodo millet was domesticated about 3000 years ago in India. It is presently being cultivated rather rarely exclusively in India. The grains contain a diverse range of high quality protein (Geervani and Eggum, 1989; Kulkarni and Naik, 2000). It shows high antioxidant activity (Hegde and Chandra, 2005; Hedge *et al.*, 2005; Chandrasekhara and Shahidi, 2011). It is rich in fibre and grows well in a variety of soil types from gravelly to clay (de Wet *et al.*, 1983). It is, however, weedy in nature and problematic throughout Old World farming systems (de Wet *et al., loc. cit.)*. It is believed that the Kodo millet was probably first harvested as a weed alongside other cereals like rice, perhaps leading to multiple domestication events of the millet across its current range. This practice continues in some parts of Africa where the weed is also sometimes harvested during famine (de Wet *et al., loc. cit.)*. In Africa, it is referred as black rice or bird's grass (M'Ribu and Hilu, 1996). Limited molecular marker analysis has shown that Kodo millet genotypes cluster by African vs. Indian origin (M'Ribu and Hilu, *loc. cit.)*. Three races of Kodo millet are recognized *viz.*, regularis, irregularis and variabilis based on panicles morphology (de Wet *et al., loc. cit.)*. In South India, two varieties *viz.*, small seeded (Karu Varagu) and large seeded (Peru Varagu) are recognized often growing together in the same field (de Wet *et al., loc. cit.)*.

8. Further Developments: Some varieties are recognized in India on the basis of characteristics such as duration of crop, number of rows of grains and grain colour. In India, nine varieties have been released during 1889 to 2002 based on features like high yielding, early maturing period, drought resistancy and resistancy to smut, etc. (Seetharam *et al.*, 2007; Anonymous, 1952). General morphological variability is high especially in tiller number, yield and time before flowering (Subramanian *et al.*, 2010; Upadyaya *et al.*, 2014). Some authors call Kodo millet 'pseudo-cultivated' (Blench, 1997). Systematic breeding of Kodo millet is largely neglected. However, the limited efforts on this line are promising. Various metrics of plant productivity include dry fodder yield, plant height and grain yield. Improvements of these traits have been observed through breeding, with four highly productive genotypes so far identified (Upadhyaya *et al., loc. cit.)*. Pathogen resitance has been noted as a good breeding target particularly to smut and rust. Both are major hindrances of Kodo yield (Upadhyaya *et al., loc. cit.)*. It is interesting to note that some cultivated landraces have maintained the perennial nature of their wild ancestor and continue to initiate culms following the maturity of older shoots (de Wet *et al., loc. cit.)*. No genetic or molecular maps of the Kodo millet genome are available (Dwivedi *et al.*, 2012), possibly because of the problem of persistent cross-hybridisation with its wild relatives. There has been some preliminary work in miRNA target site prediction using ESTs from Kodo millet (Babu *et al.*, 2013). ICRISAT conserved 656 accessions of Kodo millet, and a core collection has been established that reflects the phenotypic diversity of the entire collection (Upadhyaya *et al., loc. cit.)*.

Nirubana *et al.* (2017) assessed the genetic divergence among the 103 Kodo millet germplasm accessions using Mahalantis D^2 statistic collected from Coimbatore. The 103 germplasm lines were grouped into 11 different clusters based on D^2 analysis. Cluster I had a maximum of 63 genotypes, followed by cluster II and III with 14 each, cluster X with four genotypes and cluster VIII with two genotypes. Remaining clusters IV, V, VI, VII, IX and XI were mono-genotypic indicating wide divergence from other clusters. The highest intracluster distance was observed in cluster X indicating differences in genotypes within cluster. The highest intercluster distance was observed between cluster III and X followed by cluster IV and VII suggesting the use of genotypes from these clusters to serve as potential parents for hybridixation. The characters days to 50 percent flowering contributed maximum towards divergence followed by Fe content, Zn content and grain yield per plant.

Subramanian *et al.* (2010) evaluated 188 germplasm accessions of Kodo millet (*Paspalum scrobiculatum* L.) to evaluate genetic variability, heritability and genetic advance for eight yield component traits. The ANOVa revealed that there were significant differences among the accessions for all the traits studied. High genotypic variance, phenotypic variance, GCV and PC were observed for dry fodder yield, plant height and grain yield per plant. Broad sense heritability ranged from 19.18 percent for dry fodder yield to 56.63 percent for number of tillers. The highest genetic advance as percentage of mean was recorded by dry fodder yield followed by grain yield per plant and plant height. High heritability coupled with high genetic advance observed in this study for grain and dry fodder yield per plant indicates that these traits are under the influence of additive gene action. These traits can be considered as favorable attributes for crop improvement through selection.

9. Epilogue: Wild local varieties require 5 to 6 months to mature (Purseglove, 1988). Therefore, it is better to adapt short duration varieties released in India. This change in duration will help them as catch crop when the rice fail under cultivation. Kodo Millet grows well in rainfed area and better yield can be also obtained after less rain and even sown in poor soil. It helps sustain farmers for food as well as fodder for his cattle. Surplus grains can be also good feed for cattle.

3.6 LITTLE MILLET

Panicum sumatrense Roth ex Roem. & Schult.

(Syn. *P. miliare* Lamk.) (Poaceae)

Resume: *It is believed to have originated in India or South-East Asia. It occurs wild in India. It is thought derived from Panicum psilopodium Trin., a wild relative. It is a minor food crop. Its flour is useful for bread making. It is reliable catch crop in view of its earliness and resistance to adverse conditions.*

1. Introduction: It is also called 'Sawa' or 'Kutki' in Himalayas, especially found in northern India and also in Burma, South-East Asia and

Malaysia. It is considered to have originated in India as South-East Asia from *P. psilopodium* Trin., which is also found wild.

2. Plant Characteristics: Erect annuals, growing to a height of 03 ft.; leaves linear 15-20 cm long, glabrous; panicles contracted, ovoid-elliptic, 15-20 cm long, spikelets elliptic, upper glumes 7-9 nerved, ovate, lower glume orbicular; caryopsis glabrous striated, brown.

3. Useful Parts: Grains, straw.

4. Chemical Constituents: Protein (7.2%), carbohydrates (63.8%), fat, fibre, minerals (calcium, manganese, magnesium potassium, zinc, phosphorous, iron), thiamine, niacin (*cf.* http:www.hort.purdue.edu/newcropla tcm/ millet html).

5. Uses And Applications: Grain flour is used for bread. They can be boiled and cooked like rice and consumed with other vegetable. Straw is useful as fodder for cattle.

6. Karyotaxonomic Evidence: Little millet (*Panicum sumatrense*) is said to be allotetraploid and have 18 regular bivalents. Cytogenetics of *Panicum* millets revealed the basic chromosome number ×=9. Polyploidy is common in the genus. It is also the case of *Panicum psilopodium*. Morphologically both species are similar and have sympatric distribution especially in the Eastern Ghats (Andhra Pradesh) of India. Preponderance of 18 bivalent formation in the hybrids strongly indicate that genomes of *P. sumatrense* and *P. psilopodium* are basically similar and fully homologous. Thus, *P. sumatrense* might have originated from *P. psilopodium* through selection and further cultivation. Genomic notation of AABB is proposed for these two species (Hiremath *et al.*, 1990).

7. History: It occurs in wild state in India and south eastern Asia. It is domesticated in India. Several types exist and are under cultivation in India especially in the states like Bihar, Maharashtra, Karnataka, Tamil Nadu. It is also cultivated in Sri Lanka, Pakistan, Myanmar and other South East Asian countries. The differences are due to grain colour, plant pigmentation, shape of panicle and crop duration.

8. Further Developments: In India, nine varieties have been released during 1987 to 2001 based on characteristics such as high yield, short period of maturity and wide adaptation.

Seetharam *et al.*, (2007) documented 25 landraces from Eastern Ghats in India. Little millet (*P. sumatrense*) is thus a domesticated form of P. psilopodium (de Wet *et al.*, 1969). The chromosomes of hybrids obtained from these two species pair almost perfectly (Hiremath *et al.*, 1990). This hybridization ability combined with its wide range of cultivation across India suggests that little millet was domesticated independently several times (de

Wet *et al., loc. cit.)*. Its nutrients are comparable to other cereals. It is also drought resistant, pest and salt tolerant (Sivakumar *et al.,* 2006).

Two races *viz.,* nana and robusta are recognized based on panicle morphology. Race nana matures faster and produces less biomass than robusta (de Wet et. al., *loc. cit.)*. Locally grown landraces exhibit fair diversity in Indian Kolli hills (Arunachalam *et al.,* 2005). Nirmalkumar *et al.,* (2010) in their collections of 109 landracs revealed high diversity, heritability and genetic advancement in terms of yield and productive tillers. A different collection of 460 accessions of little millet by ICRISAT displayed genetic variation for most of the traits (Upadhyaya *et al.,* 2014). A core collection of 56 genotypes was identified which was representative of the entire seed bank.

Ashwin Kumar *et al.* (2017) studied relationship of grain smut incidence with agro-morphological characters for 20 little millet cultivars. Significant positive association of grain smut incidence with grain smut severity (0.771) and gran smut susceptibility (0.926) was recorded. Negative association of grain smut parameters *viz.,* grain smut incidence, smut severity and susceptibility index with day to 50% flowering and days to maturity were reported, which indicated that early maturing cultivars are more susceptible to grain smut. Retarding effect of grain smut incidence on morphological characters was observed. Average reduction in plant height varied 9.3 to 12.5% with a mean of 11.2%, number of tillers per plant 5.3 to 7.1% with a mean of 60%, ear head length 6.6 to 14.7% with a mean of 10.2% and average loss in grain yield per plant was 16.5 to 35.8% with a mean of 26.9% was recorded in the cultivars of little millet.

Anuradha *et al.* (2017) extended investigation on little millet to study the genetic variability, heritability and correlation of grain yield and yield related traits. Analysis of variance of 30 genotypes revealed significant variation for all the traits investigated. Grain yield recorded genotype coefficient of variation of 23.52, while phenotypic coefficient of variation was 26.61 q/ha . Heritability ranged from 0.67 (plant height) to 0.99 (day to 50% flowering, days to maturity), while heritability for yield was 0.78. Number of tillers per plant, panicle length and fodder yield were observed to be highly and positively associated with grain yield. This investigation indicated the presence of large variability and high heritability and yield and other related traits which can be used for crop improvement in case of little millet.

9. Epilogue: It can be cultivated in low rainfed area with even poor soil. The new high yielding varieties developed in India should be popularized amongst farming communities. They can be good substitute for rice when rice crop fails especially in drought period. It is good famine food. It can sustain cattle also. Other areas especially in India are desired to look for landraces with better or diverse characteristics.

3.7 PEARL MILLET

Pennisetum americanum K. Schum. (Poaceae)

Resume: *It is native of tropical Africa and probably originated in western tropical Africa where its cultivated and wild forms occur. India is thought to be its secondary centre of variability. It is a diploid crop with n=7. It is nutritious food crop and also constitutes an important forage crop. It is grown in semi-arid regions of Africa and India.*

1. Introduction: Bulrush millet, *Pennisetum americanum* (*P. typhoides, P. glaucum*) also known as pearl millet, spiked millet, cattail millet and, in India, bajra. It is usually grown as a rain-fed crop in semi-arid regions of Africa and India, although it is sometimes irrigated in the latter country. It is the most important crop in the Sahel zone approaching the Sahara in tropical Africa. The shorter maturing cultivars can be grown with less rainfall than *Sorghum*. Late-maturing cultivars become more important as the rainfall increases. The crop is grown to a lesser extent in eastern and central Africa. In Sudan Zone, it is of equal status with *Sorghum*. According to Deshprabhu (1966), bulrush millet in India occupies the fourth place after rice, *Sorghum* and wheat. The great merit of bulrush millet is that it can be grown in low rainfall areas on poor, sandy soils and that it will give economic, albeit low, yields on soil too poor.

2. Plant Characteristics: Tall annuals, growing a height of 6-15 ft.; stem 10-20 mm thick; leaves linear, long; flat, dark green; ±8 cm wide; inflorescence a dense cylindrical, cylindrical, spike-like head or panicle, 6-15 inch long; grain small, green, 3-4 mm, wedge-shaped.

3. Useful Parts: Grains, straw.

4. Chemical Constituents: Protein (22%), carbohydrate (65%), fat (1. 2%), minerals (calcium, magnesium, potassium, iron, copper, manganese, phosphorous) thiamine, riboflavin, and niacin, vitamin E. K.

5. Uses And Applications: The flour made from grains is nutritious and used for bread and cake. Green roasted ears are parched and consumed. Grains are also useful for porridge and malt. Straw or green plants constitute good fodder for cattle and also for thatching. Food products, according to Seetharam *et al.,* (2007) can be categorized into five categories *viz.,* (i) Traditional foods (10 products), (ii) Non-traditional foods (09 products), (iii) Baked food (13 products), (iv) Weaning mixtures (06 products), and (v) Diabetic products (06 products).

6. Karyotaxonomic Evidence: *Pennisetum americanum* is a diploid with 2n=2×=14 and have affinity with *P. purpureum* (Napier grass) in that part of the genus with ×=7; other species have ×=9.

Stapf and Hubbard (1934) recognized 18 species of cultivated bulrush millet. Bor (1960), Purseglove (1972) and others have recognized only a single collective species. All these forms cross readily with each other and also with related wild weedy forms growing in the area. The cultivated races have broad-tipped persistent spikelets, usually with protruding grains. In the wild species, the spikelets are narrower, pointed and readily deciduous. The grains are smaller and completely covered by the lemma and palea. The wild and weedy forms have an efficient dispersal mechanism, which is lacking in the cultivated forms. This would have been selected for freedom from shattering. All forms are thus capable of genetic exchange with related wild forms living in the same area.

7. History: Undoubtedly, despite the name *Pennisetum americanum*, tropical Africa is the native of bulrush millet. It originated probably in western tropical Africa, where the greatest number of cultivated and related wild forms exist. Chevalier (1932) opined that the cultivated bulrush millet in the Sahel zone of West Africa are very close to some of the spontaneous and subspontaneous forms. He presumed that the cultivated species originally arose in this region. The crop was taken at an early date to East Africa and thence to India probably at least 3000 years ago. At the periphery of distribution there, a second centre of variability developed and improved cultivars were evolved. African origin of pearl millet was claimed first by Koernicke and Werner (1885) based on archaeobotanical evidence from Essouk-Tadmakka (Mali) in Africa. Wild relatives are found on the southern edge of the Shahara (Chevalier, 1932; Brunken *et al.,* 1977) and it is usually considered that this was the locality of domestication. The collection of cultivated races analysed by Oumer *et al.,* (2008) showed very limited microsatellite diversity and their accessions do include a wider range of locality including both high-humidity zones and Saharan oases. Archaeobotanical study suggested pearl millet domestication on the margins of the Sahel more than 3000 years ago (D'Andrea *et al.,* 2001). Recent proposals situate this event in the 3rd millennium BC. in the far western Sahel, perhaps in Mauritiana and/or northeast Mali (Fuller *et al.,* 2007). One of the sites of Central Nigeria, the Nok region, has produced a large amount of millet dated to 800-450 cal BC (Kahlber *et al.,* 2009). The ceramics from Dhar Tichitt and Oualata revealed that domesticated *Pennisetum* was grown at least as early as 3500 BP (Amblard, 1996). Beyond Africa, the evidence is weak.

The spikelets are markedly protogynous and mostly cross-pollinated. The crop is very variable and most fields have a mixture of types. When there is some overlapping of female and male phases in the same inflorescence, or in different inflorescences on the same plant, self-pollination can occur to some extent and this can be used for the production of inbred lines for the breeding of hybrid seed. Male sterile lines were available in India for the production of hybrid seed reports (Deshprabhu, 1966). More diverse the geographical origins of the inbreds, the greater is the degree of hybrid vigour on crossing. F_1 hybrid were available which give higher production from for the local cultivars (Murty, 1969).

Thus the crop evolved by local selection for large grains and non-shattering heads from wild forms. These persisted in continued genetic exchange with their wild relatives. India became secondary centre of diversity.

8. Further Developments: As far as India is a concerned, there is a decrease in area but production has increased (Seetharam *et al.*, 2007). It is generally cultivated in Hariyana, Uttar Pradesh, Rajasthan, Gujarat, Tamilnadu, Andhra Pradesh and Maharashtra in India. Although, it is a rain fed crop, forms or varieties have been developed that can be grown after irrigation and particularly in summer. Some dwarf and early maturing varieties have been developed. Varieties which can not be suitable or accessible for bird are also developed. More than 70 hybrids have been officially released in India. Recent researches are concerned with germplasm collection, characterization, conservation, and worldwide seed distribution. India has the largest germplasm (7775) at ICRISAT. Variability occurs *w.r.t.* plant height, stem thickness, tillering, spike length, seed characters (size, shape, etc). About 21 landraces are reported from Eastern Ghats and 14 landraces from Rajasthan (Seetharam *et al.*, 2007). Bashir *et al.*, (2005) investigated genetic diversity of Sudanese pearl millet landraces using SSR markers. A collection of 214 accessions from different geographical regions of Sudan and 11 accessions from West Africa were assessed. The diversity of landraces was demonstrated by a large number (400) of detected SSR alleles. Weak genetic differentiations were observed among the geographical regions suggesting a high seed and pollen-medicated gene flow among the regions. Analysis of molecular variance revealed that the pearl millet accessions within the regions was much higher than among the regions. The large divergence of the landraces is thought promising by the authors to develop new cultivars and for development of heterotic groups which can be used to develop population and hybrid varieties with higher degrees of heterozygosity.

Bashir *et al.*, (2016) studied genetic diversity of the crop in Namibia by SSRs and morphology analysis. A total of 1441 genotypes were available from National Gene Bank representing the Namibian landraces. A sample of 96 genotypes was analysed by SSRs molecular markers. It was revealed that the Namibian *P. glaucum* germplasm has a low genetic diversity. No noticeable diversity pertaining to the regions of the accessions was observed. The authors further stated that efforts must be made to widen the available gene pool and germplasm to increase the genetic diversity of *P. glaucum*.

Hanson *et al.* (1981) made observations on genotypic variation in pear millet (*Pennisetum americanum* L.) for ability to accommodate abscisic acid (ABA) in response to water stress. A set of 16 genotypes was grown in a controlled environment. Two genotypes which contrasted in their accumulation capacity *viz.*, BJ104 and Severe 39 (the latter accumulating most ABA), maintained the difference over a range of leaf water contents and potentials. Some genotypes were grown in the semi-arid tropics with and without irrigation and sampled for ABA content. Substantial genotypic variation in ABA accumulation was recorded, which could not be attributed to differences in leaf water potential. In a field experiment comparing three genotypes Severe, 39, BJ104 and B282) differences in ABA accumulation were shown to be largely independent of genotypic differences in turgor potential. For a set of six of the genotypes, the amounts of ABA accumulating in leaves of intact, droughted plants in the field when adjusted for differences in water potential, were found to be significantly ($P<0.05$) correlated with amount of ABA accumulated in detached water-stressed leaves The author inferred that the detached leaf test adequately indicate the ability of pearl millet genotypes to accumulate ABA under field conditions.

9. Epilogue: Pearl millet is the 6th most important crop in the world, It is important in India after wheat, maize, rice and Sorghum. Its oldest record is dated to 1200 BC from Rajasthan (Vishnu-Mittre, 1974). It is suitable to grow in less fertile soil and area with low rainfall. It is drought-resistant. Early maturing varieties have been developed.

3.8 SORGHUM

Sorghum bicolor (L.) Moench. (Poaceae)

Resume: *Its primary centre of origin is tropical Africa. Cultivated Sorghum were developed in Africa. India is thought to be a secondary centre for diversification. It is the fourth important major cereals after wheat, rice and maize. In India and China, it was cultivated at an early date at least 4000 years ago. Species of the genus Sorghum are diploids or tetraploids. It is a large heterogenous genus. It provides basic nourishment for much of the human population in the semi-arid tropics of Africa and Asia. In developed countries, it is used mainly as stockfeed.*

1. Introduction: Sorghum is the fourth important and major cereals of rain-fed agriculture in the semi-arid tropics. Its various species/varieties are among the first of the wild taxa to be domesticated by man. In Egypt, it was cultivated prior to 2200 BC. In China and India, it was grown at a early date for at least 4000 years. Its native home is in tropical Africa. Botanically, the primary centres of origin is Africa. The extensive differentiation that has taken place in India justifies the regarding of India as a secondary centre for its cultivars (Visnnu-Mittre, 1974). It dispersed in temperate and tropical regions. It is called *'Juwar-i-Hindi'* in Persia. This indicates its introduction in Persia from India. There are different types *viz.,* broom corn, sweet sorghums, grain sorghums, etc.

2. Plant Characteristics: Coarse erect, annuals, growing to a height of 3-15 ft, culm cylindric, noded prominently, solid; leaves alternate, large lanceolate; inflorescence dense head or panicle; spikelets paired, sessile or pedicellate, each spikelet 2-flowered but one functional; grains (caryopsis) rounded, pinkish red, yellow to white;

3. Useful Parts: Grains, stalks.

4. Chemical Constituents: Grains contain carbohydrates (68-74%), fat (3%), protein (± 20%), vitamin B_1, minerals (calcium, phosphorous, iron). niacin, riboflavin, pyridoxine

5. Uses And Applications: White-grained varieties yield white flour preferred for human consumption in the form unleavened bread, cake, porridge, beer, syrup, etc. Sorghum starch is used as an adhesive and for sizing. Sorghum oil is used in salad or for general cooking. Some varieties are used as stock-feed and forage, in the manufacture of brushes, paper, etc.

6. Karyotaxonomic Evidence: The Sorghastrae is one of the sub-tribes of the tribe Andropogoneae. This sub-tribe contains as main genera *viz., Sorghum, Sorghastrum* and *Cleistachne*. Within the genus *Sorghum* are the sections *viz.*, Sorghum, Parasorghum, Chaetosorghum and Heterosorghum, all with chromosome numbers based ultimately on $\times$=5.

Sorghum section Sorghum has a polymorphic diploid population (2n=2×=20) in tropical Africa (which should be referred to as *S. arundinceum*) and a distinctly different diploid population in southeast Asia, Indonesia and the Philippines, *S. propinquum*. The latter has well developed rhizome, a distinctive leaf shape and very small seeds. Although hybrids of these two diploids are completely fertile, the present populations are sufficiently different in morphological characters and distribution to be thought distinct species. In between the distributions of *S. arundinceum* and *S. propinquum* and overlapping with them, the tetraploid *S. halepense* (2n=4×=40) occurs. This species distributed continuously from southern and eastern Indian through Pakistan and Afghanistan across Asia Minor to the Levant and the Mediterranean littoral. One form is widely known as Johnson Grass. *Sorghum halepense* almost arose from the doubling of the chromosomes in a cross between the *S. arundinaceum* and *S. propinquum* groups.

Section Sorghum overlaps in distribution to those of Parasorghum and Heterosorphum, as well as those of the genera Cleistachne and *Sorghastum*. Crosses between members of section Sorghum with the other genera and sections have been unsuccessful. Therefore, the origin of the cultivated crop lies within *Sorghum* section Sorghum. It was developed primarily from the wild *S. arundinaceum* of Africa but has probably been influenced by some introgression from *S. propinquum* and *S. halepense* in the areas in which these species occur (Doggett, 1970). In nature, cross between a diploid *Sorghum* and

S. halepense in Argentina gave rise to Columbus grass, *S. almum*, a tetraploid resulting from restitution in the diploid parent. The diploid chromosome number of *Sorghum* is 2n=20 and since the number for the sub-tribe Sorghastrae is ×=5, *Sorghum* is assumed to be, ultimately, of tetraploid origin. No cytological evidence for homologous segments in the chromosomes of its haploid are found. *Sorghum* behaves as a good diploid all respects.

Man selected characters *viz.,* non-shattering heads, large seeds and heads, threshability and suitable maturity. Natural selection but favourd wild-type characters. Thoday (1972) demonstrated that divergence into polymorphic populations occurs under disruptive selection, without isolation and in the presence of considerable gene flow between the components of the population. Many isolation mechanism have contributed to the diversification of the crop, in addition to the polymorphisms created by disruptive selection. Harlan and de Wet (1972) demonstrated the essential correctness of the conclusion that disruptive selection was the major influence in the development of the sorghum crop systematics. Harlan and de Wet *(loc. cit.)* were able to divide the cultivated sorghums into five basic races and a series of intermediate forms between all combinations of these races. They recognized five basic cultivated races *viz.,* Bicolor, Guinea, Caudatum, Kafir and Durra.

7. History: Undoubtedly, cultivated sorghums were developed in Africa. It is debatable whether they were developed in Ethiopia and the adjacent countries, or in West Africa by the 'Nuclear Mande' (Murdock, 1959) or in a long belt, a non-centre, across Africa. The main area of variation of the crop lies in eastern central Africa (Hutchinson, 1965). The diversity of both the wild and cultivated sorghums in this region is widespread rather than localized in the mountainous areas. Harlan and De Wet (1972) noted a lack of botanical diversity in the Ethiopian cultivated forms, many of which belong to Durra, the most highly developed of the cultivated races, or to the Durra-Bicolor intermediate type. He regarded this as an argument against Ethiopia as the area of origin of the crop. However, there appears no intrinsic reason why a crop plant should not reach its most developed form in its centre of origin, where it has been under selection for the longest time. Certainly, there is much diversity of head and grain type and it is probable that the excellence of these Durras resulted exclusion of other forms. The guinea race was diversified in West Africa and Durra in the Ethiopia-Sudan area. Cultivated sorghum meanwhile reached East Africa. Many varieties with white, high quality grains have the Guinea race characteristic of the grain turning 90° between gaping involute glumes in these areas. This character association may well have been present in the original material which moved into East Africa as well as in that which moved to West Africa. Race Bicolor has an uncertain position; it is low yielding, with generally poor grain quality and is widely scattered. It is probably preserved for its sweet, juicy stem in Africa and Asia and for its broom-making potential in the Mediterranean area (Harlan and de Wet, 1972).

Race Durra was developed in Ethiopia and the Sudan and reached to the Near East and to India. The wild sorghum varieties were carried along as weeds of the crop and diversified. Much variation may have existed in wild *S. arundinaceum* before the arrival of the cultivated plants. The wild variety *arundinaceum* in West Africa may well have been there before the arrival of the primitive crop; but wild *verticilliflorum* could well have developed and travelled as a crop weed because it shows affinities with all the cultivated races.

Thus the wild variety *arundinaceum* is associated with the race Guinea in West Africa, *aethiopicum* with bicolor and durra and *verticilliflorum* with races Kafir and Caudatum. Independent origins of the cultivated sorghums from the wild form have been proposed (Snowden, 1936). It appears more probable that the early cultivated forms of the crop were transferred from place to place, hand in hand with the new technology.

The crop was not in contact with wild diploid forms outside Africa until it reached southeast Asia, where the diploid *S. propinquum* occurs and this population may have attained the characteristic appearance of the kaoliangs of China. Some genetic exchange is possible between tetraploid *S. halepense* and the crop. The spread along the coast of southeast Asia and so to China was probably rather later, perhaps around the beginning of the Christian era. However, an identification from the Chinese neolithic period, if confirmed, would put the arrival there much earlier. This would suggest that sorghum was moved overland along the silk Road. Grain sorghum first went to America from West Africa with the slave trade. Brown and white durra were also brought from North Africa about 1874, milo about 1880, feterita in 1906 and hegari in 1908. The kafirs were introduced from South Africa about 1876.

Sorghum most probably reached to India overland from East Africa. Movement along the coast must have occurred at an early date, as the trade between the two countries via Arabia is ancient. The first reliable archaeological record of the crop in India comes from spikelets mixed with potting clay at Ahar, Rajasthan, in 1725 + 140 BC.; possibly, however, an older picture on a potsherd from Mohenjo-Daro, 2300-17 50 BC. is of sorghum. The crop, therefore, probably reached India soon after 2000 BC.

8. Further Developments: The objective of Sorghum programme in the USA was stock feed. A series of mutants affecting both height and length of maturity were selected by farmers and adaptation steadily improved. Giant milo had given rise to standard milo and dwarf yellow milo by 1906. Early white milo was grown in 1911 and double dwarf yellow milo appeared in 1918. Milos belong to the race Durra. Also series developed in the Kafirs, gave rise to a dwarf form, dawn and Club kafir. The basis of sorghum improvement in the USA has been the kafir × milo cross and natural crosses selected by farmers before 1914 included types such as header milo.

The first period related to selection of natural mutants and crosses by farmers was followed by deliberate plant breeding work. Segregates from kafir × milo, kafir × feterita and from three-way crosses involving hegari, were released as varieties in the 1920s. These included Chiltex, Bonita, Beaver and Wheatland. The last two cultivars had straight, erect peduncles and Wheatland could be harvested by standard wheat machinery thus establishing the future for the large scale growing of 'combine' sorghums. Wheatland gave rise to a number of other varieties, including Martin.

Martin was resistant to stalk-rot and occupied the greater part of the US sorghum acreage before the release of hybrids. Meanwhile, kafir × milo crosses gave rise to many varieties including 7078 and Combine kafir 60. The latter were resistant to clinch bug. Waxy starch types were developed and exploited during the Second World War and some industrial starch is still produced from sorghum.

Combine kafir-60 was sterilized and 7078 proved to be a good restorer, giving rise to the widely grown hybrid 610. Many other hybrids and parent lines were soon developed. Hybrid seed distribution to growers began in 1957 and almost all the US sorghum acreage was covered by hybrids four years later. Hybrid development represented another stage in the exploitation of the cross between Kafirs from South Africa and milos (race Durra) from North Africa. The Kafirs provided the male-sterile parents when the appropriate genome in milo cytoplasm and race Durra furnished the restorer lines.

Sorghum improvement programme in South Africa led to the production of good combine type hybrids and *Striga*-resistant varieties. Here, the market was primarily for a brewing grain to make 'Kafir beer'.

In India, the hybrid programme produced high yielding material, especially CHS I and CSH 5 and some promising new varieties for food grain. Problems are great, especially insect pests (such as shoot-fly, stem-borer and midge), birds and the witchweed, *Striga*. Occasional hybrids from the USA, such as NK 300, prove productive over wide range of conditions.

Tesfaye (2017) studied genetic diversity of Sorghum [*Sorghum bicolor* (L.) Moench.] genotypes from Ethiopia. He used total 117 accessions and two standard checks (Geremew and Baji) for yield and drought resistance. He paid attention to the traits such as basal tiller, nodal tiller, leaf number at maturity, plant height, panicle length, panicle width and days to 50% maturity. He concluded that landraces are the main source of genetic diversity for gene banks and breeding programmes. He sought attention to many studies which have shown that genetic erosion is occurring on farmer varieties because of the utilization of high-yielding varieties. He also pointed out that the landraces selected by farmers for certain desirable traits are likely to survive on a farm alongside improved varieties. According to him, the tested genotypes showed genetic variability for the traits considered. Total 119 genotypes are grouped into 13 cluster and a single accession for four cluster groups (Clusters 10, 11, 12 and 13) having ranging from 13.5 to 56, which clearly showed that there exists a diversity of the sorghum genotypes.

Bernal *et al.* (2014) studied effects of the genotypes and environment interaction on sugar accumulation in sweet sorghum varieties [*Sorghum bicolor* (L.) Moenc.] in Colombia. They evaluated ten genotypes in six contrasting environments. The results indicated that the stem dry weight, juice sugar concentration, stem sugar content and juice volume were controlled by the genetic contribution of the genotype with a large environmental contribution to their expression. Their results also indicated some genotypes having a high potential for the biofuel agroindustry due to their high sugar contents.

9. Epilogue: Thus greater attention was given in the USA to grain quality, directed purely at the better utilization of the grain in stock-feed and partly at improvement in protein quality by utilizing high lysine genes. The crop has a C_4 photosynthetic pathway and a good translocation system. The crop much benefits in a good production potential. The genetic base was broadened by a programme to convert part of the world collection to dwarf, photoperiodically-insensitive types. Disease resistances are receiving more attention, especially to downy mildew (Wall and Ross, 1970). The establishment of the International Crops Research Institute for the Semi Arid Tropics (ICRISAT) in developing nations concentrated research effort to improve the sorghum crop. Grain quality is much improved by breeding, especially for ripening under rainy conditions. Better levels of resistance to pests and *Striga* is obtained. In the developing world, these have coupled with better agricultural practices.

Dwarf plants should be capable of a better harvest index, but the best grain yields come from plant about two meters tall. Genetic male-steriles have made possible recurrent selection in random-mating populations, a procedure which holds out promise of steady crop improvement. Study of an adequate world collection were made available to breeders for better range for material to exploit. Hybrids showed ability to perform well under adverse conditions. Obviously, population pressure in the semi-arid tropics will require the development of sorghum to the greatest possible extent.

Molthaodi *et al.,* (2017) assessed genetic diversity in cultivated Sorghum [*Sorthum bicolor* (L.) Moench] from southern Africa in 22 accessions from five countries *viz.,* Botsawana, Namibia, Swaziland, Zambia and Zimbabwe. Significant variation among these accessions in both qualitative and quantitative morphological traits. They indicated promising potential as breeding material. Analysis of molecular variance revealed a high level of genetic variation. In their opinion, the genetic diversity patterns and the relationships observed should provide insights for genetic resource conservation and utilization of sorghum germplasm.

Burow *et al.*, (2012) characterized genetic diversity in a collection of 159 landraces of *Sorghum bicolor* L. collected from the colder region of China. A set of 41 microsatellites of SSR markers distributed throughout the 10 chromosomes of *Sorghum* was utilized in the analysis. Their revelation provide evidence that Chinese landraces of *Sorghum* exhibit moderate genetic diversity. The results from cluster analysis may well aid identification of diverse accessions that can serve as parental lines for efficient utilization and application of these germplasm into Sorghum breeding programmes.

References:

Amblard, Sylvie (1996) Agricultural evidence and its interpretation on the Dhars Tichitt Oualata. *In:* (Ed. G. Pwiti and Soper) Aspects of African Archaeology. Papers from the 10th Congress of the Pan African Association of Prehistory and Related Studies. University of Zimbabwe, Harare, pp. 421-427.

Anonymous (1952) The Wealth of India: Raw Materials. Vol. II D-E. CSIR, New Delhi, India.

Anuradha, Patro TSSK, Divya M., Sandhya Rani, Y. and U. Trivedi (2017) Genetic variability, heritability and correlation of quantitative traits of little millet genotypes. *Journal of Pharmacognosy And Phytochemistry* 6(6):489-492.

Arunachalam, V., Rengalakshmi, R. and M.S.K. Raj (2005) Ecological stability of genetic diversity among landraces of little millet (Panicum sumatrense) in South India. *Genet. Res. Crop. Evol.* 52:15-19.

Ashwini Kumar, Jain, A.K., Ashish Kumar and R.L. Sharma (2017) Agro-morphological characters of little millet (Panicum sumatrense) associated with grain smut incidence. *Journal of Etomolgy And Zoology Studies* 5(5):356-359.

Bapyu, R.N., Jyothi, M.N., Sharadamma, N., Sah, S., Rai, D.V. and V.R. Devraj (293) Computational identification of conserved microRNAs from Kodo millet (Paspalum scrobiculatum) Afr. Crop. Sci. J. 21:75-83.

Bashir, Elfadil M.A., Abdelbagi, M.Ali, Adam, M.Ali, E.Tahir I.Mohamed, Albrecht E.Melchinger, Heiko, K. Parzies and Bettina I.G. Haussmann (2015) Genetic diversity of Sudanese pearl millet [Pennisetum Glaucum (L.) R.Br.] landraces as revealed by SR markers, and relationships between genetic and agromorphological diversity. *Genetic Resources And Crop Evolution* 62(4):579-591.

Bernal, J.H., Gustavo A.L.M. and R.S.Hernendez (2014) Effects of the genotype and environment interaction on sugar accumulation in sweet sorghum varieties [Sorghum bicolour (L.) Moenc.] grown in the lawland tropics of Colombia. *Agron Colomb.* 32(3):307-314.

Billy McBenedict, Percy Chimwamurombe, Ezekeil Kwembeya and Gillian Maggs Kolling (2016) Genetic diversity of Namibian Pennisetum glaucum (L.) R.Br. (Pearl Millet) landraces analysed by SSR and morphological markers. *The Scientific World Journal* http:11.dx.doi.org/10.1155/2016/1439739.

Blatter, E.J. and C. McCann (1934) The Bombay Grasses. Science Monograph No. 5. Imp. Counc. Agri. Res. Mem. India.

Blench, R. (1997) Neglected species, livelihoods and biodiversity in difficult areas: how should the public sector respond? Net. Resour. perspect. 23:1-10.

Bor, N.L. (1960) The Grasses of Burma, Ceylon, India and Pakistan, Oxford, UK.

Brunken, J., J.M.J. de Wet and J.R.Harlan (1977) The morphology and domestication of pearl millet. *Economic Botany* 31:163-174.

Burow Gloria, Clere, D. Franks, Zhanguo Xin and John J. Burke (2012) Genetic diversity in a collection of Chinese sorghum landraces assessed by microsatellites. *Amer. J. Plant. Sci.* 3:1722-1729.

Chandrasekhara, A. and F. Shahidi (2011) Determination of antioxidant activity in free and hydrolyzed fractions of millet grains and characterization of their phenolic profiles by HPLC-DAD-ESI-MS. *J. Funct. Foods* 3:144-158.

Chevalier, A. (1932) Etude sur les Prairies de I'Ouest African. *Rev. Bot. Appl. Agric, trop.* 14:17-48.

Clayton, W.D. (1975) The Paspalum scrobiculatum complex of tropical Africa. *Kew Bull.* 30(1):101-105.

D'Andrea, A.C., Klee, M. and J. Casey (2001) Archaeological evidence for pearl millet (Pennisetum glaucum) in sub-saharan West Africa. *Antiquity* 75:341-348.

Darlington, C.D. and A.P. Wylie (1955) Chromosome Atlas of Flowering Plants. Allen and Unwin, London, UK.

Darmency, H. and J. Pernes (1985) Use of wild Setaria viridis (L.) Beauv. To improve triazine resistance in cultivated S. Italica (L.) P. Beauv. by hybridization. *Weed Res.*25, 175-179. doi 10. 1111/j.1365-3180-1985. tboot 33.x

De Candolle, A.P. (1886) Origin of Cultivated Plants (Translated from 2nd Ed. in French, 1959), Hafner, New York, USA.

De Wet, J.M.J., Prasada Rao, K.E. and D.E.Brink (1983) Systematics and domestication of Panicum sumatrense (Graminae). *J. d'agriculture Tradit. Bot. appliquee* 30:159-168.

De Wet, J.M.J., Rao, K.E.P., Mengesh, M.H. and D.E.Brink (1983) Diversity in Kodo millet, Paspalum Scrobi Culatum. *Econ.Bot.*37:159-163.

Deshprabhu, S.B. (1966) (Ed) The Wealth of India: Raw Materials, 7, New Delhi, India.

Dida, M.M., Wanyera, N., Harrison, Dunn M.L., Bennetzen, J.L., Gale, M.D. and K.M. Devos (2008) Population structure and diversity in finger millet (Eleusine coracana) germplasm. *Trop.Plant.Biol.*1:131-141.

Doggett, H. (1970) Sorghum. Longmans, London, U.K.

Dwivedi, S., Upadhyaya, H., Senthilvel, S., Harsh, C., Fukunaga, K. Diao, X. *et al.*, (2012) Milets: genetic and genomic resources. *In:* Plant Breeding Reviews (Ed. J.Janick) Hoboken, N.J.:John Wiley and Sons. Pp.247-374.

Fuller, D.W. Macdonald, K., and R.Vernet (2007) Early domesticated pearl millet in Dhar Nema (Mauritania): evidence of crop processing waste as ceramic temper.

In: (Ed.Cappers, R.) Fields of Change. Proceedings of 4th International Workshop for African Archaeolbotany. Borkhuis & Groningen University Library, Gronongen, pp.71-76.

Geervani, P. and B.O. Eggum (1989) Nutrient composition and protein quality of minor millets. *Plant Foods Hum. Nutr.* 39:201-208.

Graybosch, R.A. and D.D. Baltensperger (2009) Evaluation of the waxy endosperm trait in proso millet (Panicum miliaceum). *Plant Breeding* 128:70-73.

Gupta, A., Joshi, D., Mahajan, V. and H.S. Gupta (2009) Biodiversity in barnyand millet (Echinochloa frumantaceae Link. (Poaceae) germplasm in India. Genet.Resour. Crop. Evol. 56, 883-889. doi:10.1007/510722-009-9462-y.

Harlan, J.R. and J.M.J. De Wet (1972) A simplified classification of cultivated sorghum. *Crop. Sci.* 12:172-176.

Hedge, P.S. and T.S.Chadra (2005) ESR spectroscopic study reveals higher free radical quenching potential in Kodo millet (Paspulum scrobiculatum) compared to other millets. Food Chem. 92:177-182.

Hedge, P.S. Raja Sekaran, N.S. and T.S. Chandra (2005) Effects of the antioxidant properties of millet species on oxidative stress and glycemic status in olloxan induced rats. Nutr. Res. 25:1109-1120.

Henson, I.E., Mahalakshmi, V., Bidinger, F.R. and G.A.Alagarswamy (1981) Genotypic variation in pearl millet [*Pennisetum americanum* (L.) Leeke] in the ability to accumulate abscisic acid in response to water stress. *Journal of Experimental Botany* 32(5):899-910.

Hilu, K.W. (1994) Evidence from RAPD makers in the evolution of Echinochloa millets (Poaceae). Plant Syst. Evol. 189, 247-257.doi.10.1007/BF00939730.

Hiremath, S.C., Patil, G.N. and S.S.Salimath (1990 Genome homology and origin of Panicum sumatrense (Graminae). *Cytologia* 55:315-319.

http:www.hort.purdue.edu/newcroplatcm/millethtml

Hunt, H.V., Farah Badakshi, Olga Romanova, Christopher J. Howe, Martin K.Jones and J.S.Pat Helslop-Harrison (2014) Reticulate evolution in Panicum (Poaceae): The origin of tetraploid broomcorn millet, P.miliaceum. *J. Exp. Bot.* 65(12):3165-3175.

Hunt, H.V., Vander Linden, M., Liu, X., Motuzaite-Motuzeviciute, G., Colledge, S., and M.K. Jones (2008) Millets across Eurasia: Chronology and context of early records of the genera Panicum and Setaria from archaeological sites in the Old World. Veg. Hist., Archaeobot. 17, 55-518, doi:10.1007/500334-008-0187-1.

Hutchinson, J.B. (1965) (Ed.) Essays on crop plant evolution. Cambridge, UK.

Hutchinson, J.B. (1974) Crop plant evolution in the Indian subcontinent. *In:* Evolutionary Studies In World Crops (Ed. Joseph Hutchinson) Cambridge University Press, London, U.K. pp.151-160.

Jusuf, M. and J. Pernes (1985) Genetic variability of foxtail millet (Setaria italica P. Beauv.): electrophoretic study of five isoenzyme systems. *Theor. Appl. Genet.* 71:385-391. Doi:10.1007/BF00251177.

Kahlber, Stefanie, Koen Boston and Katharina Neumann (2009) Early plant cultivation in the Central African Rain Forest: First Millenium BC Pearl Millet from South Cameroon. *Journal of African Archaeology* 71(2):253-272.

Kennedy-O'Byrne. J. (1957) Notes on African Grasses: XXIX: A New Species of Eleusine from Tropical and South Africa. *Kew Bull.* 1:65-72.

Kim, C.S., Alamgir, K.M., Mutsumoto, S., Tebayashi, S. and H.S. Koh (2008) Antifeedants of Indian barnyard millet (E.frumantacae Link.) against brow plant happer, Nilaparvata lugens (Stal). Z.Naturforsch 63:755-760.3 De Wet (1983).

Kochhar, S.L. (2006) Economic Botany In The Tropics. (Repr. Ed.) Macmillan India Ltd., Delhi, India, pp.110.

Kulkarni, L.R. and R.K. Naik (2000) Nutritive value, protein quality and organoleptic quality of Kodo millet (Paspaum scrobi culatem) *Karnataka J. Agric. Sci.* 13:125-129.

Li, C.H., Pao, W.K. and H.W. Li (1942) Interspecific croses in Setaria. J. Hered. 33:351-355.

Lu Houyuan, Jianping Zhang, Kam-iu Liu, Naiain Wu, Yumei Li, Kunshu Zhou, Maolin Ye, Tianyu Zhang, Haijiang Zhang, Xiaoyen Yang, Licheng Shen, Deke Xu and Quan Li (2009) Earliest domestication of common millet (Panicum milliaceum) in East Asia extended to 10,000 years ago. Proceedings of the National Academy of Sciences of the United States of America (Ed. Dolores R. Piperno) Vol. 106 (18) 7367-7372.

M'Ribu, H.K. and K.W. Hilu (1996) Application of random amplified polymorphic DNA to study genetic diversity in *Paspalum scrobiculatum* L. (Kodo Millet, Poaceae). *Genet. Resour. Crop. Evol.* 43:203-210.

Malleshi, N.G. (2002) Malting of finger millet and specialty food based on the millet malt. Souvenier. XV Indian Convention of Food Scientists And Technologists, Mysore, pp.17-23.

Malleshi, N.G. and C.F. Klopfensteini (1997) Nutrient composition, amino acid and vitamin contents of malted sorghum, pearl millet, finiger millet and their rootlets. *J.Food.Sci.Technol.* 35:247-249.

Mehra, K.L. (1963a) Differentiation of the cultivated and wild Eleusine species. *Phyton* 20:189-198.

Mehra, K.L. (1963b) Considerations of the African origin of Eleusine coracana. *Curr.Sci.* 32:300-301.

Motlhaodi, T., Malatu Geleta, Stephen Chite, Moneim Fatih, Rodomiro Ortiz and Tomas Bryngelsson (2017) Genetic diversity in Sorghum [Sorghum bicolor (L.) Moench] germplasm from Southern Africa as revealed by microsatellite markers and agromorphological traits. *Genetic Resources And Crop Evolution* 64(3):599-610.

Murdock, G.P. (1959) Africa, its peoples and their culture history. New York, USA.

Murty, B.R. (1969) New Hybrids of Bajra. *Indian Fmg*, 19:13-15.

Nirmalkumar, A., Salini, K. and P. Veerabadhriran (2010) Morphological characterization and evaluation of little millet (Panicum sumatrense Roth ex Roem, & Schultz) germplasm. Electron J.Plant Bred 1:148-155.

Nirubana, V., Ganesamurthy, K., Ravikesvan, R. and T. Chideshwari (2017) Genetic diversity studied in Kodo millet (Paspalum scrobuculatum L.) germplasm accessions based on biomatrical and nutritional quality traits. *Int. J. Curr. Microbiol. App. Sci.* 6(10):832-839.

Phillips, S.M. (1972) A survey of the genus Eleusine (Gramineae) in Africa. *Kew Bull.* 27:251-270.

Purseglove, J.W. (1972) Tropical Crops: Monocotyledons. Longman, London, UK.

Purselglove, J.W. (1988) Tropical Crops: Monocotyledons. English Language Book Society, Longman, UK.

Ramachandra, G., Virupaksha, T.K. and M. Shadaksharaswamy (1977) Relationship between tannin levels and in vitro protein digestibility in finger millet (Eleusine coracena). *Food Chem.* 25:1101-1104.

Ramakrishnan, M., Ceasar S. Authony, Duraipandiyan, V. Al-Dhabi, N.A. and S.Ignacimuthu (2016) Assessment of genetic diversity, population structure and relationships in Indian and non-Indian genotypes of finger millet [Eleusine coracana (L.) Gaertn.] using genomic SSR markers. https://www.ncbi.nim.nih.gov/pmc/articles/PMC 4749511 8.

Salini, K., Nirmalkumari, A., Muthiah, A.R. and N. Senthil (2010) Evaluation of proso millet (*Panicum miliaccum* L.) germplasm colections. *Electronic Journal of Plant Breeding* (C4):489-499.

Seetharama, V., Elongovan, M., Khairwal, I.S., Krishna Gowda K.T., Seetharama A., Vilas, A., Topnapi and Sivaraj (2007) Sorghum And Millet genetic resources management. *In:* Agrobiodiversity, Vol.I: Crop Genetic Resources And Conservation (Ed.S.Kannaiyana and A.Gopalam) Associated Publishing Co. New Delhi, India, pp.42-74.

Singh, Ajay (2008) Plants *In:* Ancient Indian Civilizations. Agam Kala Prakashan, Delhi, India.

Sintho, W.A., Nurul, K., Narul, F., and Yudiansyah (2017) Biodiversity assessment of foxtail millet (Setaria italica L.) genotypes based on RAPD marker. *Journal of Tropical Crop Science* 4(1):21-25.

Sivakumar, S. Mohan, M. Franco, O.L. and B. Thayumanavan (2006) Inhibition of insect pest α-amylases by little and finger millet inhibitors. *Pestic. Biochem. Physiol.* 85:155-160.

Smith, P.M. (1976) Minor Crops. *In:* Evolution of Crop Plants (Ed. N.W. Simmonds) Longman, New York, USA, pp.301-324.

Smith, P.M. (1976) Minor crops. *In:* Evolution of Crops (Ed.N.W.Simonds). Longman, New York, USA. pp.301-324.

Snowdcn, J.D. (1936) The cultivated races of sorghum. Adlurd, London. UK.

Starf, O. and C.E. Hubbard (1934) Pennisetum. *In:* D.Prain (Ed.), Flora of Tropical Africa, 9, London, UK.

Subramanian, A., Nirmalakumari, A. and P. Veerabadhiran (2010) Trait based selection of superior Kodo millet (*Paspalum scrobiculatum* L.) genotypes. Electron. J.Plant Breed. 1:852-855.

Subramanian, A., Nirmalkumari, A. and P. Veerabadhiran (2010) Trait based selection of superior Kodo millet (*Paspalum scrobiculatum* L.) genotypes. *Electronic Journal of Plant Breeding* 1(4):852-855.

Tesfaye Kasahun (2017) Genetic diversity study of Sorghum [Sorghum bicolor (L.) Moenc.] genotypes, Ethiopia. *Acta Universities Sapientiae Agriculture And Environment* 9:44-54.

Thoday, J.M. (1972) Disruptive selection. *Proc.R.Soc. Lond.B.* 182:109-143.

Trivedi, A.K., Arya, L., Verma, S.K. and R.K. Tyagi (2018) Molecular profiling of foxtail millet (Setaria italica (L.) P.Beauv.) from central Himalayan region for genetic variability and nutritional quality. *The Journal of Agricultural Science* 156(3):333-341.

Ugre, R. (2008) Health benefits, storage quality and value addition of barnyard millet (Echinocholoa frumantacea L.) M.H.Sc.Thesis, Univ. Agric. Sci., Dharwad, India.

Upadhyaya, H.D., Dwived, S.L., Singh, S.K., Vetriventhan, M. and S.Sharma (2014) Forming core collections in barnyard, Kodo and little millets using morphoagronomic descriptors. *Crop. Sci.* 54, 1-10. doi:10:2135/ cropsci 2014.03.0221.

Veena, B., Chimmad, B.V., Naik, R.K. and G.Shantakumar (2005) Physico-chemical and nutritional studies in barnyard millet. *Karnatak. J. Agric. Sci.* 18(1):101-105.

Vishnu-Mittre (1974) The beginnings of agriculture: Palaeobotanical evidence in India. *In:* Evolutionary Studies In World Crops (Ed.Joseph Hutchinson) Cambridge University Press, London, UK, pp.3-30.

Wall, J.S. and W.M. Ross (1970) (Ed). Sorghum production and utilization. Connecticut, USA.

Wang, C., Jia, G., Zhi, H., Niu, Z., Chai, Y. and W.Li (1912) Genetic diversity and population structure in Chinese foxtail millet [Setaria italica (P.) Beauv.] landraces. G.32, 769-777. Doi:10.1534/g3.112.002907]

Wang, T., Fleury, A. Ma, J. and H.Darmency (1996) Genetic control of dinitroaniline resistance in foxtail millet (Setaria italica). *J. Hered.* 87:423-426. Doi 10.1093/ oxfordjournals.jhered.002303.

Watt, G. (1889-1893) A Dictionary of Economic Products of India. Vol 1-6. Periodical Expert, Shahadara, Delhi, India.

Yabuno, T. (1962) Cytotaxonomic studies on the two cultivated species and the wild relatives in the genus Echnirochloa. *Cytologia* 27:296-305.

4

Pseudocereals

INTRODUCTION

True cereals belong invariably to the grass family, Gramineae (Poaceae). There are few grain or food yielding plants which have a similar grain, chemical composition and nearly similar usage as food source. But these belong to other families of plants, and not to the Gramineae proper. For example, (i) Buckwheat (*Fagopyrum esculentum, Polygonaceae*), (ii) Quinona (*Chenopodium quinoa*, Chenopodiceae and (iii) Grain Amaranths (*Amaranthus* sp., *Amaranthaceae*). They are botanically unrelated to the cereals but having similar role to play in human food requirement, however small, and restricted mostly to their country or region of origin. The state of knowledge explained above for millets also parallels for the pseudocereals. The case of Quinoa in Bolivia and Peru mirrors that of millets in India in terms of competition with imported foods and major cereals.

Pseudocereals are thus plants that produce seeds or fruits which are used and consumed as grains, though botanically they are neither grasses nor true cereals. They are typically rich in protein and other nutrients but gluten-free. They are used in the same way as cereals. Their seeds can be ground into flour and otherwise used as cereals. However, commercialization of these grains (pseudocereals) is still quite limited. It is interesting to note that protein content in pseudocereals is usually higher than the true cereals. Moreover, they contain fair amount of vitamins. Secondly, they also provide certain important minerals. They even contain fair amount fat.

In recent times, a need is being highlighted for an improvement in the nutritional quality of gluten-free products. Pseudocereals can provide such products and satisfy the need. Celiac disease is one of the most common lifelong disorder worldwide. The most acceptable treatment to such sufferers is the strict lifelong elimination of gluten from their diet. Consumption of pseudocereals by them is advisable, nay essential. Their availability in the market is quite limited. More research is necessary to familiarize their nutritional and medicinal values. The present author has, therefore, purposefully included this neglected group of grains to study and understand their importance and even save their agrobiodiversity.

4.1 BUCKWHEAT

Fagopyrum esculentum Moench. (Polygonaceae)

> ***Resume:*** *Temperate eastern Asia is its place of origin more particularly northern India and China. It belongs to the family Polygonaceae, and not to grass family as the wheat. The earliest records are from Chinese during 10-11th centuries. Three species of the genus Fagopyrum viz., F.esculentum, F.tataricum and F.cymosum are all diploid with 2n=16 chromosomes. The last species F.cymosum is probable ultimate source of the former two species. All these three species are used in the same way for human consumption and livestock.*

1. Introduction: Although buckwheat is a crop of secondary importance, it survived over many centuries. It is brought under agriculture in each country in which grain crops are presently cultivated. It is not a member of the grass family to which generally all cereals and millets belong. But the seeds (achenes) are usually classified among the cereal grains because of similar usage.

The earliest records are from Chinese writings during 10 to 11th centuries. It was introduced in Europe during the Middle Ages and first cultivated in 1436. It is widely used in former USSR. In USA, it was brought by the early settlers. In India, it is cultivated in hilly regions of northern India and Nilgiris. It is called 'Kuttu' in India. It grows well in poor soils.

2. Plant Characteristics: Plants branching annuals, stem smooth and succulent; leaves alternate and hastate; flowers small, white or pinkish, in racemes; fruits (achenes) 3-pronged.

3. Parts Used: Grains.

4. Chemical Constituents (Grains): Protein (10.3%), fat (2.4%), carbohydrate (starch 65.0%), minerals (calcium, phosphorous, copper, manganese, magnesium, iron), vitamin B_1, nicotinic acid, riboflavin, quercetin, phytin, vitexin, rutin.

5. Uses And Applications: Flour is used in biscuits, noodles and pancakes and in the form of porridge. In India, flour is made into many food preparations and used particularly during fast days. The grains are usually used as feed for poultry and cattle. A glucoside, called rutin, is used in the treatment of capillary fragility associated with high blood pressure or hypertension. Buckwheat is gluten-free and provide a wide range of health benefits, for example, plasma cholesterol level reduction, neuroprotection, anticancer, anti-inflammatory and antidiabetic (Gimenez-Bastida and Zienski, 2015).

6. Karyotaxonomic Evidence: Three species *viz., F. cymosum, F. esculentum* and *F. tataricum* are all diploid with 2n=2×=16. These are called perennial, common and tartary respectively. Notch-seeded buckwheat (*F. emarginatum*) is cultivated in northeastern India and China. Its grains differ from Tartary and common by having the angles or edges of the hull extended into wide, rounded wings. It is similar to common buckwheat in all other respects and is completely compatible with it. It is considered as a form of *F. esculentum* (Bailey, 1917). The perennial species, *F. cymosum*, has rhizomes and differs from common and Tartary buckwheat in its shoots and branching racemes. The karyotaxonomy of buckwheat is little investigated but it is generally believed that *F. cymosum* was the original parent of both common and Tartary buckwheat (Krotov, 1963).

7. History: Temperate eastern Asia is the birth place of the buckwheats. The perennial species, *F. cymosum*, the probable ultimate source of both common and Tartary buckwheat, is native to northern India and China. Wild forms of *F. esculentum* inhabit China and Siberia. Tartary buckwheat grows in the Himalaya of northeastern India and China under cooler and harsher climatic conditions. It is better adapted than common buckwheat in such conditions.

The putative common ancestor, *F. cymosum* and the two cultivated forms suggests that, in nature, selection for annual habit, non-shattering inflorescence and low seed dormancy was operative. Tartary buckwheat has more fragile heads and more dormant seed than the common form. It appears to be the less advanced of the two cultigens. The breeding system of *F. cymosum* is unknown. It may be assumed to share with *F. esculentum* a heteromorphic incompatibility system, but *F. tataricum* is self-compatible and homomorphic. It seems as if, one cultigen, tataricum, has moved towards inbreeding under domestication, though least specialized morphologically than the other.

Buckwheat is known to have been cultivated in China for 1,000 years. It is not thought to be very ancient (Hunt, 1910). The first written records are Chinese scripts of the fifth and sixth centuries. It was introduced into Europe in the Middle Ages, reaching Germany early in the fifteenth century (Hughes and Henson, 1934). It was cultivated for several centuries in England, France, Spain, Italy, Germany and Russia. It has been also under agriculture in Africa and Brazil. It was introduced early into the American Colonies, having been relatively much more important then, than now.

8. Further Developments: Both common and Tartary buckwheat has generally been declined. Both have declined in USSR, USA, France and Canada. Canada's acreage but have increased after early 1960s as a result of demand for export. Common buckwheat is a self-incompatible, sexually-propagated crop and as such, does not lend itself readily to improvement by breeding. Sharma and Boyes (1961) have studied the heteromorphic sporophytic type of incompatibility involved and showed that the inheritance of style length is controlled by a single gene. However, short-styled flowers also differ from long-styled ones in having longer stamens, larger pollen grains, the thrum stylar incompatibility reaction and the thrum pollen incompatibility reaction. Sharma and Boyes (*loc. cit.*) postulated an S supergene for this complex of characters. There is evidence of genetic change at the S locus. Thus, homomorphic, highly self-compatible diploid lines have been isolated.

Workers in the USSR have tackled this crop. Attempts to transfer the self-compability of *F. tataricum* to *F. esculentum* have failed. A study of pollen tube growth in the interspecific cross showed that pollen tubes of *F. esculentum* reached the base of *F. tataricum* styles, where as those of *F. tataricum* were prohibited in *F. esculentum*. No hybrid seed was produced (Morris, 1951).

Seed size can generally be increased by raising the ploidy level to tetraploid since seed size and yield are important factors considered in most breeding programmes. A number of autotetraploid lines have been produced. Such lines generally have much lower fertility than their diploid counterparts and the considerable effort put into breeding at the tetraploid level has resulted in few new varieties. Buckwheat is mostly free from damage by insects or diseases. No programmes of resistance breeding have been conducted. The major emphasis in most breeding programmes was high yield.

Emphasis can be expected to continue on increasing grain yield, as there has been little improvement in this respect from the most breeding programmes. Any improvement in disease or insect resistance will probably be accomplished by introgression from desirable lines.

Fegopyrin has anticarcinogenic effect. It is isolated from the blossoms of the red flowering variety of *F. esculentum*. Such capacity makes it useful medicine in photodynamic therapy (Ebermann *et al.*, 1996). Samel and Witte (1994) suggested that fagopyrin is a potent inhibitor of protein tyrosine kinase which can control proliferative process of cancer cells. Rutin is a flavonoid with high antioxidative power. Inhibition of lipid peroxidation was observed by rutin (Yang *et al.,* 2008). Together with antioxidant activity rutin also possesses additional pharmacological efforts (Kreft *et al.*, 2003).

Juan and Henryk (2015) studied buckwheat (*Fagopyrum esculentum*) as a functional food and its effects on human health. It is a gluten-free pseudocereal. Its grain are highly nutritional food component, which in their opinion, provides a wide range of beneficial effects. Health benefits attributed to buckwheat are: plasm cholesterol level reduction, neuroprotection, anticancer, anti-inflammatory, antidiabetic effects, and improvement hypertension conditions. Also, it is reported to have prebiotic and antioxidant properties. *In vitro* and animal studies suggested that buckwheat contain bioactive compound *e.g.* D-chiroinositol (DCI), proteins and flavonoids (mainly rutin and quercetin) which may be partially responsible for the observed effects.

Duran *et al.* (2016) determined some morphological, mineral, amino and organic acid characteristics of buckwheat, bread wheat, triticale, durum wheat and triticale genotypes. They pointed out similar nutritional quality of buckwheat to other sources mentioned. However, the genotypes of buckwheat *viz.*, Aktas (A) and Gunes (G) found rich in minerals, amino acid and organic acid. It is hence buckwheat may be an important component as well as other cereals in the food industry. Being gluten-free, these genotypes could be widely used for producing gluten-free products beneficial for persons with gluten intolerance (Celiac disease).

9. Epilogue: It is indicated that the photoperiod influences the time and rate of flowering as well as plant habit (Shustova, 1965). Varieties responding to different photoperiods have been developed and grown commercially in Japan and are being utilized in other programmes in adapting lines to various regions. This is as it should be. Secondly, local cultivers have not been reported from anywhere. This study will be helpful for breeding programmes in future. Its protein is reported to be of excellent quality (Coe, 1931). We can supplement our food article if they are protein-poor. It grows well in poor soil and needs a little care. When other crops fail, one should adopt it.

One aspect that must be considered in the future is the collection and maintenance of genetic material. Few expeditions have taken place but there is good international collaboration on maintenance and distribution of seed stocks. Further collecting, to obtain a more representative sampling of native genetic resources. should result in a fairly satisfactory situation.

Sdej *et al.* (2012) reported buckwheat possessing higher antioxidant activity (AOA) mainly due to high rutin content. They found significantly higher contents of total phenolics and flavonoids in hull than in whole grain and groat. Protocatechuic, syringic and sinapic acid, rutin and quercetin were found in all tested fractions, whereas vanilic acid was found in whole grain and hull. The content of total tocopherols ranged from 23.3 mmol/g for hull to 61.8 mmol/g for groat. Hull was superior in scavenging activity on 1, 1-diphenyl-2-picrylhydrazy (DPPH), hydroxyl (HO) and superioxide anion radicals, reducing activity, antioxidants by beta-carotene bleaching method and chelating activity on Fe (2^+) as evidenced by its lower IC(50) value. The results obtained can broaden the utilization of buckwheat, especially a share of hull in whole grain flour production.

4.2 GRAIN AMARANTHS

Amaranthus Spp. (Amaranthaceae)

Resume: *Three species of the genus Amaranthus (Amaranthaceae) are native to north-western, central or southern Mexico, Central America and the Argentine Andes. They were domesticated prehistorically and are most* *ancient crop species as early as 4800 BC. The earliest archeological records are: (i) A. cruentus from Tehuacan, Puebla (Mexico) about 4000 BC. and (ii)A. hypochrondriacus about 500 AD. and (iii) A. caudatus from north-western Argentina about 2000 year old tombs.*

Although, they are overshadowed by the cereals, they are still cultivated in America, Asia and Africa. They are generally used as subsidiary food, but in the Western Himalayas they are cultivated as a regular crop and constitute the staple food in place of wheat. Their nutritional quality is comparable to the cereals. It is one of the efficient food crops. The genus is dibasic with chromosome number ×=16 and 17.

1. Introduction: The studies in archaeology in South and Central America divulged that grain amaranths have been associated with man since prehistoric times. They are among the most ancient crop species that have been cultivated as early as 4800 BC. (Agogino, 1957). Obviously, they are overshadowed by the cereals, but are still cultivated in many countries in Asia, Africa, besides America. They are still favoured since their protein quality is better and richer in amino acids as well. Moreover, they are hardy and suitable for growing in areas with a short season and poor soils. They play a better role in feeding the hungry world. *Amaranthus paniculatus* Lion is particularly mentioned as 'Jivani' in Vedic text *viz.,* Atharveda (Singh, 2008).

2. Plant Characteristics: Large, branched, leafy annuals; leaves simple, large, green or red coloured; inflorescence terminal, large panicle, red, yellow or green, monoecious, each glomerule (basic unit of inflorescence) containing one male flower and about 250 female flowers; seeds ivory.

3. Parts Used: Grains.

4. Chemical Constituents: (i) Leaves: protein (3.5%), carbohydrate (6.5%), fat (0.5%) and minerals (calcium, phosphorous, ascorbic acid, vitamin-A, miacin, thiamin, riboflavin. (ii) Grains: protein (15.0%), carbohydrate (63.0%), fat (7.0%) minerals (calcium, phosphorous), thiamin, riboflavin, niacin, acerbic acid (Krishnakumary, 2011).

5. Uses And Applications: In India, grains are used in the same way as wheat . Flour is used for making cakes or chapatis after baking. Parched seeds are ground to flour and eaten as gruel with milk or water. The grains are consumed after popping. The popped grains are put in milk and the porridge is taken on fast days. Commonly, it is used in the form of sweetmeats *e.g.* sweet balls or cake. In other countries, the grains are parched or popped or milled to flour and then consumed. Leaves are consumed as vegetable.

6. Karyotaxonomic Evidence: The cultivated species and their probable home or native regions are: *viz.,* (i) *A. hypochondriacus* (=*A. frumen, A. leucocarpus*, etc.) of north-western and central Mexico, (ii) *A. cruentus* (= *A.*

paniculatus, etc.) of southern Mexico and Central America. (iii) *A. caudatus* of the Andes. In the Argentine Andes, the typical form is grown together with a conspicuous mutant that produces club-shaped inflorescence branches with determinate growth, a trait unknown in wild Amaranths. This mutant has commonly been given specific rank (as *A. edulis*) but may better be thought as *A. caudatus* sp. *mantegazztanus* (Hanelt, 1968). The wild species that appear most closely related to the above are, respectively: the first, *A. powellii*, a pioneer of canyons, desert washes and other open habitats in the western Cordillera of the Americas. An aberrant form with indehiscent utricles has sometimes been given specific rank as *A. bouchonii*, the second, *A. hybridus* (= *A. chlorostachys, A. patulus,* etc.), a riverbank pioneer of moister regions of eastern North America and the mild highlands of Central America; and the third one *A. quiensis*, a riverbank pioneer of highland and subtropical South America.

All six species are diploids (2n = 2× = 32). They are consistently reported except that counts of both 32 and 34 are reported for *A. cruentus* and *A. potoellii*. Pal and Khoshoo (1974) suggest derivation of *A. cruentus* from *A. powellii* on the basis of their counts of 2n = 34 for both. Comparative morphology of the species does not support this standpoint. Counts of both 32 and 34 are also encountered in other sections of the genus, varying between closely related species and within certain species (Grant, 1959). Many interspecific hybrids, both spontaneous and artificial, have been observed among the grain species and their wild relatives (Murray, 1940a; Grant, 1959). Some experimental hybrids show heterosis and nearly normal meiosis but are partially sterile. Others have abnormal growth and are totally sterile, including *A. caudatus* × *A. hypochondriacus* (Pal and Khoshoo, 1972, 1974).

No spontaneous polyploids are known among the grain Amaranths but colchicine-induced autotetraploids and amphidiploids have been bred from some of them. Seed weight in the tetraploids is about double that of the diploids, suggesting agro-economic potential (Murray, 1940b; Pal and Khoshoo, 1968). Amaranths are characteristically wind-pollinated. This has exception of the grain species with colourful inflorescences, which are occasionally visited by bees (Khoshoo and Pal, 1970). The grain species and their close relatives are monoecious and self-fertile. Arrangement and sequence of anthesis of the unisexual flowers favour a combination of self- and cross-pollination.

7. History: The prehistoric American Indian peoples gathered commonly wild amaranth seeds. The wild seeds are nutritious and as large as those of the cultivated species. Archaeological proof of domestication come with the appearance of pale white seeds, contrasting starkly with the dark brown wild types. The mutation producing this change has never been recorded historically. Associated with the change in colour are improved popping quality and flavor. The earliest record of the pale-seeded crop is from

Tehuacan, Puebla, Mexico, where *A. cruentus* appeared about 4000 BC. and was joined by *A. hypochondriacus* about AD. 500 (Sauer, 1969). By the 14th century AD., pale-seeded *A. hypochondriacus* was also cultivated by Arizona cliff-dwellers (Bohrer, 1962). The earliest record of *A. caudatus* is from 2,000-year-old tombs in north-western Argentina. Its pale seeds were found there mixed with those of *Chenopodium quinoa* and with dark seeds of weed amaranths and chenopods (Hunziker and Planchuelo, 1971).

The three crop species may have been independently domesticated. However, there is an alternative possibility, namely that there was a single primary domestication of *A. cruentus* from *A. hybridus*, with the other two domesticates evolving secondarily by repeated crossing of *A. cruentus* with weedy *A. powellii* and *A. quitensis* as the crop spread into their respective territories. Pal and Khoshoo (1974) discount the role of hybridization in evolution of the grain amaranths because of low fertility in their experimental hybrids. All three domesticates has involved increased size whole plant and particularly of the inflorescence, resulting in greatly increased seed yield during evolution. All three domesticates also exhibit the effects of selection for striking anthocyanin pigmentation of leaves, stems and inflorescences.

8. Further Developments: Grain Amaranth cultivation was regarded as a symbol of paganism and repressed in Spanish America after the 16th century. The crop nearly disappeared from history. However, by the early 19th century, grain Amaranths had appeared as a staple food crop in the Nilgiri Hills of south India and in the Himalaya. They have since been observed over an increasingly wide area of India and also in interior of China to Manchuria and eastern Siberia. Pale-seeded *A. hypochondriacus* constitutes the bulk of the Asiatic crop; dark-seeded *A. hypochondriacus* and pale-seeded *A. caudatus* are minor components. In the 1940s, cultivation of *A. hypochondriacus* was begun in East Africa to supply grain to the local Indian population. The wide latitudinal spread of these species in the Old World presumably required evolutionary changes, since their flowering is controlled by photoperiod. In the Old World, *Amaranthus cruentus* has not become established as a grain crop. However, dark-seeded, deep red forms of this species have been widely planted in tropical Africa and Asia for over a century as ornamentals, dye plants, fetishes and potherbs. All three of the domesticated species may have been introduced to the Old World via Europe. They have been grown in European gardens as ornamentals and curiosities for at least 250 years. Only dark-seeded forms of *A. hypochondriacus* were known to have been present in Europe until Henelt (1968) found a pale seeded specimen in a 16th century German herbarium collection. NBPGRC India released 'Annapurna' as one of the top yielders (Joshi, 1985).

Pandey and Singh (2010) conducted genetic studies for biochemical and quantitative characters in grain amaranth (*Amaranthus hypochondriacus* L.) and evaluated 26 accessions. Leaf protein content was noted significant in four accessions *viz.*, AG-67/1 (3.152 mg g^{-1}), AG-21 (2.452 mg g^{-1}), AG-306 (2.101 mg g^{-1}) and AG-1175 (2.101 mg g^{-1}). These can be used for vegetable purpose. Biochemical characters have no significant genetic association with grain yield per plant. Biochemical traits can be improved without altering grain yield.

Xingbo and Blair (2017) studied diversity in grain amaranths and relatives and distinguished them by genotyping sequencing (GBS). A total of 10668 SNPs were discovered in 94 amaranth accessions with *APCKI* complexity reduction and 10X genome coverage Illumina sequencing. The majority of SNPs were species specific with 4568 and 3082 for the two grain amaranth originating in Cental America *viz.*, *A. cruentus* and *A. hypochondriacus*, and 3284 found amongst both *A.caudatus*, originally domesticated in South Amarica, and its close relative *viz.*, *A.quitensis*. The distance matrix based on shared alleles provided information on the close relationships of the two cultivated Central American species with each other and of the wild and cultivated South American Species with each other, as distinguished from the outgroup with two wild species *viz.*, *A.powellii* and *A.retrofleus*. The GBS data also distinguished admixture between each pair of species and the geographical origins and seed colours of the accessions. The SNPs discovered can be exploited for marker development for further amaranth study.

9. Epilogue: Some of the grain Amaranths possess the unusual and highly efficient 4-carbon photosynthetic pathway (Tregunna and Downton, 1967; El-Sharkawy, Loomis and Williams, 1968), which may account for the excellent yields claimed. However, India offered the only signs of significant expansion of the crop and sustained scientific breeding. The protein quality is richer in amino acids including lysine and methiomine. Grain amaranths are suitable for a short season and deficient soils. They are substitute for cereals from nutritional point of view. They are easy to cultivate. Grain Amaranth, whether grain or leaf, constitutes a variable solar factory. Being a C_4 plant, it is one of a few dicots, in which the first product of photosynthesis is a four carbon compound. It is most efficient to use CO_2 under a wide range of both temperature and moisture stress environments. Grain Amaranth was selected as one of the four crops from among 400 selected by NAS, USA (NAS, 1980). Theisen *et al.*, (1980) identified amaranth as one of the five crops which should receive immediate attention as a food crop better adapted to environmental stresses. It is, therefore, needed to identify broader varieties which have potential as major food sources.

4.3 QUINOA

Chenopodium quinoa Willd. (Chenopodiaceae)

Resume: *Three species of the genus Chenopodium (Chenopodiaceae) viz., C. quinoa, C. nuttalliae and C. pallidicaule are American crops, called quinoa, huauzontle and canahua, developed in pre-columbian times. The earliest archeological record seem to be quinoa grain in an Argentine tom dating ca 2000 B. P. Chenopodium hircinum is suggested as their wild progenitor. Chromosome number 2n=18 for C. pallidicaule, and 2n=36 each for other species. As far as their nutritional quality is concerned they are comparable to wheat. They constitute a staple food of South American natives, and also as poultry feed.*

1. Introduction: Quinoa makes a significant contribution to local food production. Quinoa is an annual herb cultivated for grain which is roughly comparable with wheat in nutritional quality. Canahua is a very minor grain (Gade, 1970) and the immature infructescences of huauzontle provide a vegetable which is usually eaten fried in butter. Exceptionally the last is eaten in Mexico as a grain (Sauer, 1950). The three crop have not been cultivated outside its native area and none has been accorded much scientific investigation.

2. Plant Characteristics: Annual herb, growing to a height of 4 to 6 ft. seeds round, white, red or black.

3. Parts Used: Grains (seeds), leaves.

4. Chemical Constituents: Carbohydrates (64.2%), protein (14.1%) fat (6.10%), minerals (calcium, iron, magnesium, manganese, potassium, phosphorus, sodium, zinc), vitamin-A, thiamine (B1), riboflavin (B2), niacin (B3), vitamin (B6), folate (B9), cholins, vitamin-C, vitamin-E.

5. Uses And Applications: Whole seeds are used in soups. Flour is used for toasts, biscuits, porridge, bread, cake and beer. It constitutes a good feed for poultry. Its ash is often mixed with coco leaves to increase flavor. Leaves, in USA, are used as a substitute for spinach. It is staple food of millions of South American natives.

6. Karyotaxonomic Evidence: The three species *viz.,* (i) *C. quinoa* (Quinoa) 4x=36 of Andes, (ii) *C. nuttalliae* (huauzontle) 4×=36 of Mexico, (iii) *C. pallidicaule* (Canahua), 2×=18 of Andes are cultivated in their home place. The first two are certainly closely allied and may be conspecific; one strain of huauzontle examined appeared to be no more than a very large, late-maturing form of quinoa (Simmonds, 1965). Unpigmented forms of quinoa bear a general resemblance to *Chenopodium album. Quinoa* is extremely variable in stature, rate of maturity, habit, plant and seed pigmentation and seed size. Stature and rate of maturity are negatively correlated. Variability has been little revealed. Segregation of such genes as are known is disomic, suggesting allopolyploidy (Simmonds, 1971). The variability of huauzontle and, canahua is unknown, both being rare and declining crops. Canahua, a small bushy herb with leafy inflorescence branches, has quite a different facies from the other two. Information on the karyotaxonomy of putative wild relatives is unknown.

7. History: These three indigenous American crops are developed by Indian agriculturalists in pre-Columbian times. Quinoa and huauzontle at least were important food plants prior to the Conquest. There are some archeological records but they provide no substantial clues with regard to time of evolution. The earliest record seems to be of quinoa grain in an Argentine tomb, ca 2000 B. P. (Hunziker and Planchuelo, 1971). The crop must have, however, been domesticated long before then. The ancestors of all three must have been native American *Chenopodium* spp. yet unknown. Hunziker (1943) suggested that *Chenopodium hircinum* might have been the wild progenitor of quinoa.

Comparison of quinoa with wild chenopods (*e.g. Chenopodium* album) suggests that human selection for large seeds (strictly fruits), non-shattering infructescences and low seed dormancy was operative. Quinoa seeds are usually about twice as large in linear dimensions as those of *Chenopodium album*. Canahua appears to be more primitive than the other two in having small seeds which shed freely at maturity and show some dormancy. The quinoa and huauzontle resemble in grain characters, and the strong probability that they are conspecific, suggests that the Mexican crop represents an early migrant quinoa population selected locally for a different end-use, namely vegetable instead of grain and no Mexican tradition of using chenopods for grain.

Plants are gynomoniecious (flowers mixed hermaphrodite and female) and self-pollination predominates. The breeding system of quinoa is thus distinctive. Natural crossing has been estimated as 2-9 per cent and there is no evident inbreeding depression on selfing (Gandarillas, 1967a, b). However, one population that carries a cytoplasmic/genetic male sterility has been identified in Bolivia, so enhanced outbreeding is sometimes favoured (Simmonds, 1971). Canahua differs in having hermaphrodite, more or less cleistogamous, flowers and natural crossing is very rare (about 0.1 per cent in the glasshouse). These three crops may be inferred to have originated in the areas to which they are still restricted and/or in which they are variable.

8. Further Developments: Quinoa alone retails any agricultural significance but very little breeding has been done. The recent history of all three crops is one of decline. Although, Quinoa has played a key role in food security and farmers' livelihoods for centuries in Peru and Bolivia, farmers now cultivate fewer varieties on a reduced land area. Quinoa is being substituted by imported foods. Since 1994, the government of Peru has authorized the national food assistance program (PRONAA) to purchase quinoa for use in school break-fasts and popular canteens. Its production and sales, therefore, increased (Heallin and Higman, 2009). Products made with quinoa contains more antioxidents as it contains quercetin. It is gluten-free. It contains high protein, minerals and fibre. It has a higher satiating efficiency index. It produces lower free fatty acid levels and triglyceride concentrations. It contains lysine which is mainly essential for tissue growth and repair. Quinoa is rich in magnesium content which helps to relax blood vessels and thereby alleviate migraine and reduce Type-2 diabetes by promoting higher healthy blood sugar control. It also contains a high content manganese. Manganese, being an antioxidant, helps to prevent damage of mitochondria during energy production. Presence of riboflavin improves energy metabolism within brain and muscle cells (Graf *et al.*, 2015).

Singh *et al.* (2016) reviewed Quinoa (*Chenopodium quinoa* Willd.) as a functional superfood for modern world. They concluded that quinoa being a good source of complete protein (contains all amino acids), unsaturated fatty acids, minerals, vitamins, fibre and antioxidants, it is considered as "superfood". Quinoa also helps to reduce the risk of various diseases such as cardiovascular diseases, type-2 diabetes, some cancer, high blood pressure, obesity and is also good option for people who are allergic to certain food groups. Its biodiversity and ability to sustain in adverse climatic conditions makes it an ideal crop to cultivate worldwide especially in under-developing countries of Asia and Africa, where food production is threatened by global climatic changes.

Lesjack and Calderin (2017) evaluated the effect of increased night temperature on quinoa yield grain number, individual grain weight and processes involved in crop growth under the environmental conditions (Control treatment) and night thermal increase at two phases *viz.*, flowering (T_1) and grain filling (T_2) in southern Chile. They experimented on commercial genotype *viz.,* Regalona, and a quinoa accession (Cod. BO_5, N°191, grain bank from Semmillas Baer). They recorded that temperature increase reduced grain yield by 31% under T_1 treatment and 12% when under T_2 in Regalona and 23 and 26% in the occasion mentioned earlier. Biomass, according to them was negatively affected by the thermal treatments and a positive linear association was found between grain yield and aboveground biomass across treatments. The harvest index was unaffected either by genotype or by thermal treatments. Grain number was significantly affected between treatments and this key trait was linearly associated with grain yield.

9. Epilogue: No future for these three crops other than continued decline is visualized yet. Quinoa, however, might have a place if it could be shown to have any significant advantage over the old world cereals, barley and oats. It is certainly quite attractive in respect of nutrition. There would be plenty of room for improvement by breeding. The evidence suggests that pure line cultivars would be feasible and the best first step in breeding. It is relatively easy to grow, the harvesting and post-harvesting processing of the grain labour-intensive. When dried, quinoa can be stored upto ten years. It is appreciated for its nutritional value.

References:

Activities. *J. Food. Sci.* Sep. 77(9) C954-9. Doi:10.1111/j.1750-3841.2012 02867xEpul 2012 Aug.13.

Agogino, G. (1957) Sci.News Lett. Washington. 72:345.

Bailey, L.H. (1917) Cyclopedia of American agriculture, 5th Ed, 2. New York, U.S.A. and London, UK.

Bohrer, V.L. (1962) Ethnobotanical materials from Tonto National Monument. *In:* (Ed.L.R.Caywood), Archaeological Studies At Tonto National Monument, Arizona, *Globe*, Arizona. pp.75-114.

Coe, W.R. (1931) Buckwheat milling and its by-products. U.S. Dep. Agric, Circ., 190, pp. 11.

Duran Katar, Murat Olgun and Matin Turan (2016) Analysis of morphological and biochemical characteristics of buckwheat (Fagopyrum esculentum Moench.) in comparison with cereals. *CyTA-Journal of Food.* 14(2):176-185.

Ebermann, R., Alth, G. Kreitner, M. and A. Kubin (1996) Natural products derived from plants as potential drugs for the photodynamic destruction of tumor cells, *J.Photochem. Photobiol. B.* 36 (2):91-97.

El-Sharkawy, M. A., Loomis, R. S. and W.A. Williams (1968) Photosynthetic and respiratory exchanges of carbon dioxide by leaves of the grain amaranth. *J. Appl. Ecol.* 5:243-251.

Gade, D.W. (1970) Ethnobotany of canahua (Chenopodium pallidicaule) rustic seed crop of the altiplano. *Econ. Bot.* 24:55-61.

Gandarillas, H. (1967a) Efecto de la autofecondacion sobre la quinua. *Sayaiia* 5:16-17.

Gandarillas, H. (l967b) Observaciones sobre la biologia reproductiva de la quinua. *Sayaiia* 5:26-29.

Gimenez-Bustida, J.A. and H. Zielinski (2015) Buckwheat as a functional food and its effects on health. *J. Agric. Food. Chem.* 1663 (36):7896-7913.

Graf, B.L., Patricio Rojas-Silva, Leonel E.Rojo, Jose Delatorre-Herrea, Manuel E. Baldeon and llya Raskin (2015) Innovations in health value and functional food development of Quinoa (Chenopodium quinoa Willd.) Comprehensive *Reviews in Food Science and Food Safely* 14(4):431-445.

Grant, W.F. (1959) Cytogenetic studies in Amaranthus-III. Chromosome numbers and phylogenetic aspects. *Canad. J. Genet. Cytol.* 1:313-328.

Hanelt, P. (1968) Beitriige zur Kulturpflanzenflora. I. Bernerkungen zur Systematik und Anbaugeschichte einiger Amaranthus-Arten. *Kulturpflanze* 16:127-149.

Hellin, J. and S. Hingman (2009) Underutilised plant products and market access: Challenges and opportunities. Proceedings of the International Symposium (Ed. Jaenicke *et al.*) *Acta Hort.* 806 ISHS, pp.393-406.

Hughes, H.D. and E.R. Henson (1934) Crop production principles and practices. New York, USA.

Hunt, T.K. (1910) The cereals in America. New York, USA and London, UK.

Hunziker, A.T. (1943) Las especies alimenticios de Amaranthus y Chenopodium cultivadas por los indios de America. *Rev. Argent. Agron.* 10:297-354.

Hunziker, A.T. and A.M. Planchuelo (1971) Sobre un nuevo hallazgo de Amaranthus caudatus en rumbas indigenas de Argentina. *Kurtziana* 6:63-67.

Joshi, B.D. (1985) Annapurna-a new variety of grain amaranth. Indian Farming 25(8):29-31.

Juan A., Gimenez-Bustida and Henryk Zielinski (2015) Buckwheat as a functional food and its effects on health. *J. Agric Food Chem.* 63(36):7896-7913.

Kreft, I., Fabajani, N. and M. Germ (2003) Rutin in buckwheat-protection of plants and its importance for the production of functional food. *Fagopyrum* 20:7-11.

Krishnakumary, K. (2011) Grain amaranth. *In:* Future Crop. Vol.I. (Ed. Prof. K.V. Peter). pp.40-54. Daya Publishing House, Delhi, India.

Krotov, A.S. (1963) Grechikha. Selkhozizdat Moscow and Leningrad, USSR.

Lesjak Jurka and D.F. Calderini (2017) Increased night temperature negatively affects grain yield, biomass and grain number in Chilean quinoa. *Front. Plant Sci.* https://doi.org/ 10.3389/fpls.2017.00352.

Morris, R.M. (1951) Cytogenetic studies on buckwheat. J.Hered. 42:85-89.

Murray, M.J. (1940a). The genetics of sex determination in the family Amaranthaceae. *Genetics* 25:409-431.

Murray, M.J. (1940b) Colchicine induced tetraploids in dioecius and monoecious species of the Amaranthaceae. *J.Hered.*31:477-485.

NAS (National Academy of Sciences) (1980) Underexploited tropical plants with promising economic value. Washington DC., USA, pp.80.

Pal, M. and T.N. Khoshoo (1972) Evolution and improvement of cultivated amaranths. V. Inviability, weakness, and sterility in hybrids. *J. Hered.* 63:78-82.

Pal, M. and T.N. Khoshoo (1974) Grain amaranths. *In:* J. B. Hutchinson (ed.), Euoluuonary studies in world crops: Diversity and change in the Indian subcontinent, Cambridge University Press, pp.129-137.

Pandey, R.M. and Rekha Singh (2010) Genetic studies for biochemical and quantitative characters in grain amaranth (*Amaranthus hypochondriacus* L.) *Plant Omics Journal* 3(4):129-134.

Samel, D. and de Witte P. (1994) Isolation and PTK inhibitory activity of fagopyrins from buckwheat. Abstracts of Papers And Posters 19th LOF- Symposium on Pharmacognosy and Natural Products chemistry. *Pharm. World Sci.* 16(9):13-113.

Sauer, J.D. (1950) The grain araranths: a survey of their history and classification. *Ann. Mo hot. Gdn.* 37:561-632.

Sauer, J.D. (1969) Identity of archaeologic grain amaranths from the valley of Tehuacan, Puebla, Mexico. *Amer. Antiqu.* 34: 80-81.

Sedej, I. Sakac, M., Mandic, A., Misan, A, Tumbas, V. and J. Canadoric-Brunct (2012) Buckwheat (Fagopyrum esculentum Moench.) grain fractions antioxidants compounds and

Sharma, K.D. and J.W. Boyes (1969) Modified incompatibility of buckwheat following irradiation. *Can. J. Bot,* 39:1241-1246.

Shustova, A.P. (1965) Effect of light conditions on the growth, development and yield of buckwheat. *Fiziologiya Rast.* 12:782-788

Simmonds, N.W. (1965) The grain chenopods of the tropical American highlands. *Econ.Bot.* 19:223-235.

Simmonds, N.W. (1971) The breeding system of Chenopodium quinoa. *I. Heredity*, 27:73-82.

Singh, Ajay (2008) Plant in Ancient Indian Civilizations. Agam Kala Prakashan, Delhi, India.

Singh, S., Singh, R. and K.V. Singh (2016) Quinoa (Chenopodium quinoa Willd.) functional superfood for today's world. *A review. World Scientific News* 58:84-96.

Theisen, A.A., Knox, C.G., Sprague, H.B. and F.L. Mann (1980) Possibility of introducing food crops better adapted to environmental stress. Washington DC. National Science Foundation, Report No.NFS/RA 780038.

Tregunna, E. B. and J. Downton (1967) Carbon dioxide compensation in members of the Amaranthaceae and some related families. *Canad. J. Bot.* 45:2385-2387.

Xingbo Wu and Mathew W. Blair (2017) Diversity in grain amaranths and relatives distinguished by genotyping by sequencing (GBS). *Front. Plant Sci.* https://doi.org/10.3389/fpls.2017.01960.

Yang, J., Guo, J. and J. Yuan (2008) In Vitro antioxidant properties of rutin. *LWT, Food Sci. Technol.* (Campinas.) 41(6):1060-1066.

5

Grain Legumes (Pulses)

INTRODUCTION

Next in importance to cereals as sources of human food are the legumes. Their protein content is higher than in any other vegetable product. The legume seeds or pulses also called 'grain legumes' provide the much needed proteins to predominantly vegetarian population. The relatively high protein content is attributed partially to the presence of nodules on the roots of many legumes. They contain nitrogen-fixing bacteria capable of using free atmospheric nitrogen to produce nitrates or nitrites. These are readily used by the plants and increase their supply of nitrogenous material.

The legumes belong to the family Leguminosae and belong to three tribes, *viz.,* the Vicieae, the Hedysareae and the Phaseoleae of subfamily Papilionatae. Included in the first tribe are pulses such as lentil (*Lens culinaris*), pea (*Pisum satiuvm*), chickpea (*Cicer arietinum*), grass pea or khesari dhal (*Lathyrus sativus*) and broad or horse bean (*Vicia faba*). The tribe Hedysareae includes the groundnut (*Arachis hypogaea*), a crop which is more usually grown as an oilseed crop than as a pulse. Among the Phaseoleae are the soybean (*Glycine max*), lablab bean (*Lablab purpureus*), common beans (*Phaseolus* spp.) cowpeas (*Vigna unguiculata*), pigeon pea (*Cajanus cajan*), horse gram (*Dolichos uniflorus*), the ground beans, (*Vigna subterranea*) in addition to minor pulses such as the sword and jack beans (Canavalia spp). Another tribe Galegeae (Indigofereae) includes the cluster bean or guar (*Cyamopsis tetragonoloba*) that is used for the extraction of guar gum or as a green manure. To some extent the tender pods are the source of vegetable.

Pulses or grain legumes have been an important crop ever since man began domesticating plants and have been integral part of cultural heritage. Chinese literature records the cultivation of soybean between 3000 and 2000 BC. Legumes are also mentioned in the cropping systems of the early Egyptian dynasties. The occurrence of pea and lentils have been reported by Helbaek (1966) at various archaeological sites such as Hacilar, Beidha and Jarmo, dating back 7500 to 6500 BC. Archaeological remains of *Pisum, Vicia* and *Lens* have been recovered by Renfrew (1966) from Neolithic sites in Greece. The common bean may have been the first to be domesticated in the New World with remains dating as far back as 4975 BC. from the Tehuacan Valley in Mexico. Fragments of lima bean (*Phaseolus lunatus*) or common bean, as also the beans of the genus *Canavalia* have been reported in dried human excreta (desiccated faeces or coprolites) in coastal Peru, dating back to 3000 BC. Kaplan *et al.,* (1973) have reported the occurrence of common bean and lima bean in deposits in the inter-montane Peruvian Valley dating to about 6000 BC.

The legumes have been favoured by mankind for the domestication. The importance of pulses are many, for instance: (a) grain legumes have a low water content and impervious seed coats–features valuable during transportation and storage, (b) they can be easily cultivated and mature rapidly, (c) they have an important place in the crop rotation, (d) the per acre production of proteins is higher for pulses than for cereals and (e) they provide protein-rich food for mankind and his domestic animals and birds.

Nutritive Value Of Pulses: Legumes or pulses contain 17 to 30% protein on a dry weight basis, which is nearly twice or thrice the value found in cereals and 20 times that of cassava. Soybean with as high as 49% protein occupies praiseworthy position amongst the protein-rich legumes. Besides proteins, pulses contain up to 60% carbohydrates, mainly in the form of starch. The fat content of most pulses is low, the only exception being soybean where it may reach a level as high as 20%. Also, they contain reasonable amounts of niacin and thiamine but only small amounts of carotene, riboflavin and ascorbic acid. The pulses contain five times the riboflavin (vitamin B2) of most cereals, and ten times more thiamine. Pulses are relatively rich in calcium and iron. Although dried legumes do not contain vitamin C, however sprouted seeds compensate excellently.

All the eight amino acids in the necessary amounts are present in animal proteins. They are expensive. The lower income groups cannot afford them. Each person in the highly industrialised nations gets an average of 84 g of protein daily, 39 g of which comes from animal sources. In the developing countries, on the other hand, the average intake is 52 g with only 7 g coming from meat It has been increasingly recognized, in recent time, that the production of high yielding and high-quality varieties of cereals, millets and pulses would, therefore, offer the least expensive and most immediate practical way of diminishing the threat of intellectual dwarfism in the developing underdeveloped countries.

The proteins from legumes are usually incomplete. One or more of the amino acids are present in inadequate amounts but the amino acids present neatly complement those of the cereals. They are sources of lysine, tryptophan and threonine but are low in the sulphur-containing amino acid *e.g.* thionine, cystine and cysteine. While cereal grains are low in lysine and threonine and corn is also deficient in tryptophan, they have adequate levels of the sulphur-containing amino acids. A simple way to eliminate the dangers of amino acid deficiency is to use legume and cereals together or to eat the two kinds of proteins within a short time.

Pulses are consumed as food in different ways: whole or dehusked splits (dhal), fried or puffed grain products, fermented preparations, germinated seeds and flour. In India, more than 75% of the grain legumes are milled to produce dehusked splits. These are cooked in boiling water for varying periods and eaten after adding salt and spices, after frying in vegetable oil, along with cereal preparations or as soup. Sprouted grain legumes are excellent source of vitamin-C. Pre-cooked dehydrated legumes or splits are advised for the defence forces at higher altitude. Some pulses contain toxic and biologically active substances such as trypsin inhibitors, haemaglutinins, cyanogenetic or cyanogenic glycosides, saponins, goitrogens or goitrogenic factors, anticoagulants, toxic histones, substances producing lathyrism and favism and astringent flavourings. Most of these toxic constituents are eliminated by simple preparative procedures and methods of cooking. Pulses are necessary to keep human population in good health. They are also essential to the very large number of dairy cattle and poultry. If the nation wants more milk, more eggs and more meat, then it is desirable to feed livestock with sufficient quantities of legumes. The legumes are widely cultivated in the tropics as green manures. Some of the pulses constitute fodder crops for draught and dairy animals as well.

5.1 BAMBARA GROUNDNUT

***Vigna subterranea* (L.) Verdc.**

[(Syn.) *Voandzeia subterranea* (L.) Thours]
(Leguminosae-Papilionatae)

Resume: *It is an indigenous crop of tropical Africa. Its wild form viz., subsp. spontanea is found in Northern Nigeria to Cameroun and to the Central African Republic. Chromosome number is n=11 reported for its related subspecies. The cultivated crop is a landrace grown in sub-Saharan Africa. Its pulse is mainly used for human consumption. It is advisable for preventing protein energy malnutrition.*

1. Introduction: It is also called Nyima Bean, Bogor, Peanut. Ground Bean, Kaffir Pea, etc. It is an indigenous tropical African crop mainly grown as supplementary food. The wild form *viz.,* subsp. *spontanea* is found from the Yos Pleateau and Yola (northern Nigeria) to Cameroun and possibly to the Central African Republic. It was taken to Madagascar, perhaps by Arab traders. It reached Brazil in the 17th century and then introduced to the Far Eastern tropics and the Philippines (Doku and Karikari, 1971).

2. Plant Characteristics: Annual bunchy herbs, much branched, rooting at nodes; stem pink, purple or green; leaves pinnately trifoliate, petioles long; flowers yellowish-white, spreading out of ground level; pods bivalve, subterranean, seeds 1-2 per pod, smooth, hard, white, brown, purple, red or black.

3. Parts Used: Seeds (nuts).

4. Chemical Constituents: Protein (24.5%), carbohydrates (59%), fat (6-12%), minerals (Ca, Fe, P, K, Na, Mn), thiamin, riboflavin, niacin, carotene.

5. Uses And Applications: Pulse is mainly used for human consumption. The seeds are consumed either immature or fully ripened and dry. The nuts may be eaten fresh after boiling or roasting before they are dried. The dried nuts are milled to obtain flour which is used to prepare a variety of dishes including cakes, biscuits and dumplings.

6. Karyotaxonomic Evidence: There are various differences from the cultivars *viz., subterranea* which have a clustered habit with short nodes, while subsp. *viz., spontanea* has diffuse growth and more scattered pods. Both have 2n=2×=22. It is self-pollinated and sometimes cleistogamous. Ant pollination of the more diffuse varieties is recorded in Ghana. Casual selection following hybridization seems to have been the basis of most cultivars (Hepper, 1970).

7. History: Its most likely ancestor is considered as *Vigna subterranea* var. *spontanea* (Doku and Karikari, 1971). This relationship has been confirmed through isozyme and other molecular analysis (Posquet *et al.,* 1999; Amadou *et al.,* 2001; Singrun ad Schenkel, 2004). The cultivated crop exists as landraces and is grown extensively in sub-saharan Africa.

8. Further Developments: A group of institutions from Africa, India and Europe researched molecular genetics through field evaluation to product development incorporating end-user needs. They investigated changes in gene expression under drought. They also described microsatellite development, genetic mapping, and QTL analysis in a cross derived from a cultivated landrace and a wild relative. Their genetic research integrated physiology, agronomy, nutrition, modeling and end-user preferences to identify and overcome bottle-necks to further exploitation of this crop so that the full food security potential of bambara groundnut (Mayes *et al.,* 2009). Fasasi *et al.,* (2009) evaluated into the proximate and functional properties of modified Bambara isolate-'Ogi' mix in order to improve the functional properties that affect the behavior of protein in food systems during processing, manufacturing, storage and preparation.

Odongo *et al.* (201) investigated genetic relationships among 105 Bambara groundnuts [*Vigna subterranea* (L.) Verdc.] accessions from Kenya using 12 microsatellite markers. A total of 24 alleles are revealed with a mean of 2 alleles per locus. The polymorphic information content and gene diversity values averaged 0.28 and 0.35, respectively indicating low genetic diversity among the evaluated germplasm. Genetic distance on Jaccord's similarity coefficient from the SSR marker analysis ranged from 0.08 to 1.16 among the landraces. Cluster analysis distinctly grouped the 105 accessions into 3 major clusters. The analysis of molecular variance revealed that 98% of the total genetic variation was within accessions, whereas the genetic variation among accessions accounted for 2% of the total genetic variation. The genetic diversity provides the basis for selection of appropriate parental genotypes for breeding programmes and mapping populations to further broaden the genetic base of Bambara groundnut germplasm in Kenya.

Linnemaan *et al.* (1995) studied influence of constant photoperiods of 10, 12, 14 and 16 h on development and growth in two barmbara groundnut [*Vigna subterranea* (L.) Verd.] genotypes in greenhouse experiments. Photoperiod influenced the onset of flowering in genotype *viz.*, Ankpa 4 and the onset of padding in both Ankpa 4 and Tiga Nicuru genotypes. Under 14- and 16-h photoperiods plants of Alopa 4 produced no pods. Photoperiod did not influence total above ground dry matter production per plant in Ankpa4 and had only slight effect on Tiga Nicuru. In both genotypes assimilate distribution changed after the photoperiod-induced onset of podding. In addition, a direct influence of photoperiod on partitions observed. Tiga Nicuru plants under 10-and 12h photoperiods accumulated more dry matter as leaf blades and less as stem material than plants under 14-and 16-h photoperiods. In case of Ankapa 4, the increase in pod dry matter weight per plant under the 10-h photoperiod was nearly double the increase under the 12-h photoperiod. Photoperiod thus apparently strongly effects the number of developing sinks.

9. Epilogue: The recent investigation on Bambara by Fasasi *et al.*, (2009) attributed to the addition of modified Bambara protein isolate. The mix therefore would be adequate to prevent protein energy malnutrition in people who consume it as their main protein source. Bambara is regarded 'complete food'. The mix developed can, therefore, be consumed as a sole diet and can also find relevance as ingredients in other food formulations. It grows well in poor soils and is drought-resistant. Its nitrogen-fixing roots help replenish soil nutrients. It is most suitable for intercropping systems.

5.2 BEANS

Phaseolus Spp. (Leguminosae- Papilionatae)

Resume: *The genus Phaseolus originated in the New World (the Old World species of phaseolus merged under the genus Vigna). This is also evidenced by the archaeological findings. A group of four species of this genus viz., P. vulgaris, P. lunatus, P. coccineus and P. autifolius are all diploid with 2n=2×=22 chromosomes. They have a wide range of distribution in New World (America) from cool humid uplands to warm temperate conditions. They are important source of protein and calories in human diets in both the New and Old worlds. They are usually consumed as dry beans; Phaseolus vulgaris, however, is used for fresh pod consumption.*

1. Introduction: The genus *Phaseolus* has contributed crop plants to agriculture in both the Old and the New Worlds. *Phaseolus* beans are valued grain legumes or pulse crops in many tropical countries and are usually consumed as dry beans, whereas in temperate countries, although some of the species are grown as dry beans (notably *P. vulgaris*), varieties have also been developed for fresh pod consumption and for processing as frozen vegetables. They are important sources of protein and calories in human diets in tropical Africa and America where they supplement the carbohydrate staple foods like maize, plantains and cassava. With the increased cost of animal proteins in Europe and North America, some species may become important protein-producing crops in these regions too.

2. Plant Characteristics: In general, bean species erect or twinning annuals; stems slender, herbaceous; leaves alternate, trifoliate; flowers, papilionaceous, borne in racemes; pods straight or slightly curved, generally 4 to 6-seeded, seeds being variously coloured.

3. Parts Used: Pods, seeds.

4. Chemical Constituents: In general seeds contain proteins, carbohydrates, fats, fibres, ash, vitamins and minerals.

5. Uses And Applications: (i) *Phaseolus vulgaris*: Unripe or dry seeds are used for human consumption as vegetable. The green immature pods are eaten as vegetable. They are also used frozen or canned. (ii) *Phaseolus lunatus*: Seeds are consumed fresh or dried. Dried seeds are boiled or baked for table use. In USA, the young beans are canned or frozen, sprouted seeds are consumed boiled, used as vegetable or in stews. Seeds are used as pulse and young pods as vegetable. (iii) *Phaseolus coccineus*: The fresh or dried seeds are consumed. The young pods are used as vegetable in British cuisine. Dried beans are consumed in Germany. The beans have important role in Greek cuisine. (iv) *Phaseolus acutifolius*: Seeds are consumed fresh or dried and also the young pods as vegetable.

6. Karyotaxonomic Evidence: *Phaseolus* originated in the New World and that Old World species previously classified in *Phaseolus* are now assigned to *Vigna*. Westphal (1974) reviewed the evidence for the revision of these two genera. There are four cultivated species in the New World *viz.,* (i) *P. vulgaris* (the Common, Haricot, Navy, French or Snap bean), (ii) *P. coccineus* (the Runner or Scarlet Runner bean), (iii) *P. lunatus* [the Lima (large seeded)], Sieva (small seeded), Butter or Madagascar bean), (iv) *P. acutifolius* var. *latifolius* (the Tepary bean).

All the species are diploid with 2n=2×=22. Interspecific hybridization in the genus *Phaseolus* has been carried out (Smartt, 1970). Smartt (*loc. cit.)* showed that viable hybrids of varying fertility can be produced between *P. vulgaris* and *P. coccineus*. There is some affinity between these two species and *P. acutifolius*, as shown by the formation of viable but sterile hybrids, but less affinity between *Phaseolus lunatus* and the other three species. Hybrids of *P. vulgaris* ♀ X *P. coccineus* ♂ can be produced easily while the reciprocal is produced only with difficulty. Miranda and Evans (1973) suggested that the unilateral incompatibility has developed under cultivation. *Phaseolus coccineus* are generally cross-pollinated by bees but there appears to be no genetic barrier to self-pollination in this species (Hawkins and Evans, 1973). The other three species are self-pollinating with only a small amount of cross-pollination.

7. History: Smartt (1969) showed that the archaeological record in Middle and South America is impressive compared with that of some other crop plants. Radiocarbon datings of 7680 BP. are on record for *P. vulgaris* from Callejon de Huaylas, Peru (Kaplan *et al.*, 1973) and of 7000 BP. from the Tehuacan Valley in Mexico. It is evidenced that domestication of *P. vulgaris* also occurred in Brazil and northern Argentina from a wild form of which *P. aborigineus* is a modern survivor. The large limas of *P. lunatus* were domesticated by 4,500 years ago in Huaca Prieta, Peru. The small-seeded limas were domesticated by 1,400 to 1,800 years ago in Mexico. *Phaseolus acutifolius* domesticated dated to 5,000 and *P. coccineus* to 2,200 years ago in the Tehuacan Valley in Mexico (Kaplan, 1965).

The four species of *Phaseolus* have a wide range of habitats in the wild. They range from the cool humid uplands of Guatemala at altitudes of about 1,800 m, where *P. coccineus* grows in the arid semi-desert conditions in which *P. acutifolius* is found. Warm, temperate conditions are suitable for *P. vulgaris* which occurs in an ecological transition belt at 500-1,800 m altitude, with the highest frequency of wild types at ca. 1,200 m in the Mexico Guatemala area (Miranda, 1974). *Phaseolus lunatus*, is found in the tropics and sub-tropics (Mackie, 1943). Archaeological finding of *P. vulgaris* and *P. lunatus* suggest separate domestications in Central and South America from conspecific geographic races. The small seeded types are native to Central America and the large seeded types to South America. Whether the Central American is the primary centre and the South American a secondary one, with migration from the primary to the secondary, or whether domestication was polycentric is still a mystery. Spanish galleons took lima beans across the Pacific to the Philippines and thence to Asia, also from Peru to Madagascar (Westphal, 1974). The slave trade took the lima and the common bean from Brazil to Africa and later inland to the trade routes. The latter reached Europe in the early 16th century. The French bean spread rapidly in the 16th and 17th centuries and reached England by 1594 (Purseglove, 1968). Beans also spread eastwards through the Mediterranean basin and into middle Wope. Beans were cultivated all over Italy in the 17th century and in Greece, Turkey and Iran. In North America, these beans spread through California (5,000 to 2,000 BP.) and into western areas of the USA. There were numerous introductions from Europe in the late 19th century in Eastern USA.

8. Further Developments: The most obvious effect of domestication and of subsequent plant breeding in *Phaseolus* is the modification of growth habit.

To take *P. vulgaris* as an example, in the germplasm collection of 5,000 accessions of wild material and cultivars examined at Cambridge, five races have been described under field conditions. The pattern of evolution appears to be as: (i) Middle America: Small seeded, indeterminate climbers-Race 1(a); Indeterminate semi-climbers-Race 2; Indeterminate bush types-Race 3; Determinate, multi-noded Race-4. (ii) South America: Large seeded, indeterminate climbers-Race 1(b); Determinate bush types Race 5 (Evans, 1973).

No indeterminate bush types occur in the runner and lima beans, but the latter has many determinate bush varieties, while the former has only one (Hammond's Dwarf). In the tepary, all cultivars are indeterminate bush types (Smartt, 1969). There have also been reductions in the number of branches and leaves as a result of domestication and an increase in leaf size and stem diameter in all species *P. acutifolius*.

Important changes have taken place in pod and seed size. There has also been an appreciable increase in flower size in *P. vulgaris* and *P. coccineus*. Seed numbers per pod have changed in *P. vulgaris* and *P. acutifolius*, where up to nine seeds are found in wild forms but rarely more than five in most cultivars. *Phaseolus coccineus* showed a tremendous increase in pod and seed size from 6 to 30 cm in pod length and from 100 to 1,000 mg for individual seed weight. An increase in permeability of the seeds to water has taken places under domestication which is an important property for uniform germination and for ease of cooking (Miranda, 1974).

Testa colours range in the seeds of *Phaseolus* species. In *P. vulgaris*, five main colour groups have been identified, *viz.,* white, black, red, ochre and brown. Superimposed on these ground colours may be various patterns of spots, flecks and stripes. The genetics of seed coat and flower colour in *P. vulgaris* has been studied by Prakken (1970).

Three distinct pod textures are found in the common bean: (a) the parchmented, which are very fibrous and dehisce strongly at maturity; (b) the leathery types which are less dehiscent but split readily along the sutures; (c) the fleshy or stringless pods which are indehiscent and do not split readily. Varieties with parchmented pods are used only for dry seed productions. Leathery podded varieties can be used for green pod production when young, or haricot production when fully mature. Fleshy podded or stringless types are used entirely for the production of green pods (Smartt, 1969). Since the genus *Phaseolus* is native to the tropics, many cultivars from the region are photoperiod sensitive and responsive to short days. There are also day-neutral types available and all the temperate cultivars have been selected for day-neutrality or tolerance of long days.

9. Epilogue: The first aim in breeding must be to breed for high yield potential. There is general agreement that the determinate bush varieties with pods held high off the ground are the most desirable for mechanical harvesting. There may, however, be a place for the indeterminate bush types in the tropics, since they provide an assurance of continued growth after temporary environmental stress (Leakey, 1971). Some emphasis has been placed in past on protein content and quality. The beans are rich in protein. The proteins are deficient in the sulphur amino acids. There is no unequivocal evidence that there is a significant negative association between protein percentage and yield. Although correlations between yield and protein are negative, the coefficients are generally low and there is every hope that high protein and high yield can be combined in these crops.

Crystine and methionine are the first limiting amino acids, followed by tryptophan. These crops are high in lysine and so can supplement cereal proteins, which are generally low in lysine. The presence of anti-nutritional factors such as trypsin inhibitors, haemagglutinins and flatus factors present another problem and work should be directed to reduce these components by genetic means or to ensure adequate cooking to destroy the toxic factors. Hutchinson (1970) points out the need for the breeder to work closely with the biochemist and the nutritionist.

The symbiosis whereby legumes gain N compounds from the rhizobia in the root nodules is not always efficient. The difficult breeding project of selecting an efficient combination of legume host and rhizobial symbiont for the production of effective root nodules is one of importance, especially in the tropical world. This may, at least in part, be attributed to the N reserves of these large seeded legumes, in contrast to the lack of reserves in the small seeded leguminous crops. It may be that response is shown in forage legumes grown for their vegetative parts but not in grain legumes grown for seed. There is now a greater awareness of the need for the conservation of the genetic variation available in these crops. Germplasm seed banks are being developed and utilized. International institutes will help considerably in this venture to ensure that breeding programmes do not reach an evolutionary endpoint.

5.3 BLACK GRAM

Vigna mungo (Linn.) Hepper

[(Syn.) *Phaseolus mungo* L.] (Leguminosae: Papilionatae)

Resume: *India is the primary centre of origin. Its carbonized grains are discovered from the Chalcolithic site, Navdatoli-Maheshwar in India dating 1660-1440 BC. It is considered derived from Phaseolus trinervis Hayne or P. sublobatus Roxb. which are wild in India. Chromosome number n=11 and 12 are reported for it. In India, it is a highly prized major pulse rich in phosphoric acid and good source of protein.*

1. Introduction: It is thought to be a native of India. It is under cultivation in India since ancient times. Its carbonized grains have been recovered from the Chalcolithic site, Navdatoli-Maheshwar in India dating 1660-1440 BC. (Vishnu-Mittre, 1968). It is introduced recently by Indian immigrants into other tropical countries. Presently, it is grown in Iran, Malaysia, East Africa, Southern Europe, West Indies, South and Central America, besides India.

2. Plant Characteristics: Erect, suberect or trailing annuals, growing upto 0. 8 m in height, reddish-brown hairy; leaves large, 3-foliate, leaflets ovate-lanceolate, 5-10 cm long, stipulate; flowers small, yellow, clustered at the ends of penduncles; pods narrow, cylindrical, erect or suberect, 4 to 7 cm long, seeds 6 to 10, oblong with flat ends, black, smooth.

3. Parts Used: Seeds, leaves.

4. Chemical Constituents: Protein (23.4%), fats (1.0%), carbohydrates (57.3%), fibre (3.8%) ash (4.8%), minerals (calcium, phosphorous, iron), vitamin A, carotene, thiamine, riboflavin, niacine, phosphoric acid, etc.

5. Uses And Applications: Grains are used for human consumption in the form of 'dal' or pulse either as whole, split, husked or dehusked or parched. In India, it is mainly used in making 'papad' and in preparation of idli, vada, dosa, etc. Flour is baked into bread and biscuit. Green pods are consumed as vegetable. Fried and salted seeds are eaten as a snack.

6. Karyotaxonomic Evidence: Chromosome have been reported n=11 and 12. It is although native of India, it is not known in wild. It is thought probably originated from *Phaseolus trinervis* Heyne or *P. sublobatus* Roxb. which occur wild in India. (Species of *Phaseolus* in India are now merged under the genus *Vigna*).

7. History: De Candolle (1886) and Vavilov (1926) thought India as a primary centre of its origin. Zukovaskii (1962) considered *Vigna sublobatus* as the progenitor of blackgram which grows wild in India. Ancient Sanskrit literature *e.g.* Charak Samhita, Kautilya's Arthashastra are replete with references to blackgram. It is also mentioned in vedic literature *viz.,;* Yajurveda, Atharveda, besides Brahmana, Kalpasutra, Panini's Astadhyayai and Vartik, and Patanjal's Mahabhasya (Singh, 2008).

8. Further Developments: In India, it is cultivated in majority of states. Improvement program of blackgram was started in 1943 in different states under ICAR suitable to grow in different seasons and agroclimatic zones. Varieties have been developed with emphasis on high yield, short duration, non-shattering habit, resistance to pests and diseases, good cooking qualities and high protein content.

In Asian Vigna, to which black gram belongs, the genetic relationship between cultivated and wild forms of mungbean, azuki bean and rice bean is investigated. Clear genetic differentiation between the wild and cultivated gene pools was shown in mungbean (Sangiri *et al.,* 2007), but no azuki bean (Xu *et al.,* 2008) or rice bean (Tian *et al.,* 2013). These studies revealed that wild and cultivated accessions of black gram were not clearly differentiated as shown by PCAO plot and phylogenetic tree. This means domestication from wild to cultivated black gram is relatively recent and/or is not intensive. However, archaeological evidence suggests that domestication of black gram may be as long as 3500-4500 years ago (Fuller and Heaey, 2006). Black gram and mungbean both originated in India and share several common morphological characteristics, and even cultivated utilized in similar ways (Tomooka *et al.,* 2002). Interestingly, wild black gram has much narrower distribution in comparison to wild mungbean. Black gram is found in limited areas of Asia, mainly India, Thailand and Myanmar. However, mungbean is present in Africa, Australia and Asia.

In recent times, black gram was introduced in America and Africa by Indian immigrants (Jain and Mehra, 1980) to grow as food and manure crop. Gene diversities of cultivated black gram in Africa and America are higher than those in other regions. This suggested that black gram was introduced into Africa and America from multiple regions of India, which is the centre of domestication.

Vyas *et al.* (2018) carried out investigation on 22 blackgram [*Vigna mungo* (L.) Hepper] genotypes to reveal the genetic diversity using 20 ISSR primers. Screening of the primers revealed that only 15 out of the 20 primers produced amplification. A total of 84 amplified bands were obtained, out of which 72 were polymorphic (85.71 percent polymorphisms). The total number of amplified bands varied between 1 (UBC-813 and UBC-878) and 9 (UBC-826) with an average of 5 bands per primer. The overall size was 0.285 across all genotypes. Five unique bands were detected in four genotypes, out of which the genotype U-9 gave maximum number of distinct bands. The 22 genotypes could be divided into 4 main clusters. Cluster I included 16 genotypes and the cluster II, III and IV included two genotypes each. Genotypes IC-16511 and UTTARA, UH-177 and IPU2K-21, STY-2834 and UH-177 are found to be genetically distant from each other. The authors suggested that the ISR marker could prove to be a versatile toll in further screening of the *Vigna* germplasm pool for study of genetic divergence and the establishment of phylogenetic relationship amongst accessions.

9. Epilogue: It is fast growing crop of fairly warm countries. It grows well in heavy soil such as water retentive black cotton soils. It appears that there is still scope for its improvement looking at the features and some varieties developed. It is ancient legume crop and rich in protein source. Being a short duration crop, it can be grown when other crops fail.

5.4 CHICKLING PEA

Lathyrus sativus L. (Leguminosae: Papilionatae)

***Resume**: Its probable origin is in Southern Europe and Western Asia. Archaeological remains are recovered from Jarmo (Northern Iraq) dating back ca 6000 BC. Similar records are available from Neolithic sites in Hungary and Switzerland. In India, similar finds are recorded from Navdatoli-Maheshwar dating 1600-1440 BC. and Atranji Khera dating 2000-1500 BC. Chromosoe number n=7 is reported for it. It is thought poor man's pulse. Its prolonged and excessive consumption cause 'Lathyrism' in man, being neurotoxic and hence its careful consumption is advisable.*

1. Introduction: It is also called Grass Pea, Chickling Vetch, Lathyrus Pea or Khesari in India. It probably originated in Southern Europe and Western Asia. Its archaeological remains have been recovered from Jarmo in Northern Iraq dating as far back as ca 6000 BC. (Helbaek, 1965). Similar finds have also been recorded from Neolithic sites in Hungary and Switzerland. It is grown in India as a winter crop and also in Iran, the near East, Southern Europe, parts of Africa and South America. The earliest record of this pulse comes from the Neolithic site Chirand (Bihar). It is also known from Navdatoli-Maheshwar about 1660-1440 BC., and subsequently from the early historical period at Ter (Maharashtra) and Kaundinyapur (Madhya Pradesh). It is also reported from Atranjii Khera dated to about 2000-1500 BC. (Vishnu-ittre, 1974).

2. Plant Characteristics: Much-branched, berbaceous annual creeper, stem winged, growing upto 1. 2m in length; leave alternate, pinnate with a single basal pair of sessile linear-lanceolate leaflets, ending terminally in 1 to 3 tendrils, stipules foliar and auricled; flowers blue or purple, borne singly in leaf axils; pods flattened, oblong, dorsally 2-winged, 2. 5 to 5. 0 cm long, 4 to 5 seeded, seeds wedge-shaped angular, white, yellow or brown or mottled.

3. Parts Used: Seeds, foliage.

4. Chemical Constituents: Protein (28.2%), fat (0.6%), carbohydrates (58.2%), minerals (calcium, phosphorus, magnesium, iron, copper, manganese, zinc), vitamin-A, beta-cerotene, thiamine, riboflavin, niacin, pantothenic aid, pyridoocine, folic acid, vitamin-C.

5. Uses And Applications: It is used in the form split or pulse, especially by poor classes. Seeds are carefully boiled and then consumed. It is also used as an adulterant of pulses viz., pigeon pea, lentils and chick pea. The foliage and seeds are fed to cattle. Its excessive and prolonged consumption is avoided as it is reported to cause latharism i.e. paralysis of the lower limbs especially in man. It has also a neurotoxic effect in animals. Leaves are eaten as pot-herb.

6. Karyotaxonomic Evidence: Various genotypes of *Lathyrus sativus* revealed diploid chromosome 2n=14. They showed significant differences among genotypes in long and short arms of chromosomes, total chromosome length, arm ratio, r-value, total chromosome volume and centromeric index (*cf.* Karimzadeh *et al.,* 2011).

7. History: In India, the crop is classified on the basis of flower colour, pod markings, size and colour of seeds. As many as 56 types have been differentiated based on size (small, medium, large), colour (grey, black, mottled). The smaller forms are known as 'Lakhori', while the larger one as 'Lakh'.

8. Further Developments: Lathyrism was believed to be caused by the high content of solenium and low content of methionine in its seeds. Hippocrates recorded it in ancient times. It is now known to contain an oesteotaxon and a neurotoxic amino acid, β-N-Oxalyl amino alamine (BOAA), the latter has been implicated as the cause of human lathyrism. BOAA is water soluble and easily leached out when the seeds are boiled in water. To overcome this disease, in India, it is successfully evolved varieties that are relatively free from the toxic constituents and hence safe for human consumption *Lathurus aphaca* Linn. *viz., L. imphalensis* Watt., *L. pratensis* Linn. *L. sphaericus* Retz. are cultivated for fodder (Anonymous, 1962).

Chickling Pea or Grass Pea suffers from a reputation of being toxic as its overconsumption under certain circumstances has caused neurolathyrism, a neurotoxic discase. This led us to the decision that crop should be abandoned as human food. Even seed sales were banned in some countries (Enneking, 2011). Apart from this fact, improvement in this crop is still considered a priority by national and international research centres. In the last 50 years, major efforts have been aimed at reducing the ODAP content which resulted in main cultivars of low ODAP.

Asmussen and Liston (1998) studied, for the first time, phylogeny using molecular data on both Eurasin and New World *Lathyrus* species. They used chloroplast DNA (cpDNA) characters to test the monophyly and relationships of Kupicha's *Lathyrus* sections, suggesting that some of these sections should be combined in order to form monophyletic group.

Later on, Badr *et al.,* (2002) confirmed the monophyly of the section *Lathyrus*, but they excluded other sections in their study. Pattern of protein electrophoresis (Emre, 2009) reflected the profound interspecific hybridization barriers in the genus *Lathyrus*, although *L. sativus* and *L. cicera* seem to be closer phylogenitically (Emre, *loc. cit.).*

There is a need for detailed characterization of the existing diversity to achieve effective conservation and enhance the use of germplasm collections. Jackson and Yunus (1984) showed that *L. sativus* is differentiated into several distinct forms, primarily on the basis of flower colour, size of seeds and leaves. They identified a clear distinction between the blue-flowered forms from South-West Asia, Ethiopia and the Indian subcontinent, and the white and blue-flowered forms with white seeds that have a more western distribution from the Canary Islands to the Soviet Union. This array of variation is thought due to geographical separation and selections by mankind. There is recent paradigm shift in the perception of the *L. sativus* research about its content of homoarginine, which is a alternative substrate for nitric oxide biosynthesis (Rao, 2011). A daily dietary intake of homoarginine through small quantities of *L. sativus* may have advantages and deserves to be exploited.

Conserved plant genetic resources are essential to meet the current and future needs of crop improvement programmes. Progress in *Lathyrus* breeding is slow. More efficient and faster breeding approaches are needed on this neglected but promising underutilized species.

Arslan (2017) studied fatty and compositions of 173 different grass pea (*Lathyrus sativus* L.) accessions. This study revealed total saturated fatty acids, total monounsaturated fatty acids, total polyunsaturated fatty acids and total fatty acids. In addition, unsaturated fatty acids, oleic acid, linoleic acid, gamma-linolenic acid and alpha-linolenic acid that are the main components of fatty acids. The author concluded that large variation which is found among these genotyps in terms of fatty acids, can be used as a selection criterion in future breeding trials.

Rahman *et al.* (2010) assessed genetic diversity in grass pea (*Lathyrus sativus* L.) of 25 genotypes for 13 morphological traits through analysis of variance and multivariate analysis. This analysis of variance for all the characters revealed highly significant variation among the genotypes of grass pea except for day to 50% fruiting. BD3618, BD3620 and BD3621 were found as superior on the basis of the mean performance of desirable traits. The genotypes were grouped into five clusters through principal component analysis (PCA). The inter cluster distance was maximum (17.038) between the cluster V and III and the lowest (2.6) distance between IV and I. the genotypes of cluster III might be selected as a suitable parent for future hybridization programme. The value of Vector I and II revealed that both Vector had positive values for plant height, pods/plant and seed yield/plant indicating the highest contribution of these traits towards the divergence among 25 genotypes of grass pea genotypes.

9. Epilogue: It is the hardiest of cool-weather pulse crop. It withstands drought and also waterlogging. It is grown in dry areas where cereal crops fail to establish. It is, therefore, an important famine crop. It flourishes even on heavy clay soils and matures in 5 to 6 months.

5.5 CHICKPEA

Cicer arietinum L. (Leguminosae-Papilionatae)

Resume: *Cicer has the centre of diversity in Western Asia, probably in the Caucasus region and/or Asia Minor. The crop is diploid. The earliest records go to a site in Turkey radiocarbon-dated to 5450 BC. It is being cultivated in the Mediterranean area since 3rd to 4th millennia BC. It meets the quantitative and qualitative requirements of a large part of humanity. The pulse is nutritious.*

1. Introduction: It is thought originated in southwest Asia, although wild forms are unnoticed from this region. It is one of the oldest crops cultivated in Asia and Europe. It was known to the ancient Greeks, Egyptians and Herbrews. Helbaek (1966) noted archaeological remains of Chickpea dated 5450 BC. at Hacilar in Turkey ca 4000 BC. The generic name *Cicer* is derived from the Roman word 'Cicer' alluding to resemblance of a seed to the head of a ram. The specific epithet is also derived similarly. Its introduction into India is of more early date as it has Sanskrit name 'Chanaka' and several names in other Indian languages. The western Aryans probably carried it into India.

2. Plant Characteristics: Branched, erect annual reaching 50 cm in height; leaves imparipinnate with 9 to 15 pairs of leaflets, leaflets ovate, elliptic to obovate, all parts studded with glandular trichomes; flowers pink to white and borne singly; pods inflated small, with 1 or 2 seeds, seeds wrinkled yellow, brown to nearly black.

3. Parts Used: Seeds.

4. Chemical Constituents: Proteins (17.1%), fats (5.3%), carbohydrates (61.2%), fibre (3.0%), minerals (calcium, phosphorous, iron, magnesium, potassium), carotene, vitamin A & C, thiamine, riboflavin, niacin, etc.

5. Uses And Applications: It is being nutritious, consumed raw, parched, boiled, salted or roasted. Flour is used in various forms in Indian confectionery. Germinated seeds are also consumed to prevent scurvy. Pulse is a good feed for horses. Green leaves are used as vegetable. Stalks and dry leaves constitute fodder for cattle. Flour is also used in textile sizing and adhesives.

6. Karyotaxonomic Evidence: *Cicer* is a relatively small genus, with 39 known species distributed mainly in Central and Western Asia. The most common somatic chromosome number reported in the genus so far is 2n=2×=16 (*Cicer arietinum, C.chorossanicum, C.pinnctifidum, C.anatolicum, C.bijugum, C.cunertum, C.incisum, C.judaicum, C.microphyllum* and *C.montbretii*). A different number (2n=14) has been reported for *C.pungens* as well as for *C.arietinum* and *C.microphyllum* (as *C.songaricum*). The latter reports are thought erroneous, one of the chromosomes in the complement being quite small (about 1µ). The basic number in the genus thus appears to be

×=8. This does not agree well with basic numbers (6, 7) of the tribe Viciae to which *Cicer* belongs. (Darlington and Wylie, 1955). It would agree with the tribe *viz.,* Trifoliae, where *Ononis* is placed. However, *Ononis* species counted so far are at a higher ploidy level (2n=4×=32) than *Cicer* and cannot have contributed to the development of *Cicer*. No information is available regarding the interrelationship between the cultivated *C.arietinum* and the wild species. Morphologically, it seems that *C.pinnatifidum*, *C.echinospermum* and *C.bijugum* are closest. Crosses between *C.arietinum* and *C.microphyllum* (as *C.songaricum*, cultivated in the Himalayan region) as well as the wild *C.pinnatifidum* have not been successful because of pre-fertilization barriers.

7. History: The genus *Cicer* has the centre of diversity in western Asia, probably in the Caucasus region and/or Asia Minor (Maesen, 1972). The progenitor of the present day *C. arietinum* must have been spread by the Aryans both westwards, along the Mediterranean, and eastwards, overland to India. The earliest records go to a site in Turkey radiocarbon-dated to 5450 BC. Its cultivation in the Mediterranean area is evidenced in the 3rd and 4th millennia BC. The earliest record of chickpea in India, at Atranji Khera in Uttar Pradesh (Chowdhury *et al.,* 1970), is from 2000 BC. and presumably represents an overland introduction. Early records from peninsular India are dated AD. 200-150 BC., but this may have to be extended at least to the 3rd to 5th century BC. (Vishnu-Mittre, 1974). De Candolle (1886) thought area between Greece and Himalayas as the place of origin. Vavilov (1926) suggested two centres of diversity, one in the southwest Asia and Mediterranean, and second in Ethiopia. It is called ‘Chanak’ in vedic literature *viz.,* Yajurveda and ancient Sanskrit literature Kalpasutra. Etymological evidence suggests that chickpea could have also been introduced into India independently by an oversea route.

The names applied to the crop in the Dravidian languages are completely different from the Sanskrit-derived Chana, prevalent in northern India. There was extensive maritime trade between Egypt, Israel and Southern India, with flourishing Indian colonies in Memphis. Ethiopia, the third centre of diversity, could have; received the crop from the Mediterranean and where chickpea was grown at least as early as the first millennium BC.

Chickpea seems to have diverged in two directions during the course of its dispersal. In the western part of its distribution, where it is grown under an equable Mediterranean spring-summer environment, cultivars developed with characteristic large, owl-head shaped, light-coloured seeds with little wrinkling of the seed coat. The tall plants bore leaflets and flowers are large. In the eastern and southern parts of its distribution, particularly In India, where chickpea is raised in the cool dry season, and in Ethiopia, forms with early maturity and with slender stems having small leaflets and flowers have developed. The seeds are small, typically wrinkled, ram-head shaped and dark coloured. The Indian Desi (eastern ecotype) and Kabuli (western) groups differ and probably appear more modified by human selection. The former and can be thought as more advanced (Rohewal *et al.*, 1966). The two types are reproductively isolated. Several varieties have been developed by plant breeders combining their characteristics.

8. Further Developments: It is now introduced into a few other areas. From the Iberian peninsula, it was carried by 16th century Spanish and Portuguese travellers to South and Central America, especially to Mexico. It is also introduced into East and South Africa and the West Indies by the expatriate Indian population, The large, smooth-seeded types (known as Kabuli) appear to represent a third recent (eighteenth century) wave of introduction into India from the Mediterranean region overland.

Little attention has been paid, until recently to the improvement of the crop by breeding. It was always thought as a poor man's food and grown under conditions of limited inputs where other crops cannot be grown. However, substantial breeding efforts have been made in India (Ramanujam, 1972) and in Pakistan in the past few decades. Success in improving yielding capacity over the land races has not been considerable. In India, important approach in its breeding has been to introduce large seed size and responsiveness to improved cultural conditions by crossing indigenous types with the Mediterranean ecotype. Several medium-sized types have been developed. Breeding for disease resistance or response to fertilizers have not, however, received any attention.

Sharifi *et al.* (2018) evaluated variations in chickpea (*Cicer arietinum* L.) yield and yield components by multivate technique. He used 25 chickpea genotypes. The principal component analysis recognized number of days to flowering, flowering period, number of days to maturity, canopy height, canopy width, biological yield and number of pods per plant as characters. These describe the variation with the chickpea genotypes. The cluster analysis classified the genotypes into four separate clusters, exhibiting that hybridization of genotypes across clusters could lead to increase in heterosis in cross progenies. The results obtained can be used in breeding strategies for future hybridization programmes for yield improvement and are appropriate for classification of diversity among chickpea germplasm.

Sachdeva *et al.* (2018) studied molecular and phenotypic diversity among 40 chickpea (*C. arietinum*) genotypes as a function of drought tolerance. In total, 90 alleles were detected with 3.6 alleles per locus. Polymorphism information content (PIC) values ranged from 0.155 to 0.782 with an average value of 0.4374 per locus. The size of amplified products ranged from 160 bp to 390 bp. Primer TA 136 with eight alleles showed the highest PIC value of 0.7825, indicating its ability to differentiate the genotypes at molecular level. DARwin neighbor-joining tree analysis based on dissimilarity estimates was done for the molecular data and sequential agglomerative hierarchial non-overlapping (SAHN) grouping for the morphological data. It could clearly discriminate the tolerance and the sensitivity of genotypes. Two-dimensional principal coordinates analysis (PCOA) plot indicated good diversity for drought tolerance. The genetic similarity coefficients ranged from 0.115 (genotypes BGD72 to ICCV5308) to 0.828 (genotypes ICCV10316 to ICCV92337).

9. Epilogue: There has been considerable international concern about world-wide shortage of protein foods and the energy crisis leading to fertilizer shortage. This has fostered a new interest in grain legumes, including chickpea. The major objectives was to develop a plant type, with high yield potential and responsive to agronomic inputs, which can compete on a more nearly equal footing with high yielding cereals (Swaminathan, 1972). To achieve this would require a refashioning of the plant type with greater photosynthetic efficiency and reduced photo-respiration loss and more advantageous partitioning of the photosynthate into a larger number of sinks (seeds) rather than into vegetative parts. The development of suitable legume-rhizobium combinations which can fix the large quantities of nitrogen needed for high yields were thought important. Protein contents, amino-acid profile, disease resistance and suitability for milling or processing were worthwhile objectives in the later attempts. It is well adapted to arid and semi-arid regions with low to moderate rainfall and a cool and dry climate. It matures early in 3 to 3. 5 months.

5.6 COWPEAS

Vigna unguiculata (L.) Walp. (Leguminosae-Papilionatae)

Resume: *The crop although thought to be cultivated in Central Africa, West Africa and India are both modern centres of diversity of cultivars. West African farming systems suggested that it was associated with Sorghum and Pearl Millet especially in Neolithic times (3rd millennium BC.). The crop is polyploid. It is chiefly important source of protein (particularly of lysine) and also predominant starchy diets.*

1. Introduction: The genus *Vigna* contains nearly 150 species in the world, of which the greatest number of species are particularly found in African continent. Of these, just two or three species are cultivated for food and forage. Cowpeas are an ancient crop grown in the tropics and subtropics. It is now believed that cultivated cowpea originated in central Africa from where it spread, in early times, through Egypt or Arabia to Asia and the Mediterranean. The Spaniards introduced it to America in about 1700 AD.

2. Plant Characteristics: Low-growing, bushy or trailing annuals, reaching a height of 0.9 to 1.5 m; leaves 3-foliate, long-petioled, leaflets large, cordate; flowers few, in pairs, crowded near the tip of long axillary peduncle, pink or purple or light blue in colour; pods long, cylindrical, constricted between seeds, seeds globular or kidney-shaped, smooth or wrinkled, white, creamy yellow to red to brown.

3. Parts Used: Seeds, pods, foliage.

4. Chemical Constituents: Protein (24.6%), carbohydrates (55.7%), fats (0.7%), fibre (3.8%), minerals (calcium, phosphorous, potassium, iron), viramin-A, B6, B12 and C.

5. Uses And Applications: Mature seeds are an important pulse in India and Africa. Green seeds or young pods are used as vegetable. The immature pods are eaten frozen or canned. Ripe seeds are boiled and eaten. The seeds are ground into flour, used for cakes, sweetmeats and porridge. Sprouted seeds are also consumed. Young shoots and leaves are consumed as spinach. The crop is used as a hay, silage, soil cover and green manure. Starch obtained from seeds may be used in textile-sizing and as a thickening agent in calico-printing.

6. Karyotaxonomic Evidence: Vigna is a pantropical genus of about 170 species, 120 in Africa (66 endemic), 22 in India and southeast Asia (16 endemic), and a few in America and Australia (Faris, 1965). Verdcourt (1970) recognized clear five subspecies of *V. unguiculata*. Two subspecies are wild *viz.,* (i) subsp. *dekindtiana* in the African savanna zone and Ethiopia and (ii) subsp. *mensensis* in forests, with scabrous, dehiscent pods and seed dormancy not found in the cultivars. The cultivated cowpea is subsp. *viz., unguiculata*; the other cultivated subsp. *viz., cylindrica* and *sesquipedalis*, are widespread in India and the Far East. These are now recognized as independent species within the genus *Vigna*.

The cytotaxonomy of Vigna is relatively simple, being uncomplicated by polyploidy (2n=2×=22) and with apparently little genetic and no chromosomal divergence of the cultivars from their putative ancestor. The five subspecies of *V. unguiculata* are interfertile. Attempts to hybridize cultivars with other *Vigna* species, have failed (Faris, 1965). The origin of the cultivars in Asia cannot explain the distribution and truly wild status of subsp. *dekindtiana*. Instead, the evidence all suggests that subsp. *unguiculata* was domesticated from it in Africa.

7. History: West Africa and India are both modern centres of diversity of cultivars. Circumstantial evidence arising from the study of cowpeas in the ancient West African cereal farming system strongly suggests that their early history in Africa was closely associated with that of sorghum (*Sorghum bicolor* and pearl millet *Pennisetum americanum*). Doggett (1970) thought that these cereals were domesticated in Ethiopia 5-6 thousand years ago, and then spread throughout the African savanna zone and to Asia. Archaeological evidence from West Africa suggests that sorghum was grown there in neolithic times (3rd millenium BC.).

The diversity of wild cowpeas in Ethiopia is impressive and contrasts with their homogeneity in West Africa. Faris (1965) concluded that the cultivars were domesticated from subsp. *dekindtiana* in West Africa. This is so because this is their only centre of diversity within its range. It is also probable that the West African centre of diversity is secondary and that cowpeas were domesticated with sorghum and pearl millet in Ethiopia; or that they had a 'diffuse' origin in the savanna zone after dispersal of the cereals.

Diversity probably accumulated and cultivars were adapted to the cereal farming system evolved. The expansion of the crop would have accorded opportunities for the survival of chance mutants for natural hybridization of inbred cultivars and for introgression from sympatric populations of subsp. *dekindtiana*. Though cowpeas are usually noted as nearly cleistogamous inbreeders, up to 2% outcrossing has been reported from Senegal.

Early West African cereal farmers reached the fringe of the rain forest zone in the 3rd millenium BC. (Alexander and Coursey, 1969). Their contact with hunting/ gathering communities who ate wild yams led to the development of yam cultivation in West Africa. The day-neutral cultivars of subsp. *unguiculata* are common there now, grown alone or interplanted with yams. It is probable that they evolved from short-day types and then the demand for protein favoured their spread in the yam zone.

The subsp. *unguiculata* arrived in Asia with sorghum and pearl millet which, Doggett (1970) suggested more recently than 1500 BC. There could be no further gene exchange in Asia with wild cowpeas. The expansion of the crop in India would have accorded chances for the accumulation of genetic diversity from which conscious selection probably established subspecies *cylindrica* (for forage) and *sequipedalis* (for its long, immature pods eaten as a vegetable). *Vigna cylindrica* is mentioned in ancient Sanskrit literature *viz.,* Kalpasutra and Patanjal's Mahabhasya. The cultivars from India spread into southeast Asia and the Far East and ultimately reached Europe from Asia or Egypt (Chevalier, 1944). They were referred to in Greek texts of 300 BC. and by the end of the Roman era, were established as a minor food legume in southern Europe. It was introduced in West Indies in the 16th century by Spaniards and in America in about 1700 AD.

The diversity of West African subsp. *unguiculata* contrasts with the relative homogeneity of modern populations of wild subsp. *dekindtiana*. These consist of prostrate or twining, short-day plants with short, scabrous, dehiscent pods and small, smooth, dark speckled seeds with testas sufficiently impermeable to water to confer a degree of dormancy. The loss of pod dehiscence and seed dormancy must have evolved during domestication.

The later change of greatest agricultural significance in West African cultivars was the divergence. Under the influence of human selection characteristics *e.g.* upright, day-neutral types, adapted to the climate and agronomy of the root and tuber farming system; spreading, short-day cultivars, etc. were adapted to the cereal farming system of the semi-arid savanna zone. A similar but more extreme change in Asia gave rise to the distinct subsp. *cylindrica* consisting of strictly erect, day-neutral plants.

Variation in West African cultivars of subsp. *unguiculata* includes all of the flower, pod and testa (Saunders 1960). Most day-neutral cultivars have relatively small, smooth, cream or brown seeds and pods shorter than 13 cm. Most short-day cultivars have large, white seeds with rough testas and pods up to 22 cm long.

Gyan Ansah *et al.* (2016) evaluated 100 genotypes of cowpea [*Vigna unguiculata* (L.) Walp.] for phosphorus use efficiency. The authors conducted pot experiment to screen these genotypes for their phosphate (P) use efficiencies at two levels of phosphorus: (i) triple superphosphte, 60 kg P 205 ha-1, applied to the pots (8.2 kg soil pot^{-1}) and (ii) no application (control). Tap root length, total root length, root dry weight, short dry weight, total dry weight, shoot P content, P utilization and P uptake efficiency were measured. These traits ere highly heritable based on the ratio of genotypic coefficient of variation (GCV) to phenotypic coefficient of variation (PCV). Data for the 14 variables measured from these accessions were subjected multivariate analysis using principal component analysis (PCA) and clustering criteria. Characters that contributed most to variability based on the first five principal components (Eignvalue>1) were shoot dry weight, shoot P content, lateral root number and nodule dry weight. Two main groups of genotypes were revealed based on the similarity matrix. Group (A) had 10 accessions, which are P-use efficient. These are being introgressed in adapted cutivars.

Thouseem *et al.* (2018) employed genetic parameters in seed yield components of cowpea [*Vigna unquiculata* (L.) Walp.]. They found phenotypic coefficient variation (PCV) higher than the genotypic coefficient variation (GCV) for all the traits studied. The highest magnitude of GCV was observed for pot weight followed by number of pod clusters plant^{-1}, seed yield plant^{-1}, 100 seed weight, pod length, number of pods plant^{-1}. All these characters considered except number of primary branches plant^{-1}, number of pods plant^{-1} and seed yield plant^{-1} recorded with high estimates of heritability. High heritability coupled with high genetic advance was observed for days to 50 percent flowering, length of main stem, number of pod clusters plant^{-1}, number of food cluser^{-1}, pod weight, pod length, pod girth, number of seeds pod^{-1} and 100 seed weight.

8. Further Developments: The Spanish introduced cowpeas to the New World tropics in the 17th century and day-neutral cultivars of subsp. *unguiculata* must have been taken there with the slave trade from West Africa. The crop was first grown in the southern USA in the early 18th century. Diverse cultivars have been bred there both for seed and forage production.

There are opportunities for sampling a wide range of genotypes in Indian and West African centres. Large collections are maintained in West Africa, India and the United States (Summerfield *et al.,* 1974). Genetic erosion does not seem to be operated. Many cultivars have been developed in South East Asian countries. The largest collections of cultivars is found IITA in Ibadon, Nigeria, India and USA (Grubben, 1977).

Breedinig experiments indicated that most of the varieties are easily crossed and seed prolifically. Remarkable phenotypic variability has been noticed for solution and improvements of characters *e.g.* length of peduncle, pods per plant, yield per plant, yield of grain per plant, colour of seeds. Several varieties are developed in India *e.g.* Pusa 419, Pusa Barsati, E. B. 1, 2, 3 & 4, Walior K. 3B, K11, K14, Cowpea No. 1, 2, 7, 10, 40, Cowpea Type I & II, and some others. In USA, some varieties are released *e.g.* Whippoorwill, Ner Era, Groit, California Black and Victor, etc. (Anonymous, 1976).

The cultivars exhibit diversity *w. r. t.* growth habit, leaf shape, flower colour, male sterility, high protein, drought resistance, grain yield, period of maturity, etc. (Swamy and Sadashiva, 2007).

9. Epilogue: The objective of research in Africa and Asia was always to increase the use of cowpeas in human consumption. Traditional farming systems will have to be altered to achieve this aim. Many 'land races' of cowpeas will cease to be grown. There are still hopes for breeders to produce determinate cultivars which are highly productive as densely sown sole crops, utilizing determinate growth from subsp. *cylindrica* and desirable seed types from subsp. *unguiculata*. Attention is also being given to resistance to locally important pathogens and to the very destructive insect pests which are the chief factor limiting the. yield of sole crops, at least in Africa. The protein content of seeds can be increased and its nutritive value will no doubt be improved by increasing the proportion of sulphur-containing amino acids. Laborious intercrossing programmes are required to establish the genetic base for the application of this technique to an inbreeding crop. The search for new sources of germplasm includes attempts to obtain interspecific hybrids and the inclusion of subsp. *dekindtiana* in breeding for resistance to pathogens and insect pests.

5.7 GREEN GRAM

Vigna radiata (Linn.) Wilczek

[(Syn.) *Phaseolus aureus* Roxb.)] (Leguminosae: Papilionatae)

Resume: *It is native of India. Carbonised grains are discovered from the Chalcolithic site Navdatoli-Maheshwar in India dating 1660 to 1440 BC. Phaseolus sublobatus is thought its progenitor as it grows wild in India. Chromosome number n=11 is reported for it. It is an important pulse crop in India and rich source of protein and carbohydrates.*

1. Introduction: It is also called 'Golden Gram'. It is a native of India. It has been cultivated in India since ancient times. Its carbonized grains have been recovered from the Chalcolithic site, Navdatoli-Maheshwar in India dating 1660 to 1440 BC. (Vishnu-Mittre, 1968). It is also introduced in southeast Asia, some parts of Africa, the West Indies and the United States.

2. Plant Characteristics: Erect or suberect annuals, growing upto 1. 3 m in height; leaves alternate, 3-foliate, yellowish-green; flowers purplish-yellow, in clusters of 10-22 on a long peduncled axillary recemes; pods narrow, cylinidrical, 5-10 cm long, seeds 10-15, green or sometimes yellow.

3. Parts Used: Seeds, pods, foliage.

4. Chemical Constituents: Protein (23. 6%), fats (1. 2%), carbohydrates (58. 2%), fibre (3. 3%), ash (4. 0%), minerals (calcium, phosphorus, iron), vitamin A and C, thiamine, nicotinic acid, niacin and riboflavin.

5. Uses And Applications: In India, the pulse is used for human consumption in the form of whole or split, after decortication. The tender young pods and green shelled beans are eaten as a vegetable. The sprouting beans are consumed in China and America. Sprouts are also candied. They are eaten raw as a salad or boiled. The split or dehusked gram fried in fat, after salting is consumed with tea or drinks as a snack. Parched beans, after seed coat removal, are ground into flour which is used in preparation of various Indian and Chinese dishes. Green plants are good fodder for cattle. Broken seeds are fed to livestock. It is used as green manure, being a leguminous crop, or also as a restorative crop.

6. Karyotaxonomic Evidence: The somatic chromosome number is 2n=22. Two pairs of large chromosomes are associated with nine pairs. Other chromosomes are variously classified as medium and small chromosomes.

7. History: Green gram had been growing in India since ancient times and also in Nile Valley (De Candolle, 1886). Vavilov (1926) also thought as centre of origin of cultivation. It is supposed native to India and Central Asia (Anonymous, 1976). Zukovski (1962) considered *Phaseolus sublobatus* as the progenitor of green gram as it grows wild in India. Prain (1903), Piper and Morse (1914) lent support to such deviation of green gram. It is called 'Mudga' in vedic literature *viz.,* Yajurveda and other ancient Sanskrit literature *viz.,* Brahmana, Panini's Astadhyayi and Vartik, and Patanjal's Mahabhasya.

8. Further Developments: Breeding programmes in India were started since 1925 evolving a plant type with erect habit, short duration, high yield, synchronized maturity of pods, non-shattering habit, bold and shining seed, good cooking qualities. A large number of improved varieties are released (Majumdar, 2011). Recently, AICRP developed 13 varieties suitable for cultivation in different states in India. Germplasm collections have been deposited at Pusa (Bihar, India) and have been studied for some economic importance *viz.,* period of maturity, pod characters, quality and yield of seeds. Improved strains have been recognized suitable to various states in India (Anonymous, 1976).

Pandey *et al.,* (2011) detected 42 alleles in 120 green gram accessions by using 27 STMS markers. This study indicated overall high genetic similarity among the green gram accessions. Lakhanpal *et al.,* (2000), Afzal *et al.,* (2004) and Karuppanapandian *et al.,* (2006) reached to the same conclusion based on their RAPD analysis of green gram cutivars. Bhat *et al.,* (2005) also reported narrow genetic base in their AFLP analysis of Indian green gram cultivars.

No relationship between geographic distribution and genotypic diversity is exhibited by the clustering pattern as the genotype of different geographic origins were grouped in the same cluster as observed earlier by Manivannan *et al.,* (1998). This was ascribed to the fact that genetic drift and selection in different environment could cause greater diversity than geographic distances. Bisht *et al.,* (1998) reached to the same conclusion.

Das *et al.,* (2014) studied genetic diversity in some Indian cultivars of green gram. They recorded overall moderate degree of genetic diversity among them. In their opinion, this reasonable diversity observed may be exploited for further yield improvement. They further stated that ISSR marker system can be effectively used to identify and differentiate the mung bean cultivars, to provide more evidence on the extent of genetic diversity among the selected cultivars, to determine the phylogenetic relationships among them and most importantly to evolve systematic breeding strategies to improve the yield components of the genotypes.

Lamichaney *et al.* (2017) investigated verification in preharvesting sprouting (PHS), fresh seed germination (FSG) and activity of alpha-amylase enzyme in diverse mungbean genotypes. There was a wide variation in PHS tolerance and FSG among 163 genotypes examined and 14 genotypes were found to be tolerant (<20%) to PHS. Seed germination in a pod, a measure used to evaluate PHS, varied from 7.14% in in germplasm accession Chamu4 to 82.52% in cultivated variety IPM 2-3. There was increase in alpha-amylase activity in genotypes showing high FSG and PHS, especially at 48 and 2h after germination as compared with PHS tolerant genotypes. Therefore, alpha-amylase can be used as an effective biochemical marker to evaluate a large number of mungbean genotypes for FSW and PHS. The variation in seed germinability revealed could be used for improvement programme of mungbean.

9. Epilogue: It is rather fast-growing legume crop of especially of rainy season. It requires short period of maturity. It is rich in protein and used in various preparations. The crop has received little attention in comparison to other pulse crop, although it is also grown outside India in other African countries and even in Africa. It is cultivated as mixed crop.

5.8 HORSE GRAM

Macrotyloma uniflorum (Lam.) Verdc. (Leguminosae: Papilionatae): [(Syn.) *Dolichos uniflorus* Lam.]

Resume: *It is native of Southeast Asia particularly in tropical regions. It is drought resistant crop. Chromosome number n=12 is reported for it. It is thought the poor man's pulse in India and mostly consumed as whole seeds. It is rich in protein and carbohydrate.*

1. Introduction: It is native of southeast Asia and distributed now throughout the tropical regions particularly India, Malaysia, Mauritius, Sierra, Leone, Transvaal and the West Indies.

2. Plant Characteristics: Slender, suberect, low-growing, succulent, bushy annuals, branched from base, branches intertwining among themselves; leaves 3-foliate, leaflets membranous, entire; flowers borne in bunches in axillary racemes; pods sickle-shaped, upto 5. 0 cm long, containing 5 to 7 seeds, seeds small, flattened, rhomboidal, brown to red, black or mottled, seed coat hard and shiny.

3. Parts Used: Seeds, foliage.

4. Chemical Constituents: Protein (24.0%), fats (1.3%), carbohydrates (57.3%), fibre (5.3%), ash (3.1%), 1/3 weight of ash being phospdoric acid; minerals (calcium, iron, phosphorous), vitamin A, carotene, thiamine, riboflavin, niacin, etc. (Gopalan *et al.,* 1977).

5. Uses And Applications: Seeds are generally parched and then boiled or fried. It is usually eaten as it is instead as a split-pulse. In Myanmar, the dry seeds, after boiling, are pounded with salt and then fermented to produce a sauce. The seeds are fed to cattle and horses as a concentrated food, hence the name 'horse gram'. The green plants and husk also constitute good fodder for them. It is also used as a green manure crop.

6. Karyotaxonomic Evidence: Karyotype analysis revealed 2n=20 chromosomes. The karyotype showed a prevalence of chromosomes with median primary constrictions. Satellites are associated with the short arm of 2B-type chromosomes. Secondary polyploidy has been attributed as the possible cause of secondary association and the basic chromosome number for the species is suggested to be ×=7 (Halder *et al.*, 2012).

7. History: Several varieties differing in colour of seed coat and period of maturity are known under cultivation. The seeds are brown, light red, grey, black and mottled. The black seeded type is of shorter duration than the others. Horse gram is a multipurpose, rich protein crop cultivated in India, Pakistan, Bangladesh, Sri Lanka, Nepal and West Africa. It is, however, farmer's neglected crop. It is sown in India particularly by tribal farmers especially in drought prone area. So less attention has been paid to varietal improvement in past.

8. Further Developments: A root rot caused by *Rhictonia* sp. affects the plant especially when the soil is water-logged due to heavy rains or irrigation. Anthracnose caused by *Glomerella* sp. effects stem, leaves and pods causing black sunken spots. Spraying of Bordeaux mixture is advised. Use of disease-free seeds, burning of all infected plants and cultivation of resistant strains are some measures to be implemented (Anonymous, 1948).

Kulkarni and Mogle (2011) carried out experiment to evaluate 22 horse gram genotypes for analysis of quantitative and qualitative characters. They studied characters such as days to 50% flowering, plant height, number of primary branches and pods per plant, number of seeds per pod, colour and shape of seeds, shattering ability and average yield. They concluded that sufficient genetic diversity is available at phenotypic level which, in their opinion, provide opportunity for selecting designed traits. They also suggested to evaluate these germplasm for diseases, insects and pest susceptibility and fertilizer requirement.

Latha *et al.,* (2013) evaluated seed polymorphism in south Indian horsegram taking into consideration length, breadth, thickness, volume, plumule length, radicle length, seed colour, etc. They recorded tremendous variability in horse gram seeds which can be utilized in crop breeding programmes. In their opinion, the variability can be linked to their geographical origin of the collections so that spatial distribution can also be ascertained. The narrow difference between PCV and GCV revealed less environmental influence on these characters providing scope for direct selection based on phenotype. They concluded that the extent of seed polymorphism in terms of seed size and colour do not appear to be that detrimental or decisive in seed viability.

Prasad and Singh (2015) reviewed horse gram as a underutilized neutraceutical pulse crop grown in wide range of adverse climatic conditions. They concluded that horse gram is a rich source of nutrient content. It has high levels of antioxidant and radical scavenging activities in addition to their traditional role of providing proteins and carbohydrates. It is also a rich source of natural bioactive substances *e.g.* phytic acid, phenolic acid, etc. These have tremendous potential for curing common cold, fever, throat infection, asthma, urinary stones, obesity, etc. It is, therefore, needed to exploit horse gram as a source of neutraceutical and food industries.

Sharma *et al.,* (2015) analyzed genetic diversity and population structure of horse gram using RAPD and ISSR markers. While analyzing, they also included two other species of the genus *viz., M. axillare* and *M. sargharwalensis*. They opined that the latter species is distantly related to the cultivated species than *M. axillare*. They further stated that high levels of genetic diversity of some accessions can be exploited in crop breedinig programmes for its genetic improvement.

9. Epilogue: It is cultivated as a dry-land crop in areas of moderate rainfall. It is hardy and drought resistant, grows well on all soil types, even on poor soils with very little rain. It requires minimum preparatory tillage. It attains maturity within 4 to 6 months. It is often grown as a preparatory crop on reclaimed lands.

5.9 LENTIL

Lens culinaris Medic.

(Syn.) *L. esculenta* Moench. (Leguminasae-Papilionatae):

Resume: *It is the oldest grain legumes of the Old World associated with the start of the 'Agriculture' revolution. Carbonised seeds have been recovered from early farming villages in the Near East dating 7th and 6th millennia BC. Its archaeological remains are discovered from Hacilar, Beidha and Jarmo dating 7500 to 6500 BC. The crop is diploid. The seed protein content is high and thought an excellent substitute for meat. It is reputed as food since biblical times.*

1. Introduction: Lentils are reputed as a food since biblical times. It formed a common food of the ancient Romans, Greeks, Jews and Egyptians. Lentil ranks among the oldest and the most appreciated grain legumes of the Old World, cultivated from the Atlantic coast of Spain and Morocco in the west to India in the east. It is a characteristic companion of wheat and barley cultivation throughout the belt of Mediterranean agriculture. Its archaeological remains dated 7500 to 6500 BC have been discovered from Hacilar, Beidha and Jarmo (Helbaek, 1966). Its seeds have been recovered from Neolithic deposits in Greece (Renfrew, 1966). It is believed to be indigenous to southwestern Asia and the Medeterranean region. From these areas, it then spread to northward to Europe, eastward to India and China and southward to Ethiopia.

2. Plant Characteristics: Plants slender, much-branched, erect annuals, reaching 25-40 cm tall; leaves pinnate, with 5 to 7 ovate leaflets, generally ending into a terminal tendrils; flowers borne singly or in clusters of 2 to 3 flowers at the end of flowering axis; pods short and contain only two seeds; seeds lens-shaped, brown, grey and red.

3. Parts Used: Seeds.

4. Chemical Constituents: Protein (25.1%), fat (.7%), carbohydrates (59.7%), mineral (2.1%) (calcium, phorphorous, iron, magnesium, potassium), carotene, vitamin A, thiamine, riboflavin, niacin, etc. (Majumdar, 2011).

5. Uses And Applications: Unripe young pods are used as vegetable in India. The pulse (dal) is mainly used for soup and porridge. It is an excellent substitute for meat, especially in many catholic countries. Lentil soup is popular in USA. The dried leaves, stalks, empty pods and broken pieces of pulse are fed to cattle. Seeds may be employed as a source of commercial starch for use in textile and calico printing industries.

6. Karyotaxonomic Evidence: *Lens* is a relatively small genus. It is distributed in wild only to the Mediterranean basin and southwestern Asia. In addition to the crop (*L.culinaris* = *L.esculenta*), the genus also includes four wild species (Barulina, 1930; Zohary, 1972). *Lens* thought intermediate taxonomically between *Vicia* and *Lathyrus*. Its closest affinities seem to be with *Vicia* section Ervum. The four wild species of *Lens* are all slender, ephemeral, small-flowered annuals with characteristic broadly rhomboid, conspicuously compressed pods with 1-2(3) flattened seeds. All members are apparently predominantly self-pollinated.

Comparative morphology and observations on natural hybridization implicate one wild species, namely *L. orientalis*, in the ancestry of the cultigen (Zohary, 1972). In its vegetative and reproductive morphology, *L. orientalis* appears as a miniature *L. culinaris*. The wild plant differs from the cultigen in the characteristic seed dispersal device. The small pods burst at maturity and effectively release the characteristic lenticular seeds. The chromosomes of *L.orientalis* and *L. culinaris* are similar. Both lentils are diploid, with $2n=2x=14$.

The wild progenitor of the cultivated lentil is a Near Eastern plant. This is suggested by its specific epithet. *Lens orientalis* is distributed mainly in Turkey, Syria, Israel, northern Iraq and western and northern Iran. Interestingly, this is the area wherein the earliest archaeological evidence of lentil cultivation is met with.

Lens culinaris exhibits a wide range of morphological variation both in vegetative and reproductive parts. Several distinct varieties and lines have been described (Barulina, 1930). These cultivars are grouped in two intergrading clusters: (a) small-seeded lentils (subsp. *microsperma*), with small pods and small (3-6 mm) seed. This group overlaps in seed size with wild *L. orientalis*, (b) large-seeded lentils (subsp. *macrosperma*), with larger pods and with seed attaining 6-9 mm. As in other pulses, increase in seed size is one of the most conspicuous trends under domestication. Macrosperma forms are considered as the more evolved.

7. History: Lentils are thought associated with the start of the 'agricultural revolution' in the Old World. This revolution was initiated by the domestication of einkorn and emmer wheats, barley, pea, flax and also lentil. Carbonized lentil seeds have been identified from early farming villages in the Near East (dated 7th and 6th millenia BC.). They are small (2.5-3.0 mm in diameter) and do not occur in large quantities, Larger amounts of seeds were revealed in somewhat later phases of the Neolithic settlements of the Near East. Some of these 5th millenium BC. remains are already somewhat larger than the wild form and attain 4.2 mm in diameter This was the developmental progress during domestication.

Lentils were intimately associated with the spread of Neolithic agriculture to Greece and adjacent Bulgaria. Later, they accompanied the expansion of wheat and barley in the Neolithic and Bronze-Age agriculture of the Near East, the Mediterranean basin and central Europe. Lentil also was an important element in the assemblage of Near Eastern grain crops which were introduced to Ethiopia by the Hamitic invaders. Its introduction in the Indo-Gangetic plains dates back to ca 2000 BC. Possibly, it may have been introduced even earlier with the arrival of Aryans as suggestive from the contacts between Mohenjo-Daro and the Sumerians and Akkadians of Mesopotamia (Anonymous, 1962; Zohary, 1972).

Lentil maintained itself from the Bronze Age onwards, as an important companion of wheat and barley throughout the expanding realm of the Mediterranean type agriculture with the exception of the cold and humid northern fringe. In classical times, the large seeded *macrosperma* forms had already been attained and later the crop has changed but little. It was provider of major source of protein to the entire belt of Mediterranean agriculture, from northwestern India to Spain and North Africa. Centres of genetic diversity recognized are Afghanistan, Turkey and Ethiopia (Barulina, 1930).

8. Further Developments: Lentils have been but little changed by modern plant breeding. Variation has been structured in the form of numerous true-breeding lines aggregated in land-races endemic to restricted geographical areas. Works in lentils have centred on extensive collection and subsequent evaluation of existing varieties selecting stress on yield, seed-size and resistance to diseases. In several breeding stations, this introduction-and-evaluation work was supplemented by intervarietal crosses and selection and release of new varieties. Negi *et al.,* (2000) assessed and characterized 497 accessions for 30 qualitative and quantitative traits at regional station Bhowali (Nainital, Uttar Pradesh, India). They scored attributes such as number of branches, seeds size, number of cluster per plant, pigmentation and length of pod, seed and flower colour, number of day to flower and maturity, net yield per plant, leaflet size, tendril length, etc. Many varieties have been developed for different states of India under AICRP (All India Co-ordinated Pulse Improvement Project). Two types *viz.,* bold seeded and small seeded are recognized. The former is grown in central zone, whereas the latter in the Indo-Gangetic plains.

Sehgal *et al.* (2017) investigated the individual and combined effects of heat and drought stress during seed filling and determined the response of lentil genotypes with contrasting heat and drought sensitivity. They also assessed cross tolerance in contrasting genotypes. Both heat and drought, individually or in combination, damaged cell membranes, photosynthetic traits and water relations. The effects were more severe with the combined stress. Stomatal conductance increased with heat stress. Leaf and seed sucrose deceased with each stress in conjunction with its biosynthetic enzyme, while its (sucrose) hydrolysis increased under heat and drought stress, but was inhibited due to combination of stresses. Starch increased under heat stress in leaves but decreased in seeds. It drastically declined in seeds under drought alone or in combination with heat stress. At the same time, starch hydrolysis in leaves and seeds increased resulting in an accumulation of reducing sugars. Heat stress inhibited yield traits (seed number and seed weight per plant) more than drought stress, while drought stressed reduced individual seed, seed weights more than heat stress. The combined stress severely inhibited yield traits with less effect on the drought and heat tolerant genotypes. Drought stress inhibited the biochemical process of filling more than heat stress and the combined stress had a highly detrimental effect. A partial cross tolerance was a recorded in drougt and heat tolerant lentil genotypes against the two stresses.

9. Epilogue: Lentil is unparallel in its tastiness and high quality protein. It is attractive candidate for future development. Its success depends basically on the attainment of higher yields and on the development of varieties better adapted to mechanical harvest. Reduction of flatulence factors in the seed is another goal. The crop offers a considerable amount of variation yet unexploited by the breeder. Development of pure-line cultivars remains the safest and also promising breeding strategy. Utilization of hybrid varieties would depend on achieving drastic changes in reproductive biology.

5.10 MOTH BEAN

Vigna aconitifolia (Jacq.) Marechal (Leguminosae: Papilionatae)

[Syn. *Phaseolus aconitifolius* Jacq.]

Resume: *It is native to India. It grows wild in India and cultivated forms are also found in India. Chromosome number n=11 is reported for it. It is pulse crop and rich in protein and carbohydrates.*

1. Introduction: It is also called 'Mat Bean' especially in USA on account of its mat-like spreading habit. It is probably a native of India, Pakistan and Myanmar where it grows wild and appears to have been domesticated. It is being cultivated in Sri Lanka, China, Pakistan, Texas and California in USA, besides India. De Candolle (1886) remarked that the absence of a Sanskrit name, and the variety of names in modern Indian languages, points to a recent cultivation.

2. Plant Characteristics: Slender, trailing, pubescent annuals, 10-30 cm in length; leaves 3-foliate, terminal leaflet being larger and segmented into five acuminate lobes, lateral ones 3-lobed; flowers yellow, grouped on long axillary peduncles; pods small nearly cylindrical, beak curved, 4 to 9 seeded, seeds small nearly cylindrical or rectangular, yellow, brown or mottled black.

3. Parts Used: Seeds, pods, foliage.

4. Chemical Constituents: Protein (23%), fat (0.7%), carbohydrates (59.0%), fibre (4.0%), ash (4.0%), minerals (calcium, phosphorus, iron), vitamin A and C, carotene, thiamine, ciboflavin, niacin (Majumdar, 2011; Gopalan *et al.,* 1977).

5. Uses And Applications: In India, young tender pods are consumed as vegetable. Ripe seeds are eaten cooked either whole or split (dal). Whole grains, after frying with fat, is mixed with savoury dishes to make 'dal moth'. The seeds are boiled or parched and mixed with condiments to eat. Seed flour mixed with flours of other grains are used for making unleavened bread. In USA, it is grown for pasture, fodder and green manure.

6. Karyotaxonomic Evidence: *Vigna aconitifolia* (Moth Bean) has been revealed for karyo-morphological studies. It showed 2n=22 chromosomes without any evidence of numerical variation and the overall karyotype symmetry in chromosome morphology. The chromosome complement resolved into 11 pairs which formed a graded series from longest to shortest within the idiograms. The longest chromosome of the haploid complement was almost 2. 5 times longer than the shortest one in *V. aconitifolia.* It revealed metacentric, submetacentric and subtelocentric chromosomes (Shamurailatpam *et al.,* 2015).

7. History: It grows wild in India (De Candolle, 1886). Vavilov also noted it wild in India and also cultivated forms under Indian centre of origin. It is cultivated as a crop only in India. Wild forms also occur in Sri Lanka, Pakistan, Myanmar and Thailand (Piper and Morse, 1914). Thus, India is the original home of moth bean. This is recorded by Vavilov (1926).

8. Further Developments: Recently, it has been introduced in southern China, Burma, Malaysia and USA. Improved varieties have been released suitable to different states in India. Some promising varieties *viz.,* Baleshwar 12, G moth1, India, Jawel, Type 1 and No. 88 have been developed emphasizing characteristics such as period of maturity, seed weight and chocholate colour, resistance to yellow-mosaic virus, shining seeds, green fodder yield, etc. (Majumdar, 2011).

Yogesh *et al.,* (2016) investigated morphological diversity and genetic variability in moth bean representing 121 genotypes from different geographical regions. Analysis of variance showed the presence of significant variability among the genotypes. The estimates of PCV and GCV were high for plant height, primary branches per plant, secondary branches per plant, nodules dry weight, root nodules per plant, 100 seed weight, pods per plant and seed yield per plant. Genotypes exhibited wide range of values for all characters. In general, the PCV and GCV were quite high indicating that there is greater scope for selection for improvement of these characters. High heritability coupled with high genetic advance, in the opinion of authors, reveals the presence of lesser environmental influence and prevalence of additive gene actions in their expression. The authors opined that simple selection may not be effective in improvement of the traits studied.

Harsh *et al.,* (2016) studied effect of short-term heat stress in moth bean. They conducted experiment to find out the effect of short-term heat stress on osmo-protectant and antioxidants in 37 genotypes (32 mutants and 05 varieties). A significant over accumulation of total sugar and proline along with an increased activity of CAT, GPOX and SOD was observed in most of the genotypes under heat stress. However, none of the traits studied were associated with the level of thermo-tolerance except GPOX activity. The authors opined that thermo-tolerance and biochemical parameters can be efficiently altered and improved through mutagenesis.

9. Epilogue: The crop is the most drought resistant and grown in kharif season. It is largely grown on dry, light sandy soils in the arid and semi-arid regions. Its wild forms still grow in India. The crop still needs improvements taking into consideration the wild and as well as cultivars developed. The crop, in general, is more suitable for desertic condition. It is also a short duration crop and generally suitable as a mixed crop.

5.11 PEAS

Pisum sativum L. (Leguminosae-Papilionatae)

Resume: *Peas are native of Southern Europe. Excavations of Neolithic settlements (ca 7000 BC) in the Near East and in Europe recovered carbonised seeds. Remains of peas dating back to 7500-6500 BC have been found at Hacilar, Beidha and Jarmo. The peas were cultivated before the beginning of Christian era. Peas are diploid (2n=14). They are important source of protein for human consumption.*

1. Introduction: Peas are thought native of southern Europe. It has been cultivated prior to the beginning of Christian era. They were well known to the Romans and Greeks. They were consumed as a dry pulse. Green peas were not consumed until the 16th century. Until the middle of the 17th century, they were not grown in Europe although an old crop. The earliest colonists introduced peas into America. De Candolle (1886) opined that the Aryans introduced it into Europe. Perhaps, it existed in northern India before the arrival of Eastern Aryans.

2. Plant Characteristics: Glaucous, tendrillar climbing or trailing annuals; flowers white or coloured in axillary racemes; pods swollen or compressed, straight or curved;, seeds round, smooth or wrinkled.

3. Parts Used: Seeds, foliage.

4. Chemical Constituents: Protein (22.5%), fats (1.0%), carbohydrates (58.5%), fibre (4.4%), vitamins, minerals (calcium, iron, magnesium, manganese, zinc), vitamin-A, thiamine (B1), riboflavin (B2), niacin (B3), vitamin (B6), folate (B9), vitamin-C, vitamin-E, vitamin-K.

5. Uses And Applications: Seeds are used for human consumption in the form of pea meal, split peas, as green or dry. They are available canned, frozen and dehydrated. Pea soup is also a favourite dish. The seeds may be also useful for stockseed. Foliage can be used for cattle feed.

6. Karyotaxonomic Evidence: Although *Pisum* have been classified into a number of species, its taxonomic validity is now questioned. Ben-Ze'ev and Zohary (1973) intercrossed the cultivated *P. sativum* with *P. elatius*, *P. humile* and *P. fulvum*. They confirmed first three as anything but members of a single species, *viz.*, *P. sativum*. The genus presently includes two species., *viz.*, *P. fulvum* and *P. sativum* (Davis, 1970). All members of the genus are self-pollinating diploids (2n=14). All intercross freely. The F_1 seeds of the cross *P. fulvum* X *P. sativum* are shrunken. In the reciprocal cross, although the F_1 seeds are normal, the seedlings exhibit a stunting growth and other developmental abnormalities. Meiotic analyses of such F_1 hybrid plants suggest that the two species have diverged chromosomally. This is because quadrivalents, trivalents and univalents are observed at metaphase-I. *Pisum elatius* (and certain sub-populations of *P. humile*) differ from *P. sativum* by a chromosomal translocation (Ben-Ze'ev and Zohary, 1973). The so-called perennial pea, *P. formosum*, is now ascribed to a separate genus *viz.*, *Vavilov* (Davis, 1970). Where cultivated peas are found in close proximity to *P. humile* and *P. elatius*, hybrids are observed.

7. History : The wild progenitor or even the early history of the pea crop is unknown. Excavations of Neolithic settlements (ca. 7000 BC.) in the Near East and in Europe reported carbonized pea seeds. These had a smooth surface indicating that the cultivation of the pea is as old as that of wheat and barley (Zohary and Hopf, 1973). Remains of peas, dating as far back as 7500-6500 BC., have been also discovered at Hacilar, Beidha and Jarmo (Helbaek, 1966). Vivilov (1949) conceived the probable centres of origin to be Ethiopia, the Mediterranean and Central Asia, with a secondary centre of diversity in the Near East. No certainity as to which of these is the primary source and which are merely centres of diversity. *Pisum elatius* is a tall climber with long pods and small seeds. It is found in the more humid regions of the Mediterranean. *Pisum humile* has a smaller habit and similar to *P. sativum*. It is widely distributed in the Near East and occupies drier habitats than *P. elatius*.

Intergrading forms, both between these two groups, and between them and the cultivated forms, are however found. Modern cultivars differ from the wild forms in generally having larger seed and a shorter compact habit. It must be remembered that there are also both tall and small-seeded cultivars. Some primitive forms have slightly bitter seed and a tough seed coat which allows them to remain dormant in the soil for longer period. On the basis of morphological and cytological evidence, Zohar and Hopf (*loc. cit.*) suggest that those sub-populations of *P. humile* which show no chromosomal divergence from *P. sativum* could be the ancestral form from which the cultivars were evolved. It is called 'Kalanj' and 'Kalaya' in ancient Sanskrit literature, *viz.,* Kalpasutra and Patanjal's Mahabhasya respectively.

8. Further Developments: Almost certainly, the crop has been grown by man for several thousand years. Greek and Roman writers mentioned peas, but not until the 16th century were any varieties described. A distinction was first made between the field peas bearing coloured flowers, small pods and long vines, and the garden peas, usually with white flowers and large seeds. Later, the edible podded peas and tufted or Scottish peas were mentioned, followed rapidly by more detailed descriptions of a large number of varieties (Berger, 1928). In 1787, Knight began his controlled intercrossing of varieties, peas being the first crop in which controlled breeding was undertaken for the production of new varieties. He improved two varieties of peas. He could not explain the performance of what we know as 'dominant' and 'recessive' alleles (http://plantbreeding. coe. uga. edu/indore. php?title_2). In the 19th century, English breeders produced some classic pea varieties. These persisted in horticulture for so long.

Breeding objectives were constant for considerable decades. Particular varieties were selected for the fresh market, others for harvesting as dried peas and yet others for canning. For the last of these three uses, either dried peas are soaked or immature seeds are processed. For dried and canned peas, the colour and quality of the product are important. Good quality are being readily recognized by agronomists but difficult to define in biochemical terms. Round seeded forms have been traditionally used for dried peas; for canning or for seeds harvested in the immature state, either round or wrinkled forms are used. Winkled seeds have higher sugar and lower starch contents.

A new industry has developed in past few decades for the production of immature peas for freezing and this has imposed new demands on breeders. This crop was grown on a horticultural scale but now production has reached agricultural dimensions. The primary requirement is for peas of a maturity stage. Too young or too old are rejected. The criterion of maturity stage is strictly defined, harvesting being controlled to within a matter of hours. The demand is, therefore, for a crop in which as many peas as possible are at the correct developmental stage at the time of harvest.

9. Epilogue: Improved resistance to diseases caused by *Fusarium species, Ascochyta pisi* (pod spot) and *Peronospora viciae* (downy mildew), as well as to those caused by viruses, is of importance, particularly in more humid areas. Several single gene mutations are available in peas which affect the development of the leaf or stipule. One of these (af), when homozygous, converts the leaflets to tendrils. Such a phenotype could have agronomic advantages (Snoad and Davies, 1972). The reduced amounts of foliage could lead to a more rapid and easier processing of the crop for freezing and, particularly in wetter regions, to easier drying of the crop grown for seed. With the increasing cost and demand for protein, there is scope for re-examining the potential role of the pea crop as stock feed. This will involve developing new criteria in breeding and selecting forms with high protein and improved contents of methionine and cystine. Efforts to conserve genetic variability have been left rather late and, though collections of old cultivars and breeding material exist in many countries, few thorough collections of germplasm have been assembled in centres of origin or diversity (Gentry, 1971). These should be subjected to experimentation for further improvement of crop.

5.12 PIGEON PEA

Cajanus cajan (L.) Millsp. (Leguminosae-Papilionatae)

Resume: *Cajanus has a centre of origin in India and also centre of diversity of both Cajanus and closely related genus Atylosia in India. The crop is diploid (x=11). Archaeological findings from Neolithic sites dating 3400-3000 years ago lend support to its Indian origin. It is mainly used as pulse and staple food.*

1. Introduction: About 32 species of the genus Cajanus are distributed all over the tropics. De Candolle (1886) contended that the crop was introduced to the West Indies from the coasts of Africa by the slave trade. It was also taken to Brazil, Guiana and into the warm parts of the American continent by the slave trade (*Carney* and *Rosomofl*, 2009). De Candolle also (1886) came to know that missionaries introduced it to the Fiji Islands. He concluded that Africa was the centre of origin. However, the genus does not show the expected diversity in Africa. Engler (1915) found no evidence in favour of its African origin. Instead, Vavilov (1926) noted considerable variability of this crop plant in India. He concluded that *Cajanus* originated in India (Hindu Centre). Archaeological findings from Neolithic sites in Odisha (Gopalpur and Golbai) and South India (Sangarkullu and Tuljapur) dating back 3400-3000 years ago supports the Indian origin of the pigeon pea. Mahta and Dave (1931) found 36 types only from Madhya Pradesh state of India. Burkill (1953) contended that *Cajanus* was ennobled in India. Based on etymological study of Sanskrit name, it is believed that Cajanus was introduced in Western Malaya probably in the last centuries BC. by Indian traders. The crop was carried forward to Canton in China in the 6th century AD. British colonists added it in Australia. Ecklow and Zeyher (1836) did not mention *Cajanus* in their work on the African and Austrian flora. Murdock advocate a view that *Cajanus* was introduced in Africa through interchange of crop plants took place from western India across the Arabian desert by way of Arabean sea and the Sabaean lane. He further believed that *Cajanus* was introduced in this way in the 5th century BC.

2. Plant Characteristics: A woody, perennial shrub, much branched, 1-4 m tall; leaves 3-foliately compound, leaflets lanceolate; flowers yellow, borne in axillary or terminal racemes; pods 5-10 cm long, seeds brown-black to yellowish-red.

3. Parts Used: Seeds, tender pods, leaves.

4. Chemical Constituents: Protein (22.30%) vitamin B, fats (1.7%), carbohydrates (57.2%), vitamin B6, E&K, minerals (calcium, phosphorous, iron, magnesium, sodium, zinc, etc. (Gowda *et al.,* 2011), thiamine, riboflavin, niacin, flacin, vitamin-A, vitamin-C, vitamin-E.

5. Uses And Applications: In India, it is used as pulse and staple food. Tender pods are used as vegetable. Green leaves are excellent fodder for cattle. Dried stalks are used for baskets and thatching huts and partition walls. Seeds, pericarp and husk are good feed for poultry and livestock. In Ethiopia, pods, young shoots and leaves are consumed (Zemede, 1995). It is used as stew and porridge. It is used as either fresh, frozen or canned vegetable (Gowda *et al.,* 2011).

6. Karyotaxonomic Evidence: Although more than 30 species have been recognized for *Cajanus*. It has almost certainly one polymorphic species, *Cajanus cajan* (=*Cajanus indicus*). The basic chromosome number of *Cajanus* is $\times$=11. The crop is diploid but some tetraploids and hexaploids have been obtained. *Cajanus* has been crossed with the diploid Indian species, *Atylosia lineata, A. sericea* and *A. scarebaeoides* which have also the basic number $\times$=11. A comparison of somatic karyotypes has showed that several chromosomes are similar in the two genera. *Atylosia lineata*, the species tested is the nearest to *Cajanus* (*cf.* De, 1974). He noted that there are several other *Atylosia* species which deserve study in this connection. It is concluded that the crop originated from a yet unidentified species of *Atylosia* by selection for size and vigour of plant, for nonshattering pods and for larger seeds.

7. History: The centre of origin of the crop long been rather disputed (De, 1974). There is, so far, no archaeological evidence. References to *Cajanus* in northern Indian and Deccan literature of 1600 and 1400 BP. are cited and concluded that it was a well-known crop in India by the latter date. De (*loc. cit.)* favours an origin in South India, recalling the fact that India is the present centre of diversity for both *Cajanus* and of the closely related genus *Atylosia*.

However, Purseglove (1968) and Herklots (1972) cited earlier work by Burkill (1993) as mentioning the presence of *Cajanus* seed in an Egyptian tomb of the XII Dynasty (earlier than 4000 BP.). The existence of wild varieties in Africa is not sufficient to accept De Candolle's support of Africa as the centre of origin since 'wild' varieties exist in the Caribbean into which all authors agreed that the crop was introduced in post-Columbian times. Furthermore, *Atylosia* is a predominantly Indian genus, rare and poorly represented in Africa.

The genus *Cajanus* is mentioned in ancient Sanskrit literature. It is referred in Sanskrit in a lexicon and in a medical text around 1400 BP. as Tuvari or Tuvarica or Adhaki. Two main plant types are recognised today, Tur (flavus) and Arhar (bicolour) have Indian names originated from the Sanskrit words. This indicates that both types were developed at an early stage of evolution of the crop. If one accepts an Indian origin, it is probable (De, 1974) that the crop reached to Malaysia from India around 2000 BP. on the basis of the Sanskrit names there. Later, it was taken to China (about 1500 BP.) and quite likely reached Australia through the East Indies. The generic name *Cajanus* is derived from the local Malaysian name Katjang. Between 1500 and 1000 BP. *Cajanus* was carried to East Africa, probably in the region of Zanzibar. From there it moved northwards to the Nile valley and to West Africa. The crop probably reached to the New World from Zaire or Angola by the Portuguese in early post-Columbian days, prior to the main African slave trade. Menezes (1944) suggested that 'Guando' or 'Guandu' the Brazilian name for the crop, is a Portuguese transformation of *Cajanus* by Africans in Brazil. This name has spread to Caribbean Latin America as 'Guandole' in the south, 'Gandole' in Puerto Rico, the Dominican Republic and Cuba, and 'Gungo' in Jamaica. The New Plantation colonies of the eastern Caribbean islands refer to *Cajanus* as 'Pois Angola' and 'Pigeon Pea' while Congo pea in Jamaica is a false rationalization of Gungo. The crop is now called 'Pois de Congo' and 'Ervilha do Conga' in French and Portuguese-speaking non-equatorial Africa.

8. Further Developments: The crop exhibits a great diversity and wide adaptation in India. This has arisen partly from unsophisticated selection by farmers and partly from the occurrence of fairly frequent cross-pollination. *Cajanus* is often crossed by insects. Within India, there appears some differentiation between the peninsular Tur varieties (with small early maturing plants, yellow flowers and few light-coloured seeds) and the northern Arhar types (larger, late maturing plants, variegated flowers and more numerous, darker seeds). There is also much intergradation. The two are fully interfertile and the validity of the distinction appears partly doubtful. The two types after representing the relics of two geographically independent domestications.

Work at the Indian Agricultural Research Institute (Deshpande *et al.*, 1963) was centred around combining resistance to *Fusarium* wilt with high seed yields using material from the New World and obtained useful lines. Breeding has continued and has been gathered momentum in the last decades as attention had directed gradually on protein shortage.

Its potential as a forage led Kraus (1926, 1927) to breed it for this purpose in Hawaii as early as 1922. He recorded the creation of a high yielding variety, New Era. The collection and selection of forage types was carried on later in northern Queensland. The objectives of such programmes were types which are indeterminate in growth habit, late-maturing and perennial. This provided browsing on arid land all the year round,

Gooding (1962) conducted a programme of collection, hybridization and selection aimed at a change in the nature of the crop that would allow economic green pea production. This required varieties which showed earliness and determinacy, with concomitant capacity for multiple cropping, dwarfness and suitable pod characteristics such as large seeds, many seeds per pod and good seed pod weight ratios. This resulted in varieties acceptable both for the production of green peas for direct consumption and for canning.

Odeny and Gebhardt (2009) studied molecular breeding of Pigeon pea. They demonstrated that Pigeon pea 'Sample Sequence Repeats (SSRs) (also known as microsatellites) constitute efficient sources of molecular markers for cultivated pigeon pea genotypes. According to them, these markers will be beneficial, in future, for marked assisted breeding and comparative generous analysis between the different *Cajanus* species for more efficient exploitation of the desirable characteristics therein.

Adebooye (2009) investigated reducing the oligosaccharides and anti-nutritional factors in Pigeon pea. This author showed that decortication could play a role in substantial reduction of the cooking time of *Cajanus cajan*. He also opined that this will be of immense economic benefit as it reduces the cost of fuel or fire wood. The brown colouration of the cooking water and bitter taste were also eliminated by decortication followed by cooking. He further stated that this technique of reducing the anti-nutritional factors and bitter taste should be extended to sensitize the consumers.

In recent times, the major objectives for cultivation of this crop are: (a) early podding, (b) round-the-year flowering and fruiting, (c) dual purpose varieties, (d) high multiple-harvest pod yield, (e) long green pods with fully grown ovules, (f) large attractive seeds with good taste, (g) long shelf-life of pods and seeds, etc. ICRISAT Gene Bank has 13548 germplasm accessions from 76 countries (Gowda *et al.*, 2011).

9. Epilogue: Using this germplasm, further investigations are desired. The rate of increase in the breeding and other work carried out on *Cajanus* is striking. The massive improvement needed to transform the species into a highly developed crop is appreciably attained. The necessary variability is employed with conscious preservation. Variation existing in *Atylosia* is also studied for its possible use. A much improved knowledge of *Alylosia* is made available by recent studies. The breeding of pure lines produced grain, green pea and fodder from plants that are amenable to mechanization and respond to high levels of husbandry in large fields.

References:

Adebooye, O.C. (2009) Reducing the Oligosaccharides and anti-nutritional factors contents of two underutilized grain legumes of southwest Nigeria. Proceedings of The International Symposium (Ed. Jaenicke *et al.,*) *Acta Hort.* 806 ISHS pp. 355-358.

Afzal, M.A., Haque, M.M. and S. Shanmusasundavam (2004) Randam amplied polymorphic DNA (RAPD) of selected mungbean [Vigna radiata (L.) Wilczek] cultivars. *Asian J. Plant Sci.* 3(1):20-24.

Alexander, J. and D.G. Coursey (1969) The origins of yam cultivation. *In:* (Ed.) Ucko, P. J. and Dimbleby, G. W., The domestication and exploitation of plants and animals. Aldine Publishing Co., Chicago, USA. pp. 405-425.

Amadou, H.I., Bebeli, P.J. and P.J. Kaltsikes (2001) Genetic diversity in bambara groundnut [Vigna subterranean (L.) Verdc.] germplasm revealed by RAPD markers. *Gemome* 44:995-999.

Anonymous (1948) The Wealth of India: Raw Materials. CSIR, New Delhi, India.

Anonymous (1962) The Wealth of India: Raw Materials. Vol. VI:L-M. CSIR, New Delhi, India.

Anonymous (1976) The Wealth of India: Raw Materials. Vol. X:SP-W. CSIR, New Delhi, India.

Arslan, M. (2017) Fatty acid characteristics of grass pea (Lathyrus sativus) in an East Mediterraneal environment. *Cogent Chemistry* 3C176, 1296748. httpps:/doi.org/10.1080/ 23312009.2017.1296748.

Asmussent, C.B. and A. Liston (1998) Chloroplast DNA characters, phylogeny and classification of Lathyrus (Fabaceae) *American Journal of Botany* 85:397-401.

Badr, A., El. Shazly, H., E.I. Rabey and L.E. Watson (2002) Systematic relationships in Lathyrus section Lathyrus (Fabaceae) based on AFLP data. *Canadian Journal of Botany* 80:962-969.

Barulina, E.I. (1930) [Lentils of USSR and other countries; a botanico-agronornical monograph.] *Trudy prikl. Bot. Genet. Selek. Supplem.* 40 265-319.

Ben-ze'ev, N. and D. Zohary (1973) Species relationships in the genus Pisum. *Israel. J. Bot.* 22:73-91.

Berger, A. (1928) Peas. *In:* Hendrick, Hall, Hawthorn and Berger (Eds.), Vegetables of New York 1:1-132.

Bhat, K.V., Lakhanpal, S. and S. Chandha (2005) Amplified fragment length polymorphism (AFLP) analysis of genetic diversity in Indian mungbean [Vigna radiata (L.) Wilczek] cultivars. *Indian J. Biotechnol.* 4(1) 54-64.

Bisht, I.S., Mahajan, R.K. and T.G. Kawalkar (1998) Diversity in green gram [Vigna radiata (L.) Wilczek] germplasm collection and its potential use in crop improvement. *Ann. Appl. Biol.* 132:301-312.

Burkhill, I.H. (1953) Habits of man and the origins of the cultivated plants of the Old World *Proc. Linn. Soc. London* 164:12-42.

Carney, J.A. and R.N. Rosamoff (2009) In the shadow of Slavery. Africa's Botanical Legacy In The Atlantic World. University of California Press, Berkeley, California, USA.

Chevalier, A. (1944) Le dolique de Chine en Afrique. *Rev. Bal. appl. Agric. trop.* 24:128-151.

Chowdhury, K.A., Sarswat, K.S., Hasan, S.N. and R.G. Gaur (1970) 4000-3500 years old barley, rye and pulses from Atranjikhera. *Sci, and Cult.* 37:531-533.

Darlington, C.D. and A.P. Wylie (1955) Chromosome Atlas Of Flowering Plants. Go. Allen and Unwin Ltd., London, UK.

Das, Swati, Surya Shekhar Das and Parthadeb Ghosh (2014) Assent of molecular genetic diversity in some green gram cultivars as revealed by ISSR analysis. *Advances in Applied Science Research* 5(2):93-97.

Davis, P.H. (1970) Flora of Turkey, 3, 370-373. Edinburgh, Scotland.

De Candolle, A.P. (1886) Origin of cultivated plants (Translated from 2nd Ed., in French 1959), Hafner, New York, USA.

De, D.N. (1974) Pigeon pea. *In:* J.B., Hutchison (Ed.), Evolutionary Studies In World Crops, Cambridge pp. 79-87.

Deshpande, R.B., Jeswani, L. M. and A. B. Joshi (1963) Breeding of wilt resistant varieties of pigeon pea. *Ind. J. Genet.* 23:58-63.

Doggett, H. (1970) Sorghum. Longmans, London, UK.

Doku, E.V. and S.K. Karikari (1971) Bambara groundnut. *Econ. Bot.* 25:255-262.

Ecklon, C.F. and C. Zeyher (1834-1837) Enumeratio plantarum Africae australis extra-tropicae. 3. Hamburgi. 3 Fasc. 8vo.

Emre, I. (2009) Electrophoretic analysis of some Lathyrus L. species based on seed storage proteins. *Genetic Resources And Crop Evolution* 56:31-38.

Enneking, D. (2011) The nutritive value of grasspea (Lathyrus sativus) and allied species, their toxicity to animals and the role of malnutrition in neurolathyrism. *Food And Chemical Toxicology* 49:694-709.

Evans, A.M. (1973) Genetic improvement of Phaseolus vulgaris. *In:* Nutritional improvement of food legumes by breeding. P.A.G. Symposium, F.A.O. Rome (1972) 107-115.

Faris, D.G. (1965) The origin and evolution of the cultivated forms of Vigna sinensis. *Can. J. Genet. Cytol.* 7:433-452.

Fasasi, O.S., Oyebode, E.T. and D. Olusanya (2009) Evaluation into the proximate and functional properties of modified Bambara [Vigna subterranean (L.) Vardc.] isolate 'Ogi' mix. Proceedings of the International Symposium (Ed. Jenicke *et al.,*) *Acta Hort.* 806 ISHS, pp. 301-307.

Fuller, D.Q. and E.L. Harvey (2006) The archaeobotany of Indian pulses: identification, processings and evidence for cultivation. Environ. Archaeol. 11:241-268.

Gentry, H.S. (1971) Pisum resources, a preliminary survey. Plant Genet. *Resources Nemsl*. 25:3-13.

Gooding, H.J. (1962) The agronomic aspects of pigeon peas. *Field Crop Abstr*. 15:1-5.

Gopalan, C., Ramsastri, B.V. and S.C. Subramanian (1977) Nutritive value of Indian foods (Compiled by R. S. Saini) ICAR, New Delhi, India

Gowda, C.L.L., Srivastava, R.K., Saxena, K.B. and R.V. Kumar (2011) Vegetable Pigeonpea [*Cajanus* cajan (L.) Millsp.] *In:* Future Crops (Ed. K. V. Peter). Daya Publishing House, Hew Delhi, India, pp. 260-276.

Grubben, G.J.H. (1977) Tropical Vegetables And Their Genetic Resources. International Board For Plant Genetic Resources, Rome, Italy.

Gyan Asah, G., Adu-Dapaah, H.Kumaga, F., Gracen, V. and F.K.Nartey (2016) Evaluation of cowpea [Vigna unguiculata (L.) Walp.] genotyes for phosphorus use efficiency. *Acta Hortic*. 2016. 1127.58.

http://doi.org/10.17660/ActaHortic.2016.1127.58.

Haider, Sandip, A.K. Datta, A. Mandal and B.K. Ghosh (2012) *Macrotyloma uniflorum* (Lam.) Verdc. (Leguminosae) – A note on chromosomal studies. *Cytologia* 77(4):447-451.

Harsh, A., Sharma, Y.K., Joshi, U., Rampuria, S., Sinigh, G., Kumar S. and R. Sharma (2016) Effects of short-term heat stress on total sugars, proline and some antioxidant enzymes in moth bean (Vigna aconitifolia). *Annals of Agicultural Sciences* 61(1):57-64.

Hawkins, C.F., and A.M. Evans (1973) Elucidating the behaviour of pollen tubes in intra- and interspecific pollinations of Phaseolus vulgaris and P. coccineus, *Euphytica* 22:378-385.

Helbaek, H. (1965) Isin-Larson and Horian food remains at Tell Bazmosian in the Dokan Valley Summer (Baghadad) 19:27-35.

Helbaek, H. (1966) Pre-pottery Neolithic farming of Beidha. *Palest. Explor. Q*. 98:61-66.

Hepper, F.N. (1970) Bambara groundnut (Voandzeia subterraneu) *Field Crop Abstr.* 23:1-6.

Herklots, G.A.C. (1972) Vegetables in Southeast Asia. Allen and Unwin, London, UK. pp. 232-233.

History of Plant Breeding-Plant Breeding. https://plantbreeding. coe. uga. edu/index. php? title:2

Hutchinson, J.B.J. 970) The evolutionary diversity of the pulses, *Proc. Nutr. Soc.* 29:49-55.

Jackson, M.T. and A.G. Yunus (1984) Variation in the grasspea (Lathyrus sativus L.) and wild species. *Euphytica* 33:349-559.

Jain, H.K. and K.L. Mehra (1980) Evolution adaptation relationships and uses of the species of Vigna cultivated in India. *In:* (Ed. Summerfield R.J.) Advances In Legume Science, Royal Botanic Garden, Kew, Richmond Survey. Pp. 459-468.

Kaplan, L. (1965) Archaeology and domestication in American Phaseolus beans. *Econ. Bot.* 19:358-368.

Kaplan, L., Lynch, T.F. and C.E. Smith (1973) Early cultivated beans iPbaseolus vulgaris) from an intermontane Peruvian valley. *Science*, N.Y. 179:76-77.

Karimzadeh Ghasem, Maryam Danesh-Gilevaei and Majid Aghaalikhani (2011) Karyotype and nuclear DNA variations in *Lathyrus sativus* (Fabaceae) *Caryologia* 64(1):42-54.

Karuppanapandian, T., Karuppudurai, T., Sinha, P.B., Maniya, K.A. and K. Manoharan (2006) Genetic diversity in green gram [Vigna radiata (L.). Wilczek] landraces analyzed by using random amplified polymorphic DNA (RAPD). *Afri. J. Biotechnol.* 5(13):1214-1219.

Kraus, F.G. (1926) Genetic analysis of *Cajanus* indicus and the creation of improved varieties through hybridization and selection. Proc. Hatoauan Acad. Sci., B. P. Bihshop Mus. Spec. Publ., 11:24-25.

Kraus, F.G. (1927) Improvement of pigeon pea. *J. Hered.* 18:227-232.

Kulkarni, G.B. and U.P. Mogle (2011) Characterization and evaluation of horse gram [*Macrotyloma uniflorum* (Lam.) Verdcourt] genotypes. *Science Research Reporter* 1(3):122-125.

Lakhanpaul, S., Bhat, K.V., S. Chadha (2000) Random amplified polymorphic DNA analysis in Indian mungbean [Vigna radiata (L.). Wilczek] Cultivars. *Genet.* 109:27-234.

Lamichaney, A., Katiyar, P.K., Vijay Laxmi and A.pratap (2017) Variation in pre-harvest sprouting tolerance and fresh seed germination in *mungbean* (Vigna radiata L.) genotypes. *Plant Genetic Resources*,

1-9.doi:1001017/S147926211000096.

Latha, M., Suma, A., Asha, K.I., Dwivedi, N.K., Mani, S. and A. Indiradevi (2013) Seed polymorphism in south Indian horsegram [*Macrotyloma uniflorum* (Lam.) Verdc.]: A comparative study. *Jour. Appl. Biol. & Biotechn.* 1(4):1-6.

Leakey, C.L.A. (1971) The improvement of beans (Phcseolus vulgaris) in East Africa. *In:* Crop improvement In East Africa. C. A. B. Farnham Royal pp. 99-128.

Linnemann, A.R., Westphal, E. and M. Wessel (1995) Photoperiod regulation of development and growth in Bambara groundnut (Vigna subterranea). *Field Crops Research* 40(1):37-47.

Mackie, W.W. (1943) Origin, dispersal and variability of the Lima bean, Phaseolus lunatus. *Hilgardia* 15:1-29.

Maesen, L.J.G. van der (1972) Cicer: A monograph on the genus with special reference to the chickpea (*Cicer arietinum*), its ecology and cultivation. Wageningen, The Netherlands.

Mahto, D.N. and B.B. Dave (1931) Studies in *Cajanus* indicus. Memoirs of The Department of Agriculture in India (Botanical Series) 19:1-25.

Majumdar, D.K. (2011) Pulse Crop Production: Principles And, Technologies. RHZ Learning Private Ltd., New Delhi, India.

Manivannam, N., Murugan, E., Vishwanathan, P.L., C.V. Dhankodi (1998) Genetic divergence in green gram. *Legume Res.* 21:131-133.

Mayes, S., Massive, F., Beena, A., Sheshshayee, M. S., Stadler, F., Mohler, V., Wenzel, C., Kilian, A. and A. Bonin (2009) BAMLINK-A cross disciplinary programme to enhance the role of Bambara groundnut [Vigna subterranean (L.) Verdc.] for food security in Africa and India. Proceedings of The International Symposium (Ed. Jaenicke *et al., Acta Hort* 806 ISHS, pp. 137-149.

Menezes, O.B. de (1944) Estudes Para a genetica do guandu. *Rev. de Agric. Pircicaba,* V. 19, n. 1-2, p. 51-64.

Miranda, C.S. (1974) Evolutionary genetics of wild and cultivated Phaseolus vulgaris and P. coccineus. Ph. D. thesis, University of Cambridge, Cambridge, UK.

Miranda, C.S. and A. M. Evans (1973) Exploring the genetical isolating mechanisms between Phaseolus vulgaris and P. coccineus. *Ann. Rep. Bean Improv.* Coop. 16:39-42.

Negi, K.S., Muneem, K.C. and V.D. Varma (2000) Preliminary evaluation and utilization of Lentin genetic resources. *Journ. Non-Timber For. Products* 7(1/2):89-97.

Odeney, D.A. and C. Gebhardf (2009) Towards molecular breeding of pigeonpea [Cajanus cajan (L.) Millsp.]: A case for increased production in Africa. Procee

Odongo, F.O., Oyoo, M.E., Wasike, V., Owuoche J.O. Karanja, L. and P.Korir (2015) Genetic diversity of Bambara groundnut [Vigna subteranea (L.) Verdc.] landraces in Kenya using microsatellite markers. *African Journal of Biotechnology* 14(4):283-291.

Piper, C.W. and W.J. Morse (1914) Five Oriental Species of Beans. U.S., Department of Agriculture, Bulletin No. 119.

Posquel, R.S., Schwedes, S. and P. Gepts (1999) Isozyme diversity in Bambara groundnut. *Crop. Sci.* 39:1228-1236.

Prain, D. (1903) Bengal Plants. Calcutta, India.

Prakken, R. (1970) Inheritance of colour in Phaseolus vulgaris: A critical review. Mededeelingen Landbouwhogeschool, Wageningen 70(23):1-38.

Prasad Saroj Kumar and Manoj Kumar Singh (2015) Horse gram-an underutilized neutracentrical pulse crop: A review. *J. Food Sci. Technol.* 52(1):2489-2499.

Purseglove, J.W. (1968) Cajanus. In Tropical crops, Dicotyledons. London-I, pp. 236-241. eniya kulturnkh rasteni. [The Centres of Origin of Cultivated Plants]. Works of Applied Botany And Plant Breeding 16(2):1-248p. [Russian, English].

Purseglove, J.W. (1968) Tropical Crops: Dicotyledons I. Longman, London, UK. pp. 284-310.

Rahman, S., Molla, M.R., Nazim-ul-Dowla, M.A.N., Rasul, M.G. and M.A.K. Mian (2010) Assessment of genetic diversity in grasspea (Lathyrus sativus L.) genotypes using morphological markers. *Academic Journal of Plant Sciences* 3(2):85-91.

Ramanjam, S. (1972) Some salient results of pulse research. *Indian Frng.* 21:17-18.

Rao, S.L.N. (2011) A look at the brighter facets of β–N-oxalyl-L-α, β-diamonopropionic acid, homoargininie and the grasspea. *Food And Chemical Toxicology* 49:620-622.

Renfrew, J.M. (1966) A report on recent finds of carbohized cereal grains and seeds from prehistoric Thessaly. *Thessalika* 5:21-36.

Rohewal, S.S., Ramanujam, S. and K.L. Mehra (1969) Plant type in bengal gram. *Ind. J. Genet.* 26:255-61.

Sachedeva, S., Bharadwaj, C. Sharma, V., Patil, B.S., Soren, K.R., Roorkiwal, M., Varsheney, R., and K.V.Bhat (2018) Molecular and phenotypic diversity among chickpea (Cicer arientinum) genotypes as a function of drought tolerance. *Crop And Pasture Science* 69(2):142-153.

Sangivi, C., Kaga, A., Tomok, N., Vaugan, D.A., and P. Srinivas (2007) Genetic diversity of the mungbean (Gigna radiate: Legumonosae) gene pool on the basis of microsatellite analysis. *Aust. J. Bot.* 55:837-847.

Saunders, A. (1960) Inheritance in the cowpea (Vigna sinensis) II. *S. Afr. J. agric. Sci.* 3:141-162.

Sehgal, A., Sito, K., Kumar, J., Kumar, S., Singh, S., Siddique K.H.M. and H. Nayyar (2017) Effect of drough, heat and their interaction on the growth, yield and photosynthetic function of lentil (Lens culinaris Medikus) genotypes varying in heat and drought sensitivity. *Front Plant Sci.* 2017 Oct. 17.8.1776. doi10.3389/fpls.2017.01776.eCollection2017.

Shamurailatpam Anju, Latha Madhavan, Shrirang Ramchandra Yadav, Kangila Venkatraman Bhat and Satywada Rama Rao (2015) Heterochromatin distribution and comparative karyo-morphological studies in Vigna umbellate Thunberg. 1969 and V. aconitifolia Jacquin, 1969 (Fabaeae) accessions. *Cytogenetics* 9(1): 119-132.

Sharifi, P., Astercki, H. and M.Pouresmael (2018) Evaluation of variations in Chickpea (Cicer arientinum L.) yield and yield components by multivariate technique. *Annals of Agrarian Science* 16(2):136-142.

Sharma, Vikas, Tilak Raj Sharma, Jai Chand Rana and Rakesh Kumar Chahota (2015) Analysis of genetic diversity and population structure in horse gram (*Macrotyloma uniflorum*) using RAPD and ISSR markers. *Agricultural Research* 4(3):221-230.

Singh, Ajay (2008) Plants In Ancient Indian Civilizations. Agam Kala Prakashan, Delhi, India.

Singrun, C. and W. Schenkel (2004) Fingerprinting of bombara groundnut germplas with molecular markers. *In:* (Ed. Massawe, F.). Proceedings of the International Symposium on Bambara Groundnut. Botswana College of Agriculture. September 8-12, 2003.

Smartt, J. (1969) Evolution of American Phaseolus beans under domestication. *In:* P.J. Ucko and G.W. Dimbleby (eds.). The domestication and exploration of plants and animals. London, pp. 451-462.

Snoad, B. and D.R. Davies (1972) Breeding peas without 'leaves. *Spatz* 15: 87-89.

Su, H.X. Tian, J. and Tommoka, N., Kaga, A., Isemura, T. and D.A. Vaugan (2008) Genetic diversity of the azuki bean [Vigna angularis (Willd.) Ohwi & Ohashi] gene pool as assessed by SSR markers. *Genome* 51:728-738.

Summerfield, R.J., Huxley, P.A. and W.M. Steele (1974) Cow peas, Vigna unguiculata: A Review. *Field Crop Abstr.* 27:301-312.

Swaminathan, M.S. (1972) Basic research needed for further improvement of pulse crops in Southeast Asia. *In:* (Ed.) M. Milner, Nutritional Improvement of Food Legumes By Breeding. Protein Advisory Group, U. N., New York, USA .

Swamy, K.R.M. and A.T. Sadashiva (2007) Tropical vegetable crops. *In:* Biodiversity In Horticultural Crops. Vol. I (Ed. Peter K.V. and Dr. Z. Abraham) Daya Publishing House, New Delhi, India. pp. 109-183.

Thouseem, N., Beena Tomas and Namitha Elizabeth (2018) Genetic parameters in seed yield components of cowpea [Vigna unguiculata (L.) Walp.] *Int .J. Curr. Microbiol. App .Sci.* 7(2):2268-2274.

Tian, J., Isemura, T., Kaga, A., Vaugan. D.A. and N. Tomooka (2013) Genetic diversity of the rice bean (Vigna umbelleta) gene pool as assessed by SSR markers. Genome 56:717-727.

Tomooka, N. Vaugan, D. A., Moss, H. and T. Maxted (2002) The Asian Vigna: genus Vigna subgenus Ceratotropsis genetic resources. Kluwer Academic Publishers, The Netherlands, pp. 270.

Vavilov, N.I. (1926) Studies on The Origin of Cultivated *Plants. Trudy ByuroPrikl.* 16:139-248.

Vavilov, N.I. (1949) The origin, variation, immunity and breeding of cultivated plants. *Chron. Bot.* 13:1-54.

Verdcourt, B. (1970) Studies on the Leguminosae-Papilionoideae for the Flora of East Tropical Africa IV. *Kew Bull.* 24:507-569.

Vishnu Mittre (1968) Protohistoric records of Agriculture in India. Trans. Bose Res. Inst. Calcutta. 31(3):87-106.

Vishnu-Mittre (1968) Protohistoric records of Agriculture in India. *Trans. Bose Res. Inst. Calcutta* 31(3):87-106.

Vishnu-Mittre (1974) The beginnings of agriculture: Palaeobotanical evidence in India. *In:* Evolutionary studies in World Crops. (Ed. Joseph Hutchinson) Cambridge University Press, London, UK. pp. 3-30.

Vyas, D., Joshi, A., Rajamani, G. and Jain, D. (2018) Assessment of genetic diversity in black gram [Vigna mungo (L.) Hepper] genotypes based on ISR. *Legume Research* 41(2):175-181.

Westphal, E. (1974) Pulses ill Ethiopia, their taxonomy and agricultural significance. Phaseolus. *Agric. Res. Rep. Wageningen* 815:129-176.

Yogesh, L.N., Vishwanatha, K.P., Madhusudhan, R. and S. Gangaprasad (2016) Morphological diversity and genetic variability in moth bean. *In. J. Sci. & Env. Techno.* 5(4):1912-1924.

Zemede, Asafaw (1995) Conservation and use of traditional vegetables in Ethiopia. (http://ww. Biodiversityinternational. org/publication/ Webversion/500/ch08. htm) Proceedings of The IPGRI Inter-national Workshop on Genetic Resources of Traditional Vegetables in Africa (Nairobi, 29-31 August, 1995.

Zohary, D. (1972) The wild progenitor and the place of origin of the cultivated lentil: Lens culinaris. *Econ. Bot.* 26:326-32.

Zohary, D. and M. Hopf (1973) Domestication of pulses in the Old World. Science, N. Y. 182:887-894.

Zukovskii, P.M. (1962) Cultivated Plants And Their Wild Relatives. Commonwealth Agriculture, Bureaux, London, UK.

6

Superfood Grains

INTRODUCTION

The current world population is presently over six billion people. Many are malnourished and shackled by poverty. Millions of people do not meet their daily required energy needs from their diets. The inability to secure sufficient and quality food in many African and Asian countries causes deaths daily because of malnutrition. They suffer chronically from the consequences of poor nutrition and deficiencies of some minerals and vitamins. The global requirement for protein and carbohydrates is mostly provided by only three food grains *viz.*, rice, maize and wheat (Jaenicke and Hochle-Zeledon, 2006). Moreover, the worldwide wave of 'Green Evolution' concentrated and promoted only these crops which resulted in inadequate nutrient-rich compounds. Due to inadequate levels of micronutrients available in diets, particularly in developing countries, many people may be unknowingly suffering from 'malnutrition' even if there is adequate food supply to meet their daily per capita energy requirements. They may not be 'hungry' but they are 'malnourished'. This is malnutrition in form of 'hidden hunger'. The lack of awareness of the existence and the effect of this kind of malnourishment on the body invariably leads to a host of health problems and premature deaths.

Information on 'useful' plants is stored in inaccessible literature to the user. Although much information is available in the international as well as 'grey' literature, very few countries have a complete overview of their 'useful' biodiversity and of the state of knowledge on these plants. Moreover, modernization of agriculture, horticulture and forestry are leading to loss of 'useful' plant species. As stated earlier again 'Green Evolution' carried the danger of marginalization of other 'useful' and underutilized biodiversity while focusing only few major food crops.

The 'superfood grains' of modern period are actually ancient, orphaned or forgotten food sources of primitive human societies worldover. Consuming superfood grains now means a 'down-to-earth' approach to the history of ancient human nutrition. They are now regaining their popularity and being used in new forms of recipes. They are called super grains because of their great flavour, taste, nutritional profile, high antioxidants, complete protein, energy-rich, gluten-free, evidenced-based health benefits, addition to nutrients, increase in existing nutrients, low glucemix index, vital trace elements, decrease in risk of diseases, etc. The superfood grains chia, teff, kamut, black barley and brown rice are dealt separately in this chapter for their very important and vital role in human health and nutrition. Few others such as amaranth, quinoa and buckwheat are classed in another chapter of pseudo-cereals.

6.1 BLACK BARLEY

Hordeum vulgare L. var. *nudum* Hook. *f.* (Poaceae)

Resume: *Barley is an ancient and one of the first cultivated crop. Black barley is a variant with hulless or naked black caryopsis. Both of these are found under cultivation but the latter has been neglected because of its comparatively lower yield. Both of these are distributed in similar localities in the world. Black barley is fairly rich in its nutrient contents including unique β-glucan. It is being recommended wordwide now as a healthy food source. Archaeobotanical evidence supports its cultivation since ancient times along with classic hulled barley. Chromosome number in both hulled and hulless (naked) barleys are similar 2n=2x=14. Interestingly, apart from traditional dietary uses, it has potential for finding place in processed super foods.*

1. Introduction: Barley, in general, was one of the first cultivated grains originated in the Nile River valley in ancient times. Black barley is also called Ethiopian Black Barley, African Black Barley or Naked Barley, Black Nile Barley or Hulless Barley. Basically, there re two types of barley *viz.,* hulled and hulless. The former must be rubbed off using a process called 'pearlizing', whereas in case of the latter, the hull comes off easily leaving the whole grain kernel intact. No bran is removed. In pre-historic times, barley grew among the wild grasses. The name black barley is derived from the black colour of its edible gloom (bran) layer, which stays attached to the kernel. Nomads from the Near East gathered the barley (and einkorn) long before the agrarian agriculture took hold.

De Candolle (1886) was the first to pinpoint the problem regarding the progenitor of barley. Two hypothesis are forwarded to explain whether two-rowed or six-rowed is primitive. It is certain that the progenitor of the genus *Hordeum* had a panicle-like inflorescence. The triplet being transformed into a disaspore, the formation of female lateral flowers must be suppressed in order to achieve a functional diaspora dispersal. In case of *Hordeum vulgare*, it can be concluded that the selection pressure suppressing the occurrence of female lateral flower as drastically changed when barley was brought into cultivation and the triplet shattered during harvest, storage and sowing. Thus an original form of *H.vulgare* would be a two-rowed type. The system of classifying into two-rowed and six-rowed is artificial since the latter would have been developed due to mutations. The latter should be regarded secondary development. There is still some disagreement as to which rank should be given to the various taxa. The subspecies and varieties of *H.vulgare* are identical based on features of rachis, spike (lax, 2, 4 or 6-stichous) awns, caryopsis (naked or not naked), etc. Cultivated *H. vulgare* is the most variable species within the genus, variations encountered are phenotypic due to local climatic or stress factors (*cf.* Aberg, 1948; Kamm, 1954; Takahashi, 1955; De Candolle, 1886; Navski, 1941; Schultze-Motel, 1986; Harlan, 1968, etc.)

2. Plant Characteristics: Plants loose tufted up to 1 m tall, tillers many; culms erect or geniculate; leaves flat scabrid or glabrous, linear; spikes green to purplish, nodal, generally six-rowed, central spikelet sessile, awned, kernel, naked, black.

3. Useful Parts: Grains.

4. Chemical Constituents : Grains: Starch (62-63.3%), protein (11-13.9%), lipids (2.1-2.35%), TDF (18.74-19.47%), SDF (1.57-3.18%), β-glucan (3.61-3.81%), minerals (molybdenum, manganese, selenium, copper, phosphorus, magnesium), serine, glutamic acid, proline, glycine alanine, valine, isoleucine, leucine, tyrosine, phenylalamine and arginine, lysine, histidine, proline, tyrosime, asparagine, etc.

5. Uses And Applications: Apart from uses *vis-à-vis* application barley grains, in general, black barley is being used for some other purposes. It holds good for banquets or in steam tables/buffets. It is incredible for soups and stuffing. Its shiny, black appearance adds to pilafs and salads. It is can be enjoyed by all vegetarians and cuisines of Mediterranean, Africans and Americans. It is slightly sweet, deliciously nutty flavour and chewy nature. It holds up well in heating and freezing without dehydration. Its dietary fibre is high in beta-glucan, which helps lower cholesterol by binding to biliary acids and excreting them from the body. It can be eaten alone as a hot breakfast cereal or added to soups, stirfries, rice-pilafs and grain salads. If ground, black barley flour can be used to make bread, pancakes and pasta. In Tibet (China), black barley grains are widely used in wine preparation.

6. Karyotaxonomic Evidence: Barley is a diploid self-fertilized with a low number 2n=2x=14 relatively large chromosomes. The karyotype is symmetrical and has a quotient (ratio) longest/shortest chromosome of 1.3 (Teketa *et al.*, 2003). Ghaffaripour *et al.* (2011) studied karyotypes of 15 genotypes of hull-less barley (*H.vulgare* L.). Karyotype in hull-less barley taxa studied is symmetric. The karyotype formula was 2n=2x=14=7m. They observed chromosomes with centromere at median region. The chromosome constitution was uniform among different genotypes of hull-less barley.

7. History: Barley is an ancient plant crop that played an important role in human nutrition in Asia, Africa and Europe for many millennia. Black barley is often flogged as 'Ethiopian Black Barley'. It is said to be, in fact, for more likely in North America to be a variety that originated in Russia and was introduced to America in December 1911. It was planted, and from it a selection with a stronger stalk was made that was called 'Lion'. This was the 'Black Barley' that came to be distributed in the United States.

Saraswat (1992) divulged archaeobotanical remains in ancient cultural and socio-economical dynamics of the Indian subcontinent in a chronological framework from the earliest Neolithic cultures of primitive village-farming communities to highly advanced Iron Age cultures around the beginning of the Christian era. They provided the earliest evidence of West Asian cereal cultivation dated to C 7000 BC. in Neolithic Baluchistan, Pakistan and the Middle Ganga Valley in India. The presence of crop remains from Harappan times to the end of first millennium BC. of West Asian, African and indigenous origin are indicative of trade contacts in diverse geographical regions. Interestingly common barley *viz.*, *Hordeum vulgare* L. and *Hordeum vulgare* var. *nudum* (commonly called Naked, Hulless or Black Barley), are revealed at different archeological sites. This clear indicates the development of Black Barley also in this ancient period in Indian subcontinent.

Barley (*H. vulgare* L.) is thought to be one of the oldest cultivated crops in the world. Earlier studies suggested that there were two original centres of cultivated barley, one in the Fertile Creseant (Ancient Egypt, the Levant and Mesopotamia) as indicated by the widespread dispersion of *H. spontaneum* Koch (the nearest wild ancestor of two and six-rowed cultivated barley) in this region (Harlan, 1976), and another in the Tibetan region of China on the Qinghai-Tibet Plateau supported by the discovery of *H.agriocrithan* Abeng, a six-rowed wild barley (Vavilov, 1926). Zhang *et al.* (2012) extended observations on paleo-environment and agriculture of ancient Loulan and Milan of the silk Rod in China. Active agriculture in ancient Loulan was documented in the historical literature (Ban, and naked barley was found there (Hou, 1985). Zhang *et al.* (*loc.cit.*) studied phyloliths from fossil samples collected at the said sites and informed them belonging to the naked barley. Their results indicated common milles, and foxtail millet as staple food and possibly naked barley as non-staple crops. These were food source for ancient Loulan and Milan residents between ADS 50 and 770.

It is noticeable that the replacement of naked barley by hulled barley is a common scenario in other Iberian and European regions throughout the Bronze Age (Buxo and Pique, 2008). Although naked barley maintained its relevance, hulled barley appeared more abundantly in Northern Portugal. Only in Buraco da Pala (Mirandela) is the naked barley variety more abundant than hulled barley, but even at this site there is a great increase of hulled barley when compared with previous levels (Sanches, 1997).

Pokharia and Srivastava (2013) provided current status archaeological studies in Harapan civilization. They recorded occurrence of naked barley

(*H.vulgare* var. *nudum*) at different sites during different Horappan times (Early, mature and Late) *viz.*, at Mohenjo-Daro, Harappa,Chanu-Daro and Pirak in Pakistan, and Rajasthan, Punjab, Haryana and western Uttar Pradesh in India. Gietre (2012) searched out the earliest appearance of domesticated plant species and their origin on the western fringes of the Eurasian Steppe. The author analysed pottery shards of glumes and palaea impressions found at Zanovskoe site, a Neolithic settlement in eastern Ukrain. These impressions belong to hulled barley (*H.vulgare*) and naked barley (*H.vulgare* var. *nudum*). It is further stated that the inhabitants of Zanovskoe were probably cultivating hulled and naked barley.

8. Further Developments: Sterna *et al.* (2017) analysed protein, lipids, starch, dietary fibre, beta-glucan and alpha-tocopherol contents in both covered and naked barley varieties. They reached to a conclusion that barley may differ greatly in chemical composition due to genetic and environmental factors. Their study showed that starch, protein and beta-glucans content was significantly influenced by variety, year and nitrogen fertilizers rates, whereas total dietary fibre lipid and aphe-tocopherol did not differ significantly by varieties or nitrogen usage. They further revealed that almost all aminoacids significantly differed by varieties. In case of naked barley, the contents of proline valine, isolucine and tyrosine were significantly higher than these in covered barley grain. They observed 'Naked Spring Barley' material has a genotypic potential. It could be used for the development of the improved barley breeding material specifically designed for food with a high nutritional value and healthy food production.

Gustafsson (2013) studied the origin of naked barley by the 'nud' gene. He stated that it is from previous Asian studies on naked barley known that a mutation or a deletion of the 'nud' gene expresses the naked seed phenotype. He studied the Nordic variant of naked barley. He stated that most of the results of the PCR genotyping confirmed the phenotype of the tested occasions, both naked and bulled barleys. However, one visually phenotyped naked barley cultivar (NGB-4580) still amplified with the known primers that would match the Asian hulled allele, meaning that the Nordic accession NGB-4580 of naked barley did not carry the known 'nud' deletion. This, according to the author, suggested that naked barley has arisen independently in Asia and in the Nordic countries.

Feng *et al.* (2006) studied genetic diversity, evolutionary relationships and geographical differentiation of cultivated six-rowed naked barley (*Hordeum vulgare* spp. *Hexastichon* var. *nudum* Hsu), an oldest cultivated barley in China. They used 35 simple sequence repeat (SSR) markers selected from seven barley linkage groups among 65 landrace accessions collected from different locations in China. They concluded that SSR analysis of the 65 accessions of the said taxon from Tibet possessed generally higher genetic diversity than these in Qinghai and Ganzi, with about 19% of the genetic diversity being accounted for by geographical differentiation among the three subgroups. This, in their opinion, is suggestive that Tibet might be a centre of genetic diversity of cultivated six-rowed naked barley on the Qinghai-Tibet plateau of China. This might have evolved in Tibet and spread to Qinghai and then to Ganzi prefecture of Sichuan province.

Wu Kunlun *et al.* (2017) analysed relationship between Wx gene polymorphisms and amylose content in hulless barley. *Hordcum vulgare* L. var. *nudum* Hook. *f.* is a major cereal grain in Tibet (China), which is often called hulless or naked barley. It is essential for the daily life of Tibetans. Starch is the main component in hulless barley grain. The starch in hulless barley is unique in its composition as it contains 74-78% of amylopectin and 22-26% amylose. The granule-bound starch synthase (GBSS) enzyme catalyzes one of the enzymatic steps of starch synthesis. This enzyme, also called the waxy protein, is responsible for the synthesis of amylose and is encoded by the Waxy gene (Wx.). Amylose content is an important quality trait of barley and an important parameter for the evaluation of waxy barley varieties. This study demonstrated that SSR markers amplified by primer pair P4 from the Wx gene were positively correlated with amylose content in hulless barley. Wang *et al.* (2005) described polymorphism involving P3-amplified products in wheat as well as positive correlation between the length of the P3 PCR products and amylose content. However, this phenomenon is not observed in case of hulless barley by Wu Kunlun *et al.* (*loc.cit.*). The primer P4 can be used as a complementary markers for the selection of hulles barley germplasm with different amylose content.

Global strategy for the *Ex Situ* Conservation And Use of Barley Germplasm, the International Centre for Agricultural Research in the Dry Ares (ICARDA) in Aleppo, Syria was commissioned by the Global Crop Diversity Trust. The inventory was primarily based on direct responses from institutional holders of major barley collections. Additional information was obtained from different national, regional and international information sources. The inventory showed that the total of barley germplasm holdings at 47 major barley collections (with more than 500 accessions) was 402000 accessions, besides 28 minor collections. Thus accession were 290, 820. Of these, 15% are wild relatives, 44% are landraces, 17% are breeding materials, 09% are genetic stocks and 15% are cultivars. In these, information for naked barley (*H.vulgare* L. var. *nudum* Hook. *f.*) is also included (Anonymous, 2008).

Yand *et al.* (2018) surveyed for phenolic contents and antioxidants activity of 12 blue highland barley varieties planted on the Qinghai-Tibet Plateau. Highland barley (*Hordeum vulgare* L. var. *nudum* Hook. *f.*) is a type of barley, also known as hulless barley due to its bareness of grain. It is the major crop in this region and the main food relied on by farmers and herdsman. Their results showed that the blue highland barley grains are rich in phenolic compounds and have a strong antioxidant activity. According to them, the significant differences among the different varieties signify that the phenolic substances of this barley varieties coexist in both free and bound forms. The seeds mainly consists of gallic acid, benzoic acid, syringic acid, p-coumaric acid, naringenin, hesperidin, rutin,catechin and quercetin. These varieties, in their opinion, can be used as a source of natural antioxidant food.

Yan *et al.* (2016) analysed mineral analysis of hulless barley (*Hordeum vulgre* L. var. *nudum* Hook *f.*) originated from the Qinghai-Tibet Plateau (China). The mineral contents of K, Mg, Ca, Na, Fe, Zn, Mn, Cu and Se in 13 cultivars of hulless barley and its beta-glucan concentrates were determined. Their results showed that there were no marked content differences among cultivars in plant origins, but some elements were enriched in beta-glucan concentrates, of which Ca, Fe, Zn, Mn, Cu and Se are elements related to blood sugar. With the help of beta-glucan, in their opinion, this concentrate may be useful functional food or a good dietary supplement for diabetic patient.

Liang *et al.* (2017) performed a transcriptomic analysis using two accessions of Tibetan hulless barley (*H.vulgare* L. var. *nudum* Hock *f.*) *viz.* Z772 (drought-tolarant) and Z013 (drought-sensitive) which grow in harsh environment of the Qinghai-Tibetan Plateau (China). They found more up-regulated genes of Z772 than Z013 under both mild and severe dehydration treatments. Under mild dehydration stress, the pathways exclusively enriched in drought-tolerance genotype Z772 included in endoplasmic reticulum, tricarboxylic acid (TCA) cycle, wax biosynthesis and spliceosome. Under severe dehydration stress, the pathways that were mainly enriched in Z772 included Carbon fixation in photosynthetic organisms, pyruvate metabolism, orphyrin and chlorophyll metabolism. The candidate gene for drought tolerance were selected based on their express patterns. These data provided pathways and a targeted set of candidate genes that might be essential for deep analyzing the molecular mechanisms of plant tolerance to drought stress.

Taketa Shin *et al.* (2008) inferred that the covered (hulled) or naked (hulless) caryopsis in Barley is controlled by a single locus ('nud') on chromosome arm 7HL. On the basis of positional cloning, they concluded that an ethylene response factor (ERF) family transcription factor gene controls the covered/naked caryopsis phenotype. This is validated by (i) naked cultivars, (ii) two x-ray induced 'nud' alleses with a DNA lesion and (iii) gene expression strictly localized to the testa.

9. Epilogue: This botanical heirloom is a part of living history. It is ancient grain. This chewy, delicious nutty treat goes from field to table with obviously no processing. No bran or germ is removed. The complete edible kernel is intact for the full nutritional benefits. It is our way of supporting centuries-old plant variety and preserving the genetic diversity of our food supply. It has passed down through generations and nearly lost. Naked black barley is generally rich in various nutrients. This potential could be used for the development of the improved barley breeding material particularly for food with nutritional value. Obviously, such improved varieties and even baked black barley itself could be wider used for healthy food production.

Black or naked (hulless) barley get its deep purple pigment from its high antioxidant level of anthocyanin. This oxidant is said to be more stable in grains than in the vegetables and fruits. It is also high in gluten-soluble fibre. This aids in lowering cholesterol and reducing glucemic levels and thereby lowering down the risk of heart diseases and diabetes.

Modern industrialized countries disregarded the naked black barleys for some reason. This was not a good thing from a nutritional standpoint. Because of pearlizing, barley are rendered devoid of bran during the process of rubbing off the hull. On the contrary, the naked or hulless (black barley, no bran is removed) leaving the complete, whole grain goodness of the barley intact.

This hearty grain resemble a coffee bean in shape and cooks up to a glossy purplish-mahogany colour. It is wonderful in soups and stuffinigs or alone as a beautiful alternative to rice. It is also best mixed with other grains when used alone. Moreover, it has a somewhat chewy texture.

It appears that cultivation of naked (hulless) barley is as old as that of hulled barley but is less common worldwide due to significantly lower yields and only minor breeding activities (Pandey *et al.*, 2006). Compared to hulled barley, the free-threshing character of hulless barley proportionally increases contents of protein and the limiting amino acids lysine and thereonine respectively (Bhatty, 1999) and well levels of beta-glucan but lowers contents of insoluble dietary fibre compounds (Baidoo and Liu, 1998).

Overall, coloured cereals receive a growing interest due to their antioxidant properties (Nam *et al.*, 2006; Philpott *et al.*, 2006) and protective functions under severe environments (Carletti *et al.*, 2014). Plants with highly pigments are more resistant to biotic and abiotic stresses (Pandey and Dhakal, 2001). Black Barley is no exception to these merits.

6.2 BROWN RICE

Oryza sativa L. (Poaceae)

Resume: *Brown rice is a unpolished rice retaining its outer bran layer and other nutrients intact. The modern method of polishing deprives bioactive nutrients and hence brown rice is being favoured as functional food source. It has the same chromosome number 2n=24 and whole genome sequencing has been completed. Being rich in nutrients, it benefits in health and suffices well in our nutritional requirements. It is used in the form of various* *processed products worldwide as superfood. Germinated brown rice (GBR) has proved more fruitful and beneficial in human health and nutrition as it increases existing nutrients and adds as well during the process of GBR. New products are also released and used in record times.*

1. Introduction: Rice is a important staple food grain and primary crop grown all over the world. It is a major part in the diet of more than half the world population in tropical region. In India, it is called 'Dhanya' meaning the sustainer of human race. There are different types of rice *e.g.* white rice, black rice, sticky (glutinous) rice. They are not equal in terms of health benefits and nutrients. Brown rice is one of the healthiest type of rice. Its health benefits are partially due to the way it is processed or prepared. In case of common white rice, the hull and outer layers of kernel containing bran and germ are removed to make it white. Only starchy endosperm is left. In case of brown rice only the hull of the rice kernel is removed during preparation and thereby the value of nutrition of rice kernels is intact. Brown rice can be turned into white rice by removing and polishing more of the kernel but this process leads to loss in nutrients. Thus brown rice is highly nutritious food.

2. Plant Characteristics: Usually a grass growing over 1m tall, culms (stems) upright, noded; leaves long, linear, alternate; flcrets borne in branched panicle, each spike compressed with solitary floret; floret shortly pedicallate; caryopsis elongate, husked.

3. Useful Parts: Grains.

4. Chemical Constituents : Carbohydrates (77.24%), protein (7.85%), fat (2.92%), vitamins *e.g.* thiamine, riboflavin, niacin, pantothenic acid (B5), vitamin B6, folate (B9), minerals (calcium, iron, magnesium, manganese, phosphorous, potassium, selenium, sodium and zinc).

5. Uses And Applications : The nutritional and functional properties of rice bran are suitable for baked products *viz.*, cookies, muffins, breads, crackers, pastries and pancakes. The stabilized rice bran is successfully incorporated in the production of yeast bread because hygroscopicity of the rice bran may improve its moisture in the baked products. Defatted rice bran is used as a substitute of the wheat flour for making cookies. Rice bran protein have been found to be of high quality and application in food and pharmaceutical industries. It is worth to note that germinated brown rice (GBR) is being used in diets in various forms which benefits in increasing the existing nutrients and also add few other nutrients during the process of GBR.

6. Karyotaxonomic Evidence: The genome of *Oryza sativa* is composed of 2n=24 chromosomes. Its AA genome has a total length of 430 Mb (megabase, a nucleotide length of 1000000 base pairs. Rice has thus a relatively small genome if compared with that of other common cereals. The whore genome of rice has been sequenced (Goffs *et al.*, 2002; Yu *et al.*, 2002).

Rice has 40000-50000 functional genes with about 15000 distinct gene families.

7. History : Bahadur (2003) reported that it is better to eat brown (unpolished) rice because the outer bran layer of the grain is rich in fiber, vitamins, and minerals. Further germination of brown rice (thereby making GBR) is necessary for enhancing nutrients required for a good health. The change of style food from polished rice to GBR maintain and promote the healthy life and improve the quality of life (Hiroshi, 2005).

GBR was developed for marketing in Japan in 1995. GBR products were developed and marketed first by Domer Co. (Veda City, Nagano Prefecture, Japan). The city government, Mino-cho of Kagawa Prefecture (Japan) was one of the earliest organizational engaged in the production of GBR. The brown rice is soaked for 1 or 2 nights. This changes the internal minerals and brown rice becomes more nutritious, easier to chew and tastier. It has been reported that the GBR may enhance brain functions and reduce the level of lipids or fats in the blood (Patil and Khan 2011).

Consumption of brown rice became popular in Japan back 1970's because of rich fiber and other nutrients contained in the brown rice. Popularity of its consumption did not last long because brown rice has to be cooked in the pressure cooker and was hard to chew and less tasty. GBR overcame the problem which can be cooked in an ordinary rice cooker and is soft enough to chew even for children (Patil and Khan, *loc. cit.*).

8. Further Developments: Issara and Rawdkuon (2016) review literature about rice bran and its application in food products. During the milling process, rough rice is milled to produce polished edible grain and removal of brownish layer. This process generates rice bran and rice hull (Friedmn, 2013). Now-a-days, rice bran is applied in food industry due to high nutritional promoting and rich of major source of phytochemical compounds. Currently, rice bran used for rice bran oil extraction and wax production and used as a food ingredients (Watchararuji *et al.*, 2008). These products are currently used for baked products, energy bars and protein fortification of powdered drink formulations (Mitchell, 2009). Several countries use rice bran for healthy beverage or functional drink. Rice bran yields rice bran oil after extraction which is used for cooking. This promotes good human health and also produce animal feeding (Friedman, 2013). The whole rice bran is used especially in beverage production, cosmetic manufacture and pharmaceuticals.

Generally, rice bran is a good source of bioactive phytochemicals *e.g.* γ-oryzonal, tocopherols and tocatrienols, which have beneficial properties and antioxidant properties (Watchararuji *et al.*, 2008; Moongngarm *et al.*, 2012). Rice bran has rich source of protein and known amino acid. Amino acids are essential to the body especially in part of muscle building and maintenance (Han *et al.*, 2014). Bioactive phytochemical in rice bran *e.g.* phytosterols, γoryzonals, tocopherols, tocotrienols, polyphenols and squalene are useful for nutraceuticals technology. The antioxidative activities of each component in rice bran deactivate the natural by-products known as free radicals (Patel and Naik, 2004).

Esa *et al.* (2013) overviewed health benefits and applications of by-products of rice such as rice bran, rice germ, broken rice, rice huet, rice straw with particular emphasis on rice bran. They pointed out their nutritional values, antioxidant properties, and role of diterary fibre, phytic acid, tricin, vitamin-E, oryzonal and hypocholesterol effect as anti-cancer. They explained suitability of rice bran in various processed food products. They pointed out the potential of bran to be converted into human food to improve food security. They inferred it as important source of ingredients of functional food and nutraceuticals.

Germinated brown rice (GBR) is considered whole food because only the hull of the kernel is removed and there is no damage to its nutritional value. Patil and Khan (2011) reviewed germinated brown rice as a value added rice product. They concluded that the intake of GBR: (i) ameliorates the hyperglycemia (ii) boosts the immune system (iii) lowers blood pressure (iv) inhibits development of cancer cells and (v) assists the treatment of anxiety disorders. They explained the role of various nutrients in case of GBR. They concluded that novel foodstuffs from extruded GBR would be acceptable to consumers and food industry as a promising foodstuff which contains more nutritional and bio-functional components than ordinary rice products.

Analyses on GBR conducted indicated that during the process of germination, nutrients in the brown rice change drastically. Kayahara *et al.*, (2001) showed that not only the existing nutrients are increased but new components are also released from the inner change due to germination. The nutrients which increased significantly are GABA, lysine vitamin-E, dietary fiber, niacin, magnesium, vitamin B1 and vitamin B6 (Kayahar *et al.*, *loc. cit.*). The other nutrients that increased in GBR were inositols, ferulic acid, phytic acie, tocotrienols, potassium, zinc, 9-oryzonal and prolendopeptidase inhibitor (Kayahara and Tsukahara, 2000). Trachoo *et al.* (2006) found that in GBR the increased nutrients were vitamin-B, reducing sugar and total protein content.

Emam *et al.* (2012) reviewed anti-diabetic properties of germinated brown rice emphasizing type 2 diabetes. Germination of brown rice (GBR) is one way of increasing bioactive concentrations to enhance its functional effects. They reached to a conclusion that diet is an important factor in worsening of type-2 diabetes. *In vitro* data, animal experiments and clinical studies suggested GBR to have a strong potentials of an anti-diabetic functional food that could provide a steady supply of bioactive compounds, for maintaining health and preventing complications in type-2 diabetes. The authors, however, desired to have further long-term studies using different

GBR products to establish the full potential of GBR in the management of type-2 diabetes.

9. Epilogue: Brown rice is obviously a highly nutritious food. It is a whole grain that is relatively high in fiber, gluten-free, a good source of magnesium, phosphorus, selenium, thiamine, niacin, manganese and vitamin B6.

Brown rice has many health benefits. High magnesium content is good for heart which help for regulating blood pressure and offsetting sodium in the human body. Thus it is associated with healthier carotid arteries and blood pressure levels. Brown rice is abundant in lignans which help protect against many other diseases. Brown rice is rich in fiber content which keeps bowel movements regular and reduces constipation. Moreover, it is gluten-free and thus a popular staple of gluten-free diets. Brown rice has potential in reducing risk of diabetes. It is protein and energy rich. Brown rice is associated with lower risk of colon cancer, prostrate cancer and breast cancer. Presence of selenium heps induce repair in damaged cells and thereby help stop cancer cells from spreading. It is also antioxidant rich which function as anti-inflammatories.

The structure and function of the rice genome has been identified. The genus *Oryza* has 22 wild rice species with nine genomes (AA, BB, CC, BBCC, CCDD, EE, FF, GG and HHJJ). This is in addition to the two widely cultivated species *Oryza sativa* (AA genome) and *O. glaberrima* (AA genome) (Vaughan, 1994). After completion of genome sequencing of *O.sativa*, genome diversity of the entire genus *Oryza* is the next target for genomic analysis. These analyses have to include problems of genome organization and comparisons at a chromosome level, sequence level, functional level and evolutionary level. Accordingly, studies for resolving mechanisms, involved in organization of genome functions, will need novel breakthroughs in molecular, biological, cytogenetical, biochemical and genetic methods.

It is interesting and noteworthy that instead of using simply whole grains of brown rice, one should opt for germinated brown rice (GBR) because this process not only increased existing nutrients but also adds by releasing other nutrients. Thus GBR is more effective in view of health and nutritional benefits.

It is worth to note that very recently another nutritious and medicinally important variety/race of rice has been revealed (Borah *et al.*, 2018). It is called 'Black Rice', just like 'Black Barley' and having four landraces *viz.*, Chakhao poireiton, Chakhao angouba, Chakhao amubi and Chakhao pungdol amubi. These are being cultivated traditionaly by farmers belonging to Meity community in the state of Manipur in India. Its consumption is important from neutraceutical point of view. It possesses antioxidents, antidiabetic and anticarcinogenic properties. Its agronomically important characteristics are:

less ploughing, stress tolerant, lower water requirement, no application of fertilizers and farmyard manure, pest and disease resistant, stable production on marginal farmlands and higher by product yield.

Thus improvement in case of rice, we have now to notice its journey from white rice to brown rice to black rice.

6.3 CHIA

Salvia hispanica L. (Lamiaceae)

Resume: *Chia (Salvia hispanica L.) is thought a native of Central America. It was once a staple food used by ancient Aztecs. Its production was depressed for some centuries but it came back being a promising and beneficial source for health and nutrition. It is now accepted as a superfood source. Its seeds are now being used for commercial industrial products in various forms. It has chromosome number 2n=2x=12. Its cultivation is being extended in other parts of the world.*

1. Introduction: Chia has a long history of plant-human interacton. A desert plant species *viz.*, *Salvia hispanica* (Lamiaceae), a member of the mint family, yields Chia seeds. It is sold under its common name 'Chia'. It is believed to be a native of Central America. The seed was a staple in the ancient Aztec diet. The seeds of a related plant *viz.*, *Salvia columbariae* Benth. (called 'Golden Chiia') were used primarily by the Native Americans in the southeastern United States (Immel, 2003).

Salvia hispanica is described and pictured in the 'Mendoza Codex', 'Florentine Codex' and 'Aztec Codices' created between 1540 and 1585. Both sources describe and picture this plant species and also the use by the Aztecs. The 'Mendoza Codex' indicated that it was widely cultivated and offered a tribute in 21 out of the 38 Aztec provincial states. It was a staple food (Cahill, 2003). It was cultivated by the Aztecs in pre-Columbian period. These are presently cultivated on a small scale in their ancestral homeland of Central Mexico and Gautemala and cultivated commercially throughout Central and South America. Economic historians suggested *S.hispanica* as important as maize in pre-Columbian. Mexico, and in some areas more important (Harrey, 1991). Its seeds were valued for food, medicine and oil (Sahgun, 1950-1982). After Spanish contact and colonization, its cultivation plummeted. However, highly productive domesticates are still under cultivation in a few areas.

2. Plant Characteristics: An annual herb, upto 1.75 m tall; leaves opposite, 4-8 cm × 3-5 cm; flowers purple or white, borne in clusters in a spike at the end of branches; seed small, oval, mottle-coloured with brown, grass, black and white.

3. Useful Parts: Seeds.

4. Chemical Constituents: Carbohydrates (42.12%), protein (16.54%), fat (30.74%), minerals (calcium, iron, magnesium, manganese, phosphorus, potassium, sodium, zinc), vitamin-A, thiamine (B_2), riboflavin (B_2), niacin (B_3), folate (Bg), vitamin-C and vitamin-E.

5. Uses And Applications: Chia seeds can be soaked in water and consumed directly or mixed with any kind of juice or milk. They may also be ground and made into a gelatin-like substance or eaten raw. Chia seeds may be sprinkled or ground up on top of other foods. They can be mixed into smoothies, breakfast cereals, energy bars, granola bars, tortillas and bread. The gel obtained from the ground seeds may be used to replace as much as 25% of the egg and oil content in cakes.

Chia seeds contained more calcium, omega-3 fatty acid, antioxidants, fiber, protein and micronutrients. These all benefit in various function of different systems of human body and boost energy. Chia seeds are antioxidant-rich which help protect the cells against free radical damage and premature aging. Antioxidants are also influential in helping skin repair from the everyday damage of sun exposure and environmental irritants. Protein-rich chia seeds are essential for collagen production, which helps skin stay-strong and supple. Chia seeds contain abundant the omega-3 fatty acid ALA to ease inflammation of the skin. This also strengthens the skin hair and nails. Magnesium helps reduce inflammation associated with acne. It also relaxes the nervous system and regulates cortisol, the stress hormone that wreaks havoc of skin.

6. Karyotaxonomic Evidence: The chromosome number 2n=2x=12 found for *S.hispanica* concides with the number reported by Estilai and Hashemi (1990). The study carried out by Claudio Palma-Rojas *et al.* (2017) showed that chromosome pairs 5 and 6 and in the limit between submetacentric and subtelocentric types. This is not the case for the other four pairs, whose morphology is clearly defined according to the chromosome nomenclature proposed by Levan *et al.*, (1964). Chromosome numbers in the genus *Salvia* vary extensively from the 2n=2x=12 in *S.hispanica* (Harley and Heywood, 1992) to 2n=8x=88 in the octoploid *S.guaranitica* (Alberto *et al.*, 2003). This variation occurs both due to polyploidy and to a wide diversity of basic numbers (X) of chromosome (Zhiyun *et al.*, 2004). Basic numbers of chromosomes such as 7, 8, 10, 11, 12 and 13 have been found for popyploid species of *Salvia* (Alberto *et al.*, *loc.cit.*).

7. History : Chia (*Salvia hispanica* L.) is a native plant of Central Mexico. The chia was classified by the Swedish botanist Carl Von Linne in 1753, who named it *Salvia* (save or cure) *hispanica* (Spanish) that in Latin means Spanish plant to cure or save (Urbino, 1887). It is evident that this name was wrongly given because Chia was collected by Lofling in Madrid (Lepez, 2007).

This species is not native from Spain, but it was carried by Cristobal Colon from Mexico to this country (Urbino, 1887). However, 178 years later it was effectively demonstrated that chia is no native from Spain (Vavilov, 1931). But its origin centre is Mexico and Central America is possible that, considering in the past, its binomial name was a mistake. He did named *Salvia hispanica*, but he named it as *Salvia chia* Fernald. On Nahua language the Chian word (today named Chia) means oily, because the Aztecs used the Chia Word to refer to all species of *Salvia* genus. Their main characteristic is their high oil content. The Spaniards were unknown to the botanical classification of plants used by the Nahua people.

8. Further Developments: Cahill and Provance (2002) studied genetics of qualitative traits in domesticated Chia (*Salvia hispanica* L.). Three F_2 segregating generations from crosses between wild and domesticated parents were scored for three qualitative traits. A single recessive gene 'scc' was found to govern the white seed characteristic. A single dominant gene 'spp' was found to control striated stem pigmentation. A complete dominance of open calyx over close calyxes were observed in F_2 generations, not conforming to Mendelian ratios. A complementary test was conducted for this non-shattering trait between two lines representative of geographically and morphologically divergent domesticated varieties. Complementary gene action was not observed in any F_1 plants, and all F_2 plants were homogenous with respect to the trait, suggesting the same genetic control for non-shattering among domesticated varieties data for linkage of 'ssp' and 'scc' indicated that two loci segregate independently.

Sandoval-Oliveres and Parades-Lopez (2013) isolated and chracterised proteins of Chia seeds (*Salvia hispanica* L.). Chia is a plant that produces seeds rich in some neutraceutical compounds with high protein contents. These authors characterized the seed storage proteins obtained from chia seed flour. The main protein fraction corresponded to globulins (52%). Sedimentation coefficient studies showed that the globulin fraction contains mostly 11S and 7S proteins. The molecular size of all the reduced fractions were about 15-50 kDa. Electro-phoretic experiments showed four bands of globulins in the range of 104-628 kDa. The denaturation temperatures of crude albumins, globulins, prolamins and glutelins were 103, 105, 85.6 and 91°C respectively. Albumins and globulins had relatively good thermal stability. A good balance and essential amino acids was found in the seed flour and globulins, especially of methionine + cysteine.

Chia, the ancient grain is becoming enormously popular in modern food regimen worldwide. It contains more super source of omega-3 fatty acid, which is associated with many physiological functions in human body. Chia seed is a potential source of antioxidants with the presence of chlorogenic acid, caffeic acid, myricetin, quercetin and kaemferol which are believed to have cardiac, hepatic protective effects, antiaging and anti-carcigogenic characteristics. Its dietary fiber is beneficial for digestive system and controlling diabetes mellitus with higher concentration of beneficial unsaturated fatty acids, gluten-free protein, vitamins, minerals and phenolic compounds. Therapeutic effects of Chia in control of diabetes, hypertension, anti-blood clotting, laxative, antidepressant, antianxiety, analgesic, vision, etc. is scientifically established (Rahman Ullah *et al.*, 2016).

Coates and Ayerza (1998) introduced Chia as a commercial crop in Northwestern Argentina. The economy of this country heavily depended on sugar and tobacco crops. Prices of these crops depressed and caused significant economic problems. Alternative crops are actively sought. These authors grow Chia (*S.hisponica* L.) as new crop in their country being a source of industrial oil for the cosmetic industry and θ-3α-linolenic acid for the food industry They expected this new crop could help diversity of the local economy. A project to develop Chia as a commercial crop was initiated in 1991. They recorded a average seed yield 606 kg/ha. Oil content and fatty acid composition varied between different fields in their country. Biomass production and seed yields, in their opinion, were affected by weather, geographical location and production practices. The authors still desired additional experiments to assess the effects of these factors on Chia yield, oil content and composition fully.

Jamboonsri *et al.* (2012) extended Chia crop in North America as there has been renewed interest in Chia because of excellent source of W3 fatty acids and dietary fibre for healthy diets. The authors set out to develop the genetic diversity in floral induction to provide germplasm for production in the US and other temperate area of the world. They demonstrated that new early flowering lines are able to flower and a photoperiod of 15h under greenhouse conditions. In field conditions, some selected new lines flowered at a photoperiod of 14h and 41 minutes during the 2009 growing season in Kentucky and produce seeds in a range of environments in temperate areas. This was the first report on induction of early flowering Chia mutants. The genetic mechanisms of floral induction varies among plant species.

9. Epilogue: Chia as a crop arrived 21st century without having incorporated the changes required to adapt for modern production. Chia did not travel on the same way its ancient mates of American origin such as corn (maize), bean, tomato did. As a result, the Chia crop was unknown to other countries and its use was confirmed to the regions which adopted its use before arriving the Spaniards. After 1990 decade, the status of Chia as a crop and food source changed. Although there are several food sources in the market, none of them is safer and more sustainable than Chia. It is the unique source that we can eat directly. Botanists predicted that a future of this species would be one important crop in the food industry to produce supplements and neutraceutical is now confirmed. In recent years, it is being assessed for its cosmetic benefits. It is doubtless that the importance of Chia as a functional food on the world will continue achieving and the next generation of functional food, supplement and neutraceutical products where the main active ingredient will have as raw material the chia seed, an ancestral crop of Mexico, that although was forgotten for more than 500 years.

6.4 KAMUT

Triticum turanicum Jakubz.

(Syn.) *T.turgidum* subsp. *turanicum* (Jakubz.) A.Love & D.Love (Poaceae)

Resume: *The origin of wheat is intriguing and multiple. Kamut or Khorsan wheat is a tetraploid (2n=28) and said to be ancient food grain from Fertile Crescent. It has many health benefits as compared to modern commercial varieties of wheat. It was once forgotten but came back with promising properties revealed by modern researches. It is being used in various forms of food articles. It is still being debated taxonomically and genetically.*

1. Introduction: Kamut (Kah-moot) is an ancient tetraploid grain closely related to wheat, characteristically closer to durum wheat. 'Kamut' is actually a brand name but the common name for it is 'Khorasan wheat' Kamut or Pharaoh Grain'.Farmers in Turkey call this grain 'Camel's Tooth' because of its shape. It is also called 'Prophet's Wheat' referring to another legend that Noah brought the grain with him on the ark. In ancient Egyptian language, Kamut means wheat. Some Egyptologists claim the root meaning of this word is 'soul of the Earth'. Kamut grain is similar in shape to a wheat grain but about twice the size of modern wheat. It has slightly sweet taste and nutty flavour.

Wheat was one of the first domesticated food crops. It has been the basic staple food of the major civilizations of West Asia, North Africa and Europe for about 8000 years. It continues to be the most important food grain source for mankind and is being cultivated on more land area than any other commercial crop (Curtis, 2002). The commonly used wheat criginated thousands of years ago through naturally occurring hybridization of their progenitors. In recent times, there has been an ever-increasing number of varieties, while ancient varieties of this cereal have been largely forgotten. Kamut or Khorasan is one such once neglected food grain but receiving its glory and fame because of its some unique properties. It is being dilated in this attempt.

2. Plant Characteristics: Plants (culms) 120-130 cm tall, upper internodes solid; leaves linear, long blade nodding, puberlent; spike lax, 9-15 ca 1 cm, rachis tough, spikelets 16-20 mm, with 3-5 florets; glumes narrow, apex acute, lemma awned, black, palea equaling lemma; canyropsis ellipsoid, usually free from lemma and palea, 10-12x ca 3mm, vitreous.

3. Useful Parts: Grains.

4. Chemical Constituents: Carbohydrates (60.64%), protein (14.33%), fat (2.2%), fiber (10.67%), minerals (silicon, iodine, zinc, boron calcium, cobalt, iron, phosphorous, manganese, nickel, potassium, selenium, sodium, molybdenum, copper), amino acids, *e.g.* tryptophan, cysteine, methionine, aspartic acid, threonine, serine, glutamic acid, proline, glycine, alamine, valine, isoleucine, leucine, tyrosine, phenylalimine, lysine, histidine and arginine.

5. Uses And Applications: Kamut can be directly consumed or is milled into flour and can be used in many recipes. Kamut flour works well in cakes, cookies, crackers and homemade flat breads. It is useful for pizza and pasta. It is also used in the form of flaked or puffed cereal, bread mixes, breakfast cereal, waffles, pancakes, bulgur, drinks, frozen meals, beer, snacks, cold salads, pilafs, soups (Stallknecht *et al.*, 1996; Kamut).

6. Karyotaxonomic Evidence: A wealth of data on the genomic structure of wheat has accumulated. By the 1920s, it was known that cultivatd wheat species of the genus *Triticum* have chromosome of 2n:14, 28 and 42 (Sakmura, 1918). This suggested a basic 1x chromosome number of 7 and the occurrence of diploid (2n=2x=14), tetraploid (2n=4x=28) and hexaploid (2n=6x=42) wheat species. Two species *viz., T.turgidum* and *T.timopheevii* have cultivated as well as wild forms. The hybrids between these two are sterile, establishing them as valid biological species. The wild forms of both tetraploid wheats *T.turgidium* spp. *dicoccoides* (also known as *T.dicoccoides*) and *T.timopheevii* ssp. *armeniacum* (also known as *T.armeniacum*) are widely distributed in the fertile Crescent. Kihava (1919) and Sax (1922) indicated that *T.monococcum* and *T.turgidum* share one genome in common, whereas *T. turgidum* and *T.aestivum* share two genomes in common. The remains of cultivated emmer (*T.turgidum* ssp.*dicoccum* have been revealed at several archaeological sites in Syria dating 7500 BC. (Zohary and Hopf, 1993). Durum wheat (*T.turgidum*) has a genome size of 10 billion bp of DNA organized into 14 pairs of chromosomes. The structure of A-and B-chromosomes of durum wheat is essentially identical to the corresponding homologous in bread wheat (*T.aestivum*).

7. History: A legend has it that grains of Kamut were found entombed in the pyramids of Egypt and then it was brought under cultivation. It is said that A U.S.airman of Portugal bought 36 kernels of Kamut grain from someone who claimed it came from the pyramids of Egypt. The airman mailed these kernels home to his father. His father was a wheat farmer. He planted the 36 grains, of which only 32 kernels sprouted out. He tended carefully this crop over subsequent six years. He grew this crop to 1500 bushels.

This unique wheat was shown at the country fair. It was named there as 'King Tut's Wheat'. But the grains ended up since it was fed to the cattle by the farmer. Bob Quinn, a founder and president of Kamut International, was a young boy in the crowd at the time. A year later, he remembered this strange grain. He scoured the country and got a pint of it. He then built up the strain and generated enough interest to market it commercially in the late 1980s (*cf.* Quinn, 1999).

The origin of Kamut is uncertain. It is believed that it probably survived the years as an obscure grain kept alive by the diversity of crops common to small peasant farmers perhaps in Egypt or Asia Minor. It is thought to have evolved contemporary with the free-threshing tetraploid wheats. Most scientists agreed that it is *Triticum turgidum* (AABB) which also included the closely related durum wheat. Although its identity is disputed, it is favoured for its great taste, texture, nutritional qualities and hypo-allergenic properties (*cf.* Quinn, 1999).

8. Further Developments: Sofi *et al.* (2013) attempted to charcterise Kamut wheat (Khorasan Wheat) to characterize and examine the effect of a replacement diet with their products on cardiovascular a risk parameters. They assigned the participants to consume products like bread, pasta and crackers made from Kamut for eight weeks in a random order. They observed: (i) a significant reduction of metabolic risk factor such as total cholesterol, low-density liprotein cholesterol and blood glucose, (ii) redox status significantly removed in both thiobarbituric and reactive substance and carbonyl levels, (iii) significant increase of serum potassium and magnesium and (iv) significant reduction in key pro-inflammatory cytokines. The authors concluded that a replacement with Kamut products could be effective in reducing metabolic risk factors, markers of both oxidative stress and inflammatory status.

Sofi *et al.* (2014) studied effect of *Triticum turgidum* subsp. *turanicum* wheat on irritable bowel syndrome (IBS). They examined the effect of a replacement diet with organic, semi-whole grain products derived from it. Participants consumed bread, pasta, biscuits and crackers made from Kamut grains. During the intervention period with these products, patients experienced a significant decrease in the sevarity of IBS symptoms *e.g.* abdominal pain, bloating, tiredness and satisfaction with stool consistency.

Khorsan (Kamut) wheat flour as a substitute for other ingredients can contribute to the improvement of the nutritional value of the final product. Bread made from it has more carotenoid and richer protein content as compared to modern wheat varieties (Pasqualone *et al.*, 2011). Also, total flavonoids, phenolics and antioxidant capacity are higher in Kamut flakes than in corresponding conventional products (Sumczynski *et al.*, 2015). Abdel-Aal el and Rablski (2008) reported a higher concentration of total phenolics in Kamut wheat than in 10 common wheat cultivars.

Veronika Michalcova *et al.* (2014) studied taxonmical classification and origin of Kamut wheat. These authors opined first that although history of

Kamut is interesting and has stable place in the market, it has been not a common subject of genetic studies. They further stated that it has not been successfully taxonomically classified yet. There are few studies which attempted to classify this tetraploid wheat as: *Triticum polonicum* L., *T. turanicum* Jakubz., *T.turgidum* L. and *T. durum* Desf. These studies are based on cytological and comparative methods. Veronika Michalcova *et al.*, (*loc. cit.*) extended molecular and statistical methods to classify Kamut wheat. They concluded that Kamut brand wheat originated as a natural hybrid between *Triticum dicoccon* conv. *dicoccon* and *T. polonicum*. It is not original ancient Egyptian wheat. They further suggested that Etruscan wheat has the same parents as Kamut.

9. Epilogue: Kamut is related and similar to wheat. It is generally more digestible and an excellent substitute for common wheat. It has many benefit nutritionally. It has more protein, amino acids, lipids and rich in Zn, Mg and vitamin-E. It is slightly lower in fiber as compared to wheat. Interestingly, Kamut is grown organically and has never been modified genetically or hybridized. It is not treated with harmful pesticides or artificial fertilizers. This is why it has achieved a status of high-quality grain.

IBPGR (1990) informed total number of accessions available in wheat collections around the world under different six types of wheat with total number=650566. Genetic resources have played a significant role in wheat improvement and will continue to do so by providing breeders with the variability they require. Most projections suggests that wheat, the most important food source, will also be the primary food staple in the developing nations. Demand of wheat in developing nations is expected to grow. In near future, we have to be aware about production-consumption gap. Genetic improvement of wheat is likely to remain as a major source of productivity gains. Genetic resources are fundamental to the world's food security and central to efforts to alleviate poverty.

Kamut (Khorasan) wheat is a ancient grain crop and had been all but forgotten. Recently, it made a comeback. This is because of its great taste, texture, nutritional value and hypoallergenic properties. Kamut has a rich and buttery flavour. It is digestible easily. It has more proteins, lipids, vitamins, minerals and amino acids than common wheat. It is thus a more nutritious substitute. It is a high energy grain because of its high percentage of lipids, which provide more energy than refined carbohydrates. The manganese in Kamut promotes strong and healthy bones. Being a high-fiber food, it aids in the regularity and functions of the digestive system. Its high content of phosphorous is important for kidney function and help the body detox by eliminating toxins and waste through urine. Protein makes up the structure of the cells, organs and muscles in human bodies. Kamut is a high source of protein that helps our bodies make harmones, coenzymes, blood cells and even DNA. The zinc content is able to prevent the common cold and symptoms of other illness. The manganese present in Kamut grains supports brain health. Consumption of high-fiber foods as Kamut is effective for lowering cholesterol as it helps to get rid of toxins and unwanted waste. Recently, there has been a great deal of research focusing an ancient wheat grain varieties. These heirloom grains are proving to be a rich source of health-promoting substances. Kamut is one such highly delicious and nutritious source.

Ancient Kamut wheat has a potential which can be transformed into a large variety of products. The health-promoting properties of Kamut are not generally evident in other modern commercial wheat varieties. Further scientific evidences are desired to consider Kamut or Khorasan based products as functional foods. The results obtained through recent researches are promising. There are some elements of interest that challenge the scientists to deepen the scientific knowledge about this ancient heritage.

6.5 TEFF

Eragrostis tef (Zucc.) Trotter (Poaceae)

Resume: *Teff (Lovegrass) is thought a native of Ethiopia and also domesticated there since ancient times. It is an allotetraploid cereal crop with 2n=2x=40. It is being hailed as a new superfood and becoming increasingly desirable in cuisines worldwide and also is valued for health benefits. It is rich in dieters, fiber, protein carbohydrates, minerals and amino acids. Its processed nutritional products have a great demand in modern markets. It is much-sought energy-crop of the day.*

1. Introduction: The common name 'Teff' is thought originated from the Amharic word 'teffa' which means lost, due to small size of the grain. It is a fine grain about the size of a puppy seed. It is called lovegrass, Annual Bunch Grass, Tef, Taf, Tteff, Thaft, etc. Ethiopia is considered its site of origin. It was domesticated in Ethiopia 4000-1000 BC. It is an ancient grain but became a modern staple. It has long been a dietary staple for Ethiopia. Health-conscious consumers have been gravitating to it since it is nutrient dense and have not been genetically modified. Most of the teff is consumed in North America, Europe and other parts of the world like the Netherlands, Australia, etc. Teff is being hailed as a new superfood. Being a naturally gluten-free, and it makes a great wheat alternative nutritionally. This tiny grain is the national pride of Ethiopia and getting global attention in health and nutritional problems. Teff is thus one of the obscure ancient grain and is becoming increasingly desirable in cuisines of many.

2. Plant Characteristics: A grass with a large crown, many tillers, shallow diverse root system, 25-13 cm tall; leaves linear, alternate; panicle loose, lax or compact, with multiple branching; florets white to dark brown, spikelets consisting 2-12 florets; seed very small 1-1.7 mm, colour variable, white to ivory white or tan to deep brown to reddish brown purple.

3. Useful Parts: Grains.

4. Chemical Composition: Carbohydrates (19.86%), protein (3.87%), fat (0.65%), minerals (calcium, iron, magnesium, manganese, phosphorus, potassium, sodium, zinc), thiamine (B_1), riboflavin (B_2), niacin (B_3), vitamin (B_6), folate (B_9).

5. Uses And Applications: In Ethiopia, teff is consumed as a flat bread called 'injera' as well as a porridge, unleavened bread and even beverages. Teff flour, water and salt are needed for making injera. It requires an overnight fermentation. It is spongy, slightly sour and pleasantly nutty. Teff seeds can be boiled like pasta. Ground into flour, it is used in baked goods. It also works wonderfully in stews, adding both thickness and texture. It is used in gluten-free recipes like pancakes, breads, crepes or waffles, snakes, wraps, etc. Food startups incorporate teff grains.

6. Karyotaxonomic Evidence: *Eragrostis tef* (Poaceae), called Tef, is an allotetraploid with 2n=4x=40 chromosomes. Previously, it was suggested a total of 14 wild *Eragrostis* species as potential progenitors. Phylogenetic analysis of sequence data from the nuclear gene 'waxy' and the plastid locus 'rps16' strongly support the widely *E.pilosa*, a wild allotetraploid. Another previously proposed progenitor *viz.*, *Eragrostis heteromera* is shown by the 'waxy' data to be a close relative of one of the tef genomes. Plastid sequences from five varieties of tef and four *E.pilosa* accessions are identifical and therefore are uninformative with respect to the question of multiple origins of these polyploids. The 'waxy' phylogeny also resolves the relationships among other allopolyploids, supporting a close relationship between the morphologically similar allotetraploids *viz.*, *E.macilenta*, *E.minor* and *E.mexicana*. The authors also concluded that *E.cilianensis*, another morphologically similar allopolyploid, to have shared one diploid progenitor with these species but derived its other genome from an unrelated diploid (Ingram and Doyle, 2003).

7. History: Teff (*Eragrostis tef*) is one of the earliest domesticated crop (Murphy, 2007). It is thought originated in Ethiopia between 4000 BC and 1000 BC. *Eragrostis pilosa* is the most probable wild ancestor of it (Ingram and Doyle, 2003). It is an allotetraploid (2n=2x=40) cereal crop grown primarily in Ethiopia. The grain is used to make a variety of food products, including 'injera', a spongy fermented flat bread that serves as the staple food for most Ethiopians. Various species of *Eragrostis* have been considered as progenitor of tef. The general concensus is that *E.pilosa* is the most likely candidate for the direct wild progenitor of *E.tef*. The former is the most common species in Ethiopia. Tef is also known as an 'emergency crop' since it is grown late in the spring when the growing season is warmer and most of the crops have already been planted. It does withstand frost. It is characterized by the US National Research Council as having the 'Potential to improve nutrition, boost food. Security, foster rural development and support sustainable landcare' (Anonymous, 1996).

Based on morphological data the following species have been identified by different researchers as the ancestors and contributors to the origin of tef or as species closely related to tef:

(i) Ancestors of tef (Costanza, 1974): *E.pilosa, E.macilenta, E.aethiopica, E.pseudotef, E.longifolia* and *E.atrovirens.*

(ii) Species contributors to the origin of tef (Endeshaw, 1978): *E.pilosa, E.curvula, E.aethiopics, E.cilionensis, E.mexicana* and *E.bicoour.*

(iii) (a) Species sufficiently closely related to tef (Ponti, 1978): *E.pilosa, E.aethiopica, E.mexicana, E.cilianensis, E.minor, E.barrelieri.*

(b) Species suspected to the close enough (Panti, 1978): *E.macilenta, E.aegyptiaca, E.papposa, E.heteromera* and *E.bicolour.*

(iv) Species closely related based on cytological evidence (Tavassoli, 1986): *E.aethiopica* (2x), *E.pilosa* (2x), *E.mexicana* (6x), *E.barrleiri* (6x), *E.minor* (2x, 4x) and *E.cilianensis* (2x, 4x, 6x).

8. Further Developments: Fan Zhu (2018) recently reviewed chemical composition and food uses of teff (*Eragrostis tef*). He pointed out the reasons for its global popularity such as: (i) attractive nutritional profile *e.g.* gluten-free and high dietary fiber content (ii) gluten-free food items, (iii) sensory quality of teff-based products, (iv) use of whole grain teff flour in healthy food market, (v) genetic diversity. He further concluded that there is great potential to adapt teff to the other parts of the world for healthy food and beverage production.

Dame (2018) analysed major and trace elements in teff (*Eragrostis tef*) which generally play a vital role in proper functioning of enzymes through modulating reactions, constitute their active sites, stabilize their structure and mediate oxidation process. A total 44 elements are detected and quantified in teff grain samples. The most abundant elements measured are K, P, Ca, Mg, and Na. The least abundant elements are Er, Eu and Sb. The concentration of Fe in white and red teff varieties are comparable to concentrations reported for wheat and corn (maize).

Gebremariam *et al.*(2014) reviewed teff (*E.tef*) as a raw material for malting, brewing and manufacturing of gluten-free food and beverages. Interest in teff has increased noticeably due to its very alternative nutritional profile and gluten-free nature of the grain, making a suitable substitute for wheat and other cereals in their food applications as well as foods for people with celiac disease. They concluded that: (i) having an excellent balance of amino acid composition, makes teft an excellent material for malting and brewing, (ii) being small in size, it is made into whole-grain flour (bran and germ included) resulting in a very high fiber content, (iii) very useful to improve the haemoglobin level in human blood and helps to prevent malaria, anaemia and diabetes, (iv) a good potential to be used in foods and beverages because of nutrient composition, (v) high levels of simple sugars and α–amino acids are essential for fermentation and beer making.

Seyfu (1991) reviewed prospects and research needs based on experiences of Ethopian farmers. To sum up in brief are: (i) can be grown in areas of moisture stress, (ii) can be grown in water logged areas and withstands anaerobic conditions better than other cereals, (iii) suitable for use in multiple-cropping systems *e.g.* double, relay and intercropping, (iv) straw a valuable feed for cattle, (v) high demand and high market in diet and hence enables farmers to earn more than with other crops, (vi) reliable and low-risk crop, (vii) in moisture-stress, farmers use it s a rescue crop, (viii) not attacked by weevils and other storage pests and (ix) has fewer disease and pest problems.

Daba Tadesha (2017) focused an nutritional and socio-cultural values of indigenous crop of teff (*Eragrostis tef*) in Ethiopia. Two distinct teff varieties 'Kuncha' and 'Key-teff' representing the white and red/brown varieties respectively, were analised for their nutrient compositions. The white varieties, in general, have higher market price, social and cultural values. The red varieties are thought more nutritious especially in rural areas. The nutrient composition of two varieties in this study confirmed that 'key-teff' (red variety) is superior to 'Kuncho' (white variety) in most minerals and essential amino acids. This crop has high socio-cultural values in Ethiopia but it is relatively low yielding. Production of teft needs improving its productivity, efficient domestic use and promoting its international market values to ensure sustainable production.

9. Epilogue: Teff is now one of the most nutritious alternative grains. If offers calcium, fiber, protein and antioxidants. It is a energy-rich and healthy. It helps protect bones and provides long-lasting satiation that can even help to support weight loss efforts. It is higher than wheat in nutrients. It is much higher in amino acids which are great for supporting energy levels. Teff is high-fiber grain and ensures regularity in bowel movements. It is gluten-free and hence, of late, gaining popularity. The iron in teff increase oxygenation to organs. Its nutrient content is higher because nutrient-packed bran and germ make up the majority of the teff grain. It is hence climbing up as superfood.

Tef is adapted to a wide range of environments and is presently cultivated under diverse agroclimatic conditions. It can be grown from sea level up to 2800 m asl, under various rainfall, temperature and soil regimes. Teff performs excellently at a altitude of 1800-2100 m, annual rainfall of 750-850 mm, growing season rainfall of 450-550 mm and a temperature range of 10°-27°C. A very good result can be obtained at a altitude range of 1700-2200 m and growing-season rainfall of 300 mm.

References:

Abdel-Aal el S.M. and I. Rabalwski (2008) Bioactive compounds and their antioxidant capacity in selected primitive and modern wheat species. *Open. Agric*.J. 2:7-14.

Abera, E. (1948) Cereals and peas from Eastern Tibet and their importance of the origin of cultivated plants. *Ann. Agri.Coll.Sweden* 15:235-250.

Alberto, C.M., Sanso, A.M., C.C. Xifreda (2003) Chromosomal studies in species of Salvia (Lamiaceae) from Argentina. *Bot. J. Linn. Soc.* 141:483-490.

Anonymous (1996) Tef. In: Lost Crops of Africa. Volume-I: Grains. Academic Press. pp.215-236.

Anonymous (2008) Globa strategy for the Ex Situ Construction And Use of Barley Germplasm (ICARDA) Aleppo, Syria.

Baidoo, S.K. and Liu Y.G. (1998) hulles barley for swine: ileal and faecal digestibility of proximate nutrients, amino acids and non-starch polysaccharides. *J. Sci. Food. Agric.* 76:397-403.

Ban, G. (1992) HanShu. Beijing: Zhongua Book Company Reprint in 1962 (*In* Chinese).

Bhatty, R.S. (1999) The potential of hulless barley: A review. *Cereal Chem.* 63:97-103.

Borah, N., Florida, D.A., Semwal, R.L. and S.C.Garkoti (2018) Chakhoao, (Black Rice: Oryzza sativa L.): A culturally important and stress tolerant traditional rice variety of Manipur. Indian J.Tradit.Knowle. 17(4):789-794.

Buxo, R. and R. Pique (2008) Arqueobotanica. Los usos de las plantas en la peninsula Iberica. Ariel, Barcelona.

Cahill, J. P. (2003) Ethnobotany of Chia, Salvia hispanica L. (Lamiaceae). *Economic Botany* 57(4):604-618.

Cahill, J.P. and M.C. Provance (2002) Genetics of qualitative traits in domesticated (Salvia hispanica L.) *Journal of Heredity* 91(1):52-55.

Carletti, G. Nervo, G. and L. Cattivelli (2014) Flavonoids and malanins: a common strategy across two kingdoms. *Int. J. Biol. Sci.* 10:1159-1170.

Claudio Palma-Rojas, Basilio Carrasco, Carla Golzalez and Herman Silva (2017) Genetic, cytological and molecular characterization of Chia (Salvia hispanica L.) provances. *Biochemical Systematics And Ecology* 73:16-21.

Coates, W. and R. Ayorza (1998) Commercial production of Chia in Northwestern Argentina. *Journal of The American Oil Chemists Society* 75(10):1417-1420.

Constanza, S.H. (1974) Literature and numerical taxonomy of teff (Eragrostis tef) M.Sc. Thesis, Cornell University, Urbana, Illionois.

Curtis, B.C. (2002) Wheat in the world (cited 2013 Feb.19)

http://www.fao.org/docrep/006/y4011e/4011eo4.htm.

Daba Tadesh (2017) Nutritional and Socio-cultural values of Teff (Eragrostis tef) in Ethiopia. *International Journal of Food Science and Nutrition* 2(3):50-57.

Dame Z.T. (2018) Analysis of major and trace elements in teff (Eragrostis tef). Journal of *King Saud University*, Science (In Press, Corrected Proof).

De Candolle A.(1886) Origin of Cultivated Plants (1959) (Translated from 2nd Ed. In French) Hafner, New York, USA.

Emam Mustapha Umar, Nur Hanisah Azmi, Muhammad Iqbal Bhanger, Norsharina Ismai and Muznah Ismail (2012) Antidiabetic properties of germinated brown rice: A systematic review. Evidence based Complementary And Alternative Medicine Vol. 2012 : 12 pages.

Endeshaw, Bekele (1978) Biochemical and morphological studies of the relationships of Eragrostis tef and other Eragrostis spp. M.Sc. Thesis, University of Birmingham, Birmingham, UK.

Esa Norhaizan Moh., Tan Bee Ling and Loh Su Peng (2013) By-products of rice processing: An overview of health benefits and applications. *J.Rice.Res.* 1:107.

Estilai, A. and A. Hashemi (1990) Chromosome numbers and meiotic behavior of cultivated Chia. Salvia hispanica (Lamiaceae). *Hort Science* 25:1646-1647.

Fan Zhu (2018) Chemical composition and Food uses of teff (Eragiostis tef). *Food. Chemistry* 239-402-415.

Feng Zong-Yun, Zhang Li-Li, Zheng Yi-Zheng and Hong-Qing Ling (2006) Genetic diversity and geographical differentiation of cultivated six-rowed geographical differentiation of cultivated six-rowed naked barley landraces from the Qinghai-Tibet plateau of China detected by SSR analysis. *Genetics And Molecular Biology* 29(2):330-338.

Friedman, M. (2013) Rice brans, rice bran oils and rice hulls: Composition, food and industrial uses, and bioactivities in humans, animals and cells. *Journal of Agricultural And Food Chemistry* 61:10626-10641.

Gebremariam, M.M., Zarnkow, M. and T.Becker (2014) Test (Eragrostis tef) as a raw material for malting, brewing and manufacturing of gluten-free food and beverages: A review. *J. Food Sci. Technol* 51(11):2881-2895.

Ghaffaripour *et al.* (2011) Karyotypes and evaluation of studies in some genotypes of hull-less barley. *Turkish Journal of Field Crops* 16(2):245-250.

Giedre Motuzaite-Matuzevciute (2012) The earliest appearance of domesticated plant species and their origios on the western fringes of the Eurasian Steppe. *Documenta Prachistoric* XXXIV:1-21.

Goffs *et al.* (2002) A draft sequence of the rice genome (*Oryza sativa* L. ssp. japonica). *Science* 296-92-100.

Gustafsson, Dan (2013) The origin of naked barley (Hordeum vulgare L. ssp.vulgare) studied by the nud gene. Bachelor's Thesis, Department of Physics, Chemistry and Biology, Linkoping University, Linkoping.

Han, S.W., Chee, K.M. and S.J. Cho (2014) Nutritional quality of rice bran protein in comparison to animal and vegetable protein. *Food Chemistry* 172:766-769.

Harlan, J.R. (1968) On the origin of barley. In: Barley. U.S.Dept. Agric. Handbook No.338.

Harlan, J.R. (1976) Barley. In: Evolution of group Plants (Ed.Simonds, N.)Academic Press, London (UK) and New York (USA).

Harley R.M. and C.A. Haywood (1992) Chromosome numbers in tropical American Labiatae. In: Advanes in Labiatae Science (Ed. Harley, R.M. and T.Renolds) Royal Botanic Gardens, Kew, London. pp.211-246.

Harrey, H.R. (1991) Land politics in the valley of Mexico: A two thousand year perspective. University of New Mexico Press, Albuqurque.

Hou, C. (1986) The discoveries of common millet, naked barley and valuable wheat inflorescence in the Loulan site. *Agricultural Archaeology* 2:225-227 (In Chinese).

Immel, Diana L. (2003) Chia, Salvia Columbariae Benth.: Natural Resources Conservation Service, US Department of Agriculture, USA.

Ingram, Amand L. and J. Jeff, Doyle (2003) The origin and evolution of Eragrostis tef (Poaceae) and related polypolids: Evidence from nuclear waxy and plastid rps 16. *Amer J.Bot.* 90(1):116-122.

Issara, U. and S. Rawkuen (2014) Organic rice bran milk: Production and its natural quality attributes. In: Proceeding of 1st Joint ACS AGFD-ACS ICSCT Symposium on Agricultural And Food Chemistry. Montein Riverside Hotel, Bangkok, Thailand pp.82-88.

Jaenicke, H. and I. Hoschle-Zeledon (Ed.) (2006) Strategic framework for under utilized plant species research and development with special reference to Asia and the Pacific, and to Sub-Saharan Africa. ICUC, Colombo, Sri Lanka and GFUUS, Rome Italy.

Jamboonsri, W., Phillips, T.D., Genere, R.L., Catill, J.P. and D.F. Hilderbrand (2012) Extending the range of an ancient crop, Salvia hispanica L.- a new W3 source. *Genet. Resour. Evol.* 59:171-178.

Kamm, A. (1954) The discovery of wild six-rowed barley and wild Hordeum intermedium in *Israel. Ann.Agri. Coll.* Sweden. 21:287-320.

Kamut-An Ancient Grain Rediscovered. https://whatscoking americа.net/charlotte Bradlley/Kamut_AncienGrain.htm

Kayahara H. and K. Takmura (2003) Surprising live germinated brown rice. Shougekukan-Square Co., Tokyo, Japan.

Kayahara, H. and K. Tsukahara (2000) Flavours, health and nutritional quality of pre-germinated brown rice. (Ed.Spanier, A. H.) Royal society of Chemistry, Cambridge.

Kayohera, H., Tsukahara, K, Takai, T. (2001) Flaror, health and nutritional quality of pre-germinated brown rice (Ed. Spanier, A.H.). Royal Society of Chemistry, Cambridge.

Kihaya, H. (1919) Ueber cytologische studien ben einigen Getreidearten. I.Spezies-Bastarde des Weizens and Weizenroggen-Bastard. *Bot.Mag.*32:17-38.

Lepez, J.N. (2007) Las plantas vasculares de la comunidad de Madrid. Catalogo floristico, claves dicotomicas y studio detallado de la familia Compositae Giseke. Tesis Doctral Universidad Complutens. Madrix Espana, 409.

Levan, A., Fredga, K. and A.Sandburg (1964) Nomenclature for Centromeric position on chromosomes. *Hereditas* 52:201-220.

Liang Junjun, Xin Chen, Guangbing Deng, Zhifen Pan, Haili Zhang, Qio Li, Kaijun Yang, Hai Long and Maoqun Yu (2017) Dehydration induced transcriptomic responses in two Tibetan hulless barley (Hordeum vulgare var. nudum) accessions distinguished by drought tolerance. Genomics 18:775.

Michalcova, V., Sabo, M. Hauptvoges, P. and Z. Ivanicova (2014) Taxonomical classification and origin of Kamut wheat. *Plant Syst. Evol.* 300:1749-1757.

Mitchell, C.R. (2009) Chapter-13: Rice Starches: In:Production And Properties (3rd Ed.) (Ed. BeMiller J. and R. Whistler) Academic Press, San Diego.

Moongngarm, A., Daomukda, N. and S.Khumpika (2012) Chemical compositions, phytochemicals and antioxidant capacity of rice bran, rice bran layer and rice germ. *APCBEE Procedia* 2:73-79.

Murphy, Denis J. (2007) People, Plants And Genes: The Story of Crops And Humanity. Oxford University Press, Oxford.

Nam, S.H., Choi, S.P., Kang, M.Y., Koh, H.J., Kozukue, N. and M.Friedman (2006) Antioxidative activities of bran from twenty one pigmented rice cultivars. *Food Chem.* 94:613-620.

Nevski, S.A. (1941) Beitrage zur Kenntniss der wildwachsenden Gersten in Zusammenhang mitder Frage uber den Ursprung von Hordeum vulgare L. and Hordeum distichon L. *Trudy. Bot .Akad Nauk SSSR.* Ser. 1(5):64-255.

Pandey, A.K. and M.R.Dhakal (2001) Phytomelanin in Compositae. *Curr. Sci.* 80:933-940.

Pandey, M., Wgner, C. Friedt, W. and F.Ordon (2006) Genetic relatedness and population differentiation of Himalayan hulles barley (Hordeum vulgare L.) landraces inferred with SSRs. *Theor.Appl.Genet.* 113:715-729.

Pasqualone, A., Piergiovanni, A.R., Caponio, F., Paradiso, V.M., Summo, C and R.Simeone (2011) Evaluation of the technological characteristics and breed-making quality of alternative wheat cereals in comparison with common and durum wheat. *Food. Sci. Technol. Int.* 17:135-142.

Patel, M. and S.N.Naik (2004) Gama-oryzanol from rice oil: A review. *Journal of Scientific & Industrial Research* 63:569-578.

Patil Swati B. and Md. Khalid Khan (2011) Germinated brown rice as a value added rice product: A review. *J.Food Sci. Technol.* 48(6):661-667.

Philpott, M., Could, K.S., Lim, C. and L.R. Ferguson (2006). In situ and in vitro antioxidant activity of sweet potato anthocynins. *J.Agric Food Chem.* 54:1710-1715.

Polharia, A.K. and C.Srivstava (2013) Current status of archaeolotanical studies in Harappan Civilization: An archaeological perspective. *Heritage* 1:118-137.

Ponti, J.A. (1978) The systematics of Eragrostis tef (Graminae) and related species. Ph.D.Thesis, University of London, UK.

Quinn, R.M. (1999) Kamut: Ancient grain, new cereal. In: Perspectives on New Crops And New Uses (Ed.J. Janick) ASHS Press, Alexandria, VA. Pp.182-183.

Rahman Ullah, M.Nadeem, A. Khalique, M., Imran, S. Mahmood, A. Javid and J. Husain (2016) Nutritional and therapeutic perspectives of Chia (Salvia hispanica L.): A review. *J. Food Sci. Technol.* 53(4):1750-1758.

Sahagun, B. (1950-1982) Florentine codex: General history of the things of New Spain. In: Monographs of The School of American Research (Ed. Anderson A.J.O. and C.E.Dibble), University of Utah Press, Salt Lake City.

Sakamura, T. (1918) Kurze Milleilung uber die chromozomenzalen and dia verwandtchaffsverhaltnisse der Triticum-Arten. *Bot.Mag.* 32:151-154.

Sanches, M.J. (1997) Pre-historia Recente de Tras-os-Montes e Alto Douro, Sociedade Portuguese de Anthropolosia e. Ethnologia Porto.

Sandoral-Oliveros and O. Parades-Lopez (2013) Isolation and characterization of proteins from Chia seeds (Salvia hispanica L.) *J. Agric. Food. Chem.* 61(1):193-201.

Saraswat, K.S. (1992) Archaeobotanical remains in ancient cultural and socio-economical dynamics of the Indian subcontinent. *Palaeobotanist* 40:514-545.

Sax, K. (1922) Sterility in wheat hybrids. II Chromosome behaviuor in partially sterile hybrids. *Genetics* 7:513-550.

Schultze-Model I. (1986) Hordeum. In: Rudolf Mansfelds Verzeichnis landwirtscha-ftlicher und gartnerischer Kulturpflanzen (Ohne Zierpflanzen) (Ed. Schultze-Motel.I).

Seyfu Katema (1991) Germplasm evaluation ad breedinig work on teff (Eragrostis tef) in Ethiopia. In: Plant Genetic Resources of Ethioia (Ed. Engels, J.G. *et al.*,) Cambridge University Press, Cambridge, UK. pp.323-328.

Sofi *et al.* (2013) Characterisation of Khorasan wheat (Kamut) and impact of a replacement diet on cardiovascular risk factors: cross-over dietary intervention study. *European Journal of Clinical Nutrition* 67:190-195.

Sofi *et al.* (2014) Effect of Triticum turgidum subsp. turanicum wheat on irritable bowel syndrome: A double-blinded randomized dietary intervention tial. *J. Nutr.* 13:1-8.

Stallknecht, G.F., Gibberston, K.M. and J.E. Romey (1996) Alternate wheat cereals as food grains: Einkorn emmer, spelt, Kamut, and triticale. In: Progress In New Crops (Ed. Janick) ASH Press, Alexandria, VA. pp.156-170.

Sterna, V., Zute, S. Jansone, I. and I.Kantane (2017) Chemical composition of covered and naked spring barley varieties and their potential for food production. *Pol. J. Food Nutri. Sci.* 67(2):151-158.

Sumczynski, D., Bubelova, 2., Sneyd, J. Erb-Weber, S. and J. Mlcek (2015) Total phenolics, flavonoids antioxidant activity, crude fibre and digestibility in non-traditional wheat flakes and museli. *Food. Chem.*174:319-325.

Takahashi, R. (1955) The origin and evaluation of cultivated barley. *Adv.Genet.* 7:227-266.

Taketa Shin *et al.* (2008) Barley grain with adhering hulls is controlled by an ERF family transcription factor gene regulating a lipid hiosynthesis pathway. *PNAS* 105(10):4062-4067.

Taketa, S., Linde-Laursen, I. and G. Kunzel (2003) Cytogenetic Diversity. In: Diversity of Barley (Ed.. Th.van Hintum, H. Knupffer and K.sato Elsevier Science B.V., Amsterdam. The Netherlands pp.97-119.

Tavssoli, A. (1986) The cytology of Eragrastis tef with special reference to E.tef and its relatives. Ph.D.Thesis, University of London, UK.

Urbina, M. (1887) Lachia y sus aplicaciones, La. Naturaleza. *Sociadad Mexicana de Historia Natural Tomo* 1:27-36.

Vaughan, D.A. (1994) The Wild Relatives of Rice. International Rice Research Institute Press, Manila.

Vavilov, I.N. (1931) Mexico Y Centroamerica como centro basico de origen de las plantas cultivadas del neuro mundo. *Revista de Geografia Agricola* (1994 reprint) 20:15-34.

Vavilov, N.I. (1926) Studied on the origin of cultivated Plants. *Bull.Appl.Bot.Genet Plant Breed USSR.* 16:1-248.

Wang, F., Zhao, H., Wang, Y., X.Z.Wang (2005) Relationship between fragment length polymorphism with waxy and amylase content in wheat. *Journal of plant Physiology And Molecular Biology And Molecular Biology* 31:269-274.

Watchararuji, K., Goto, M., Sasaki, M. and A.Shotipruk (2008) Value-added subcritical water hydrolysate from rice bran and soybean meal. *Bioresource Technology* 99:6207-6213.

Wu Kunlan, Xidohua Yao, Youhua Yar, Dezhao Chi and Zongyun Fong (2017) Analysis of the relationship between Wx gone Polymrphism and amylose content in hulless barley. *Czech. J. Genet. Plant Breed* 53(4):144-152.

Yan Wei-wel, Hao-ying-ye Yao, Shao-ping Nie and Ya-Jing (2016) Mineral analysis of hulles barley grown in different areas and its β-glucan concentrates. *Cogent Food & Agriculture* 2:1-8.

Yang Xi-Juan, Bin Dang and Ming-Tao Fan (2018) Free and bound phenolic compound content and antioxidant activity of different cultivated Blue Heighland barley varieties from the Quinghai-Tibet Plateau. *Molecules* 23:1-20.

Yu *et al.* (2002) A draft sequence of the rice genome (Oryza sativa L. ssp. indica). *Science* 296-79-92.

Zhang, J., Houyuan Lu, Naiqin Wu, Xiaguang Qin and Luo Wans (2012) Palaeoenvironemt and agriculture of ancient Loulan and Milan on the Silk Road. *The Holocene* 23(2):208-217.

Zhang, J., Lu, H., Wu, N., Qin, X. and L.Wang (2012) Palaeoenvironment and agriculture of ancient Loulan and Milan on the Silk Road. The *Holocena* 23(2):208-217.

Zhiyun, Y., Gona, X. and Y. Pan (2004) Cytological study of six Salvia species (Lamiaceae) from the hengduanshan mountains region of China. *Caryologia* 57(4):360-366.

Zohary, D. and M. Hopf (1993) Domestication of plants in the old worlds. 2nd Ed. Oxford UK, Clarendon Press.

7

Edible Fruits

INTRODUCTION

In ancient times, our ancestors lived on wild plants. Most of our common fruits have their origin in the region of the world that was inhabited by primitive man. When man initiated organised agriculture, he grew mainly grain crops and some fruit trees in his vicinity. It is possible that some fruit plants might have been cultivated even before the cereals. Archaeological findings afford certain clues about the primitive diet and the gradual change that occurred with the domestication of plants The backbone of Mesopotamian agriculture consisted of crop plants that are still important to the world's food supply. Fruits can be studied from different viewpoint as the following:

(i) History: In the New World, man cultivated the indigenous food plants that we now use such *e.g.* maize, tomato, squash, pumpkins, groundnuts kidney and lima beans, avocado, Brazil nut, cashews nut, black walnut, pineapple, cacao, etc.

(ii) Necessity: Cultivation of fruits is now a profitable enterprise. Fruits should be thought as a necessity and not a luxury. They are classed as 'protective foods', absolutely essential for the maintenance of health. They are understood to promote health and check disease. Nutrition experts recommend the consumption of fruits in our daily diet in addition to cereals, millets, pulses, etc.

(iii) Morphology: A fruit is the seed bearing portion of the plant. It consists of a ripened ovary. Other floral parts are mostly associated with fruits and a fruit thus developed is called a pseudocarp. All fruits developing from an inferior ovary are necessarily false as the floral tube is united with the ovary wall. It becomes fleshy and ripens with it. If the receptacle or some other floral parts become associated with a fruit developing from a superior ovary and the two ripen together, a false fruit is also formed by this way.

(iv) Terminology: The use of the terms *viz.,* fruit and vegetable cause confusion. Many fruits *e.g.* tomato, squash, cucumbers, corn and eggplant are popularly called vegetables, although botanically they are ripened ovaries. On the contrary, very few vegetables are known as fruits. One vegetable used as a fruit is rhubarb, but it does not satisfy the botanical definition of a fruit. Rhubarb is actually a petiole. In everyday communication, the term 'fruit' usually refers to matured ovaries that have fragrant aromatic flavor and either are naturally sweet or are normally sweetened with sugar before consumption. Like many vegetables, however, they do provide much appreciated variety and flavour to the diet. They are commonly served as a dessert. Fruits and vegetables share common composition, method of cultivation, harvesting, storage properties and processing. One is hardput to draw a sharp line between them. Realizing this point, some cucurbitaceous fruit yielding plants which are mainly consumed as dessert are presented under the chapter of vegetables for convenient treatment of the entire group of cucurbits.

(v) Chemical Composition: Fruits are quite juicy and low in proteins and fats, except few. The water content may be as high a 96% of the total weight. Fruits are inferior in food value except banana and dates which are rich in carbohydrates. Fruits and vegetables are important sources of both digestible and indigestible carbohydrates. The digestible carbohydrates occur mostly in the form of sugars and starches, the indigestible cellulose and pectic materials provide roughage (which is thought necessary for the proper functioning of the alimentary tract). Fruits are useful, good sources of minerals and vitamins (particularly A and C). Apart from these, fruits contain organic acids (malic, citric and tartaric acids), ethereal substances, tannins and pigments.

(vi) Technology: Undoubtedly, fruits are man's oldest food and must have always been a source of pleasure for him on account of their flavor, taste and nourishing quality. Most fruits are eaten fresh or are preserved (dried, canned frozen, pickled or cooked with sugary syrups). They usually perish once they have attained maturity. Some fruits *e.g.* coconut and citrus can be handled well. The advent of air transport, refrigerated storage and new packaging techniques have, of late, made a tremendous array of fruits available fresh in many cities worldwide. The most delicious fruits *e.g.* the tropical cherimoya, mangosteen, etc. are very delicate. They cannot easily be transported. Virtually they are unknown beyond their area of occurrence.

(vii) Classification: Fruits can be classified in different ways but only two important ones are dilated here.

(i) **Number of Ovaries Developing the Fruit:** Fruits are classified into three groups as: (a) Simple fruits developed from a single ovary of one flower; (b) Aggregate fruits developed from numerous ovaries of the same flower. The individual ripened ovaries may form a drupe or an achene etc., (c) Multiple fruits developed from the ripened ovaries of several flowers developed from the ripened ovaries of several flowers crowded on the same inflorescence.

(ii) **Adaptability To Climate:** From standpoint of horticulture, the fundamental distinction is made based on climatic conditions. Accordingly, three categories of fruits are recognized: *viz.,* (i) temperate, (ii) subtropical and (iii) tropical. Temperate fruit crops are deciduous and need a period of cold. Tropical fruit plants are evergreen and are most sensitive to low temperatures, while subtropical crops may be either deciduous or evergreen and are usually able to withstand a light frost. In each category three more subcategories can be differentiated as: (i) tree fruits, (ii) small fruits and (iii) nuts. Fruits borne on trees are termed tree fruits which include cherry, pear, orange, papaya, apple and date. Fruits borne on low-growing plants, *e.g.* shrubs, lianas and some herbs are known as small fruits *e.g.* grape, raspberry, cranberry, strawberry, etc. Nuts, on the other hand, are characterized by a hard shell that is separable from the inner kernel-the meat *e.g.* walnut, pecans, almond, cashewnut, etc.

(iii) **Dietary And Non-food Uses:** Fruits are good sources of flavonoids, minerals, vitamins, carotenoids, electrolytes and another bioactive compounds that have impact on human health. Fresh fruits are generally high in fibre, vitamin-c and water. Their regular consumption is generally associated with reduced risks of many diseases and functional declines associated with aging. Apart from these, fruits have also non-food uses. Some fruits provide wax to make candles *e.g.* bayberry. Fruits like cherry and mulberry provide natural dyes. Coir fibre obtained from coconut shells have a variety of domestic uses.

7.1 APPLE AND PEAR

Malus Spp. and *Pyrus* Spp. (Rosaceae)

Resume: *Apple (Malus) and Pear (Pyrus) originated primarily in Asia Minor, the Caucasus Soviet Central Asia, Himalayan India and Pakistan. They are important fruit crops especially of the cooler temperate regions. Remains of apple fruits and seeds are recovered in the ruins of pre-historic lake dwellers in Switzerland. Romans were familiar with 22 varieties of apples. Both genera viz., Malus and Pyrus have a basic chromosome number of* $x=17$*, most species being diploid, except few.*

1. Introduction: Apple is believed to have originated in the Caucasus mountains of western Asia where vast forests of wild apple trees exist even today. It has been under cultivation since the earliest times. Remains of apple fruits and seeds are recovered in the ruins of pre-historic lake dwellings in Switzerland. Neolithic finds in central Europe include the remains of sliced dry apples, indicating that even then they were used as dried fruits or stored for the winter. Twenty-two varieties were known to the Romans.

Apples were carried to Israel, Turkey and many parts of Europe nearly 5000 years ago from its homeland. From Europe and Asia, they have been introduced to all other continents. The large-scale introduction of apple varieties into India was made in 1887 by Alexander Coutts. He planted 100 varieties in his orchard (called Hillock Head) in Mashobra near Shimla, The variety Ambri is said to be indigenous to Kashmir and is the most popular. It is widely grown in the valley. Post-Columbian explorers brought apples to Mexico and South America. Later, Spanish priests carried it to the southwest United States. In the early settlements of the Ohio valley, John Chapman started many small orchards as he wandered early in the 19th century. He became one of the legendary figures of what were then the borderlands. Apples are now grown everywhere except the tropics. Pear is native of Eurasia. It is extensively grown in Asia and Europe. China is the largest pear-producing country. It is also grown in Italy, USA, Japan, Spain and Yemen.

Apples and pears are the most important fruit crops of the cooler temperate region. There is a considerable international trade in apples and pears both within major regions such as Europe and also between continents in both hemispheres. For example, there is a very significant sale of fruit from South Africa, Australia and New Zealand to Europe, particularly to the United Kingdom.

2. Plant Characteristics:

(i) Apple: Tree round-crowned, reaching to a height of 12.5 m; leaves simple and dull-green; flowers borne in terminal umbellate cymes; pinkish-white in colour; fruits spherical pome, exocarp and mesocarp in the fruit fleshy and containing few to many seeds.

(ii) Pear: Similar to apple tree in general appearance, but taller and more upright; leaves alternate, oval, toothed margin; flowers appear with the leaves, white, borne in clusters; fruit characteristically pyriform in shape, sweeter and more juicy than the apple.

3. Parts Used: Fruit.

4. Chemical Constituents:

(i) Apple: Fruit: Sugar (11%), fibre (1.0%), malic acid (90-95% of the total acid content), esters, essential oils (amyl esters of formic, acetic and caprol (acids), water (84.1%), small quantities of proteins, fats and minerals.

(ii) Pear: Protein (0.4%), fat (0.1%), carbohydrates (10.6%), minerals (calcium, phosphorous, iron), vitamin A (thiamine, riboflavin, nicotinic acid).

5. Uses And Applications:

(i) Apple: Generally fresh apple fruits are directly consumed. They are also processed in the form of apple butter, jelly, canned apple sauce and apple juice (sweet cider), hard cider (fermented product), dehydrated apples, its flour, etc. The refuse from the cider press (apple pomace) is used in the preparation of jams, jellies and marmelades. Apple murabba is also a popular product in India. The juice is a converted into vinegar. There are some other by-products *e.g.* apple concentrate, powder, pumice and syrup.

(ii) Pear: Pears are used for table fruits. They are also canned. A beverage, perry (similar to cider) is made from juice.

6. Karyotaxonomic Evidence: The genera *viz., Malus* and *Pyrus* have a basic chromosome number of ×=17. Rehder (1954) recognised 25 species distributed in five sections. *Malus* and the genus *Pyrus* have 15 species (in a single section).

Most species in both genera are diploid. Tetraploids and/or triploids are also frequently found in several species. The *Malus* species in the section chloromeles (all indigenous in North America) are the major exception being mostly either triploid or tetraploid (Wilcox, 1962). *M. ioensis*, possibly the stem species of the section, is diploid. All species contributed to cultivated varieties are diploid. A high proportion of scion varieties, especially of apple, are triploid. Selection by man has been strongly in favour of plants which resulted from the fertilization of unreduced ovules by normal haploid pollen. Cytologically and genetically, 2n=2×=34 plants usually behave as normal diploids. There are also a number of suggestions, often conflicting (Knight, 1963) about the possible origin of *Malus* and *Pyrus* by secondary polyploidy from species with lower basic chromosome numbers.

Hybridization between the two genera is difficult and derivatives rarely survive. The two genera are sufficiently distinct to suggest that they arose separately from more primitive species. Intergeneric hybrids between other genera in the Pomoideae (*Sorbus* × *Malus, Sorbus* × *Pyrus* and *Cydonia* × *Pyrus*) appear to be more readily achieved than *Pyrus* × Malus hybrids (Knight, 1963). Even the relatively rare genus *Docynia* from the Himalayan region is probably more closely related to *Malus* than *Pyrus* is to *Malus*.

It has been traditional to classify the cultivated apple as *M. pumila* (formerly *M. communis* or *Pyrus malus*), although some workers prefer to split this species into *M. domestica* (the cultivated apple), *M. pumila* (including var. *paradisiaca*, the paradise apple) and *M. sylvestris* (the wild crab apple). The apple is self-incompatible and can be hybridized readily with most other Malus species. It is probable that most European and Asian species have contributed to the evolution of present day cultivars.

The cultivated pear is frequently classified as *P. communis* but, like the apple, is self-incompatible. It crosses readily with related species. There have been many recent contributions from other species (Knight, 1963) and probably very many more unrecorded.

7. History: It is generally accepted that the primary centre of origin of both *Malus* and *Pyrus* is within the region which includes Asia Minor, the Caucasus, Soviet Central Asia, and Himalayan India and Pakistan (Wilcox, 1962). Some authors (Knight, 1963) claim a more easterly origin, even as far east as western China in case of *Pyrus*. The secondary centres of origin of the individual species within the two genera are either within the primary centre of origin (*M. pumila, M. trilobata*) or to the east (*M. baccata, M. ioensis, P. ussuriensisi* or to the west. (*M. florentina, P. longipes, P. nivalis*). No species of both genera has a natural distribution which includes areas both east and west of the primary centre. The development of species has been primarily due to physical isolation in a variety of environments.

Knight (1963) considered that *M. sylvestris* and *M. pumila* were the major ancestral species of modern apples but with *M. spectabilis, M. prauii, M. baccate, M. prunifolia and M. astracanica* as significant additional contributors. He considered *P. communis, P. nivalis* and *P. serotina* to be major contributors to the cultivated pear, with *P. syriaca, P. ussuriensis* and *P. longipes* significant minor donors. Some authors suggested several additional wild forms in the origin of modern cultivars (Knight, 1963). *Pyrus* did not spread to the New World until after its introduction by European settlers. Apples and pears have existed in Europe from prehistoric times and cultivated in both Europe and western Asia since the earliest historical times. Early records indicate that Pearmain and Costard type apples were important in England by the 13th century. Pippin types probably became important in England later, following introductions from continental Europe in the 16th century.

8. Further Developments: Pears were bred on a large scale in Belgium during the 17th century and also bred in France and England during the same period. Apple and pear cultivars and seedlings derived from them were taken to North America by the early settlers from Europe. Apples and pears were usually propagated in Europe by grafting. Initial distribution was most by seed in North America. As a result, the genetic diversity accumulated in North American orchards. Subsequent selection in North America gave rise to a range of local varieties distinct in many respects from those grown in Europe. Few European apple varieties are grown in North America, although several American varieties (Golden Delicious, Jonathan and Delicious) have become important, of late, in Europe. European pears are widely grown in North America but with no American pears are well established in Europe.

North American apple and pear breeders initiated the use of wild species in the latter half of the 19th century, for better winter hardiness and also fireblight resistance in pear. The immediate effect was to reduce fruit quality. This problem has largely been overcome by several generations of backcrossing. Varieties originating in North America or Europe are the important commercial types in the world. Thus, for example, the very promising new commercial apple and pear industry of Himachal Pradesh in northern India is based almost exclusively on such varieties.

All important apple and pear varieties are grafted on special rootstocks as they cannot be readily vegetatively propagated on their own roots. These rootstocks (especially the range of apple rootstocks) are widely used in modern orchards throughout the temperate fruit-growing areas of the world. Apart from the great success of clonal rootstocks, many growers, including a high proportion in North America also depend on invigorating but very variable seed propagated rootstocks.

Many *Malus* species have been involved in breeding rootstocks *e.g. Malus robusta* 5 (*M. baccata* × *M. prunifolia*). It is widely used in the colder parts of Canada. Wild species of pears have been used as rootstocks because of their winter hardiness and adaptation to a range of environments. The pear species rootstocks are normally seed-propagated. Such rootstocks are widely used in North America. Clonal *Cydonia* selections have been the most widely used rootstocks for pear in Europe.

9. Epilogue: A trend was set towards the use of a very few major varieties throughout the world. The loss of each local type reduced the scope for further evolution by breeding. Fruit collections made in some national institutes will, if properly supported, play an increasingly important role in helping to retain for future use the remainder of such genetic diversity. The evolution of improved cultivars will also be very seriously affected if the present trend towards the elimination of stands of wild species continues without steps being taken to preserve. Judging by the most advanced breeding programmes and the steady increase in the number of species being utilized, it seems likely that almost all of the *Malus* and *Pyrus* species will be involved in the evolution of future varieties and/or rootstocks.

7.2 AVOCADO

Persea americana Mill. (Lauraceae)

Resume: *Avocado (Alligator Pear) is a native of Mexico and Central America. Wild forms representing ancestors are found in Honduras and Coasta Rica. Its archaeological evidence is available from Tehuacan Valley.*

The species of genus Persea have chromosomes 2n=2x=24. The fruits are used more as a delicacy than a staple food. Fruits are eaten raw or in various preparation and also employed in the cosmetics.

1. Introduction: Avocado or Alligator pear is a native of Mexico and Central America. It is one of the most important contribution of the New World to human diet. It has been under cultivation for its edible fruits for about 9000 years. It was a staple food of the early Central American civilizations. Archaeological findings at Tehuacan Valley revealed that the fruit size was increased due to human selection in ancient times. First, it was introduced into Jamaica in about 1650 and to Spain in 1601. It reached Asia in the middle of the 19th century. Now, it is cultivated throughout most the world's tropics and subtropics (Purseglove, 1974). In India, it is grown in Southern States.

2. Plant Characteristics: A small to medium-sized evergreen tree, reaching 20 m in height; leaves are dark green, glabrous, elliptic to lanceolate, simple and aromatic; flowers small, fragrant, greenish-yellow and borne in a large, compact, axillary panicles; fruits pear-shaped, ovoid or even rounded, single-seeded berry with smooth exocarp and fleshy mesocarp.

3. Parts Used: Fruits.

4. Chemical Constituents: Protein (1.7%), fat (26.4%), carbohydrate (5.1%), vitamin A as β-carotene, niaan, riboflavin, thiamin (Muthukrishnan and Khadar, 1996).

5. Uses And Applications: It is used more as a delicacy than a staple food. The flesh is eaten raw or spread on bread like butter to make sandwitches. It is used as a salad or cooked. The extracted fat is employed in cosmetics.

6. Karyotaxonomic Evidence: The 80-odd Persea species are placed in two graft- and hybridization-incompatible subgenera *viz., Eriodaphne* and *Persea.* All species of both subgenera have 2n=2×=24.

Persea americana belongs subgenus *Persea.* It contains a number of varieties, previously classified as separate species (Bergh *et al.,* 1973) including the three that ranke up the avocado of commerce *viz.,* vars. *americana, guatemalensis* and *drymifolia.* These are equivalent to the horticultural races known respectively as 'West Indian', 'Guatemala' and 'Mexican'. They are also described as tropical, subtropical and semitropical on the basis of increasing cold hardiness and general climatic adaptation. There are many differences between them in other botanical traits including some of great horticultural significance. The avocado has been crossed with *P. floccoso* but with none of the other wild species that have been tried.

Wild forms representing ancestors of each of the three varieties are known. A very primitive form restricted to small areas in Honduras and Costa

Rica could be close to the progenitor of all cultivated types. Primitive *guatemalensis* types are known from isolated spots in Mexico, Guatemala and Honduras. Similarly, primitive *drymifolia* (southern Mexico) and *americana* (Colombia) forms have also been noted. The three horticultural races appear domesticated independently from wild forms in the region Colombia to southern Mexico. The wild forms all have small fruits with large seeds and little pulp so the trend of early selection is evident.

7. History: Schroeder (1968) based on his fossil studies indicated that species of *Persea* grew abundantly in what is now California about 50 million years ago. The crop apparently originated in Central America. Smith (1969) recovered seeds in connection with human camp sites in-southern Mexico as early as perhaps 7000 BC. with cultivation indicated by about 5000 BC. He showed that there was human selection for larger fruit size beginning about 4000 BC. He opined that the inhabitants of Mexican valley had made much less selective improvement. The avocado cultivars are presently grown around the world without significant improvement in horticultural quality over that gained by early Amerindians. Selection was apparently most effective in *guatemalensis*, selecting its larger fruits and smaller seed and that too with separation of seed coat from flesh.

Development of local Indian names indicated importance of avocado fruit *e.g.* the Incan 'palta', the Mayan 'on' and, especially, the Aztec 'ahuacatl' (whence the Spanish 'ahuacate' or 'aguacarc' which was corrupted into the English 'avocado'). Indian people had brought the fruit as far south as Peru on the western side of the continent. Later, the Spaniards introduced it to Chile, Venezuela, the West Indies, the Madeira and Canary Islands. It then became generally established wherever its varieties were climatically adapted. Its spread into the Old World tropics was, however, comparatively late in 19th century.

Outbreeding prevails in avocado. The resulting possibilities for selection is provided by the groves near Atlixco, Mexico. These are usually regarded as hybrids of *guatemalensis* and *drymifolia*. They may simply be the result of evolutionary adaptation to an intermediate environment. A number of superior commercial traits were selected here including 'Fuerte'. The Rodiles family owned these groves for many generations. They kept planting seeds from the finest of the trees and eventually obtained thousands of seedlings of unusually high average quality. The great variability in quality even amongst such seedlings made asexual propagation highly needed.

8. Further Developments: In Florida, near the end of the 19th century, asexual propagation was first recorded (Ruehle, 1963). Clonal cultivars were selected in many regions in this century *e.g.* California, 'Bacon', 'Zutano' and 'Hass' the 'Booth' selections in Florida; the 'Sharwil' in Australia; the 'Ettinger' in Israel; and many others. There exist sterility barriers between the three varieties. Recent breeding has primarily involved inter-varietal hybridization. In tropical and near tropical areas, only *americana* is well adapted but hybrids of it with *guatemalensis* are performing well and are valuable for extending the harvest season. In less tropical regions, hybrids of *guatemalensis* with *drymifolia* predominate. They combine the cold hardiness of the latter with the superior horticultural traits of the former and they bridge the two seasons of maturity. Genetical constitutions vary from largely *drymifolia* ('Ettinger', 'Zutano') through approximately equal proportions ('Fuerte', 'Sharwil') to largely *guatemalensis* ('Hass'). Trihybrids of all three varieties have been developed in Hawaii and in the University of California breeding programme. The result of all this varietal hybridization led to polytypic species with all gradations of traits.

All the leading cultivars developed as chance seedlings; rarely, has even the female parent been known. Moreover, all were developed by private individuals. The University of California programme emphasized self-pollination with the dual aims of direct selection and more uniform breeding lines for hybridization. Selection programmes were also undertaken by institutions in Florida. As a result, important cultivars are available *viz.*, Fuerte, Has, Lula, Pollock, Ardith, Pinkerton, Yulia, Green, Purple, etc.

9. Epilogue: Private and public selections will probably continue in avocado-growing regions around the world. There is no alternative for local selection for local adaptation. But the fact that cultivars superior in one country have generally proved to be also superior in distant countries with similar climates. This suggests continued dominance of the cultivar picture. There will probably be continued emphasis on inter-varietal hybrids, which will further reduce varietal distinctions.

Breeding objectives pertain to all those attributes of tree, rootstock and fruit (Bergh, 1975) that result in heavy production, good shipping quality, sales appeal and consumer satisfaction. The major disease problem in most regions is *Phytophthora cinnamomi* causing root rot. Resistance to the fungus is common in species of *Eriodaphne* but members of the two subgenera are graft-incompatible. The most promising approach appears to be hybridization and selection among the few avocado lines that show a little root-rot resistance with a gradual development of commercially resistant rootstocks.

7.3 BANANAS

Musa Spp. (Musaceae)

Resume: *Banana (Musa) is one of the oldest fruits to known to mankind and thought originated in the humid tropics of Southeast Asia. It is a tropical fruit but consumed throughout the world as a dessert fruit and in various forms. It was first referred in ancient Hindu writings and epics dating as far* *back as 500 BC. The wild bananas are all diploid. Various species of the genus Musa have 2n=2×=20 or 22 chromosomes. Generally,cultivated bananas are triplod. Parthenocarpic development of fruit is common.*

1. Introduction: The banana is thought to have originated in the humid tropical regions of Southeast Asia, somewhere in the mountainous regions of Assam (India). Myanmar, Thailand or Indochina (Sauer, 1952). It is one of the most popular and important of all tropical fruits It has spread all over the tropical world. It is consumed throughout the world even though it is cultivated in the tropical countries. Banana is one of the oldest fruits known to mankind. Banana was first referred in ancient Hindu writings and epics dating as far back as 500 BC. It was the forbidden fruit in the earthly Paradise, according to a Hindu legend, and it was with its leaves that the first man and woman covered their nakedness. It has also legend that the wise men of India sat underneath the shade of the banana tree, refreshing themselves for further wise thought by eating bananas. When Alexander the Great invaded India more than 2000 years ago, he found large plantation of banana growing in the Indus valley.

Arab traders carried the banana across the Indian Ocean to the Malagasy Republic in the 5th century BC. In the Koram and in Arab legends, the banana plant is described as 'The Tree of Paradise'. Linnaeus probably had this legend in mind while naming the banana as *Musa paradisiaca* (paradisiaca from the Latin, meaning 'of paradise'). It then reached to Africa, from where the Portuguese are said to introduce into the Canary islands after 1402. Thence, it was taken to Santo Domingo (present Haiti and Dominican Republic) by a Spanish priest from Santo Domingo, the banana spread the West Indies and Central America.

2. Plant Characteristics: Herbaceous tree-like tall perennial, reaching to a height of 3-9. 5 m with underground rhizome, the trunk (pesudostem) composed leaf-sheaths; leaves glossy, green, oblong, large reaching 3. 0 m in length; inflorescence a complex spike, peduncle stout, female sterile and male flowers borne on basal, middle and upper nodes of inflorescence; fruits elongated, curved or not, parathenocarpic, ripe when yellow and seedless.

3. Parts Used: Berries.

4. Chemical Constituents: Ripe Banana: sugars (glucose, fructose) (20.4%), starch (1.2%), fats (0.2%), protein (1.22%), vitamin A. C but poor in vitamin B, minerals (iron, phosphorous potassium and calcium), β-cerotene, riboflavin, niacin, ascorbic acid.

5. Uses And Applications: Ripe banana is consumed throughout the world as a dessert fruit. Many products are prepared from banana fruits *e.g.* chips, figs, puree, jam, confections, slice, soft drink, dehydrated banana, etc. Banana powder is used as baby food and in the manufacture of biscuits, chocolates, jellies, jams and a variety of confections. Green bananas are cooked and consumed as vegetable. Other products *e.g.* alcohol, cinegar are obtained by fermentation of pulp. (Anonymous, 1952, Khader, *et al.*, 1996).

6. Karyotaxonomic Evidence: The wild bananas are all diploid and are distributed in southeast Asia and the Pacific (Table-A). The related genus *Ensete* has 2n=2×=18 and is distributed in tropical Africa and southeast Asia.

Table-A : Classification of the Bananas

Section	2n=2×=	Remarks
Eumusa	22	13-15 species, southeast Asia, including *Musa acuinata* and *M. balbisiana*
Rhodochlamys	22	5-7 species, southeast Asia
Australimusa	20	5-7 species, New Guinea area
Callimusa	20	6-10 species, southeast Asia

Cultivated bananas (Eumusa cultivars) mostly evolved from *M.acuminata* and *M. balbisiana*. They have chromosome numbers based on 11 (22, 33, 44). The small and unimportant. Fe'i group of cultivars had an independent origin from species of Australimusa and are diploid with 2n=2×=20. Two other members of the group are: (i) Manila hemp *i.e.* Musa textilis. This species, a member of Australimusa with 2n=2×=20, is native to and cultivated in the Philippine Islands and Borneo. It has been but little bred. There is no evidence that cultivars are significantly removed from their wild progenitors, (ii) *Ensete ventricosum*, native to highland east Africa, has long been cultivated in the highlands of Ethiopia.

7. History: Banana evolution is related to parthenocarpy and sterility in the edible diploids. Wild banana fruits are trilocular berries. The seeds in them are surrounded by a mass of sweetish acid, starchy, parenchymatous pulp. The ovaries generally swell slightly if unpolinated and remain as empty shells. The growth of pulp depends upon a stimulus from the developing seeds. In case of edible bananas, the stimulus to pulp growth is autonomous. Seeds are unnecessary. The autonomous stimulus that culminates in vegetative parthenocarpy is due to complementary dominant genes which are present in wild forms of *Musa acuminata*. It is possible to have thinly parthenocarpic fruits which are edible if unpollinated but heavily seeded if pollinated. The typical seedless edible banana is, therefore, the product of two evolutionary processes *viz.*, parthenocarpy and sterility, both essential. Sterility in the edible diploids is thought basically genetic female sterility. Many female-sterile but male-fertile edible diploids are known. Gametic sterility also manifest in both ovules and pollen. It has been caused by the carrying over from wild ancestors and the accumulation during clonal existence of structural chromosome changes.

Edibility first evolved in wild *Musa acuminata* The species is very variable. The primary centre was the Malay peninsula. Human selection favoured parthenocarpy plus seed-sterility leading to the- production of edible seedless fruits. It is followed by male sterility consequent upon structural heterozygosity. Edible diploids are still distributed in southeast Asia. They have largely been supplanted by the polyploids. There remains a considerable centre of diploid diversity in the New Guinea area.

Generally cultivated bananas are triploid (2n=3×=33). Triploid plants are more vigorous and their fruits grow faster than diploids. It is clear that there are three major groups of triploids (*A. acuminata* genome; *B. balbisiana* genome): AAA, AAB, ABB. There are also rare hybrid diploids (AB) and tetraploids (AAAB, AABB, ABBB) (Richardson *et al.*, 1965). There is no evidence that *M. balbisiana* ever evolved edibility on its own. (Vakili, 1961), so that the hybrid groups of cultivars originated by outward migration of edible diploid, male fertile NA types into the area of IVI, balbisiana, followed by crossing and polyploidy. Probably, balbisiana genomes are responsible for hardiness and resistance to seasonal drought in the monsoon climates north of the primary centre of origin.

There are several hundred clones in existence of which AA, AAA, AAB and ABB are the most numerous; AB, AABB, AAAB and ABBB are much rarer. No natural AAAA clones are known.

The crop reached West Africa before European contact and a few clones were carried thence in the late 15th century to the New World. It is likely that bananas entered Africa from Malaysia, instead India, by way of Madagascar and/or the east coast in the hands of Indonesia migrants and travelled up the great lakes. More or less contemporaneously during the first millenium AD., the crop moved eastwards across the Pacific to the further islands. Pre-Columbian presence in the New World is not on record. Little is known about the Fe'i group of cultivars. They occur in New Guinea and Pacific islands. They are diploids (with 2n=2×=20) and phenotypically belong in Australimusa, with the wild New Guinea species *Musa maclayi* as a likely progenitor.

8. Further Developments: Bananas were widely spread throughout the tropics by the end of the 16th century. Relatively few clones had moved out of the centre of origin in southeast Asia and this pattern persists today. Diversity declines in the sequence Asia-Africa-America.

In the later 19th century, the banana export trades developed mainly in tropical America and have depended for about a century on few clones, all triploids (AAA), *viz.*, Gros Michel and various mutant members of the Cavendish group. These are well adapted to large-scale cultivation and ocean transport to distant markets but are susceptible to certain diseases. Bananas constitute perhaps the best example in the history of agriculture of the pathological perils of monoclone culture.

Gros Michel is susceptible both to Panama disease (banana wilt) caused by *Fusarium oxysporum f. cubense* and to leaf spot (Sigatoka) caused by *Mycosphaerella musicola* which appeared in the New World in 1933. It is, however, rather tolerant of the burrowing nematode, *Radopholus similis*. The Cavendish group of cultivars is resistant to wilt but very susceptible to leaf spot and nematodes. Leaf spot can be controlled by spraying and nematodes by fumigation. Panama disease is uncontrollable in a susceptible variety. This fact provided the stimulus for banana breeding. Nowadays, resistance to all three diseases is the aim.

The approach is essentially developed to breed edible diploids (AA) which were vigorous, productive, female sterile, male fertile and disease resistant. These are crossed as males on to the disease susceptible. Gros Michel (AAA) is, if pollinated, very slightly seed fertile and, with much labour and difficulty, progenies can be built up. About one third of the seedlings are tetraploid (AAAA). In practice, tetraploids from Gros Michel are too large and, since about 1960, its semi-dwarf mutant, Highgate, has been substituted for it. The mutation behaves as a dominant with good penetrance in tetraploid progeny.

9. Epilogue: Breeding banana is a difficult exercise due to complexities resulting from parthenocarpy, sterility, polyploidy and vegetative propagation. Breeding efforts have been directed towards the improvement of dessert bananas considering fruit quality, dwarfness and disease resistance. Some objectives *e.g.* commercial characters (uniform size of fruits, shelflife, thick storage, ripening period, pericarp, etc.) and local utilities should be attended.

The use of secondary triploids deserves comment. Early backcrosses to wild AA clones (predictably) yielded only rubbish. But later combinations of improved diploids by primary tetraploids have produced occasional very attractive bunches, , though of poor fruit quality. There are possibilities here, but the prospects do not seem good. The genetic base of banana breeding is narrow. The only available female parent is Gros Michel/Highgate, the Cavendish group being completely seed-sterile. On the male side, the base has been deliberately widened over the years by collecting wild and edible acuminate diploids. Rather good collections of these are available, along with large stocks of other materials both wild and cultivated. Little effort has ever gone into breeding bananas for local use in the tropics. Most established polyploid clones are rather seed sterile but fertile enough, if pollinated, that useful tetraploid progeny can be obtained.

7.4 BER

Ziziphus Spp. (Rhamnaceae)

Resume: *Ber (Jujube) is an ideal fruit for the arid and semi-arids regions. It is thought to be originated in Central Asia wherein it is also found in wild. It* *is cultivated in China and India since the last 400 years or more. It was consumed by the ancients of the Chalcolithic age (1500-100 BC.). Presently, it is eaten fresh (when ripe) and in various forms. Basic chromosome number is reported to be ×=12. High degree of polyploidy is noted under cultivation.*

1. Introduction: Jujubes (Ber) were consumed by the ancients of the Chalcolithic age (1500-1000 BC.) and they have been under cultivation for the last 400 years in India and China. Their fruits are mentioned frequently in Yajurveda, Sutras and other ancient Sanskrit literature authored by scholars like Panini, Kautilya and Patanjali. In Yajurveda, three types of jujube are mentioned *viz.*, Badara, Koal (Kuvala) and Karkandhu, the former two referred to *Z. mauritiana* Lam. and the last one to *Z. nummularia*.

Ber is an ideal fruit tree for the arid and semi-arid regions where most of other fruit crops cannot be grown either due to lack of irrigation facilities or adverse climatic and soil conditions (Sharma and Kore, 1996). The centre of origin for this fruit crop is Central Asia (De Candolle, 1886). It is found wild, semi-wild and in cultivated forms in most parts of its occurrence. Some species of *Ziziphus* are being cultivated commercially in the areas like India, China and Uzbekistan *e.g. Z. zujuba* L., *Z. mauritiana* Lam. *Z. nummularia* (Syn. *Z. rotundifolia*). Of these, *Z. jujuba* is commonly grown in China and *Z. mauritiana* in most part of India, except eastern part of India wherein *Z. nummularia* is found under cultivation.

2. Plant Characteristics: Small or medium-sized, spiny trees; leaves ovate to broadly ovate-elliptic to suborbidular-ovate and prominently 3-nerved; flowers small greenish-yellow borne in axillary clusters or cymes; drupes smooth, yellow to orange when ripe, globose to ovoid, fleshy containing stone.

3. Parts Used: Drupes.

4. Chemical Constituents: Carbohydrate fat (0.07%), ascorbic acid, protein (0.8%), vitamin B and (succinin acid, carotene, thiamin, riboflavin, malic acid, niacin. The percentage vary species to species and cultivar to cultivar, minerals (calcium, phosphorous, iron) (Morton, 1987).

5. Uses And Applications: The fruits are usually eaten fresh but also in other forms *e.g.* dried (raising), candied, pickled and some products *e.g.* squash, juice, ber butter.

6. Karyotaxonomic Evidence: Basic chromosome number for the ber (*Ziziphus*) is ×=12. It is monobasic (Sharma and Kore, 1996). Sharma and Kore (*loc. cit.*) reported high degree of polyploids, for example, the chromosome number of Umran (a cultivar) is 2n=48 (tetraploid), in other cultivars 2n=96 (octoploid) is documented. Jhar ber (*Z. rotundifolia*) showed

2n=47 or 72. Khoshoo and Singh (1963) reported 35 chromosomes for other cultivars. Out of 35 cultivars, 33 are reported tetraploids (2n=48), one pentaploid (2n=60), and two octaploid (2n=96). Polyploidy, therefore, seems responsible for self-incompatibility (Anonymous, 1948-1976).

7. History: Collection of germplasm and selection of promising types has been attempted by some institutes and agricultural universities in India. A trial with 40 cultivers was conducted at the Regional Fruit Research Station (Bahadurgarh). Umaran, Thornless, Sanaur-2, Seo, Kala Gold and ZG-3 were found to be promising (Anonymous, 1966). Likewise, at the Haryana Agricultural University (Hissar), cultivars *viz.,* Sandhura Narnaul, Gola, Kaithli, Muria, Maharana, Sanaur-5, Katha Phal and Umran were recommended for commercial cultivation. Also, at Indian Agriculture Research Institute (New Delhi), the cultivars *viz.,* Banarsi, Karaka, Katha, Pondu, Narma, Seb, Gola and Ilaichi have been found to be promising (Sharma and Kore, 1996). In India, there are 90 or more cultivars differing in the habit of tree, leaf shape, fruits form, size, colour, flavor, keeping quality and fruiting season (*cf.* Morton, 1987).

8. Further Developments: Cross-compatibility was studied in ten cultivars. The cross between Ilaichi and Kakrola Gola resulted in the highest fruit set, whereas cross between Katha Phal and Safed Selected, as also between Kaithli and Kakrola Gola showed complete incompatibility. Umaran showed the best combining ability with other cultivars when used as female or male parent. It was observed that premature drop was reduced with colchicine application (Sharma and Kore, 1996).

Variability, heritability and expected genetic gain were reported for twelve growth character in 30 genotypes of ber. High variability was recorded for the all growth characters. Heritability estimates were quite high for most of the characters except spread of the tree. Maximum heritability was obtained for the days from fruit set to ripening followed by per cent fruit set. It was suggested that for selection of high yielding genotypes attention should be paid upon selection for more fruit set and less fruit drop (Sharma and Kore, 1996).

9. Epilogue: Ber grows well in arid and semi-arid areas. Moreover, it grows under varying climatic conditions. Only temperature below freeing is injurious to the fruits and young plants. Otherwise, it is a good candidate where fruit trees or even crops do not flourish easily. Thus, it is ideal fruit for people who especially lead life in these extreme conditions of life. Although, some products from fruits can be developed in practice, these are not circulated or sold commercially on large scale. This is one area in which attention is desired. It is, being xerophytic, mostly survives disease and pest free. A large number of characteristics although revealed, they are not exploited experimentally to their fullest potential. There is good scope for more improved varieties through breeding programmes especially for commercial and even local purposes.

7.5 BLUEBERRY

Vaccinium corymbosum L. (Ericaceae)

Resume: *Blueberry is a native to eastern Canada and eastern and southern United States. Fruits find a significant place in diets of the American natives and processed products are enjoyed by modern societies of the said region. Blueberry taxa are diploid, tetraploid and hexaploid with 2n=24, 48 and 72. It plays an important role in local economy and ecology as well.*

1. Introduction: Blueberry is a native to eastern Canada and the eastern and southern United States, from Ontario east to Novo Scotia and south as far as Florida and eastern Texas. It has been naturalized later in various places in Europe. Japan, New Zealand, the Pacific Northwest of North America (Walter, 2000, 2010; Uttal, 1987; Vander Kloet, 1980). It is also commonly called Blue Huckleberry, Tall Huckleberry, Swamp Huckleberry, High Blueberry and Swamp Blueberry. It belong to a large north temperate genus *Vaccinium* (Ericaceae). The different species of this genus are of value for their attractively acid, red or blue berries. Fruits of many wild species have long been collected. Some domestication has occurred in recent time. The plants are bushes, highbush or lowbush depending upon their sizes. However, the cuiltivated blueberries outyield the wild populations but the labour of picking it is considerable.

2. Plant Characteristics: Deciduous erect, sometimes suckering shrub, twiggy, 1.8-3.7 m tall, spreading, branches angular to terete; leaves elliptic, glossy green, subcoriaccous, serrate or entire, up to 2 inches, alternate, dense, turn to a brilliant red, orange, yellow or purple in autumn; flowers white to light pink, bell or urn-shaped, in axillary clusters, fruits, blue-black berry, pea-sized, sweet when ripe, glabrous, seeds 10-20.

3. Parts Used: Fruits.

4. Chemical Constituents: Carbohydrate (0.7%), protein (0.3%) and fat, negligible amounts of micronutrients, mineral (manganese), vitamin-C, vitamin-K.

5. Uses And Applications: Fruits of blueberries are juicy and sweet. Blueberries are sold for fresh eating or processed as individually quick frozen fruit, puree, juice or infused berries. These may then be used in a variety of consumer goods *e.g.* jams, jellies, sorbets, syrups, blue berry pastries, pies, muffins, sauce, snack foods, compotes, or as an additive to breakfast cereals. Blueberry jam is made from sugar water and fruit pectin. Bluberry sauce is sweet prepared using blueberry as a primary ingredient. Blueberry wine is made from the flesh and skin of the fruits, which is fermented and matured.

The fruits can be also dried and used like raisins. They can be smok-dried, sun-dried, boiled and baked. A tea is made from the leaves and dried fruits.

6. Karyotaxonomic Evidence: Blueberry has been investigated karyologically. It includes diploids, tetraploids and hexaploids. Chromosomes are reported as 2n=24, 48, 72 (Vander Kloet, 1980). *Vaccinium corymbasum* (2n=4x=48), the highbush blueberry, and *V. angustifolium* (2n=2x=24), lowbush blueberry, are cultivated. North American species with large fruits introduced also in Europe. Cultivars of the former incorporate genes from *V.asheri* (2n=6x=72) and *V.australe* (2n=4x=48), both American species. *Vaccinium macrocarpon* (2n=2x=24, central and eastern North America) and *V.oxycoccus* (2n=4x=48, more or less circumboreal) are craneberry species (red fruits), both often being placed in the genus *Oxycoccus*. Both are cropped from the wild but also cultivated in North America and in Europe (Lamb, 1973).

7. History: It is often found in dense thickets with moist acidic soils. Every morphological variant of the bluberry has been named formally at one time or another. At lest 25 such taxa have been raised to specific rank, none is distinct throughout its putative range nor has the properties normally associated with biological species (Vender Kloet, 1980). Many species of the *Vaccinium* are thought to have been cultivated by native Americans. The impact of their agricultural endowment to the world is a testament to their time and work in reshaping and cultivating the flora indigenous to the Americas (Mann, 2005). *Vaccinium corymbosum* L. is one of the species likely used by these peoples. It has later studied and domesticated by Frederick Vernon Coville, an American botanist. After 1910, he began to work on blueberry. He was the first to discover the importance of soil acidity. He revealed that blueberries need highly acidic soil, and also the effect of cold on blueberries. He further noted that they do not self-pollinate.

Many cultivars have been selected. They vary upon in respect of size and flavour, besides growth habit. Some varieties are available *e.g.* Bluecrop, Blueray, Jersey, North Blue, Patriat, Rubel and Sunshine Blue. These vary regarding their size and sweetness of fruit, compactness of plants, dwarfness, upright habit, fruiting earliness, hardyness, midseason fruiting, need of pollination, density of colour, foliage colour, ornamented value, etc.

The blueberry is one of the native fruit-yielding plant species in North America. Native Americans harvested the wild blueberries. Their special use is its function as dye, coloring items. It is also known as a medication for stomach complaints (Vaccinium caesanense Plant Profile).

8. Further Developments: Blue berries were traditionally hand-picked with berry-picking rakes. Modern farmers use machine harvesters that shake the fruit off the bush of cultivated blueberries The fruits are then cleaned and packaged (Production-226-Progress) .

Vander Kloet and Austin-Smith (1986) reported that the seed and pulp energy differed considerably from three geographic locations. Northern plants produced fruit with low seed energy and high pulp energy, whereas southern plants produced fruits with high seed energy and low pulp energy. Pulp caloric values varied as: (i) Florida: 52 calories/berry (ii) Nova Scotia: 141 calories/berry and (iii) Ontario:184 calories/berry. Vander Kloet and Austin-Smith (1986) speculated that the fruit ripening patterns of highbush blueberries may be related to the nutritional needs of avian seed dispersers. Mass fruiting in the north occurs in summer when avian dispersers are numerous.

Highbush blueberries rarely produce vegetatively. When burnt or disturbed, the plant occasionally produce new plants from root sprouts 3 to 6 feed away from the parent plant (Vander Kloet, 1988). However, a study by LeBlanc and Leopold (1992) in a central New York shrubby swamp thicket indicated that highbush blueberry is a good sprouter following disturbance. Two years after stems were cut at ground level, highbush blueberry sprouted. They concluded that this population of highbush blueberry was being maintained through sprout recruitment.

Antonios Petridis *et al.* (2018) studied phytosynthetic limitation in highbush blueberries (*Vaccinium corymbosum* L.). They established that: (i) blueberry fruit development depends on newly assimilated carbon and plants do not store a reserve carbohydrate buffer (ii) leaves reach a photosynthetic threshold at a light intensity of approximately 500 μmol $m^{-2}s^{-1}$ (iii) leaf stomata are slow to respond to dynamic light environments resulting in extensive stomatal limitation in such environments (iv) UK growing conditions fluctuate in a range that is likely to impact photosynthetic performance. The authors opined that the data provided indicate several potential avenues for improving blueberry yield stability.

9. Epilogue : Blueberries yield berried fruits. These have to be consumed as and when fully ripened. Being berried fruits, their shelf-life is obviously shorter. A limited amounts of its yield can be exhausted fresh. These are produced also on large scale, apart from the yield of homestead gardens. In such circumstances, these fruits can be processed in various forms for dietary purpose *e.g.* sauce, syrups, raisins, etc. Thus the blueberries afford a fair chances from commercial point of view. These plants can be grown for ornamental purpose in home gardens or grown in pots. Being a polyploid taxon, it exhibits fair diversity especially related to fruit characteristics. Suitable cultivars or varieties can be employed depending upon the edaphic and environmental factors.

7.6 BREADFRUIT AND JACKFRUIT

Artocarpus Spp. (Moraceae)

Resume: *Bread fruit (Artocarpus altilis) is a native of Malaya and Pacific Islands. Jackfruit (Artocarpus heterophyllus) is supposed to be native of Western Ghats in India. Both species of the genus Artocarpus are diploid with 2n=2×=56 chromosomes. Both are now widespread in the tropics under cultivationas as dessert fruits and also used in various forms. They are thought to be of ancient domestication of Vavilov's Indomalayan centre.*

1. Introduction: Several species of *Artocarpus* are domesticated for their fruits and other uses in Malaysia and Oceania. Chief among them are : (i) Breadfruit (*A. altilis*) and (ii) Jackfruit (*A. heterophyllus*). The former is a native of Malaya and Pacific Islands. It is now widespread in the tropics. The latter is native to India and cultivated in India especially in Maharashtra, south India along the East and West Coasts. The latter is supposed to be native of Western Ghats in India. Now, it is also widespread in the tropics. In India, they are cultivated in Orissa, Andhra Pradesh, Utter Pradesh, Bihar and Bengal states.

2. Plant Characteristics:

(i) Breadfruit: A middle-sized tree; leaves 1-2 ft. long, more or less deeply pinnatifidly cut in the upper part; male flowers in cylindrical or club-shaped yellow-green inflorescences; female flowers in large, rounded, ball-shaped green inflorescences; fruits borne in small clusters, melon-sized, prickly brownish, yellow when ripe, with fibrous yellow pulp.

(ii) Jackfruit: A large, evergreen tree; leaves elliptic or obovate, entire, dark green, shining above, petiolate, male and female flowers in separate, fleshy spikes, bore on older branches and trunk; fruits 12-30 × 6-12 in., hanging on short stalk from the trunk and larger branches, prickly, yellowish-green, pulp yellow, seeds oblong or reniform.

3. Parts Used: Fruits, seeds.

4. Chemical Constituents:

(i) Breadfruit: Protein (1.5%), fat (0.2%), carbohydrate (17.9%), minerals (calcium phosphorous, iron), vitamin-A.

(ii) Jackfruit: Protein (1.9%), fat (0.1%), carbohydrate (18.9%), minerals (calcium, phosphorous, iron), vitamin A & C, riboflavin, thiamin.

5. Uses And Applications:

(i) Breadfruit: The fruits are eaten roasted, steamed, baked or sometimes made in dough. They are also preserved by desiccation or as fermented paste. They are used as vegetable.

(ii) Jackfruit: The fruits and seeds are used as vegetable. In Kerala state of India, tasty and delicious fried chips are prepared. The pulp is made into local food articles like papad and sweet bread.

There are two varieties *viz.*, Kapa and Barka. The former has sweet and fleshy pericarp, and the latter is with mucilaginous sour pericarp of inferior quality. Unripe fruits are used as vegetable and also made into pickles. It is preserved as papad or dehydrated leather. Ripe fruits are consumed fresh or preserved in syrup. Seeds are rich in starch and consumed after boiling or roasting. Flakes are canned. Rind extract is used for making jelly.

6. Karyotaxonomic Evidence: The genus *Artocarpus* includes about 50 species and it is native to southeast Asia. The wild progenitor of the breadfruit although not identified, it probably belonged to the series Incisifolii from the eastern borders of Indonesia, western Micronesia and New Guinea. (Jarrett, 1959). The other two domesticates *viz.*, jackfruit and champedak, are closely related and belong to a different series *viz.*, the Cauliflori. All three are basically diploid with 2n=2x=56. Triploidy appears to be common among seedless breadfruit clones.

The breadfruit and its relatives *viz.*, (i) *A. altilis* (breadfruit) (=*A. communis*) (2n=56, 84) are distributed in Malaysia to eastern Polynesia, (ii) *A. integer* (champedak) (*A. champeden*) (2n=56) is found in Malaysia, and (iii) *A. heterophyllus* (jackfruit) (2n=56) found in Malaysia to India.

7. History: The above three species appear be of ancient domestication in Vavilov's Indomalayan centre. The breadfruit, a multipurpose plant is used for food, fibre, latex and timber, must have attracted man's attention very early, just as it struck the first European explorers of the Pacific with wonder. The crop, was domesticated in the eastern parts of the archipelago. It may be thought to have accompanied the human migrations that peopled Oceania 1000-2000 BP. It became there a staple food plant. Local selection isolated a large number of clones. Many of them are seedless. Parthenocarpy may be inferred. It seems that the crop has been profoundly modified in cultivation. All three species are monoecious and diclinous, bearing separate male and female inflorescences on specialized laterals. They are commonly propagated by root-cuttings, breadfruit nearly always so, jackfruit and champedak also often by seed. They are outbreeders and jackfruit seedlings are very variable (Purseglove, 1968). *Artocarpus mariannensis* of western Micronesia is taxonomically similar to the breadfruit and has sometimes been implicated in the evolution of the crop.

8. Further Developments: Breadfruit and jackfruit were widely distributed in the tropics in the 18th to 19th centuries They attained little importance outside their native areas. The story of the introduction of breadfruit to the West Indies (1793) and the associated mutiny on the Bounty has often been told (Purseglove, 1968).

The three crops are not used in any significant plant breeding effort, or even clonal selection, although breadfruit's potential for food production is higher. The stimulus may yet be provided by a serious disease, Pingelap, which has appeared in the last decades in the islands of Micronesia. Characterized by either a sudden foliage wilt or by die-back, it is of unknown origin (O'Connor, 1969) but it would seem that resistant clones can be found. Investigations of variability in the breadfruit have been reported by Jarrett (1959) and Fosberg (1960). Some decades as ago there was, in Apia, Western Samoa, a collection of Oceanian breadfruit clones established there for comparative study and breeding purposes.

9. Epilogue: The aforesaid three crops have been components for millenia of varying importance, of subsistence economies. The breadfruit and jackfruit have considerable potential. Deliberate breeding is necessary to exploit it. Being cross-pollinated and seed propagated, the jackfruit has served forms differing in size, shape, quality period of maturity, flavour, acidity, sweetness, taste etc. These afford better characteristics for future breeding.

7.7 CASHEW

Anacardium occidentale L. (Anacardiaceae)

Resume: *Cashew is thought to be a native of Brazil. It reached to South-East Asia in 16th century. In India, it was introduced by the Portuguese in the Malabar coasts in 16th century and then the Amboina in Indonesia. It gained commercial importance in the early 20th century and now stands second in international trade amongst the major edible nuts. Kernels are consumed after roasting and also in various forms. Diploid chromosome number is 2n=42.*

1. Introduction: It is one of the important tropical tree crops. It stands second in international trade amongst the major edible nuts *viz.,* hazelnut, cashewnut, walnut and almonds. It is thought to be a native of Brazil. It was introduced in the Malabar Coast of India in 16th century by the Portuguese. From there, it reached to other parts of South-East Asia (De Costa, 1578) and also eastward to Amboina in Indonesia (Rumphius, 1962). Its dispersal to South-East is believed to have been carried by birds, bats, monkeys and human agents (Burkill, 1935). The earliest record to this region is that of *Loureigo* (1790). The cashew kernels gained commercial importance in the beginning of the 20th century and it is only in the past few decades that cashew gained the status of the horticultural crop.

2. Plant Characteristics: Evergreen branched tree with light black, rough shallowly longitudinally fissured bark; leaves obovate-oblong or elliptic-obovate, 6-19 × 3. 5-9. 5 cm, glabrous and coriaceous; male and bisexual, small flowers borne in terminal panicles; nuts reniform, pale greenish-yellow or dark purple.

3. Parts Used: Nuts, apples.

4. Chemical Constituents:

(i) Kernel: Protein (21.0%), carbohydrate (22.0%), fat (47.0%), minerals (calcium, phosphorous, iron), carotene, vitamin B1, riboflavin (Nambiar *et al.,* 1996).

(ii) Apple: Protein (0.2%), carbohydrate (11.6%), fat (0.1%), minerals (calcium, phosphorous, iron), vitamin-C (Nambiar *et al.,* 1996).

5. Uses And Applications: Kernels are eaten after roasting and used in curries, sweetmeats, etc. The apples (fleshy stalk or torus) is eaten as a fruit.

6. Karyotaxonomic Evidence: *Anacardium occidentale* has been examined cytologically. It showed a diploid and haploid chromosomes of 2n=42 and n=21 respectively. The karyotype was mostly symmetric, composed mainly of metacentric pairs and several submetacentric. Similarity in the morphology, number and behavior of the chromosomes in the accessions from different populations or origin attests to the degree of genetic closeness of the selections. Polymorphism in chromosome number was not recorded among the selections. Degree of genomic closeness suggests that gene exchange among the accessions could be possible (Aliyu and Awopetu, 2007).

7. History: It was De Candolle (1825) who recognized two cultivars *viz.,* (i) Americanum and (ii) Indicum based on the ratio of nut to peduncle. Later, some authors also recognized different cultivars based on: (i) red and yellow apple, (ii) varying intensities of yellow, red and pink colour of apples, and (iii) apple and nut characteristics, (iv) size of nut, thin skin, etc. (*cf.* Nambiar *et al.,* 1996). In India, some cultivars have been released *e.g.* BPP 1 to 6, Vengurla 1 to 5, VRI 1 to 2, ULLAL 1 to 2, Anakkayam-1, etc. These have been based on different characteristics considering yield, size and weight of nut, colour and shape of apple, flowering and fruiting duration, tree canopy and size, etc. Depending upon the performance of cultivars, they have been released considering suitability of region of state in India. In India and Tanzania, some success was also achieved for commercially viable vegetative propagation technique. Moreover, damages by pests, diseases and disorders have been studied and control measures were suggested from time to time (Nambiar *et al.,* 1996).

8. Further Developments: Breeding was carried out in India for improvement of cashew. Selection was based on yield, nut size, synchronized flowering phase and a high shelling percentage. Hybridization work carried out at Anakkayam (Kerala) included an exotic parent. The progeny showed better performance than crosses between local types. About ten promising hybrids were evaluated. Similar attempts were also made at Vengurla and Madakkathora. These efforts helped to evaluate germoplasm and resulted in developing supplier strains with high yield, bold nuts and good shelling percentage (Nambiar *et al.,* 1996).

9. Epilogue: The future work should be aimed at minimum pest resistance or tolerance in the developed cultivars so that the benefits of high yielding potential can be well realized. Also, rationalization of fertilizer input, developing cultivars with better nutritive quality nuts and developing a package which is within the reach of local community are desired.

7.8 CHERRY, PLUM, PEACH, APRICOT AND ALMOND

Prunus Spp. (Rosaceae)

Resume: *Various species of Prunus are natives of Old World, especially of Eurasia. Many of them originated, developed and cultivated in China long before the Christian era. They are the most important fruit crops generally grown in the cooler temperate regions of the world. All of them are used as table fruit, highly nutritious and constitute luxury food. The genus Prunus is characterized by a basic chromosome number x=8. They are diploid or tetraploid.*

1. Introduction: The cherries are natives of Old Word, especially of Eurasia and were cultivated long before the Christian era. In USA, they have been grown since colonial times. They are also grown in Pacific Coast and still grow wild there. It has been cultivated for more than 2000 years. It was known to the Lake Dwellers and the Greeks and Romans. The plum especially *Prunus salicina* Lindl. is a native of China. It is also domesticated in America. The Peach (*Prunus persica*) is a native of China, where it has been grown thousands of years. Its numerous varieties have been developed in China and many legends have sprung in regard to its fruit. It reached the Mediterranean region very early in history and the Romans knew at least six its kinds. It was brought to the USA by the early colonists. The apricot (*Prunus armeniaca*) is called 'seeds of the Sun' in Iran. It is native of Asia and still grows wild over a wide area in this region. It was cultivated in China as early as 2000 BC. and soon reached India, Egypt, Iran and Armenia. It was introduced into Europe in the 1st century. The almonds (*Prunus dulcia* (Mill.) D. A. Webb; (Syn. *P. amygdalus* Batsch.) are amongst the world's most popular nuts and have remained so since Biblical times. It seems to have originated in the Near East, where they still grow in wild and from where it spread westward to Europe. It is introduced in California (USA), South Australia and South Africa.

After apples and pears, the cherries and plums are two of the most important fruit crops grown in the cooler temperate regions of the world. The peaches and apricots overlap the warmer portion of the plum and cherry growing areas but, like the cherries and plums, also require adequate winter chilling. Almonds are increasingly important nut crop and are cultivated in regions with a Mediterranean climate. They also require at least a short period of winter chilling.

2. Plant Characteristics:

(i) Cherry (*Prunus cerasus*): A small or tall tree with a birch like bark; white or pinkish flowers produced in clusters; fruits small with a round smooth stone.

(ii) Plum (*Prunus salicina*): A tree, 30-40 ft. in height; flowers white; fruits large, globose, smooth, clustered and variously coloured.

(iii) Peach (*Prunus persica*): A low tree; attractive pink flowers produced before leaves, round fruits have a velvety skin and a compressed, pitted or furrowed stone, endocarp either separating easily or adheres closely.

(iv) Apricot (*Prunus armeniaca*): A small tree of about 20-30 ft. in height; flowers pink, produced before leaves; fruit peach-like in shape and colour, velvety when young and has a yellowish-orange flesh.

(v) Almond (*Prunus dulcia*): A small but graceful tree; leaves alternate oblong-lanceolate, serrate; flowers very colourful in clusters of 1 to 3; fruit a green drupe with pubescent exocarp; fleshy, tough and leather mesocarp, and hard shell (endocarp) enclosing 1 or 2 ovoid, flattened seeds.

3. Parts Used: Fruits in all types.

4. Chemical Constituents:

(i) Cherry: Protein (1.1%), fat (0.5%), carbohydrates (14.5%), minerals (calcium, phosphorous, iron), vitamin-A, thiamine, riboflavin, niacin, ascorbic acid.

(ii) Plum: Protein (0.7%), fat (0.4%), carbohydrates (11.7%), minerals (calcium, Phosphorous, iron, copper), vitamin-A, ascorbic and nicotinic acid.

(iii) Peach: Amino acids (arginine, lysine, lyrosine, serine, glucine), Ca, K, Mg, Fe, Cu, Zn, Mn.

(iv) Apricot: Acids (malic, citric), sugars (glucose, fructose, 5.3-8.6%), tannin, free amino acids.

(v) Almond: Kernels: fats (58.9%), protein (20.8%), carbohydrates (10.5%), fibres (1.7%) and minerals.

5. Uses And Applications:

(i) Cherry: Cherries are used as table fruits, in pies, for glace fruits, and in canning. Cherry juice is distilled to obtain cherry brandy. The juice is also used for syrup, cherry cider, jelly and jams.

(ii) Plum: Plums are used for fresh fruits, cooking, canning and jams.

(iii) Peach: It is a favourite table fruit and popular fruit for canning. It is also dried or frozen. Oil obtained from seeds is used for cooking.

(iv) Apricot: Apricots are used as table fruit. They are dried, frozen, canned, candied and made into a paste. Seed oil is a substitute for almond oil.

(v) Almond: Kernels are highly nutritious and luxury food. They are consumed either raw or in confections and puddings.

6. Karyotaxonomic Evidence: The genus *Prunus* have a basic chromosome number of ×=8. Rehder (1954) has classified *Prunus* into 77 species. The sub-genera, sections, occurrence and major crop plants are, given below with their respective chromosome numbers (Knight, 1969):

Subgen. Prunophora

Sect. Euprunus (6 spp.; 2×-6×; Europe to China)

P. domestica, European plum (6×); *P. insititia*, Damson plum (6×);

P. cerasifera, Cherry plum (2×, 3×, 4x, 6×); *P. salicina*, Japanese plum (2×, 4×).

Sect. Prunocerasus (13 spp; 2×; North America).

P. americana, American plum (2×).

Sect. Armeniaca (6 spp.; 2×, 3×; Europe to Korea).

P. armeniaca, Apricot (2×).

Subgen. Amygdalus

Sect. Euamygdalus (6 spp.; 2×, 8×; western Asia to China).

P. persica, Peach (2×); *P. amygdalus*, Almond (2×).

Sect. Chamaeamygdalus (1 sp.; 2×, 3×; southeastern Europe to eastern Siberia).

Subgen. Cerasus

Sect. Microcerasus (8 spp.; 2×; north temperate).

Sect. Pseudocerasus (14 spp.; 2×, 3×, 4×; central Asia to Japan).

Sect. Lobopetalum (2 spp.; 4×; China).

Sect. Eucerasus (3 spp.; 2×, 3×, 4×, 5×; Europe to central Asia).

P. avium, Sweet cherry (2×); *P. cerasus*, Sour cherry (4×).

Sect. Mahaleb (3 spp.; 2×; Europe, western Asia, North America).

Sect. Phyllocerasus (1 sp.; China).

Sect. Phyilomahaleb (1 sp.; 2×; Manchuria, Korea, Japan).

Subgen. Padus (11 spp.; 4×; Europe, Asia, North America).

Subgen. Laurocerasus (2 spp.; 8×, ca 22×; Europe to western Asia).

Amygdalus and *Prunophora* form one group for purposes of genetic transfer. Within this group, almonds and damson plums are isolated from the mainstream of genetic transfer. Transfer between the Amygdalus-Prunophora group and the sweet and sour cherry group is only rarely direct, being usually via the *Microcerasus* section *Cerasus*. The Amygdalus part of the group appears to be more loosely connected to the *Microcerasus* bridge than the Prunophora portion of the group.

The subgenus *Padus* is only weakly related to other *Prunus* (via *Cerasus* and *Amygdalus*) if the number of hybrids between *Padus* and other species is considered as a guide. The subgenus Laurocerasus thus appears to be even more isolated from other subgenera.

7. History: The first diploid species in the genus *Prunus* arose in central Asia. The Eucerasus section, containing the sweet and sour cherries, were early derivatives of such an ancestral *Prunus* species. The present day *Cerasus* species in the section *Microcerasus* are probably closer to this ancestral *Prunus* species than the commercial cherries in the Eucerasus section, as the *Microcerasus* form a genetic bridge for hybridization purposes between the Eucerasus and the *Amygdalus* and *Prunophora* subgenera.

The North American cherry species *P. besseyi* and *P. pumila* must have been derived from the central Asian *Prunus*. The Chinese *P. tomentosa*, also an important present-day element of the *Microcerasus* bridge, is probably a branch from the link between the North American *Microcerasus* and the *Prunus* centre of origin.

The tetraploid sour cherry, probably originated either from diploid *P. avium* or from tetraploid *P. fruticosa*. If *P. fructicosa* played role in the ancestry of *P. cerasus*, then the sour cherry must have derived before the spread from western Asia of both the sweet and sour cherry to their secondary centre of origin in Europe, as the centre of origin of *P. fruticosa* remains in western and central Asia.

The plums appear to hold the centre of the *Prunus* stage, as they have the greatest genetic diversity of any subgenus and are a link between the other major subgenera. The centres of origin for plums include: Europe for *P. domestica* (European plum); western Asia for *P. insititia* (Damson plum); western and central Asia for *P. cerasifera* (Cherry plum); China for *P. salicina* (Japanese plum); and North America for *P. americana* (American plum). In early historic times, selection by man in the various centres of origin and also in the many areas to which they were distributed in early and more recent historic times has ensured the retention of more genetic diversity than if selection had been practised at only one centre.

Prunus americana and the other plum species of the section Prunocerasus, all with a North American centre of origin, seem to be closely related to the Asian and European plums of the section Euprunus. The separation of plums into Prunocerasus and Euprunus appear to have a geographical rather than a genetical basis. It is likely that Prunocerasus plums were established in North America relatively recently as it is difficult to imagine that they could have moved eastwards from central Asia to North America at a very early stage and still remain relatively unchanged. If Prunocerasus and Euprunus are part of a single section, then *P. salicina* of China may be a branch from the link between the North American plums and the plums from central Asia. *Prunus domestica*, the European plum, is the most recent crop species to have arisen within Prunophora. *Prunus instititia, P. cerasifera* and *P. spinosa* were probably involved in its ancestry.

The apricot (*P. armeniaca*) also like the peach, has a primary centre of origin in western China. However, the apricot has secondary centre of origin in western Asia. *Prunus armeniaca* appears to be further from the centre of the genus *Prunus* than the plums. The factors which contributed to selection for self-fertility must have also thereby contributed indirectly for isolation of the species.

The almond (*P. amygdalus*) established in a separate centre of origin in an area extending from central to western Asia. The peach and the apricot both have self-fertilizing breeding systems in the harsh environment of western China. The almond, in the slightly more moderate environment of central and western Asia, remained self-incompatible (Vavilov, 1951). The ancestral species from which both peach and almond arose probably split at an early stage somewhere in central Asia, with the peach evolving to the east in western China and the almond evolving and moving westward. The almond maintained a foothold in the ancestral homeland of central Asia.

The peach, with its Chinese centre of origin close to where most of the present-day *Microcerasus* species occur, appears to be closer to the genetic centre of *Prunus*, and hence closer to the *Microcerasus* bridge to the cherry group, than the almond, which has a centre of origin to the west of all the *Microcerasus* species.

8. Further Developments: All fruit crops were grown from seed in early historic times. The best selections were propagated vegetatively on seedling rootstocks, usually of the same or closely related species. Quite recently there has been a trend towards selection and breeding of uniform vegetatively propagated rootstocks as a replacement for variable seedling types (Tydean, 1962).

It is probable that cherry was domesticated several centuries before this time in the region of Asia Minor or Greece, since Theophrastus, in his history of plants (300 BC.) mentioned the cherry. The cultivated forms of the cherry spread throughout the Roman Empire to include the area already covered by the wild forms of *P. avium* and *P. cerasus*. There was a decline in interest in cherries after the fall of the Roman Empire and (as late as 1491).

An important German Book, the Herbarius (1485) by Johannes, listed only two types of cherry *viz.,* Sweet and Sour, even though the Romans recognized numerous types many centuries earlier. Cherries were again extensively planted especially in Germany, from the 16th century onwards. Cherries were taken to North America by the early settlers from Europe and spread with them across the continent. The sour cherries were more favoured in the east and sweet cherries in the west. The cherry reproduces fairly true from seed, particularly the tetraploid sour cherry. This is partly due to the natural isolation of Eucerasus cherries from other *Cerasus* species. They developed to the west of the central Asian *Cerasus* centre of origin whilst most other *Cerasus* species evolved to the east. With so many cherry varieties so similar to each other, it is common to find cultivars grouped by type rather than by individual name.

The five main groups of plum evolved in three distinctly different areas of the world until about 200 years ago; since then they have been used increasingly as a common gene pool in breeding. All five plum groups, and many wild plum species, can be readily hybridized. They together form an outstanding pool of genetic variability. Many wild species of plum have been successfully used in breeding programmes, especially in North America. The dominant commercial group is the hexaploid European plum, *P. domestica.* It gained very significant contributions from both wild and cultivated diploid species.

Apricots were introduced into Europe during the Roman era. The production of late-flowering types and of types adapted to environments of new areas of production were major breeding objectives. Firmer flesh, higher fruit quality, disease resistance and hardiness were also been important objectives. The apricot hybridizes with the plum and so-called 'plumcot' varieties have been selected. Yellow fruited types which resemble the purple apricot (*P. dasycarpa*) can be produced by crossing apricot with the Myrobalan plum (*P. cerasifera*).

The peach spread from China to Europe during the last 2,000 years. It was subsequently taken to North America and other parts of the world such as South Africa and Australia.

Almonds (*P. amygdalus*) spread at an early stage from Asia to the Mediterranean and then to North America. There has been a very considerable increase in production especially in California. In Mediterranean and Asian countries such as Turkey, almonds are almost exclusively grown from seed. In North America, they are mostly propagated vegetatively. Breeding was aimed at kernel selection, late-flowering, self-fertility and early harvest (Kester, 1969).

9. Epilogue: A trend in *Prunus* breeding is related to studies of the inheritance of economic characters and the subsequent transfer of such characters to new commercial varieties. An integral part of this trend will be the increased use of wild species, particularly as contributors to the breeding of scion varieties where adaptation to new environments is important. Also, wild *Prunus* species will provide valuable characters needed in the breeding of new vegetatively propagated rootstocks. Many species will be tested and little known types such as *P. jacquimontii* may conceivably provide valuable rootstock parents for both the cherry and plum sides of the *Microcerasus* bridge.

The cultivated plums and related wild species of Prunophora appear to be in a particularly favourable position for the exchange of genetic material both within the subgenus and also with *Amygdalus* and *Cerasus* and are likely to be increasingly used for this purpose.

Commercial cherries are the most isolated group. If the potential for future improvement of the cherry crop is to be raised to a satisfactory level, it will be necessary to intensify the effort devoted to hybridizing with species outside the Eucerasus section of *Cerasus* (Zylka, 1970) and the prospects for producing new rootstocks. The well-known Japanese ornamental cherries (*P. serrulata*) are a possible source of resistance to the aphid, *Myzus* cerasi.

The subgenus *Laurocerasus* will have much to contribute in the immediate future to the production of better varieties. The possibilities for using the subgenus Padus in the near future are slightly better. The potential contribution of wild *Prunus* species to the production of new varieties, suited to an even greater range of environments than those available to our present *Prunus* fruit crops, is considerable. In the attempts of developing countries to mimic the advances made in developed countries by replacing indigenous types and wild stands with western varieties, there is a danger that a vital portion of the natural genetic variability will be permanently lost. The pending switch from seed to vegetatively propagated almonds in Turkey, part of the centre of origin for *P. amygdalus*, is an example of such a hazard.

7.9 CITRUS

Citrus Spp. (Rutaceae)

Resume: *Most Citrus species are thought to be native of tropical and subtropical regions of South Coast Asia, especially India and China. All of them are popular for their fruity sweet taste and sources of vitamin-C. Their fruits are esteemed for delicious dessert fruits and used in various forms of preparations, drinks, preserves, liquors, etc. They are medicinally important. The Western countries became familiar during the Middle Ages. Various species are characterized by basic chromosome number x=9, majority of species being diploid. Parthenocarpy and nucellary polyembryony prevails amongst the species of the genus.*

1. Introduction: It is believed that most *Citrus* species are native to tropical and subtropical regions of Southeast Asia, particularly India and China. Chinese literature written as early as 2200 BC should evidence of citrus cultivation in China (Ghosh, 1996). It is believed that *Citrus* was introduced in Greece between 250 and 200 BC. (Tolkowsky, 1938).

In Europe, citrus was introduced in the 16th century by the Portuguese. During 1st century AD, Arabians took it to the east coast of Africa. It is believed that Columbus took orange and lemon seeds to America (Ghosh, 1996).

The utility of citrus fruits in checking scurvy was scientifically proved in 1756 by John Lind, a surgeon in the British navy. The use of limes or lemons as such was established as a custom on British ships. The sailors were hence called limeys by people from other countries even today. All citrus fruits contain varying proportions of vitamin-C.

The genus *Citrus* contains the most important species, sweet, bitter and mandarin oranges, lemon, lime, grapefruit, citron and the shaddock or pummel, etc. All these are appreciated all over the world for their 'fruity' sweet taste and refreshing virtues. The cultivated species of *Citrus* are supposed to be natives of south-east Asia particularly eastern India, Indo-china, southern China and the Philippines. The grapefruit, however, appear originated in the West Indies as a chance cross between this species and the sweet orange, or as a bud mutation from the pummelo. Since remote times, the *Citrus* species were taken to Europe by Arab traders by the overland route via India, Persia and Syria. The citron become known to European civilisation about three hundred years before Christ and probably introduced by Alexander the Great. The Western World became familiar with the more popular citrus fruits only during the Middle Ages. Sour or seville oranges were taken to Europe in the 11th century AD. The lemon and lime seems to have arrived next by the same route about the 12th or 13th century. The sweet orange did not reach Europe until the second half of the 15th century and the mandarin or tangerine not before 1805. Portuguese and Spanish explorers quickly introduced citrus into the New World after the voyages of Columbus. From the West Indies, citrus fruits soon spread to continental America, reaching Florida until 1565 and then to Georgia and South California during the 18th century by Spanish missions. They are now cultivated in every tropical and subtropical region of the world.

2. Plant Characteristics: Species of *Citrus* much branched, evergreen, small aromatic trees or shrubs, often spiny; leaves unifoliate (compound) derived from trifoliate compound; flowers white or purplish pink, often scented, borne either solitary axillary or clustered in small terminal cymes; fruits a specialised berry (hesperidium), globose, subglobose or oblong, pointed with a thick and leathery protective covering and spongy mesocarp.

3. Parts Used: Fruits.

4. Chemical Constituents: The chemical composition of *Citrus* fruits differ with species and cultivars. Approximately it is: Juice (40-45%), rind (25-30%), pulp (20-35%), sugars (glucose, sucrose, 5-8%), pectins (citric acids, little malic, 1-2%), small amounts of different acids, proteins, essential oils and minerals.

5. Uses And Applications:

(i) Lime:

(a) *Citrus aurantiifolia* (Christm.) Swing: The fruits make a good source of vitamin-C. They are used raw and pickled. The oil distilled from the peel is used in confectionary, pharmaceuticals and toilet preparations. Fruits are extensively used for culinary purposes; for flavouring jams, jellies, marmalades, and alcoholic drinks, and as a garnish.

(b) *Citrus reticulata* Blanco: The fruits are delicious. They are rich in vitamin-C. The peel is used for marmalades. The essential oil is distilled mainly in Nagpur and Coorg and used in confectionary, pharmaceuticals and toilet preparations. It is most valued commercial orange. Fruit is used mainly as dessert and in the production of orange juice.

(ii) Sweet orange [*Citrus sinensis* (Linn.) Osbeck]: The fruits are edible. They are good source of vitamin-C. The peel is used in the source of orange oil. Fruit is sweet and juice, nutritious, highly esteemed dessert fruit.

(iii) Sour Orange (*Citrus aurantium* Linn.): The fruits are edible. They are used in the preparation of confections, marmalades, liqueurs and other drinks. It is rich source of pro-vitamin A and B1.

(iv) Shaddock [*Citrus maxima* (Burm.) Merrill.]: The fruits are edible. They are good source of vitamin-C. Fruits are esteemed for dessert, made into jams and marmalades, considered nutritive and refrigerant.

(v) Citron (*Citrus medica* Linn.): The fruits are edible. They are enriched with vitamin-C. Fruits are used mainly for pickling, also candied. Peel is made into marmalades and other preservers.

(vi) Sweet Lime (*Citrus limettioides* Tanaka.): The fruits are edible and good source of vitamin-C. It is also used as a root stock for sweet oranges and mandarins.

(vii) Grapefruit (*Citrus paradisi* Macf.): The fruits are edible, rich in vitamin-C, and also a fair source of vitamin-B. It is used mainly as a breakfast fruit. Dry and fortified wines, brandies and cordials are prepared from the fruits.

(viii) Lemon [*Citrus limon* (Linn.) Burm. *f.*]: The fruits are eaten raw and pickled. The juice is used for lemonade and other beverages as a flavouring substance, as bleaching agent and stain remover. The rind is the source of oil of lemon. Lemon oil is also used for flavouring liqueurs. Lemon juice is very useful for survey.

6. Karyotaxonomic Evidence: *Citrus* and its nearer relatives are represented by 28 genera in the tribe Citrea (subfamily Aurantioideae). Of these, six genera, including *Citrus*, comprise the true citrus fruits, which are evergreen trees with highly specialized pulp vesicles in the fruit. The genera *viz., Citrus, Fortunella* (the kumquat) and *Poncirus* are closely related. *Poncirus* is the monotypic genus. The rest other three genera are *viz., Microcitrus, Eremocitrus* and *Clymenia*, the last being probably the most primitive.

Swingle (1967) recognized 16 species in the genus *Citrus*. Tanaka (1954) proposed 145 species and later 157. A classification including 36 species has been proposed by Hodgson (1961). The *Citrus* forms of primary economic importance have many characters in common and are widely interfertile. Their hybrids are also often fertile. Sterility and incompatibility do occur. Asexual seed reproduction is prominent and it is not species-limited. It resulted in confusion in knowing recurring genotypes after seed propagation. In all *Citrus* species and related genera, the basic chromosome number is x=9. These include at least 13 genera and well over 70 forms within *Citrus* (Krug, 1943). Majority of them are diploid. A few polyploids have been identified in nature including a tetraploid *Fortunella hindsii* and a tetraploid *Triphasia*.

7. History: The subfamily Aurantioideae (Rutaceae) was confined to the Old World. *Citrus* and its relatives arose in southeast Asia. The main centre is apparently eastern India. Relationships can be traced through the East Indies, Australia, central China, Japan and even Africa, to which four genera are endemic. Tanaka (1954) marked a line of demarcation in southeast Asia. This line separated areas of probable development and spread of some *Citrus* species and relatives. It runs south-eastwards from northern India, passing north of Burma and south of the island of Hainan. To the southwest is Tanaka's subgenus Archicitrus. It includes the lemon, citron and orange. To the east and northeast is Metacitrus, which shows relationships to present-day mandarins, *Poncirus* and *Fortunella*.

Some information is available about the history of the principal horticultural forms together with data on reproductive behaviour and usefulness as rootstocks. It will be seen that one form (grapefruit) originated in the New World and that only two groups regularly reproduce sexually. The ease of hybridization among groups is indicated by the existence of many hybrids of horticultural importance.

Table 1: Principal horticulture groups of Citrus

Name*	Probable first written records		Embryony Type	Prominent rootstocks
	Europe	New World		
1. Citron	300 BC, Persia	—	Sexual	—
2. Lemon	12th-13th C.	1493m, Haiti	Partly nucellar	Rough lemon (hybrid)
3. Lime	13th C.	—	Partly nuccellar	Palestine sweet
4. Pummelo	12th C.	17th C., Barbados	Sexual	—
5. Sour orange	11th C.	—	Highly nucellar	Many
6. Sweet orange	15th C.	1493, Haiti	Highly nucellar	Many
7. Mandarin	1805	Mid 19th C., USA	Some cvs, nucellar, some sexual	Cleopatra
8. Grapefruit	—	18th C., West Indies	Highly nucellar	—

* (Webber, 1967; Swingle, 1967; Frost and Soost, 1968).

8. Further Developments: The first systematic *Citrus* breeding was begun in 1893, by the US Department of Agriculture (Florida) and the University of California (Riverside) in 1914. Efforts by both continued and the USDA extended its studies at other locations (Cameron and Frost, 1968). Although, breeding studies were conducted in Java, the Philippines, Italy, Japan and southern Russia, these were, however, interrupted by the late war. Research was extended hence in Japan (Nishiura, 1964).

Although *Citrus* is capable of self-pollination, cross pollination can also take place. The interfertility of many forms, and even of *Citrus* with *Poncirus* and *Fortunella*, makes hybridization relatively easy unless other factors interfere (Frost and Soost, 1968). Genetic sterility is present in many oranges, such as Washington Navel, and in numerous cultivars of other species. High percentages of staminate flowers in lemon and lime serve only as pollen sources. Intergeneric progenies vary greatly in vigour and fertility. Self-incompatibility is determined by a series of oppositional alleles of poorly known distribution. Only in the pummelo are the alleles known to be widespread (Soost, 1964). Parthenocarpy occurs in some cultivars *e.g.*

Washington Navel orange and Satsuma mandarin. Nucellar embryony is valuable for the reproduction, of genetically uniform plants for rootstocks. It is also obstacle in breeding. Several common forms, including the orange, mandarin and grapefruit, include cultivars which bear embryos that are nearly all asexual. Some related cultivars also produce varying proportions of sexual ones. Two species, the pummelo and the citron, are completely sexual. Inheritance, of the character has proved relatively simple. Completely sexual parents always produced offspring which are completely sexual. The asexual x sexual crosses can produce both types. Progeny which are sexual have also been obtained when both parents were at least partially asexual. Many new sexual hybrids have been produced.

Citrus is raised on rootstocks and breeding programmes for both rootstocks and scions have been carried on. Fruit size, flavour and season of ripening, as well as tree vigour, are critical for scion improvement. Fruit shape and colour, number of seeds and storage life are also important for fresh use. High solids, good colour and favour are sought for fresh use. Fruit size has often been a limiting character. Therefore the large-fruited pummelos have frequently been used as parents with, for example, mandarins and oranges. Hybrids of good size and flavour are obtained. High seed number is a persistent problem. Generally, seediness occurs in most diploid progeny and hybrids which are infertile often give low fruit set. Parthenocarpy sometimes provides for good fruit set in a seedless selection.

Fruits with different maturity are always of interest and a wide range of ripening periods can be obtained from parents ranging from early-ripening mandarins to late oranges such as Valencia. A low-acid pummelo has produced mild, early-ripening progeny. High levels of soluble solids, with excellent flavours have frequently been obtained among mandarin hybrids. Some Citrus selections tolerate higher maximum temperatures than others and certain hybrids sunburn or fail to set fruit in hot desert areas. Mutation is prominent in citrus, and can appear as limb sports, fruit sectors or whole plants from nucellar seeds. Several valuable cultivars, including the Washington Navel orange (parthenocarpic), the Marsh grapefruit and Shamouti orange, arose by mutation.

The breeding programme in Florida included crosses of *Citrus* with *Poncirus* to produce cold-hardy rootstocks. Several hardy hybrids were obtained. Additional crosses with *Poncirus* have since been made. In Japan, the genus was crossed with other cold-tolerant cultivars such as Yuzu and Natsudaidai. Crosses for cold-hardy scion types have been made, involving the Satsuma mandarin, Meyer lemon, *Fortunella* and Natsudaidai.

The most important diseases of citrus include *Phytophthora* root rot, citrus canker, the viruses of tristeza, psorosis and exocortis, and the mycoplasma causing stubborn disease. The citrus nematode and the burrowing nematode are also injurious in some areas. Since *Poncirus* is relatively resistant to *Phytophthora*, tristeza and the citrus nematode has often been used in crosses with *Citrus* for new rootstocks. Hybrid rootstocks tolerant of the citrus nematode have been produced but tolerance is complicated by the occurrence of nematode biotypes,

Stubborn disease is serious and wide-spread and injures many scion types regardless of rootstock. A study on response to salinity among F_{1S} involving Rangpur lime and certain mandarins as parents was made by the USDA.

Hybrid tetraploid seedlings seldom occur from crossing diploids. Tetraploids grow more slowly and are more compact and generally less fruitful than comparable diploids. Seed number depends on cultivars. Tetraploids have little commercial potential but they are valuable for the production of triploids. Many spontaneous tetraploids in *Citrus* and *Poncirus* have occurred as nucellar seedlings.

Triploids frequently arise as sexual seedlings. They can be systematically produced by crossing tetraploids with diploids. However, production by crossing 4x (female) by 2x (male) is often not feasible for want of sexual 4x parents. The reciprocal cross (2x × 4x) produces many tetraploid individuals (Esen and Soost, 1973). These occur because of the frequent production of diploid megagametophytes as well as the failure of most triploid embryos. The frequent occurrence of diploid megagametophytes also can be utilized for the production of triploids from 2x × 2x crosses. Triploid seeds are much smaller than sister diploid seeds. Unlike triploids from 2x × 4x crosses, their embryos and endosperms develop normally. By germinating the small seeds, recovery of triploids can be increased. Triploids are generally more vigorous than tetraploids. Fruitfulness has been highly variable. All have few or very few seeds. Triploids show promise for commercial use when selection for high yield is successful.

9. Epilogue: The range of strictly sexual parents has been increased and more can now be developed by breeding. Sexual tetraploids will greatly facilitate production of triploids. Several biochemical characters may have promise for separation of sexual and asexual seedlings (*cf.* Pieringer *et al.*, 1964). Advances in breeding and genetics by conventional methods are slow because of quantitative inheritance of most characters, long generation time and nucellar embryony.

7.10 CURRANTS

Ribes Spp. (Grossulariaceae)

Resume: *Various species of Currants belong to the genus Ribes and native to both the New and Old Worlds. They are cultivated for fruits in the erstwhile USSR, Britain, West Germany and Poland and North America. All*

species are diploid with 2n=2×=16 chromosomes. Main bulk of crop is processed for juice and for various food products. They are highly valued for vitamin-C.

1. Introduction: Currants are native to both the New and Old Worlds. They are grown in Europe since the middle ages and were early brought to America, where they are naturalized. Black and red currants are grown on a limited scale in most cool temperate countries. The major producers are the USSR and eastern and northwestern Europe, particularly Britain, West Germany and Poland. Much of the east European crop is exported to Western Europe. Currants are little grown in North America since, being alternate hosts of white pine blister rust (*Cronartium ribicola*), they are banned wherever five-needled pines are an important crop.

2. Plant Characteristics:

(i) Black currant (*Ribes nigrum* L.): Erect shrub, upto 2. 0 m; leaves 3-5 lobed, serrate, 4-8 cm long; flowers greenish purple, in pedulous racemes; berries globose, black, strong aromatic.

(ii) Red currant (*Ribes rubrum* Linn.): Erect shrubs; leaves ovate-cordate, 3-5 lobed; flowers greenish yellow, in elongate, pedulous racemes; berries red, smooth.

3. Parts Used:

(i) Black currant (*Ribes nigrum* L.): Berries.

(ii) Red currant (*Ribes rubrum* Linn.): Berries.

4. Chemical Constituents:

(i) Black currant (*Ribes nigrum* L.): Pectin (0.9-1.7%), minerals (calcium, potassium, magnesium), sugars (glucose, fructose), malic acid, flaronis (kuempferol, quercin, myricetin, cyaniding), rich in vitamin C.

(ii) Red currant (*Ribes rubrum* Linn.): Pectin (0.8%), protein (1.1%), sugar (monosaccharides 4.4%), acid (citric 1.2%), and ascorbic acid.

5. Uses And Applications:

(i) Black currant (*Ribes nigrum* L.): They are used in making jelly and jam, sauces, pies, pastilles, wine, syrup, etc. It is especially valued for the high vitamin-C content of its products.

(ii) Red currant (*Ribes rubrum* Linn.): It is popular for making tarts, jams, jellies, etc.

6. Karyotaxonomic Evidence: The genus *Ribes* includes nearly 150 species. All are diploid (2n=2×=16) distributed mainly in the temperate regions of Europe, Asia, North and South America. Most intra-sectional (and some intra-subgeneric) combinations produce vigorous, fertile F_1's, while intersubgeneric and many intersectional combinations either fail or produce sterile hybrids which invariably show meiotic irregularities. Black currants are included by Janczewski (1907) in the Eucoreosma section (10 sp.) of the subgenus Coreosma, and red currants in the subgenus Ribesia (15 sp.). Four species of Eucoreosma and four of Ribesia are important in the evolution of black and red currant cultivars respectively. Characteristics and worldwide distribution of both black and red currents are:

(I) Black currants:

(i) *R.nigrum:* The European black currant. The sub-species *eurapaeum*, *scandinacicum* and *sibiricum* recognized in Russian literature. Northern Europe, central and northern Asia to the Himalayas.

(ii) *R.dikuscha:* A winter-hardy species with bloomed fruits. Eastern Siberia, northern Manchuria.

(iii) *R.ussuriense:* Resembles *R.nigrum* but suckers. Manchuria to Korea.

(iv) *R.bracteosum:* The Californian black currant. Erect growing with bloomed fruits borne on long racemes. Alaska to northern California.

(II) Red currants:

(i) *R.sativum:* Under cultivation gave rise to the large-fruited 'cherry' currants (*R.macrocarpum*) and white-fruited forms. Western Europe.

(ii) *R.petraeum:* A variable species, donor of resistance to leaf spot, with fruits ranging from red to almost black. Mountainous regions of Europe and North Africa, Siberia.

(iii) *R.rubrum:* Pink and white forms known. Central and northern Europe, northern Asia.

(iv) *R.multijlorum:* Dense inflorescences with over 50 flowers. Mountains of southern Europe.

7. History: Currants were domesticated in northern Europe since the last 400-500 years. A red currant is first described in the German manuscript of the early 15th century. The black currant (*R. nigrum*) was first recorded in Britain in 17th century herbals. It sought attention for medicinal properties of the fruits and leaves. Three species contributed to the early evolution of red currants. *R. sativum* was the only one grown in 1542, *R. petraeum* was introduced into cultivation shortly after this and *R. rubrum* very much later. European currants were taken to North America, probably in the 17th century, by early English settlers.

Colour forms (white and pink red currants and green black currants) were listed in 18th and 19th century catalogues. Fruit size and fruiting habit of the older cultivars differ little from those of wild types. There was probably unconscious selection for self-fertility, as selfed progenies of most cultivars show marked inbreeding depression. This suggests that the wild progenitors of the crop plants were obligate outbreeders. Black currant varieties available in Europe by the 19th century included Baldwin, Black Naples and black Grape and all are extant. Baldwin is the leading British cultivar today. During the 19th century, the numbers of named varieties increased considerably because of the introduction of plants raised by private individuals or nurserymen from open-pollinated seed of the narrow range of the then existing cultivars. Hatton (1920) was able to classify 26 varieties, all of pure *R. nigrun* descent, into four main groups of similar varieties, the types being French Black, Boskoop Giant, Goliath and Baldwin .

Black currant growers in North America relied on European cultivars although, towards the end of the 19th century, a number of Canadian-bred varieties derived from earlier importations were released. The arrival of white pine blister rust by this time on pine seedlings from Europe resulted in severe epiphytotics on pine, leading to the banning of *Ribes* crops in large areas of the continent. Red currant varieties are very long-lived. Red and White Dutch were described in Mawe and Abercrombie's (1778) Dictionary Of Gardening And Botany, along with four other red currants. Still other varieties were raised later both in Europe and north America until in 1921. Bunyard (1921) described for 31 distinct types and classified these into five groups according to the species they most closely resembled. The Raby Castle group, derived from *R. rubrum pubescens*, includes Raby Castle, raised in Britain before 1860, and Houghton Castle (*rubrum* × *sativum*), introduced about 1820. The Versailles group comprises the large-fruited descendants of *R. sativum matrocarpum*, including Versailles, raised about 1835 in France, Cherry (raised before 1840 in Italy) and Fay's Prolific and a number of other American cultivars introduced in the late 19th century. The Gondouin group comprises the derivatives of *R. petraeum* such as Gondouin (*petraeum* × *sativum*), Prince Albert and Seedless Red. The Scotch groups are descendants of *R. rubrum*, while the Dutch groups have flowers of the savtium type (Knight and Keep, 1958; Keep, 1975).

8. Further Developments: Breeding was, at beginning, largely confined to intercrossing existing varieties. The main selection criteria was *e.g.* yield, fruit size, fruit quality, and season. In Britain, work of this nature on black currants resulted in a series of cultivars, usually hybrids between representatives of two of Hatton's four groups, released between 1927 and 1962. Wellington XXX, the most widely grown cultivar in Britain today after Baldwin, was introduced in 1927. These cultivars did not represent a marked advance on their parents, being equally unreliable in cropping. However, this work demonstrated the importance of hybrid vigour for commercial performance and the necessity for a broader genetic base if major advances were to be made. Subsequent breeding programmes markedly considered these findings. The main centres of black currant breeding over the past few decades have been western Europe, Scandinavia and Russia. The high cost and increasing scarcity of hand labour in Europe for this labour intensive crop, coupled with comparatively low returns, have put a premium on characters that reduce the costs of production. There was emphasis on breeding for regular, heavy cropping and for pest and disease resistance. Breeding for hardiness, with the particular aim of extending the northern limits of black currant growing in the then Russia was an additional major objective besides high vitamin-C content. On the advent of mechanical harvesters, there was influence on breeding programmes and some recent varieties (*e.g.* the Dutch Black Reward and the German Invigo) have firm fruits desirable for machine harvesting by shaking and erect habit.

Related wild species were used as donors of pest and disease resistance not found or poorly developed in *R. nigrum* itself. In Russia, *R. dikuscha* provided strong leaf spot resistance and hardiness, Primorskij Cempion, much used in later breeding, being one of the first commercial derivatives. Consort (*R. nigrum* × *R. ussuriense*), bred in Canada and deriving resistance to white pine blister rust from *R. ussuriense*, has been widely used in European black currant breeding. Other donor species, include *viz.,* the North American *R. sanguineum, R. glutinosum, R. cereum*, supplying resistance to mildew, leaf spot, leaf-curling midge (*Dasyneura tetensii*) and aphids; and the European gooseberry, *R. grossularia*, donor of strong major gene resistance to the gall mite. Red currant breeding in the 20th century has been on a relatively small scale. Varieties released since Bunyard's (1921) review include the American Red Lake, a leading cultivar in Europe, and the Dutch Jonkheer van Tets and Maarse's Prominent. These are in direct line of descent from those raised in previous century.

In a nutshell, modern currant breeding programmes have mostly involved intercrossing locally adapted with genetically dissimilar imported cultivars of complementary phenotypes. Early products of such programmes released both in Europe and Russia. Additional variability was supplied by related species which are being increasingly used as donors in backcrossing programmes.

9. Epilogue: The commercial currant growing in future will certainly depend on the use of mechanical harvesters, so that breeding is likely to be aimed increasingly at the upright growth habit and firm, even ripening fruit needed for machine harvesting by shaking. In case of red currants, these and the additional requirement for a more regular distribution of the crop over the bush are likely to favour *R. multiflorum* derivatives. Programmes involving cultivars of widely differing provenance and wild species are just beginning to produce commercial black currant selection. The introduction of reliably heavy-cropping varieties which can be machine-harvested increases the profitability of the crop. The increasing demand for fruit-flavoured ready-to-eat products such as yoghourt and other dessert confections should ensure an expanding market in future.

7.11 CUSTARD APPLE

Annona squamosa L. (Annonaceae)

***Resume:** Custard Apple is South American in origin. It is a native of West Indies. If naturalizes fast in tropical countries and hence sometimes (erroneously) thought native to such as country like India. Its fruits are delicious, pleasant flavoured but have short shelf-life. It is hence consumed mostly locally. Chromosome number are reported to be 2n=2x=14 to 16 in species of Annona.*

1. Introduction: Custard Apple (*A. squamosa* L.) is a native of South America and West Indies (Bailey, 1949). The occurrence of wild stands in the nature, Sanskrit names and carvings of the custard apple fruits and leaves and mention in Ain-i-Akbari had led many to believe that custard apple is the Indian origin. But Watt (1889) concluded that it is introduced and no species of the genus *Annona* is endemic to any Indian forest. After its introduction by the Portuguese in India, the seeds dispersed quickly and got acclimatized throughout south India. It is now cultivated throughout tropical regions. It is identified among the sculptures of the Ajanta caves and Bharhut Stupa (Watt, 1889).

2. Plant Characteristics: A small tree or large shrub with light rough black bark; leaves simple, alternate, double ranked, elliptic-lanceolate; flowers green, perianth fleshy, borne solitary or 2-4 on short extra-axillary peduncles; fruits are turbercled, globose or cordate-ovoid, 5-10 cm across, seed brown or black, shining, pulp white-creamy.

3. Parts Used: Fruits.

4. Chemical Constituents: Fruits: Carbohydrate (23.5%), protein (1.6%), fat (0.4%), minerals (calcium, phosphorous, iron), thiamin, riboflavin, niacin, vitamin-C (Rathore, 1996).

5. Uses And Applications: It is mostly consumed when ripe, being delicious. The edible pulp is granular with blend of sweetness and pleasant flavoured. It is also used in ice-creams, milk products and preserved as jam, jelly, etc.

6. Karyotaxonomic Evidence: Chromosome number for *Annona squamosa* is reported to be 2n=2x=14 to 16. It is a diploid species. RAPD showed a low percentage of polymorphism (*cf. Bhagal et al.*, 2009).

7. History: Although, a number of local cultivars are available, except few, none of them has got the status of commercial cultivar due to seed propagation which results in considerable variability. Some cultivars are: Balanasar, Barbados, British Guinea, Brandy, Islander, Red Sitaphal, Washington PI 98797 and Washington PI 107005. The cultivars differ from each other in fruit shape, size, colour and eating quality. Red Sitaphal is a distinct cultivar probably originated as a chance seedling (Venkataratnam, 1963). Its fruits are pinkish dark and pulp erythrite-red. The fruit quality is, however, similar to other cultivars. In a varietal trial of seven cultivars at Bangalore, only the cultivar Balanagar showed better fruit quality (Anonymous, 1987).

8. Further Developments: There are few commercial cultivars, which need improvement through breeding. The propagation by seeds vary greatly in productivity, size, colour and fruit quality, besides fruit weight. Some superior genotypes were selected at the MPAU, Rahuri (India) on the basis of yield, size of fruit, areoles, percentage pulp, number of seeds per fruit and tree vigour. Some of these were thought promising. Hybridization is comparatively easier, especially intraspecific and interspecific. Breeding on these line is desirable.

9. Epilogue: With a view to evolving the genotypes bearing fruits of uniform size, better keeping quality, less number of seeds, smooth surface, etc. is still a need of hour. Custard apple is generally consumed locally, and there is no international trade. Improved varieties keeping in view the local and international requirements should be developed. The fruits are early perishable and work should be extended on their shelf life which will be then suitable for international trade.

7.12 DATE PALM

Phoenix dactilifera L. (Arecaceae)

Resume: *The Persian Gulf is supposed to be its native. It inhabitant of hot sunny climate with low humidity. Probably, record of its cultivation go as back as 5000 BC. in southeastern Asia (the present day Iraq). Fruits are rich in various nutrients and often consumed fresh (when ripe) and in various forms of preparations. It is a diploid species with $2n=2\times=36$ chromosomes.*

1. Introduction: The date palm is a plant of hot sunny climate with low humidity and as such symbolic of arid desert climates. It is one of the oldest fruits known to mankind. Its origin is believed somewhere around the Persian gulf. It became naturalized in Arabia, North Africa and northern India during ancient times. It was probably first cultivated as early as 5000 BC. in or near the 'Fertile Crescent' of southwestern Asia (or Mesopotamia, the present Iraq). Clay tablet bears a picture of a date palm made in Babylon 5000 years ago. The plant was introduced in Italy long before the Christian era. Malaga in Spain is the only place in Europe where date palm flourished, and from there it reached to Mexico two or three centuries ago. Spaniards also introduced it into California in 1765, however, commercial cultivation in America started after 1890 from select varieties brought from Egypt and Persian gulf.

In the Indus valley, date palm is believed to have been introduced by the soldiers of Alexander the Great in the 4th century BC. Excavations of Mohenjo-Daro revealed that this fruit was in Indo-Pakistan as early as 2000 BC (Pareek, 1996).

2. Plant Characteristics: A slender, unbranched palm reaching a height of 70-100 ft. or even more, bears suckers at base and grown in clumps; leaves stiff, feathery, pinnate and rigid forming a crown at the top, 10-20 ft. in length; plant dioecious (male and female plants separate); fruit one-seeded berry, hard and green at first and then turns yellow or red with sugary flesh, seeds cylindrical and grooved.

3. Parts Used: Berries (fresh or dry).

4. Chemical Constituents: Carbohydrate (70.6%), protein (1.9%), fats (2.5%), vitamin A, B1, B2 and nicotinic acid, minerals (calcium, iron, potassium) small amount of copper, magnesium, chlorine, sulphur (Nixon and Carpenter, 1978).

5. Uses And Applications: Ripe dates are often consumed fresh or in the form of jams, jellies and pastes. In desert regions, it is eaten in many other forms. The Arabs remove the stone (seed) and replace with a lump of butter. They are often eaten curdled with milk. In date growing countries, various wines are made by fermenting the cull dates (inferior grade) and the product on distillation yields a strongly alcoholic liquor.

6. Karyotaxonomic Evidence: The six species examined cytologically by Beal (1937) were all diploid with 2n=2×=36 and of rather constant karyotype. Interspecific hybrids, both natural and experimental, are numerous. and fertile. Natural hybrids of *P. dactylifera* with *P. reclinata*, the, only two *Phoenix* species producing basal suckers reportedly produce fruits of good quality (Munier, 1974). *Phoenix atlantica* (spurious date) could be a possible ancestor or an intermediate form. Its variety maroccana produces edible fruits.

Date seedlings display much variability, as also do cultivars of different regions. This variability led Chevalier (1952) to postulate that, if one original progenitor of *P.dactylifera* ever existed, it must have been modified by hybridization with other species in different parts of the range, for example, with *P.sylvestris* in Indo-Pakistan, *P.reclinata* and variant types in northeastern Africa and *P.canariensis* in northwestern Africa and Spain.

Species of *Phoenix* are dioecious but, there is no evidence of a heteromorphic sex chromosome pair. The sex ratio in the date is about 1:1 and occasional female fertility on male palms has been noted. It is not known which sex is heterogametic in the absence of cytological differentiation.

7. History: The genus *Phoenix* was once widely distributed in Europe in Tertiary time but retreated southwards as the climate changed. It remained abundant in the North African region until desiccation there restricted its distribution.

Cultivation of the date possibly dates back to Neolithic times. The earliest record is a sample of stones from Egypt dated about 4500 BC. Traces from the 3rd millenium BC. from western India, records of the Sumerian civilization, and the etymology of the name (from the Hebrew, 'dacheb') jointly lend Support Chevalier's (1952) suggestion that domestication occurred simultaneously in different places in the area between the Indus and the Atlantic. It seems safe to assume, however, that selection for succulence of fruit must have been a major feature. Vegetative propagation by suckers was no doubt practised from earliest times. The interfertility of *Phoenix* species and the occurrence of natural hybrids suggests that the whole, or a large part of the genus could be regarded as the 'primary gene pool'.

Artificial pollination is another ancient cultural practice. It is mentioned in the cuneiform texts of Ur (about 2300 BC.). In seedling stands with approximately equal number of male and female trees, natural pollination would secure adequate fruit setting. Since, however, male trees produce very abundant pollen, they are retained in a proportion of one male to 25-50 females and pollination was traditionally carried out by hand. The process involving much climbing, is laborious. It is displaced by machinery that blows the pollen through long tubes (Baradi, 1968).

8. Further Developments: Populations originally consisted of chance seedlings from which outstanding individuals were selected. They were vegetatively propagated. Such clones are not only the numerous local varieties but also the important and widely distributed ones. This practice of selection is still going on in many primitive agricultural societies.

Nixon, in California about 1930, made a detailed study of metaxenia. During the phenomenon the pollen exercises a direct influence on the maternal tissues of the fruit, inducing differences in time of ripening or size and colour. Nixon (1966) defined the promising male tree as one which would flower synchronously with prospective female clones. The male tree produce abundant viable pollen. and show favourable metaxenia effects.

Male clones for use in commercial plantings are thus one objective of date breeding. Commercial clones, however; are all female and if it is desired to make intervarietal crosses, it is necessary first to obtain the male counterparts of some, at least, of the selected females. This was attempted in California by back-crossing repeatedly to the chosen female (Nixon and Furr, 1965). Inbreeding depression might be predicted to follow from backcrossing an out breeder such as the date. In Arizona, inbreeding experiments with the progeny of the variety Deglet Noor were abandoned after three generations because of disappointing results (Nixon and Furr, 1965).

Work has been done on a devastating disease called 'Bayoud', caused by *Fusarium oxysporum f.* sp. *albedinis* in Morocco and Algeria. It is transmitted by contaminated soil and plant parts. The disease engenders the existence of excellent varieties *e.g.* Deglet Noor but resistant clones (*e.g.* Tazerzait, Takerboucht) are known. Field testing for resistance is possible and clones and seedlings are screened in Morocco (Louvet and Toutain. 1973).

All commercial date palm cultivars have arisen by selection of chance seedlings based on local needs. The primary goal of date breeding is to achieve the highest fruit quality and yield consistent with local requirements *e.g.* tolerance to adverse growing conditions, disease and insect pest, modification of growth habit, uniform inbred lines, etc. (Carpenter and Ream, 1976). In the USA, date palm breeding has started in 1948 with the primary objective of producing female clones adopted to mechanical harvesting and processing, and slower in vertical growth, etc. (Carpenter, 1979; Pareek, 1996).

9. Epilogue: Long life cycle, dioecious habit and low rate of vegetative propagation in the date palm poses peculiar problems for the breeders. Little has been achieved to date and rapid progress in the future is not to be expected. The numerous clones and seedlings that exist in all date-growing countries have yet hardly been exploited by breeding.

Morphological characters of the fruit bunch, physiological characters of the palm and disease resistance will no doubt claim the breeders' attention. Besides resistance to Bayoud, resistance also to Graphiola leaf disease and to mites are reported. Selection of good polliniser is important since the pollen affects fruit size and the time of ripening. Also, fruit set is closely related to the percentage of viable pollen and temperature during pollination.

7.13 DRAGON FRUIT

Hylocereus undatus (Haw.) Britt. & Rose (Cactaceae)

Resume: *Dragon fruit viz., Hylocerus undatus and even few other edible species are native of American deserts. The species of the genus Hylocereus are diploid with chromosome number 2n=22 and they are progenitors of another useful species of the tetraploid genus Selenicereus. The dragon fruit finds place in many processed food recipes. It has also potential for use in medicine, additives in food preparations and cosmetic products.*

1. Introduction: Dragon fruit belong to the genus *Hylocereus* (Cactaceae). It is cultivated in Southeast Asia, Florida, the Caribbean, Australia and tropical and subtropical regions. The genus *Hylocereus* is originally native to Mexico, Gautemala, Costa Rica. E1 Salvador and northern South America. Generally, three types of species are hailed as dragon fruit: (i) *H.undatus*, (pink-skinned and white flesh) (ii) *H.constaricensis*, (red-skinned and red flesh), and (iii) *H.megalanthus* (yellow-skinned and white flesh). It is also called Pitahaya, Bell of The Night, Strawberry, Pear, Queen-of-the-Night Cactus, Fire Dragon fruit, Night Blooming Cereus, etc.

2. Plant Characteristics: A climbing cactus vine, terrestrial or epiphytic, stem triangular, fleshy, scarlet, skin with green spines; reaching 20 ft long, climbing by aerial roots; flowers large, white inside and greenish-yellow outside, perianth segments numerous, 30 cm long or more, bell-shaped, fragrant when in bloom; fruit oval, elliptical or pear-shaped, large vividly pink outside, when ripe pulp white, sweet, seed black, numerous, dotted.

3. Useful Parts: Fruits, Seeds.

4. Chemical Constituents: (i) Fruit Pulp: Carbohydrate (10.76%), protein (2.27%), lipids (0.16%), minerals (Aluminium, cadmium, calcium, lead, copper, cobalt, chrome, iron, phosphorous, manganese, mercury, nickel, potassium, sodium, selerium, zinc), fatty acids (palmitic, palmitoleic, heptadecanoic, oleic, linoleic, alpha-linoleic, archidic, eicostrienoic, arachidonic, eicosapentaenoic, heneicosanoic, behenic, docosahexaenoic, tricosanoic, stearic and tetracosenoic acids.

(ii) Seeds: Myristic acid, palmitic acid, stearic acid, palmitoleic acid, oleic acid, cis-vaccanic acid, linoleic acid and lenolenic acid.

5. Uses And Applications: Ripe fruit is sweet and edible. It is used in salads and can be processed in various products *e.g.* wine, juices, sherbets, jam, jelly, syrup, candy, pastries, yoghurt and ice-cream. Flowers also can be used for tea. Flowers buds can be used to make soup or mixed in salads and eaten as vegetable. It is also coming up as additive (colouring material) in food recipes, medicine and cosmetics.

6. Karyotaxonmic Edivence: Chromosome numbers and meiotic behavior for the climbing species *viz.*, *Hylocereus undatus, H. polyrhizus* and *Selenicereus megalanthus* are reported by Lichtenveis *et al.* (2000). The species of the genus *Hylocereus* are diploid with 2n=22 chromosomes. Whereas *S. megalanthus* is a tetraploid with 2n=22 chromosomes. Major crossability barriers do not exist between the species of the genus *Hylocereus*. Reciprocal intergeneric croses are successful between the species of *Hylocereus* and *Selenicereus megalanthus*. This suggested that Hylocereus species might be one of the diploid progenitors of the tetraploid *S.megalanthus*.

7. History: Dragon fruit has been consumed in its native countries since pre-Columbian times and grown specifically for its edible fruits. Sine the 20th Century, *H.undatus* is widely planted as a night-flowering ornamental plant and as a fruit crop. On the commercial scale as a fruit crop is grown in many tropical countries, particularly in Vietnam and South-East Asian countries. It was first grown in Vietnam for the king and then gained popularity among the wealthy people and today it is leading fruit export of Vietnam. Its scientific name (*H.undatus*) was published first in Britton's Flora of Bermuda in 1918. The genus includes about 15 species native to Central and South America. It is also naturalized in some countries *e.g.* Taiwan, Canary Islands, Spain, Queensland and have become also invasive in New Calendonia, Chile, Brazil, Cuba, etc. In Brazil, it was introduced as a garden plant and then escaped from cultivation. At some places *e.g.* Florida (USA), it posed a threat to native plant species.

8. Further Developments: Chakma *et al.* (2014) studied effect of NPK doses and cow dung manure on the yield of dragon fruit is Chittagong Hill tracts of Bangladesh. They made observations and concluded that: (i) there were significant differences in plant heights (ii) plant circumference increased significantly (iii) maximum spreading canopy was noted in east-west direction (iv) the day to first flowering varied from 8 to 30 May (v) number of fruits increased to certain level of fertilizer application (vi) significant differences in size and weight of fruits was observed. The fruit yield was increased.

Tenox *et al.* (2012) analysed the polyphenolic components of dragon fruit and its antioxidant activity. Betacyanin fractions exhibited the highest reducing and radical-scavenging capacities among the extracts and fractions tested by FRAP and DPPH assays respectively.

The antioxidant activity of dragon fruit is due to its ascorbic acid content (vitamin-C). Vitamin-C is a powerful antioxidant agent with free radical scavenging activity, very useful to treat photoaging. It neutralizes UVB-generated free radicals. Experimental models involving human skin demonstrated similar effects of ascorbate, applied along or in combination with vitamin-E (Chiu and Kimball, 2003).

Farriol *et al.* (1994) investigated the effects of vitamin-C on an aged human fibroblast culture under oxidative stress. They observed cytoprotective effects of vitamin-C against the oxidative stress produced by the addition of 5mMH_2O_2 to the culture medium. The result showed cytoprotective effects of vitamin-C on aged fibroblasts at every tested dose Dragon fruit is hence adequate to formulate cosmetic products for the protection of the skin and hair integrity against oxidative processes.

Collagen contributes about one third of protein in the body and plays a key role in providing the structural scaffolding for cells, tissues and organs. Exposure of cultures of human connective-tissue cells to ascorbate (vitamin-C) induced an 8-fold increase in the synthesis of collagen with no increase in the rate of synthesis of other proteins (Sharma *et al.*, 2008).

Perez *et al.* (2005) evaluated wound healing activity of aqueous extract of dragon fruit and flowers. Wound healing effects are studied on incision (skins breaking strength) and excision (percent wound contraction) and the nature of wound granulation tissues, which were removed on 7th day. The collagen, hexosamine, total proteins and DNA contents are determined, in addition to the rates of wound construction and the period of epithelization. In streptozotocin diabetic rats, where healing is delayed, topical applications of *H. undatus* produced increases in hydroxyproline, tensile strength, total proteins, DNA, collagen content and better epithelization thereby facilitating healing. Dragon fruit extract is thus recommended to formulate cosmetic products with moisturizing, soothing and reepithelizing activity.

Moshfegh *et al.* (2013) introduced a new natural product from dragon fruit in the market. Consumer's choice has been increased in recent years towards using natural food additive. They used dragon fruits, particularly outer layer, inner layer (albedo) and flesh. They concluded that the albedo powder of dragon fruit can be used for food colouring. The Dragon fruit colouring powder (DFCP) is found worth to examine while accessing longevity, nutrition level, optimal temperature and pH value, optimal length of time to heat exposure, botulin measurement and colour stability. The peal of dragon fruit is a waste material and thus can be used a additive *vis-à-vis* colouring matter in foods.

9. Epilogue: *Hylocereus undatus* (Dragon Fruit) was cultivated for the first time in Vietnam especially for the King, which was then adapted by wealthy people. The species of Dragon fruit originally are hailed from American continent. They are wild in their native place but are now cultivated in America and other countries in the world as superfood. It was/is useful as fruit source but also used as functional food source. It benefits mankind not only in his nutrition but also in health problems. It has also potential as colouring additive in food preparation. Interestingly, its fruit peel, a waste material, is also being used in modern period. It is also notable that it has also shown potential for use in cosmetics. These recent studies will certainly prove rewarding in near future.

7.14 EMBLIC MYROBALON (AONLA)

Emblica officinalis Gaertn.

(Syn.) *Phyllanthus emblica* L. (Euphorbiaceae)

Resume: *The Emblic Myrobalan (Aonla) is indigenous to tropical south-eastern Asia, particularly to India. It is hardy, prolific bearer and most enjoyed as a source of vitamin-C, besides some processed or preserved products. It is much valued medicinally especially in India since ancient times. Chromosome numbers are 2n=28. Natural cultivars are generally grown.*

1. Introduction: It is considered indigenous to tropical south-eastern Asia, particularly in central and southern India (Firminger, 1947). It is also reported from Sri Lanka, Malaysia and China. It is generally found wild in the mixed deciduous forests. It is quite hardy and prolific bearer and highly renumerative even without much care. It is valued more medicinally. It is one of the consistent fruits of popular Ayurvedic medicine 'Triphala Churna' in India and also an ingredient of 'Chyavanprash', an ancient drug remedy for good health.

2. Plant Characteristics: A tall tree with greenish-grey bark; leaves distichous, close-set linear and small; flowers small, greenish-yellow, unisexual, densely fascicled on new shoots; drupes about 1-2.5 cm across, globrous, fleshy, globose pale greenish-yellow, seeds 6, trigonous.

3. Parts Used: Drupes.

4. Chemical Constituents: Protein (0.5%), fat (0.1%), carbohydrates (14.0%), minerals (calcium, phosphorous, iron), nicotinic acid, vitamin B and C.

5. Uses And Applications: It is the richest source of vitamin-C and highly nutritive. It is consumed fresh, processed and preserved in several ways. The fruits are made into preserves (called murabba), sauce, juice, candy, dried chips, tablets, vinegar, chutney, sherbet, jelly, pickles, tophies and powder. It is valuable medicinally being diuretic, laxative, alternative and antiscorbutic. (Hazarika, 2014; Bajpai and Shukla, 2016).

6. Karyotaxonomic Evidence: Chromosome number has been reported as 2n=28, while a variation from 2n=98 to 104 has been also observed (*cf.* Bajpai and Shukla, 1996).

7. History: The cultivars of Emblic Myrobalam are recognized simply on the basis of size and colour of fruits or after the names of places. Their colour range from green-tinged, red-tinged, pink-tinged, white-streaked and Bansi Red (Singh *et al.,* 1967). Some cultivars are grown in the district of Pratapgarh (Uttar Pradesh, India) *viz.,* Banarasi, Chakaiya, Francis, NA4 (Krishna), and NAS (Kanchan). Some characteristics such as habit (erect, spreading, drooping), number of branchlets per node, fruit (size, flattened, oblong), skin (smooth, yellowish, soft, semi-transparent), colour (greenish, yellow, red, bluish), suitability for preparing pickles, flesh (fibrous), keeping quality, etc. have been employed while deciphering the cultivar types (*cf.* Bajpai and Shukla, 1996). Their chemical composition and fruit weight has been also decided.

8. Further Developments: It can be propagated by both-by seeds and vegetative method. The former method by seeds do not come true to type and produce small-sized fruits and also inferior in quality. In order to overcome the disadvantages of seed propagation, multiplication of superior types has been suggested by adopting vegetative methods (Hayes, 1957; Singh *et al.,* 1967). Various methods of vegetative propagation are being adopted *e.g.* grafting, cutting and top working. Various methods of plant protection from pests, diseases and physiological disorders are followed to have better yield. Breeding programs are not on record.

9. Epilogue: The quality of fruits depends on type of flesh (hard, soft), taste (sour, bitter), size, fibre content, etc. Although, Emblic Myrobalon (Aonla) is prolific bearer and highly renumerative, it is minor fruit crop of commercial significance. It is not directly used as food. It is enjoyed in the form of various recipes or articles or in medicine. Only natural cultivars are being cultivated. It is favoured for consumption mostly being a rich source of vitamin-C.

7.15 FIG

Ficus carica Linn. (Moraceae)

Resume: *The Fig is a native of Southern Arabia and then spread in the Mediterranean region early. It finds mention in the Bible and also noted by Theophrastus (370-287 BC.). It is thought to be under cultivation since 3000 BC. Its fruit is actually a 'syconium', also called receptacle. It is consumed as a dessert when fresh (ripe), dried or in the form of various preparations. The basic chromosome number of the genus phoenix is ×=13.*

1. Introduction: The generic name *Ficus* is derived from *ficus Ruminalis*, a sacred fig tree. According to Roman legend, this tree sheltered the infants Romulus and Remus who were to become the founders of Rome (about 746 BC.) while they were being cared for by a she-wolf and a woodpecker. The tree was named for Ruminia, a goddess who watched over suckling animals. The specific epithet carica refers to Caria, an ancient political division of Asia Minor bordering on the Aegean Sea.

The fig originated in Southern Arabia and spread early to the Mediterranean region. It has been mentioned in the Bible frequently. Theophrastus (370-287 BC.) also mentioned many varieties of fig in his 'Enquiry Into Plants' and gave a detailed account of its cultivation. It was brought under cultivation 3000 BC. During the age of exploration following discovery of America by Columbus, the fig was taken to most subtropical areas of the western hemisphere.

2. Plant Characteristics: Moderate sized deciduous tree; leaves ovate-orbicular, 3-5 lobed; syconium containing male, female and sterile small flowers, fruits axillary, rear-shaped.

3. Parts Used: Fruit (syconium).

4. Chemical Constituents: Carbohydrates (17.1%), protein (1.3%), ether extract (0.2%), minerals (calcium, copper, phosphorous, iron), carotene, riboflavin, ascorbic acid, vitamin-A.

5. Uses And Applications: The ripe fresh fruits are consumed as a dessert and so also when dried. They are also preserved or canned. They are used in baking and ground up for fig coffee. They are also used in the preparation of fig wine, brandy and jam.

6. Karyotaxonomic Evidence: The genus *Ficus* is contains about 2,000 species of tropical and sub-tropical trees, shrubs and vines, worldwide in distribution (Condit, 1969). Besides *F. carica*, the only other species cultivated for fruit is *F. sycomorus*, the sycamore of Scripture. It is grown locally, mostly in Egypt and. Israel. Several other species have edible fruits, which, however, lack palatability.

The basic chromosome number of the genus is ×=13. Somatic chromosome numbers have been reported for slightly over 100 species (Condit, 1928, 1933, 1964). Excepting five African species which have somatic numbers of 52 (2n=2×=52) and *F. elastica* var. *decora*, which is triploid (2n=3×=39), all are diploids with 2n=2×=26. Cytotaxonomic relationships in *Ficus* are unknown. However, *F. carica* has been hybridized with *F. pseudo-carica* and *F. pumila*).

7. History: During the War of the Titans, according to Greek mythology, Zeus was pursuing Ge and her son Sykeus when Ge, in order to save Sykeus, metamorphosed him into a fig tree. The city of Sykea in the ancient country of Cilicia derived its name from this myth. An Athenian myth credits the goddess Demeter with having revealed 'the fruit of autumn' (the fig), to human beings.

The fertile region of southern Arabia, where wild fig and caprifig trees are still to be seen, is thought the centre of origin of *F. carica*. It must have been domesticated millinia ago. It spread later in cultivation through all the countries of southwestern Asia and the Mediterranean basin. It reached to all parts of the world during the age of exploration and Columbus's discovery of the New World in 1492.

The differences between fig and caprifig are determined either by a recessive gene with pleiotropic effects or by two closely linked recessives as: (i) G, normal allele, short-styled pistils, (ii) g, recessive allele, long-styled pistils, (iii) A, normal allele, androecium present, (iv) a, recessive allele, androecium suppressed.

Thus caprifigs are *GA/GA* and *GAlga* and figs *ga/ga*. The various possible matings segregate as expected. Homozygous caprifigs are rare and have, recently identified. In the wild, in the presence of *Blastophaga*, one would expect caprifigs to set few seeds. The mating system so would approach dioecism.

There are no sex-linked vegetative characters distinguishing seedling caprifigs from figs. Identifying the sex of an individual growing on its own roots can only be done when the first syconia appear usually at an age of 3-7 years. The time can be lessened to 1-3 years, however, by grafting scions from 6- to 8-month old seedlings on to the branches of older trees.

During course of evolution, female trees (figs proper) have become differentiated into three groups. *viz.,* (i) Smyrna type: The breba crop is lacking or scanty and tends to be caducous. The main crop is usually adequate but drops prematurely if not caprified; the drupelets contain fertile seeds. (ii) San Pedro type: The breba crop is usually good but the main crop syconia drop unless caprified; caprified figs contain fertile seeds, while the- breba crop and uncaprified main crop figs may be 'seedless' or may contain 'seedlike bodies' (parthepocarpic drupelets). (iii) Common type: IS may or may not produce a breba crop; no caprification is necessary for the main crop; both breba and main crop figs may or may not contain parthenocarpic drupelets.

Nearly 605 valid cultivars were recognized even after the reduction of synonyms (Condit, 1955). These were distributed among the horticultural types as Smyrna, 115; San Pedro, 22; Common, 470. All important cultivars produced on a large scale commercially today, for example: Sari Lop (syns, Lob Injir, Calimyrna), Dottato (syn. Kadota), Verdone (syn. White Adriatic) and Brunswick (syn. Magnolia), were selected centuries ago by persons unknown. Some have been under cultivation for a very long past. Sari Lop, for example, was grown in the Meander Valley of Turkey for about 2000 years; Dottato of Italy was praised by Pliny (AD. 23-79); Verdone has been cultivated in the Adriatic region for hundreds of years. It was introduced into England in 1727 by Richard Bradley. Franciscana has been grown at Eastepona, Spain for last few centuries. It was introduced into Mexico and Chile by Spanish missionaries soon after the conquest. In 1768, Franciscan missionaries from Mexico established it in California where it called 'Black Mission'.

8. Further Developments: The US Department of Agriculture obtained scion-wood of the Chiswick fig collection from the Royal Horticultural Society of England in 1895. Scions of every cultivar, 66 in all, were grafted successfully on trees in the orchard of the California Nursery Company at Niles, California. This collection served as foundation stock for early attempts at breeding.

New cultivars developed for the first time by hybridization. They were probably those of Burbank, followed by those of Swingle (1908) and Hunt (1911). Condit (1928 to 1969) carried on studies of fig genetics and cytology and engaged in fig breeding at the University of California, Riverside. He developed two new Adriatic-like cultivars, *viz.,* Conadria and DiRcdo, for commercial planting as drying and paste figs, and two other cultivars *viz.,* Excel and Flanders, for home gardens. Promising new cultivar *viz.,* Tena was developed by W. B. Storey for drying and paste. The five cultivars named here seem to be the only once ever to have been introduced into the trade from a planned breeding programme.

Success in breeding was largely due to the use of the capri fig Croisic It was being cultivated near Cordelia, California by 1893. Its genetic behaviour was like that of other caprifigs, but in its progenies are at least some seedlings that produce succulent syconia.

9. Epilogue: The objectives of fig breeding is to improve tree and fruit characters of both fig and caprifig with special reference to the commercial production of dried figs and fig paste. The main objectives in case of fig cultivars are: (i) syconia equal to those of Sari Lop in attractiveness, size, shape, and quality but not requiring caprification (ii) early to mid-season maturity and dropping of fruit to avoid the spoilage which occurs after midseason, Desirable fruit characters are: golden or greenish yellow skin like that of Sari Lop which dries to a light-straw colour, white meat, amber pulp, characteristic fig flavor, medium size and weight, ovoid or pyriform shape, small ostiole with tightly closed eye to exclude disease-carrying insects.

The main characters to be bred in case of caprifigs are succulent syconia like those of Croisic, desired size, shape and colour and persistence on the tree without benefit of *Blastophaga*. For caprifigs to be used for caprifying Smyrna type cultivars, the characters are: caducous syconia to protect growers from buying numerous blanks, both early and late maturing for full coverage of the period of anthesis, dry, hollow syconia to encourage hatching and excitation of *Blastophaga*, and good pollen production.

7.16 GRAPES

Vitis vinifera L. (Vitaceae)

Resume: *Vitis vinifera, called 'Grape' is denizen of Middle Asia. It is still found in wild from northeastern Afghanistan to the southern borders of the Black and Capsian seas. Its cultivation and use in wine making was recorded as back of 5000 years. It is mentioned in the Bible. It is associated with mankind even before association of man with wheat and rice. Species of Vitis are ancient secondary polyploids involving three basic sets in the combination (6+7)+6=19 chromosomes. It has gone diploidization. Fruits are rich in sugars and some acids. They are delicious, nourishing and consumed fresh or dried. Principally, they are used in wine making.*

1. Introduction: Association of the grape with man is older than that of wheat and rice. The genus *Vitis* contains about 60 species, majority of which being native to North America. It is one of the earliest fruits consumed by mankind. Its seeds were found with the earliest human remains in the lake dwellings of Switzerland and in the oldest tombs of Egypt. Its cultivation and use in wine making was recorded as back as 5000 years. It is mentioned in the Old Testament several times. It is also mentioned in the Bible that Noah planted a vineyard. It was introduced, apparently, into Palestine from the north in about 5000-6000 BC. Documents concerning viticulture and wine-making in Egypt date back to nearly same period. It was carried from Palestine to islands in the Mediterranean sea by the Phoenicians. It reached to France in 600 BC. In India, it seems to have been introduced by invaders from Persia (Iran) and Afghanistan around 1300 AD. However, the mention of grape in

India has been made in the medical treatise written in the 1st century AD. During 17th century it was introduced to South Africa and 19th century to Australia.

Centuries before Columbus, Indians in North America used to gather wild grapes of species *viz.,* fox grape (*V. labrusca* L.), riverbank grape (*V. vulpina* L. Syn. *V. raparia* Miehx.) and scupper or muscadine (*V. rotundifolia* Michx.). Before Columbus about 500 years, Norseman visited the continent and found many wild grapes and hence named the country 'Vine Land'. Columbus brought grape cuttings to the Caribbean Islands. *Vitis vinifera* was introduced into the US as early as 1616. Attempts to grow some cultivars from Europe failed due to unfavourable climate and winters but could grow well in California.

Three main types of grape *viz.,* fox grape (*V. labrusca* L.), muscadine grape (*V. rotundifolia* Michx.) and European (Old World) or also sometimes called California grape (*V. vinifera* L.) are grown all over the world. Italy occupies first place in the grape producing in the world followed by France, Spain, USA and Turkey.

In the area of the earliest human civilization, in a region somewhere between the Black and Caspian seas, the grape still grows wild. This region is considered to be the original home of the Old World Grape (*V. vinifera* L.) (Snyder, 1937; Jindal, 1996). Grape culture first began in Asia Minor. From there, it spread both west and east (Olmo, 1976). There is no evidence of its cultivation on west of Greece until the 1st millennium BC. The westward movement carried out from Asia Minor and Greece, following the Phoenician Sea routes. It was introduced in the New World at the time of discovery and later by Spaniards and Portuguese during their voyages and conquest. According to Olmo (1976), grapes might have been introduced in tropical India somewhere in 620 BC.

2. Plant Characteristics: Woody tendrillar climbers; leaves alternate, palmately lobed and veined; flowers greenish, borne in panicles; grape clusters develop laterally near the base of leafy branches; grape varieties differ in colour (white, green, pink, red or purple), rounded or ellipsoidal in shape.

3. Parts Used: Berries.

4. Chemical Constituents: Mainly glucose and fructose; pentoses and pentosans in small quantity; organic acids *e.g.* mainly tartaric, malic, citric ascorbic, oxalic, gluconic, phosphoric and glucovonic acids in small amounts amino acids, minerals (calcium, manganes, magnesium, copper, aluminium, boron, chloride, potassium, sodium, iron), vitamin B, etc.

5. Uses And Applications: Being delicious, refreshing and nourishing, berries are consumed fresh or dried in the form of raisins, juice and canning. They are used principally in wine making. Their waste products are utilized in various ways *e.g.* stock feeds, fertilizers, acetic acid, cream of tartar, a fixed oil, tannin, Selected fruits are used for canning.

6. Karyotaxonomic Evidence: The genus *Vitis* includes about 60 species. It is unique in having 38 very small somatic chromosomes in the family Vitaceae that regularly form 19 bivalents at meiosis (Lavie, 1970). Wild grapes belong to three geographical groups *viz.,* American, Middle Asian and Oriental. North America, especially the southwestern and Gulf region of the USA, is particularly rich in *Vitis*. Most other related genera, including *Muscadinia*, have 2n=2×=40. Formerly classified as a section of *Vitis*, *Muscadinia* has only three known species, restricted to the southeastern USA and northeastern Mexico. The colonists of the Carolinas cultivated *M. rotundifolia* directly and domestication dates from the latter part of the 19th century. Since this species does not hybridize naturally with sympatric species of *Vitis*, it is a case of the domestication of a single species in *situ*. *M. rotundifolia* can be hybridized experimentally with *V. vinifera*. The cross helped for improving the disease resistance of *V. vinifera* and the fruit quality of rotundifolia. The ancient basic chromosome numbers in the family are probably 5, 6 and 7. *Vitis* species are thus ancient secondary polyploids involving three basic sets in the combination, (6+7) + 6 = 19. Muscadinia species, on the other hand) are (6+7) + 7 = 20. Both have undergone diploidization to give regular bivalent pairing. All known Vitis species are easily crossed experimentally and the F_1 hybrids are vigorous and fertile. Researchers of natural populations suggests that hybridization has occurred and continues. Even though species cannot be delimited on degree of genetic isolation, the species classification remains useful because of its practical significance in separating norms of variation that are important in breeding *e.g. riparia* is highly resistant to phylloxera but *vinifera* quickly succumbs; and *riparia* roots readily from dormant cuttings but *aestiualis* does not.

7. History: All about 10000 cultivars of the Old World area considered derived from the single wild species, *V. vinifera*, of Middle Asia still found from northeastern Afghanistan to the southern borders of the Black and Caspian Seas. Legend and tradition both warrant ancient Armenia as the home of the first grape and wine culture. Small refuge areas isolated by the glacial epochs are found scattered in southern Europe, but their role in domestication is questionable. Three principal groups of cultivars are proposed *viz., occidentalis*, the small-berried wine grapes of western Europe; *orientalis*, the large oval-berried table grapes; and *pontica*, intermediate types of Asia Minor and eastern Europe (Negrul, 1938).

The berries of wild *vinifera* (*sylvestris*) are palatable and the wine is of a quality comparable to that made from present cultivars. Domestication started when migratory nomads marked forest trees that supported particularly fruitful vines. This was mostly near watering holes serving their herds. Sparing these vines was associated with a spiritual taboo respected by other tribes. Sedentary agriculture was developed and the mixed deciduous forest was cleared. Fruit trees and vines were spared along boundary lines where irrigation ditches were developed. The vines were thus out of reach of grazing animals. Vineyards as such developed later when they could be protected by high mud walls from the ever-present sheep and goats.

Wine grape cultivation was under way in the Near East as early as the 4th millennium BC. The products of the vine were exported westwards from very early times (Helbaek, 1959). The westward movement extend from Asia Minor and Greece through the Phoenician sea routes. The spread of the vine was associated with that of the Christian faith during the Roman period. In the Middle Ages, the Catholic monasteries throughout Europe were the guardian of select vineyards. A vineyard covered by the eruption of Vesuvius in AD. 79 illustrates the practices recommended in the early Roman agricultural manuals of that era (Jashemski, 1973). The vine followed the main river valleys, the Danu, Rhone, Rhine; Tiber and Douro (AD. 55). The northernmost vineyards were established along the Moselle Valley in Germany. The *vinifera* grape was introduced to the New World at the time of discovery. It was then accompanied practically all the Spanish and Portuguese voyages of discovery and conquest. The first recorded introduction to the east coast of the United States was in 1621 by the London Company. This was probably preceded by the Spanish landings in Florida. The most recent incursions of *vinifera* are in tropical countries. In 1958, *V. vinifera* was introduced into the Phillipines from California.

The *vinifera* grape was introduced into different zones and often hybridized with native Vitis to produce new races better adapted to local environments *e.g.* American hybrids the Atlantic seaboard. The Alexander, Concord and, Delaware, etc. In the Caribbean islands and Venezuela, introduced *vinifera* has introgressed with the tropical caribea to produce new vigorous races of Criollas that are disease resistant and tolerant of the climate. Some Japanese and Chinese cultivars are oriental species introgressed with *vinifera*.

The wine grapes are seeded, acidic and juicy. Seedlessness evolved in a wide range of expression (Stout, 1936). The cultivars were developed with very small, parthenocarpic berries such as the ancient Greek variety Black Corinth, dried to produce currants. The Kishmish or Sultanina is said to be steno-spermocarpic, abortion of the seeds occurring soon after fertilization. Seedless varieties arise by somatic mutation. These are of different types and involve different genetic backgrounds (Olmo and Baris, 1973).

8. Further Developments: *Vitis* species are dioecious or occasionally subdioecious when some male flowers transform to hermaphrodites. The sexual type is determined by three alleles. All cultivars of Europe and the New World are hermaphrodites and self-pollinating. In Middle Asia, however, many are female. In a warm, dry climate and with close planting of mixed cultivars, cross-pollination is effective. As these female cultivars are moved to new areas and isolated, they are rapidly selected against poor fruitfulness. Female vines are useful in breeding, eliminating the need for emasculation (Levadoux, 1946).

Only aneuploids (2n+2) appear as rare aberrants. Autotetraploids were first described in 1929 and have been found to arise spontaneously in most cultivars. They are often periclinal chimeras; only two layers are involved in meristem differentiation, so that three types of tetraploid chimera have been found: 2-4, 4-4 and 4-2. The larger berry size was favoured for the possibility of producing improved table grapes. In general, the autotetraploids have poor cultural characteristic *e.g.* being less fruitful, irregular in berry size and more fragile. Moreover, the root system is weak. Tetraploids are better grafted on diploid rootstocks. A few tetraploid varieties have long been grown commercially in greenhouses. Thereby special attention to pollination and cultural factors is possible. Allotetraploids are more promising (Jelenkovic and Olmo, 1969) since selection can proceed in a wider genetic base. Triploids are highly sterile but may be useful for vigorous rootstocks, especially if a wide range of resistance to soil pests is desired. Pentaploids are weak and unuseful.

From earliest times, *vinifera* grapes were propagated by cuttings or layering and remained mostly free from pests and diseases. However, vines began to die in French vineyards in 1860 and, in 1868 due to a root aphid, Phylloxera. The French government sent specialists to the United States to discover the best sources of resistance. Selections of *riparia, rupestris* and *berlandieri* proved most useful. From about 1880 onwards, interspecific hybrids were deliberately bred as rootstocks. Thus *berlandieri* was the best adapted to calcareous soils but rooted with difficulty. It was, therefore, crossed with *vinifera* to improve propagation. The breeding of rootstocks was the first massive improvement programme. The germplasm of the *vinifera* scion, however, remained intact.

Grafting, however, is expensive. Many breeders therefore set out to breed new vines. Some 'French hybrids' proved valuable, *e.g.* in having better resistance to fungus diseases and greater hardiness. They were better adapted in some areas than ordinary *vinifera* and produced wine of passable quality. They form the base of new wine industries in many parts of the world where *vinifera* is not well adapted.

9. Epilogue: Programmes of cultivar improvement was increasingly directed toward disease and insect resistance. A high priority was virus resistance. Native species must be more thoroughly studied, screened and compared as sources of resistance. Allopolyploidy as a tool to produce larger berry size and better cultural feature are ignored in the quest for improved table varieties. We should see the use of native tropical species as a base in greatly extending the zone of commercial grape culture. Hardy clones, having short growth cycles should be further extended. A beginning has been made with oriental *amurensis* but American species can also be used. Selected female cultivars to obtain mass hybridization can be useful. Increasing need for mechanization of harvesting, pruning and other cultural methods are new demands. The long generation time from seed to fruit could, perhaps, be shortened by the use biochemical methods of selection at the earliest possible stage of seedling development.

7.17 GUAVA

Psidium guajava L. (Myrtaceae)

Resume: *Guava has origin in tropical America. It is now widely spread in the tropics and subtropics and regarded as an 'apple' of the tropics. It is rich in sugars and relished when ripe or consumed in various forms of preparation. It is diploid with 2n=22 chromosomes. Its seedless cultivars are triploid in nature. Its shelf-life being shorter, it is usually consumed locally.*

1. Introduction: Guava is regarded the apple of the tropics. It originated in tropical America and seems to have been growing from Mexico to Peru (De Candolle, 1886). It is now grown widely all over the tropics and subtropics. Presently guava producing countries are Asian Countries, the Hawaiian Islands, Cuba and India. It is believed to be introduced in India in the early 17th century by the Portuguse (Watt, 1892).

2. Plant Characteristics: A medium-sized tree with greenish-grey smooth bark; leaves opposite, sessile, ovate-lanceolate or elliptic-oblong; flowers borne in axillary pairs of singly and are white in colour with numerous stamens; berries usually pyriform, sometimes globose, variable in size and shape, containing numerous seeds in white or pink-coloured pulp.

3. Parts Used: Berries.

4. Chemical Constituents: Sugars: (i) reducing (4. 45%), (ii) non-reducing (5. 23%) acid (2. 45%), vitamin-A, C, thiamine, riboflavin.

5. Uses And Applications: Fruits are relished when mature or ripe. They are usually plucked fresh from the tree. Excellent salad and pudding are prepared from the shell of the ripe fruit. Guava jelly is well-known. It can also be canned in sugar syrup or made into fruit butter. Fruit juice is used to prepare sherbet and ice-cream. Dehydrated guava is a source of vitamin-C. It can be used frozen.

6. Karyotaxonomic Evidence: In case of guava, most of the cultivars are diploid with 2n=22 chromosomes. The seedless cultivar is triploid in nature and fruit bearing is less in it. Crosses between diploid and triploid were made to evolve cultivar with less seeds and more productivity. As a result, diploid, trisomic and tetrasomic plants were raised. The fruits of tetrasomics were normal in size, shape and less number of seeds. Interest has been shown on aneuploidy breeding. The chromosome imbalance thus obtained in aneuploidy, imparted ovular sterility which resulted in reduced seed formation and size (*cf.* Mitra and Bose, 1996). Growth of plant was observed adversely affected due to the trisomic condition and the plants were usually less vigorous than diploid. The trisomic, in general, produced smaller fruits with fewer seeds as compared to diploid cultivars. The fruits of tetrasomic plants were smaller with fewer seeds than the trisomics. Some of the tetrasomics were extremely dwarf.

7. History: The genus *Psidium* contains about 150 species (Hayes, 1970). Of these, few species are related with the common or cultivated guava species. These are *viz., P. guineese, P. guajava, P. pomiferum, P. pyriferum, P. montanum, P. friedrichsthalianum* and *P. cattleianum*. The common guava was earlier classified as *P. pyriferum* and *P. pomiferum*. In all these, fruit size, shape, habit and colour fruit are characteristics for each of them. In short, there is a fair diversity of characteristics which can be utilized in breeding programmes.

Likewise, there are some cultivars haveing their own characteristics described by earlier workers (Watt, 1908; Macmillan, 1914; Popenoe, 1920). Certain characteristics have been studied for some cultivars *e.g.* plant growth, yield, physicochemical composition, bushy habit, pulp colour, seeded or seedless, suitability for canning and cultivation in dry region, cooking, idly making, texture (smooth or rough) (*cf.* Mitra and Bose, 1996). Apart from these, plant and fruit features of guava has been also studied in relation to soil, climate, propagation, cultivation, intercropping, flowering, fruiting, etc. from agricultural point of view (Mitra and Bose, *loc. cit.*).

8. Further Developments: Selection of improved individuals has been done by the growers and pomologists. Several selected cultivars have been identified as better producers of having better quality fruits. Many of these selected plants were subsequently propagated vegetatively for commercial cultivation for example, (i) L-49, a selection from Allahabad, Safeda. (ii) Safed Jam, (iii) Kohir Safeda, etc.

9. Epilogue: Guava is propagated by seeds and vegetative propagation. The latter is sometimes used for uniform production in view of commercial purposes. Parthenocarpy is not a common feature in guava even in commercially seedless cultivar. Various chracteristics noted above of cultivars still offer opportunities for breeding programmes in view of better fruit quality and fruit bearing. Shelf-life of guava, being a berry, is short and hence technology to overcome this difficulty should be developed.

7.18 JAMUN

Syzygium cumini (Lamk.) Skeel (Myrtaceae)

(Syn.) *Eugenia jambolana* Lamk. *Syzygium jambolana* Lam.

Resume: *The genus Syzygium sensu stricto) is segregated from the large protean genus Eugenia L. Jamun is a native of India and East Indies. It is found under cultivation in other tropical and subtropical countries. The basic chromosome number is x=11 for the genus. Polyploids are of common occurrence and hence it is variable in respect of fruit characteristics. Shelf-life of its fruits is short being a seasonal one. The fruits are mostly enjoyed as a dessert. Few processed products such as wine beverage, jellies, jam, juice, etc. are available.*

1. Introduction: Jamun is sometimes mistranslated as 'blackberry', which is a different fruit (i.e. *Rubus laciniatus*). It is said to be a native of India and the East Indies (Singh, 1969). It is found under cultivation in Thailand, Philippines, Madagascar and few other countries. Recently, they are introduced in some tropical countries *viz.*, West Indies, East and West Africa and some subtropical countries as well as such as Florida, California, Israel and Algeria (Popenoe, 1920; Morton, 1963-1964). In India, apart from its utility as edible fruit, it is planted for shade and windbreak on roads and avenues (Bajpai and Chaturvedi, 1990). It is found throughout India in the tropical subtropical regions. It is also reported from the lower ranges of Himalayas up to an elevation of 1300 m and also in the Kumaon Hills up to 1600 m. It is commonly found under cultivation in the larger parts of India from the Indo-Gangatic plains in the North to Tamil Nadu in the South (Bajpai and Chaturvedi, 1990; Singh *et al.*, 1967). It is one of the most hardy fruit crop and can be grown successfully even in marshy areas (Bajapai and Chaturvedi, 1990).

2. Plant Characteristics: A large tree with ash-white bark, dead bark flaking off, branches terete and glabrous; leaves simple, opposite, decussate, elliptic, ovate-lanceolate or oblong-lanceolate, chartaceous, acute or acuminate; flowers pale greenish-white, borne in cymes arising from old leaf-scars; berries dark purple, smooth, oblong or subglobose, seed one, large, globose or oblong-cylindric.

3. Parts Used: Fruits.

4. Chemical Constituents: Fruits: Carbohydrates (19.7%), protein (0.7%), fat (0.1%), minerals (calcium, potassium, magnesium, phosphorous, sodium, iron), thiamine, riboflavin, niacin, vitamin B_6 and C. (Singh *et al.*, 1967).

5. Uses And Applications: Jamun fruit is tasty, pleasantly spicy flavoured and sub-acidic. It is mostly liked for dessert purposes. Apart from its fresh consumption, it is used for making delicious jam, jellies, squash, wine beverages, vinegar and pickles (Ochse *et al.*, 1961). Its use as squash is a very refreshing drink, particularly in hot season.

It is also medicinally significant. A little quantity of fruit syrup is beneficial for curing diarrhea. Jamun juice mixed with mango juice in equal proportion is useful for quenching thirst for diabetic patients. Vinegar made from juice obtained from slightly unripe fruits is carminative, diuretic and stomachic. This has also cooling and digestive virtues (Thaper, 1958). Jamun seed, being rich in protein, carbohydrate and calcium, forms a good feed for domestic arnimals (Thaper, 1958).

6. Karyotaxonomic Evidence: Chromosome studies in the family Myrtaceae, in general, are still scarce. Recently, Costa (2004) and Costa and Forni-Martins (2006 a,b, 2007) reported a predominance of 2n=22. These authors confirmed the basic chromosome number x=11 for the Myrtaceae. They reported 2n=22 chromosomes for four species of the genus *Eugenia*. Vijaykumar and Subramanian (1985) conducted cytotaxonomical studies in South Indian Myrtaceae. They investigated chromosome numbers in three species of the genus *Syzygium*. They reported chromosomes 2n=22 for *S.calophyllifolium, S.jambalanum* and *S.montanum*. Ohri (2015) provided the examples of intraspecific polyploidy series for *Syzygium cumini* (2x, 3x, 4x, 5x, 6x), *Eugenia jambolana* (2x, 3x, 4x, 5x, 6x) and *Eugenia uniflora* (2x, 3x). These reports are clearly indicative of variability of species especially in respect of fruit characters of present interest.

7. History: The large protean genus *Eugenia* L. contains many species from both the Old and New World. The Old World species of it have been reduced to *Syzygium* P. Browne ex J. Gaertner. Airy Shaw (1966) restricted *Eugenia* to largely American species and favourd the separation of the most Old World species into a genus *Syzygium* (*sensu lato*). The reproductive anatomy provided evidence for separation into the genus *Eugenia* and *Syzygium* (Schmid, 1972). The genus *Syzygium* Gaertn. previously formed a part of the genus *Eugenia* L. which contains several species of evergreen trees and shrubs. Many species yield edible fruits, and also useful as ornamental and medicinal values. A wild species *viz.*, *S.fruticosum* bears small but edible fruits. A popular and most liked fruit is the rose-apple (called Jamun) is *S.jambos*. It occurs in South India and West Bengal. It is also very ornamental, fruits being yellow, insipid in taste with high protein content. A small tree with edible fruits *viz.*, *S.fruticosum* bears small but edible fruits. It is grown as windbreak. A small tree with edible fruits *viz. S.zeylanica* occurs in the Western. Ghats of India (Naik, 1963). Another species *viz.*, *S.malaccensis*, called Malay Rose-apple, is found in South India. Surinam or Pitanga Cherry is *S.uniflora*. It occurs in South India which bears bright red fruit with aromatic flavour. Apart from this, *S.javanicum*, called Water-apple, is also found South India.

The common cultivar grown in North India is known as 'Ra Jamun' which yields big oblong fruit with deep purple or bluish-black colour. Its pulp is purple-pink juicy and sweet, the stone (seed) being small one. It is generally sold in markets of rural and urban places during June to July. There is also a small-sized, late maturing cultivar. It exhibits round shape and deep purple to blackish colour, pulp being purple. It contains less juice and lesser in sweetness as compared to the cultivar 'Ra Jamun'. It generally ripens during the month of August (Bajpai and Chaturvedi, 1990). There are very few standard cultivars *viz.*, Early Wild, Late Wild, Pharendra, Small Jamun, Dabka, Krian Duhat and Suepr Duhat, etc. It must be remembered that, in general, jamun is extremely variable in tree habit, fruit quality and season of fruiting.

8. Further Developments: Some experimental studies have been carried out to reveal basic features and for the sake of improvement and better yield. A high percentage of germination is observed when the seeds are sown afresh within 2-3 weeks (Mowry *et al.*, 1941). However, they loose their viability if stored seeds are sown (Shanmugavelu, 1967). Shanmusavelu (1970) treated the seeds with GA3 and improved the seed germination and growth of seedlings. This author in 1966 reported 45% rooting with 100 ppm IBA and 405 with 100 ppm IAA treatments. Bose and Sadu (1974) reported tip cuttings of *S.javanicum* rooted well under intermittent mist and treatment with IBA and 500 ppm. This produced 100% rooting. Singh *et al.* (1967) budded one-year old 10-12 mm thick root-stocks in July-August. Forket method of budding was found better than shield or T-budding (Singh *et al.*, 1967).

The new shoots emerge in two distinct phases *viz.,* (i) from February to May and (ii) August to October. Their growth is completed in two phases *viz.*, (i) from first week of March to second week of April and (ii) from second week of August to last week of October or second week of November. It is noted that the flush opening in February shows maximum growth and flowering (Misra and Bajpai, 1971). Flower bud differentiation on 5-10 month old branches start from last week of January and continue for 43 days. The floral parts arise in acropetal succession (Misra and Bajpai, 1973).

The maximum anthesis (18.71 to 43.08%) and dehiscence were reported between 10 am and 12.00 noon. The pollen fertility was noted higher in the beginning of a season. The maximum receptivity of stigma was recorded one day after anthesis (Misra and Bajpai, 1975).

The Jamun requires cross-pollination by honey-bees, houseflies and wind. The maximum fruit-set was observed when pollination was completed one day after anthesis. A sharp decline was then observed in fruit-setting (Misra and Bajpai, 1975). Misra and Bajpai (1971) observed flower and fruit drop just after opening of flowers and continue up to maturity. Their drop is found at three different stages *viz.*, (i) during bloom or shortly thereafter, (ii) after 35-40 days of full bloom and (iii) after 42-50 days of full bloom. These authors reduced flower and fruit drops by spraying 60 ppm GA twice after a gap of 15 days.

Percy Lancester and Bose (1965) succeeded in obtaining F_1 hybrid of a cross between the 'alba' variety of Water-apple (*S.javanicum*) and Rose-apple (*S.jambos*). These hybrids possessed prolific fruit-set and the fruits were larger in size than of either parent.

Fruit weight is dependent character in Jamun. It is governed by many factors *viz.*, length and breadth of fruit, volume, size, pulp weight, pulp thickness, pulp percentage, pulp to seed ratio, seed length, seed size and seed volume etc. Interestingly, these characters also showed variability (Prabhuraj, 2002).

9. Epilogue: Very few standard cultivars of Jamun are under cultivation. There are, however, many seedling strains which provide a fair scope for selection of better cultivars. While selecting cultivars, one must lay emphasis on (i) oval or oblong shape of fruit, (ii) deep purple or bluish-black colour of fruit, (iii) larger size, (iv) ample pulp, (v) juice content, (vi) sweetness, (vii) earliness and (viii) smaller stone (seed). Its cultivation needs avoidance of very heavy or light sandy soils. It requires dry atmosphere during flowering and fruiting. Early rains benefit for proper growth development and ripening of fruits. Polyembryony exists in Jamun and hence apomictic seedlings should be opted for plantation. The old, unproductive trees or even damaged young plants can be rejuvenated easily by top working, employing scion wood or buds from the superior clones. The existing cultivars have resulted from selection and cloning of superior trees.

Jamun is a seasonal fruit. It is difficult for harvesting and perishable early. It is hence doubtful for becoming a commercial fruit crop. Its harvesting is a tedious job as the tree has to be climbed to hand-pick mature fruit. Single or few trees as generally planted and not found in orchards. However, extension of the markets for processed products *e.g.* wine, beverage, juice, etc. can offer the best chances.

Devi *et al.* (2002) also opined recently that there use no significant differences among the genotypes about pulp to seed ratio and fruit to seed ratio. They further commented that major emphasis in selection should be higher pulp weight and fruit size.

7.19 KIWIFRUIT (KIWI)

Actinidia deliciosa (A.Chev.) C.F.Liang et A.R. Ferguson (Actinidiaceae)

Resume: *Kiwifruit (Kiwi) is also called 'Chinese Gooseberry'. It is a native of China and also its national fruit. It belong to the genus Actindia (Actinidiaceae). Different species contribute for Kiwifruit yield. The species* *are polypolid and chromosome numbers vary species to species from 2n=58, 116 to 174. Actinidia deliciosa is inferred to be allopolyploid. Their fruits are enjoyed in various processed forms.*

1. Introduction: Kiwifruit is a national fruit of China. Kiwifruit (Kiwi) or Chinese Gooseberry bears edible berried type of fruit. It belong to the genus *Actinidia* (Actinidiacee). New Zealand growers adopted its name as 'Kiwifruit' for commercial purpose since 1974, whereas the Europeans referred it as 'Chinese Gooseberry' (Morton, 1987). The name 'Kiwifruit' is also shortened simply as 'Kiwi' since about 1966 especially when the fruit was first exported from New Zealand to the United States (Deverson and Kennedy, 2005). Kiwifruit is native to North-Central and Eastern China (Morton, *loc.cit.*). It was described for the first time dating as back as to the 12th century China during the Song Dynasty (Ward and Courtney, 2013). Its cultivation spread from China in the early 20th century to New Zealand, wherein it was commercially planted for the first time (Morton *loc.cit.*). It was popularized during the period of World War-II especially when British and American servicemen stationed in New Zealand.

2. Plant Characteristics: A twining or climbers woody, vigorous shrubs; leaves oval to circular, cordate, alternate, long-petioled, red hairy when young, mature leaves dark green, hairless adaxially hut downy-white abaxially; flowers dioceious or bisexual, borne singly or in threes in leaf axils, white, turning to buff-yellow later, fragrant; fruits oval, ovoid or oblong up to 6.25 cm long, skin russet-brown, densely studded with short, stiff, brown hairs, flesh firm until fully ripe, glistening, juicy, bright green, sometimes yellow, browning or off-white, seeds minute, dark purple or black.

3. Useful Parts: Fruits.

4. Chemical Constituents: Fruits (Edible Portion): Carbohydrate (17.5%), protein (0.79%), fat (0.07%), minerals (calcium, iron, magnesium, phosphorous), thiamine, niacin, riboflavin, vitamin-A, ascorbic acid.

5. Uses And Applications: Ripe fruits are consumed. They are served in salads, in fish, fowl and meat dishes and also as appetizer. They are used in pies, puddings, cake-filling, ice-cream and sauce. Fruits are used in breads and beverages. They can be canned after softening partly and peeled. They are preserved after peeling and slicing in the solution of sucrose (12%), ascorbic acid (1%) and malic acid (0.25%). They are used to flavour ice-cream and juices of commercial production. Slightly under ripe fruits are used for making jellies, jam and chutney, being rich in pectin. Peeled whole fruits are pickled with vinegar brown sugar and spices. Cull fruits are made into wine. Using slices of fruit, meat can be tenderized (Morton, 1987).

6. Karyotaxonomic Evidence: Watanabe *et al.* (1990) investigated chromosome numbers in Kiwifruit (*A.deliciosa*) are related species. They observed chromosomes in the *A.deliciosa* CVS 'Matua' (Staminate Clone), 'Abbott' and 'Bruno' (Pistillate clones) and two native wild species *viz.*, *A.polygama* and *A.arguta*. The commercial cultivars revealed 2n=174, whereas those of *A.arguta* were 2n=58, 116 and 174 and those of *A.polygama* were 58. These data confirm earlier findings that the basic chromosome number X=29 (11) and that with increasing polyploidy.

Atkinson *et al.*, (1997) investigated genetic origin of Kiwifruit (*A.deliciosa* var. *deliciosa*) using phylogenetic analysis of DNA sequences derived from the polygalacturonase gene. Their results indicated that hexaploid kiwifruit had an allopolyploid origin with the diploid *A.chenensis* contributing one genome (genome A) and another (as yet unidentified) diploid series contributing a second genome (genome B).

7. History: The genus *Actinidia* (Actinidiaceae) includes nearly 60 species. However, the most common kiwifruit comes from the species *viz.*, *Actinidia deliciosa*. There are few other Kiwifruit yielding species such as *A.chinensis* (Golden Kiwifruit), *A.coriacea* (Egg Gooseberry), *A.arguta* (Hardy Kiwifruit), *A.kolomikta* (Arctic kiwifruit), A.melanandra (Purple Kiwifruit), *A.polygama* (Silver Vine) and *A.purpurea* (Hearty Red Kiwifruit) (Mortan, *loc.cit.*).

8. Further Developments: Studies have been on record revealing many medicinally useful compounds, among which antioxidants and serotonin may be beneficial to treat sleep disorders. Hsiao *et al.* (2011) assessed the effects of Kiwifruit on patterns, onset and quality of sleep. They reached to the conclusion that Kiwifruit consumption may improve sleep onset, duration and efficiency in adults.

Liu Y. *et al.* (2017) searched for rapid radiations of both Kiwifruit hybrid lineages and their parents. They sequenced 25 taxa representing natural diversity in the genus *Actinidia* with an average mapping depth of 26x on the reference genome to reconstruct their reticulate history. They found evidence, including significant gene free discordance, cytonuclear conflicts, and changes in genome-wide heterozygosity across taxa, collectively supporting extensive reticulation in the genus. At least two separate parental species pairs were found involved in the repeated origin of the hybrid lineages, in some of which a further phase of syngameon was triggered. They found no evidence of hybrid origin. The backbone taxa have distinct demographic histories and are the product of recent rounds of rapid radiations via sorting of ancestral variation under variable climatic and ecological conditions. Their results suggested a mode for conservative plant diversification through two layers of radiations, consisting of the rapid evolution of backbone lineages and the formation of hybrid swarms derived from these lineages.

Jin-Hu Wu (2012) carried out manipulation of ploidy for Kiwifruit breeding and study of *Actinia* genomics. The genus contains about 60 species, only two of which are commercially cultivated on a large scale and these have been only recently introduced into cultivation. Dioecism and the range of ploidy levels, coupled with a high basic chromosome number, seriously hinder the study of the genetics of *Actinidia* and breeding using traditional gene introgression. *In vitro* manipulation of ploidy levels, production of haploids and/or polyhaploids by culture of pollen or unfertilized embroys, chromosome doubling and protoplast fusion could bypass these barriers, which hinder traditional cross breeding. They described the application of these technologies and the contribution they can make to the development of new and novel types of kiwifruit and to investigations on the genomics and evolution of *Actinidia*.

Richardson *et al.* (2011) studied fruit development of the diploid Kiwifruit, *Actinidia chinensis* 'Hort 16A'. They found that many aspects of fruit morphology, growth and development similar to those of the model fruit tomato, except for a striking difference in fruit ripening progression. The early stages of fruit ripening occur as the fruit is still growing and many ripening events are not associated with autocatalytic ethylene production Autocatalytic ethylene is produced late in the ripening process as the fruit begins to senesce. This study provided a reference framework for subsequent physiological and genomic studies, and will allow cross comparison across fruit species, leading to a greater understanding of the diversity of fruits found across the plant kingdom.

Lai *et al.* (1989) investigated fruit growth in Kiwifruit (*Actinidia deliciosa*). The authors demonstrated different patterns of assimilate translocation in the Kiwifruit vine, such that any shortage in the local supply of carbohydrate for fruit growth within a lateral was readily met by surplus carbohydrate from other fruiting or non-fruiting laterals. They recorded assimilate transport in either a basipetal or acropetal direction, depending upon the relative positions of the regions of carbohydrate surplus and shortage. They observed translocation over seven nodes distance in a 1-year-old stems. Fruit on laterals with low leaf-fruit ratios (1:1) imported the most ^{14}C label from neighbouring laterals. The partitioning of ^{14}C assimilate towards these fruit increased with the physical proximity of the laterals. While fruit growth was increased by the presence of non-fruiting laterals in girdled canes, fruit size was reduced by large leaf numbers on non-girdled canes.

Mcpherson *et al.* (2001) studied flower quality and fruit size in Kiwifruit (*Actinidia deliciosa*). Authors grew vines at five sites that represented the climatic range of the major kiwifruit growing regions of New Zeland. They applied hydrogen cyanamide to half the vines as three of the sites. They observed a progressive increase in ovary fresh weight and ovary dry weight from warmer to cooler sites. Hydrogen cyanamide increased ovary fresh weight by between 6 and 20% and fruit weight by up to 11% at the two warmest sites. Variation in ovary fresh weight and dry weight accounted for some of the variation in fruit weight 170 days from flowering in the first season, but over both seasons, by itself, had no predictive value. The highest correlations between from fruit weight and flower quality across both seasons was with the ratio of pedicel length to seed number (r^2=0.56).

9. Epilogue: Kiwifruit is not derived only from a single taxon but it is a group of few species of *Actinidia* genus that add to its yield. They are varied in respect of their vegetative, floral or fruit characters and hence exhibit fair diversity of the fruit. Nutritionally, it is very useful, apart from human health benefits. They are polyploids adapted to varied habitats. Apart from their fresh consumption, a wide range of processed products can be had from their fruits. Few species are also exploited from wild. They are being studied for their fruit quality and yield and still have scope to widen research activity regarding especially processed products and breeding programmes for more improved varieties for the welfare of mankind.

7.20 KUMQUAT

Fortunella Spp. (Rutaceae)

Resume: *Species of Fortunella are native of South Asia and the Asia-Pacific region. It earlier formed a part of the larger genus Citrus. They are commonly referred as Kumquats and popular for their edible fruits. Interestingly, their entire fruits are consumed. All species of this genus are characterized by the chromosome number 2n=18. Apart from their edibility, the plants are employed in Bonsai making.*

1. Introduction: It is commonly called 'Kumquat'. It is a native of South Asia and the Asia-Pacific region. It has long been under cultivation in Japan, India, Taiwan, the Philippines and South-East Asia. Its earliest reference appears in Chinease literature. It is introduced to Europe in 1846 by Robert Fortune who was a collector for the London Horticultural Society and later on into North America. Its taxonomy is dilated under the genus *Citrus* elsewhere, in which formerly it was placed (*cf.*Kumquat-Wikipedia).

2. Plant Characteristics: Slow-growing, evergreen shrubs or small trees, 2.5 to 4.5 m in height, branches dense, rarely bear thorns; leaves dark glossy green; flowers white, similar to *Citrus* flowers, borne solitarily or in clusters in the leaf-axils; fruits resembling the orange (i.e. *Citrus sinensis*) but much smaller, as shaped and sized as a large olive, brilliant orange coloured (Kumquat-Wikipedia).

3. Parts Used: Fruits.

4. Chemical Constituents: Edible portion contains carbohydrate (72.1%), protein (3.8%), fat (0.4%), minerals (calcium, phosphorous, iron, sodium, potassium), vitamin A, thamine, riboflavin and niacin (*cf.* Kumquat).

5. Uses And Applications: Fruits are a versatile ingredient in both sweet and savory preparations. Slices are cooked with sugar and water and used to top pound cake or ice- cream. Slices are added to pomegranate juice, champagne and club soda for punch. Fruits are mixed with diced avocado, red onion, cilantro and lime for a fresh salsa. In China, Japan and Taiwan, fruits are often preserved as jams or in salt. They are added to teas in Taiwan. Interestingly, the plant is used as an ornamental Bonsai tree in South-East Asia. The seeds are a source natural pectin (*cf.* Kumquat Information-online).

6. Karyotaxonomic Evidence: All species of the genus *Fortunella* are confirmed diploid with 2n=2x=18 chromosomes. The chromosomes of *Fortunella* are classified into six types depending on CMA-banding. Almost of the species of the genus had either one or two characteristic type F chromosomes. *Fortunella hindsii* have the simplest karyotype without types A, E, and Dst chromomes (1B+1C+14D+2F). As a whole, the D type chromosomes are predominant in this genus (Yasuda *et al.*, 2016).

7. History: *Fortunella* species, commonly known as 'Kumquats' have been called 'the little gems of the citrus family'. Until about 1915, species of *Fortunella* were included under the genus *Citrus*. Dr.W.T.Swingle segregated these species in the genus *Fortunella*, which includes six Asiatic species. The common name 'Kumquat' or 'Cumquat' or 'Comquot' means 'gold orange' in China. Kumquats are said to be native of China. They have been grown in Europe and North America since the mid-19th century, particularly as ornamental dooryard trees and as potted specimens in patios and greenhouses. In South India, they are raised only at high elevations. They are also cultivated to some extent in South Africa and Australia. Most species (called here varieties) are used for food *e.g.* (i) Hong Kong (*F.hindssi* Swing.), (ii) Marumi (*F.japonica* Swing.), (iii) Meiwa (*F.crassifolia* Swing.) and (iv) Nagami (*F.margarita* Swing.) (*cf.* Kumquat).

8. Further Developments: Xiaoyong *et al.* (2005) analysed intergeneric somatic hybrid plants between Changshou Kumquat and Dancy tangerine by using flow cytometry (FCM), simple sequence repeats (SSR) and cleaved amplified polymorphic sequence (CAPS). This was aimed at characterizing the ploidy level, and the nuclear and cytoplasmic compositions of two integeneric somatic hybrids of Changshou Kumaquat (*Fortunella obovata*), leaf patent, and Dancy tangerine (*Citrus reticulata*), suspension parent. FCM showed that both somatic hyrids were tetraploids. One out of two SSR primer pairs was able to detect polymorplism between the fusion parents, and the somatic hybrids showed additive bending profiles, showing them to be allotetraploid hybrids with inherited nuclear DNA from both parents. CAPS showed that 4 and 1 polymorphic loci were obtained for parental chloroplast and mitochondrial genomes respectively. This research helped in understanding the relationship between nuclear and cytoplasmic compositions and the likely field performance of somatic hybrids in the future.

Miranda *et al.* (1997) obtained somatic hybrids by fusion between *Poncirus trifoliata* (2X) and *Fortunella hindsii* (4X) protoplasts. Southern blot analysis of three regenerants revealed that they carried specific rDNA fragments from both fusion partners, thereby confirming their hybridity. In contrast, mitochondrial DNA (mtDNA) and choroplast DNA (cpDNA) were unidirectionaly transmitted from the callus parent without any evidence of recombination. No differences in the restriction fragment patterns of rDNA, mtDNA or cpDNA could be detected among the regenerants. Flow cytometry showed that two regeneration were hexaploids and one was pentaploid. Takami *et al.* (2004) also obtained intergeneric somatic hybrids between round Kumquat (*Fortunella japonica* Swingle) and 'Morita' navel orange (*Citrus sinensis* Osbeck) to create breeding parents for interploid crosses aiming to release new seedless Kumquat like cultivars for Japanese markets.

9. Epilogue: Although the species of *Fortunella* were included previously under the genus *Citrus*, their fruits stand out from rest of the *Citrus* species. In case of *Citrus*, one can eat only the pulp and make use of juice. But in case of Kumquats (*Fortunella*) one can consume entire fruits with skin and all. The skin is substantially the best part. It has a sweet, spicy flavour. Fruits are to be harvested when fully ripe and orange in colour. Fruits can be stored well for 1-2 months at 36-39° F and as such are better for commercial purposes. It is to be noted that they last ony a few days at room temperature. They are a good balanced source of vitamins and antioxidants.

7.21 MANGO

Mangifera indica L. (Anacardiaceae)

Resume: *It is native of Southeast Asia and has been cultivated in India for more than 4000 years. It is deeply integrated in Indian culture. It is regarded 'King of Fruits' being the most delicious, nutritious and widely used throughout world in various forms. It is rich in sugars, vitamins and minerals. Wild Mangifera indica grows in north-western India. Various species of mango have 2n=2x=40 chromosomes. Polyembryony is common in mango cultivars.*

1. Introduction: The mango is one of the most important tropical fruits and, in the opinion of many, one of the most delicious fruits in the world. It is cultivated throughout the tropics but is most abundantly produced in India, where it occupies about 0. 9 Mha. The mango is a tree which sometimes attains great size (up to 30-40 m tall). It is adapted to seasonal tropical climates at low altitude; a dry season is essential for fruit setting. Purseglove (1968) has reviewed the economic botany of the crop.

One of the oldest, popular and choicest and highly prized as a dessert fruit is the Mango. It is regarded 'King of Fruits' in India and certainly deserves the title of 'national fruit' of India. It is native of Southeast Asia and has been cultivated in India for more than 4000 years. Its wild trees are recorded from northeast India to the Malay Archipelago. It is a sacred tree in India and holds prominent place among the best fruits of the world. It is closely connected with folklore, religion, ritual, customs, Hindu mythology, festivals and observances. It is said that a mango grove was presented to Lord Buddha by Amradharika so that he could relax in the shade of the mango tree. It is represented on the Buddhist stupas of 110 BC., including the famous Barhut stupa.

Mango is one of the fruits so closely associated with the history of agriculture and civilization. There are over 1000 named mango varieties throughout the world. This is, in fact, a testament to their value to mankind. These names reflect origin and spread of mango tree along the spread of mankind. It has been under cultivation in the Indian subcontinent for over 4000 years (De Candolle, 1904). Legend has it that a mango grove was presented by Amradarika to Lord Buddha so that he might use it as a place of repose (Popenoe, 1920). It finds mention in the travellogues of ancient travellers from Hiuentsang (632-645 AD) who was the foreigner to bring mango to the notice of the outside world. Amir Khusrau (1330 AD), the great Sufi saint and poet, sung its praise during the time of Mohammed Tuglhaq (1325-1351 AD). The greatest tribute to this esteemed plant was paid by the Emperor Akbar (1556-1605). He established the 'Lakh Bagh' (a mango orchard having 100000 plants) in Darbhanga (Bihar, India). In Ain-Akbari, Persian literature, was written by the great chronicler Abul Fazl (1590 AD) during the reign of Emperor Akbar. He discussed in detail the cultivars, cultivation and quality of mangoes.

Mango was introduced into East Africa by the Persians by the 10th century AD. From its original home, the Portuguese colonists introduced it into Brazil around 1700 from India, and from there it reached to the West Indies in 1742 and Mexico early in the 1800s. It was introduced in Florida in the 15th century. Presently, it is being grown in all the warmer regions of the world.

2. Plant Characteristics: A magnificent, evergreen, large spreading tree reaching a height of 25; leaves glossy, dark green, elliptic-lanceolate, leather, nearly 8-40 × 2-10 cm in size; flowers small greenish, or reddish, borne in large panicles; fruits large ovoid-oblong drupe, pulpy and juicy.

3. Parts Used: Fruits, kernels.

4. Chemical Constituents: (i) Ripe fruits contain sugars (sucrose, glucose, fructose), protein, fats, minerals, fruit acids, vitamin A, B and C. Seeds are rich in carbohydrate, fat and protein. Unripe fruits contain citric, malic, oxalic and succinic acids, vitamin A and C, rich in potassium. (ii) Unripe fruits contain: protein (0.7%), fat (8.8%), carbohydrates (0.01%), minerals (calcium, phosphorous, iron), carotene (as vitamin-A), vitamin B1, B2 and C riboflavin, ascorbic acid (Anonymous, 1948).

5. Uses And Applications: Young and unripe fruits are picked and used in chutneys, amchur and culinary preparations. Ripe fruits are used in the manufacture of juice, squash, jams, jellies, preserve (murabba) and canned. Ripe fruits are mostly consumed as fresh dessert. Amchur prepared from unripe fruit is used as souring agent for soups and vegetables. Kernels are salted and consumed. Fresh mangoes are processed and preserved into a wide range of products including pulps, frozen slices, pulp fruit leathers, canned in syrup, and sliced in brine. In Thialand, they are sliced or grated in fresh salad, soaked in water and sugar, salted, dried, sliced in vinegar or fish sauce or eaten as crunchy fruit. In India, drink called 'Panah' is prepared from boiled unripe fruit by adding sugar. It is consumed during regular meal.

Mango essence is used to flavor many food products such as drinks, ice-creams, wines, teas, breakfast cereals, biscuits and muesli bars. In India, the seed is eaten as a boiled or baked vegetable or ground into a starchy flour (*cf.* Bally, 2006).

6. Karyotaxonomic Evidence: Mukherjee (1972) recognized 41 species distributed in two sections. All are southeast Asian. They occur from northern India to the Philippines and new Guinea, with a marked centre of diversity in Indochina-Malaysia-Indonesia. Wild *M. indica* occurs in north-eastern India. It is related to *M. sylvatica, M. caloneure, M. zeylanica* and *M. pentandra*. These species, and all other Mangifera species and cultivars so far examined, have $2n=2\times=40$ (Mukherjee, 1950; Roy and Visweswariya, 1951). *Mangifera indica* and *M. sylvatica* are karyotypically related. Secondary meiotic associations suggested secondary polyploidy based on $\times=8$. The polyploidy is presumably ancient for no basic number as low as 8 is known in the Anacardiaceae (where $\times=12$, 14, 15, 16). Recently tetraploidy ($2n=4\times=80$) is reported in a potyembryonic seedling of an Indian cultivar (Roy and Visweswariya, 1951).

7. History: According to Mukherjee (1972), the local origin of the mango from wild *M. indica*, without the intervention of other wild species is pointed out. The fruits of several other species are eaten locally. Wild mangoes in general have fruits which are unacceptably fibrous-resinous and may even be poisonous. Succulence, low fibre, small stone and low resin content have been human selection. The dates of 4000 B. P. have been mentioned for its domestication. The crop spread out of India eastwards, in the first instance, as early as 400-500 BC. in the hands of Buddhist monks. It did not become generally diffused through the islands and into the Pacific until very much later westwards. Mukharjee (1972) opined that it moved as late as the 16th century, in the hands of Portuguese travellers, from Goa, by way of East Africa to West Africa and the New World. He suggested, however, an alternative possibility that it spread by way the Sabaean Lane as early as the 10th century. Mangoes were first grown in the West Indies in the 18th century and reached Florida in 1833.

The best mango fruits are hailed from selected clones. The great majority of trees are unselected seedlings often more or less wild. The crop is outbred and hence there is usually great variation between seedlings. Polyembryony prevails and appear true breeding (in effect, apomictic) cultivars known in southern India. In general, good cultivars should be propagated vegetatively and one form or another of budding is usual. The practice reached India in the 16th century, introduced by Europeans, and has since been adapted for the world-wide dissemination of chosen clones by means of bud wood. Male and female flowers are separate within the inflorescence and insect pollination is essential. Outbreeding is encouraged by dicliny and entomophily. In general, mangoes are be self-compatible. Sharma and Singh (1970) however, found evidence of some incompatibility.

The mango population generally consists of vast numbers of sexual seedlings, a few polyembryonic lines and many selected clones. India alone is estimated to have 1,000 clones established: elsewhere.

8. Further Developments: Breeding was started in India early in 20th century and the work has been reviewed by Mukherjee *et al.,* (1968) and by Singh (1969). Hand-crossing is remarkably unrewarding, success rates of the order of three fruits per 1,000 pollinations having; been recorded. The life cycle of mango is long. It is shortened by well-chosen grafting procedures. Polyembryony pose problems and hence none of the seedlings produced have proved to be outstanding. The basic objectives of breeding have been regular bearing and good fruit quality.

Some useful somatic mutants have been also picked up *e.g.* the cv. Haden in Florida (Young and Ledin, 1954). Induced mutation studies have not been encouraging.

9. Epilogue: There is undoubtedly a need for export varieties as well as for improved cultivars for local consumption. No improvement is achievable in a crop in which abundant variability is available, especially if new selection criteria are presented to the breeder. Probably, replacement of the laborious and ineffective hand-pollination by some mass-pollination technique is essential if adequate populations are to be raised. Dwarfness might be achieved by mutation, natural or induced, in otherwise desirable clones Indeed, some clones with trees markedly smaller than usual are established. Still, much remains to be done in regard to developing cultivars suited to different regions and differing needs. There are many wild species and in the developing countries of South Asia. Attention should be paid on their fundamental aspects before they are lost. Mangoes are selected by their characteristics such as eating quality, keeping, growing characteristics, fruit qualities like flavor, aroma, flesh texture, fibre, fruit size, colour, etc. The adapting of a variety to environmental pressures of drought, wet weather during flowering, temperature, pests and diseases are also important selection criteria as these decide the cropping consistency and fruit quality.

7.22 PAPAYA

Carica papaya L. (Caricaceae)

Resume: *It is a native of eastern Central America. It is under cultivation even before Columbian time. Presently, it is grown throughout the tropical countries and warm part of the world. Ripe fruits are nutritious and cherished fresh and also in the form of various preparations. Papain obtained from it is used in the manufacture of chewing gum and in the leather industry. Various species of the genus Carica have 2n=2x=18 chromosomes, majority of species being dioecious. Sex-reversals are also observed in case of Papaya.*

1. Introduction: The 'Papaya' or 'Papaw' has the origin in the lowlands of eastern Central America (*cf.* Muthukrishnan and Irulapean, 1996). In pre-Columbian times, it was cultivated by the Indian and thence it reached to Africa, India and the Caribbean islands by the early Portuguese and Spaniards. It is now grown throughout the tropical countries and warmer regions. It was taken by the Spanish to Manila in the middle of 16th century and reached Malacca afterwards. From there, it was brought to India in the 16th century.

2. Plant Characteristics: A giant, large, laticiferous fast-growing and hollow-stemmed, mostly unbranched tree, with a crow of leaves at top; leaves long-petioled and palmetely-divided; plant dioecious (separate male and female plants), flowers yellow, males in panicled inflorescence, female flowers large, borne either singly or in few-flowers axillary cymes; fruits large, ovoid or oblong hollow berries, turn yellow or orange when ripe, contain soft, fleshy mesocarp with buttery consistency.

3. Parts Used: Berries.

4. Chemical Constituents: Ripe fruits: carbohydrates (9.5%), protein (0.5%), fat (0.1%), vitamins (A, B1, B2 and C), minerals (calcium, phosphorous, iron), riboflavin, nicotinic acid.

5. Uses And Applications: It is an excellent breakfast fruit. It has sweet musky taste. It is used for salad, jellies, preserves, pies, sherbets, juice and confections. Unripe fruit is used in a preserved. Papain obtained from its latex is used in canned meats and in meat-tenderising preparations and also in pharmaceuticals as a milk laxative. Papain is also used in manufacture of chewing gum and in the leather industry for treating hides. It is employed in the textile industry to prevent shrinkage of wood and silk. It is a component of some toothpastes for curing pyorrhea (a gum disease). In USA, it is used to produce 'chill-proff' beer.

6. Karyotaxonomic Evidence: The Caricaceae, a small dicotyledonous family consists of four genera and 31 species, The genera and the number of species included are *viz., Carica* (22), *Facaratia* (6), *Farilla* (1), *Cylicomorpha*

(2). The first three are indigenous to tropical America, the last to equatorial Africa (Badillo, 1971). To date, chromosome numbers have been counted in only eight species of *Carica*; none in the other three genera. All species have 2n=2×=18. Species having edible fruits are found only in *Carica*. Besides *C.papaya*, they are *C.chilensis*, *C.goudotiana*, *C.monoica*, and *C.pubescens*.

Two putative natural interspecific hybrids are known *viz., C. monoica* × *pubescens* and *C.* × *heilbornii* of which pistillate trees only are known. Four phenotypic variants of *C.* × *heilbonii* are known, all of which yield large, seedless, parthenocarpic, edible fruits. They are grown by Andean people who maintain them clonally by cuttings, Badillo (1971) classified them as nm. Heilborni, nm. Chrysopetala and nm. Fructifragrans. Horovitz and Jiminez (1967) reported producing hybrid plants identical to those of *C.* × *heilbornii* by making reciprocal crosses between. *C. pubescens* and *C. stipulata*. Cytological evidence does not lend support to the supposition that *C. papaya* arose as an interspecific hybrid of unknown parentage.

All members of Caricaceae are dioecious except for three species of *Carica viz., C. monoica, C. pubescens* and *C. papaya*. They have sexually ambivalent forms which go through 'sex-reversals' in response to climatic and/or photoperiodic changes during the year. *Carica monoica* is strictly monoecious but at certain times of the year may lack pistillate flowers. Trees of *C. pubescens* exist in three basic sex forms *viz.,* pistillate, staminate trees are unresponsive to seasonal climatic changes. The andromonoecious trees are sexually ambivalent, bearing staminate, perfect and pistillate flowers in varying proportions at different times of the year. *Carica papaya* exists in the same three basic sex forms as *C. pubescens*. The pistillate tree is stable. Staminate and andromonoecious trees may be: (i) phenotypically stable; (ii) phenotypically ambivalent, going through seasonal sex-reversals, during which they bear varying proportions of staminate, perfect and pistillate flowers.

Papaya trees are classified by Storey (1958) in 31 heritable phenotypes depending on peduncle length and ramification and seasonal sexual responses. Fifteen of these are variations belong staminate trees, and fifteen variations to andromonoecious trees. The remaining phenotype is the pistillate tree.

Matings among the basic sex forms yield segregating progenies as:

Sr. No.	Mating	P	S	A
(i)	P × S	1	1	—
(ii)	P × A	1	—	1
(iii)	S × S	1	2	—
(iv)	A × A	1	—	2
(v)	A × S	1	1	1
(vi)	S × A	1	1	1

(P = pistillate, S = staminate and A = andromonoecious)

Staminate and andromonocecious plants are heterozygotes and pistillate ones are recessive homozygotes. A zygotic lethal factor eliminates dominant homozygotes. There are three different hypotheses on the genetics of sex determination in the species:

(1) Hofmeyr's (1967) hypothesis considers genic balance. The symbols M1 and M2 represent inert or inactivated regions of slightly different lengths on sex chromosomes which are inert or from which vital genes are missing. This is responsible for the zygotic lethality of M1M1, M2M2 and M1M2 genotypes. The homologous region, m, is normal. The viable genotypes are: M1m, staminate; M2m, andromonoecious; m m, pistillate.

Concentration of genes for femaleness is on the sex chromosomes and the greater concentration for maleness on the autosomes. Thus, the genotype m m, is pistillate and its homozygosity confers phenotypic stability. Since M1 is the longer of the inert regions, it is manifested phenotypically as staminate because of the greater influence of the autosomal factors. The shorter M2 region is less influenced by autosomal genes so the M2m genotype is expressed phenotypically as andromonoecious. The heterozygosity of M1m and M2m renders them susceptible to alteration in phenotypic expression by external influences.

(2) The hypothesis of Horovitz and Jiminez (1967) proposes a reversionary process for sex determination and expression. It is assumed that dioecism is the primitive state in Caricaceae and sex determination is of the classical XX-XY type with heterogametic male and YY lethal to the zygote. At some point in time, a sexually ambivalent form appeared in the genus from which the three present-day exceptions to dioecism (*C.monoica*, *C.pubescens* and *C.papaya*) arose. In *C.papaya*, the ambisexual mechanism built up on the Y chromosomes, giving rise to a modified homologue, the Y2 chromosome which, in the heterogametic genotype XY2, is expressed as the sexually ambivalent andromonoecious form. This occurred without any alteration of the X chromosome. This hypothesis holds that andromonoecism and polygamy followed the evolution of an XX-XY system and are of fairly recent origin.

(3) Storey (1953, 1967, 1969) putforth his hypothesis on progressive evolution of dioecism in the family. An unknown ancestor is assumed to have had perfect flowers which evolved as: The hypothetical flower must have been much like the perfect, perigynous flowers of *C.pubescens* and C. papaya, having a superior, 5-carpelled pistil, a gamopetalous corolla terminating in a 5-parted limb and 10 (2×5) stamens inserted in the throat of the corolla tube. This form is known as elongata in *C.papaya*. The staminate flower was originated by phylogenetic elimination of the functional pistil. The pistillate flower was derived in a succession of extraordinary changes which included: carpellody and connation of the upper cycle of stamens accompanied by abortion of the original carpels to

constitute the pistil of a new 5-stamened (pentandria) flower; carpellody and connation of the stamens of the pentandria flower accompanied by abortion of the pistil to become the pistil of a new pistillate flower; almost complete freeing of the petals; and change of insertion of the ovary from superior to partially inferior (pleurogyny). The pistillate flower was derived as a morphological anomaly in that its pistil consists of a whorl of stamens transmuted into carpels at the expense of the original elongata carpels and the upper set of stamens.

The symbol in this hypothesis (SA) represents the sum of the factors involved in transmuting the ancestral androcium into the present-day gynoecium; (sa) stand for normal androecium development; (SG) stand for factor or factors responsible for suppression of the gynoecium in the staminate flower; (sg) permits the (SA) factors to function ontogenetically in developing the replacement gynoecium. The symbol I presents the recessive sex-linked zygotic lethal factor that enforces heterozygosity on the staminate and andromonoecious plants; and C represents the factor that inhibits crossing-over between the sex-determining factors and the lethal factor, responsible for the non-existence of pistillate plants carrying the I factor. As in the other two hypotheses, heterozygosity permits ambisexuality. The sex-determining genotypes are expressed as:

(i) staminate and andromonoecious: (sa) 1C (SG)/(SA)+ +(sg)

(ii) pristillate: (SA)++(sg)/(SA)++(sg)

The sex homologues may be represented as: Mh, andromonoecious or hermaphrodite; MS, androecious, staminate or male; In, gynoecious, pistillate or female.

Length and ramification of the inflorescences are secondary sex characters. The alleles which give them expression are on the sex chromosomes but linkage with the sex-determining factors is not absolute. In effect, all staminate and andrornonoecious forms are identical. The classification of trees as the one form or the other depends largely upon the nature of the inflorescence. In this regard, Storey (1958), and Hamilton and Izuno (1967) do agree.

7. History: The papaya originated in the lowlands of eastern Central America. No record of its history is available prior to discovery of the New World. However, it must have been cultivated for a very long time by early civilizations for it was known from Mexico to Panama and occurred in a wide range of diverse types. The earliest mention is by De Oviedo in his book 'The General And Natural History of The Indies' (1535). It spread rapidly along early tropical trade routes and by 1800, was being grown throughout the tropical world, even in remote islands of the South Pacific.

Studies of floral anatomy indicated that the prototype of *C. papaya* was perfect-flowered (Storey 1967). Probably it had small fruit like those of the dioecious form of *C. papaya* presently naturalized throughout the Caribbean region and known variously by *C. cubensis, C. jamaicensis, C. partorricensis* and others. Present-day dioecism evolved from the prototype with the derivation of unisexual flowers (Storey 1967). From the viewpoint of the evolutionary trend towards dioecism in the family, it appears that the present-day forms of *C. papaya* owe their existence to continuous selection and cultivation by mankind.

8. Further Developments: Production of Papaya in Hawaii is based on the cultivar Solo, with pyriform fruit weighing 350-500 g borne by the hermaphrodite trees of an inbred gynodioecious strain. Parental lines are cultivated in isolation and only seeds from hermaphrodite trees are planted. These produce hermaphrodites and females in the ratio of 2:1. If seedlings were planted singly at the rate of 1,080 per hectare, there would be about 720 hermaphrodites and 360 females. There is no known means of separating the sexes in the early seedling stage. Seedlings are hence planted in threes of which two are later eliminated. With each place reduced to a single tree about six months after planting when the seedlings begin to flower and the sex of each can be deciphered. The final stand is 1,020 hermaphrodites and 40 females.

An important cultivar in South Africa is Hortus Gold, of which the fruit is borne by female plants of a dioecious strain. Seedlings of this strain separate as 1 female: 1 male so there would be 540 of each in a one-hectare planting. About 10-16 per cent of males in a dioecious planting is sufficient for pollination. Seedlings are planted in threes and later reduced to a final stand of 945 females: 135 males, or 12.5 per cent males.

9. Epilogue: The problems that need to be attended are: (i) enhance the probability of increased fruit and latex production, and development of virus-resistant cultivars, (ii) elimination of ambisexual andromonoecious forms that tend to become female sterile at certain times of the year or show a tendency toward stamen carpellody, (iii) development of homozygous andromonoecious forms by possible elimination of the zygotic lethal factor or by *in vitro* culture of embryos, (iv) inducing a sex-linked vegetative character for eliminating unwanted sex forms in the early seedling stage, (v) early and low bearing, (vi) selection for inflorescences of moderate length (about 7.5-10.0 cm) bearing single fruits, to circumvent crowding with resulting misshapen fruit, (vii) selection for an ovarian cavity circular in transverse section and with an easily separable placenta, (viii) breeding of dioecious cultivars for places in which andromonoecious types are excessively sexually ambivalent, (ix) breeding to expand regions of production by hybridizing with species that are more cold tolerant (*e.g. C. pubescens*); and breeding for increased latex yield.

7.23 PINEAPPLE

Ananas comosus Merr. (Bromeliaceae)

Resume: *It is a native of Northern South America. It still occurs in wild in southern central Brazil and Venezuela. Pre-Columbian oceanic contacts based on presence of pineapple in parts of the old World is still an enigma. Columbus is supposed to introduce it to Europe. The large so-called 'fruits' are syncarps formed from the entire inflorescence. Fruits being rich in sugars, citric and malic acids are favoured as a highly valued dessert fruit. They are added in various food preparations. Ananas is diploid with* $2n=2x=50$ *chromosomes.*

1. Introduction: The pineapple, also called 'Ananas' is a native of Northern South America. It is one of the few tropical fruits that has been exploited commercially and has a more rapid rise in international trade because of its own intrinsic qualities. Many of its cultivars are distributed from its home to Central America and the West Indian Islands. Columbus and associates were the first Europeans who tasted the fruit in 1493. Later Spaniards and Portuguese introduced it in Africa and by 1548 it was being cultivated in India and the Dutch East Indies. The British started growing it in hot houses. They selected forms with large and seedless fruits and introduced them into the America, Africa, India, Malaya, Australia and the Azores. Wild forms are still found in its native place. The 'pineapple' is used as it resembles to the pine cones. It is now grown in both hemispheres in frost-free areas. Within America, with tribal migration and trades between tribes, the pineapple cultivars were distributed throughout most of tropical America and the species developed into a cultigen in the process. It spread in America by the time of Columbus, who took it to Europe. It reached in India in 1548 (*cf.* Sen, 1996).

2. Plant Characteristics: A short-lived perennial, with a short stem; leaves stiff, dark-green, in rosettes, fleshy, 3 ft or more in length with prickly margin and spine-tipped; 3-merous flowers borne in dense heads and crowded by a tuft of leaves; large fruits are syncarps, multiple fruits formed from the whole inflorescence.

3. Parts Used: Fruits.

4. Chemical Constituents: Sugars (13%), fruit acids (citric and malic), vitamins (A, B1, B2, C), protein (0. 6%) minerals (calcium, phosphorous, iron).

5. Uses And Applications: It is a popular and highly valued dessert fruit because of unusual sweet-sour taste. When fresh, it is a source of vitamins A, B and C. The slices of pineapples and fruit juice are canned for export and domestic consumption. The dried waste is used for making vinegar. Pineapple is a popular addition to fruit cakes, bread, pies, biscuits, fritters and puddings.

Its juice is fermented into wine, brandy and liqueur and the bran constitutes a good cattle feed. White fibres obtained from leaves, called 'Pina' are used for making 'pine cloth which is popular in China, Taiwan and the Philippines. Waste material after extraction is used for paper-making. Other by-products are alcohol, calcium nitrate, citric acid and vinegar.

6. Karyotaxonomic Evidence: *Ananas* is closely related monotypic genus *Pseudananas*. Both distinguished from other Bromeliaceae by their syncarpous fruit. *Ananas* bears a terminal crown of reduced leaves, whereas *Pseudananas* has no such crown. Both genera share the basic chromosome number $\times$=25 common throughout the Bromeliaceae. *Ananas* is diploid (2n=2$\times$=50) while *Pseudananas* is tetraploid (2n=4$\times$=100). They are distributed in parts of lowland South America. Among the wild species of *Ananas, A. bracteatus, A. fritzmuelleri* and *A. parguazensis* appear closely related. Both *A. bracteatus* and *A. fritzmuelleri* produce fleshy, edible but seeded fruits and *A. bracteatus* has in fact been cultivated in the Parana river area (Collins, 1960). Fruits of *A. ananassoides* and *A. erectifolius* become nearly dry at maturity, with little flesh. *Ananas erectifolius*, which has long, almost spineless leaves has, however, been considered a potential fibre crop. *Ananas bracteatus, A. ananassoidcs* and *A. erectifolius,* and the F_1 hyqrids are completely fertile. When *A. comosus* was crossed with *Pseudananas*, the F_1 plants were mostly tetraploid, with only a few of the expected triploids, and again fully fertile (Collins, 1960).

7. History: 'Wild' pine-apples occur in southern central Brazil and differ from 'wild' pineapples in other parts of Brazil and Trinidad. It was but not clear whether they were recent escapes from local cultivars or links between an unknown wild comosus and modern cultivars (Collins, 1960). 'Wild' pineapples have also been reported from Venezuela.

According to Collins (1960), the pineapple originated in the Parana-Paraguay basin wherein *A. bracteatus, A. ananassoides* and *Pseudananas* also occur in wild. He further stated that the pineapple was domesticated by the Tupi-Guarani Indian and then spread north and west. Prior to the discovery the New World, pineapples were widespread in South America east of the Andes and had reached the West Indies, Mexico and Central America (Collins, 1960; Patino, 1963).

Ananas comosus is the only self-incompatible species known in the genus. *Pseudananas* and their wild species of *Ananas* are partially self-fertile. Brewbaker and Gorrez (1967) have shown that the pineapple possesses a gametophytic S-allele type of incompatibility. In pineapple, inducing polyploidy does not break down the incompatibility system. It seems unlikely that self-incompatible *A. comosus* could be derived from self-compatible *A. bracteatus* or *A. annnassoides*. The only archaeological record for the pineapple consists of minute fragments, claimed to represent pineapple seeds and bracts, in two coprolites from caves in the Tehuacan valley of Mexico dated about 200 BC. -AD. 700. The pineapple is almost certainly a lowland South American domesticate. It seems to have been a rather late addition to the cultivated plants of the region, possibly because it occurred outside the main centre of domestication. The Amerindians are least concerned for selecting against spiny leaves in the pineapple. 'Spineless' forms, in which spines occur at the top few inches of the leaf, are dominant to wild-type spiny (Collins, 1960). Spineless types reportedly predominate amongst endemic forms in the upper Amazon basin (Patino, 1963). However, other Indian varieties are spiny. The Indians had also developed varieties which differed in flesh colour, acidity and flavour. There seems to be no particular centre of diversity for the crop.

The pineapple supposed to lend support to theories of pre-Columbian trans-oceanic contacts. Heyerdahl (1947), the Norwegian explorer, claimed that pineapples were present in the Marquesas before Europeans discovered Polynesia. Merrill (1954) opined that the first record of the pineapple in Polynesia was 1769, when Captain Cook sowed pineapple seeds in Tahiti. Pineapples are also noted on wall carvings in ancient Assyria, on pottery models in Egyptian tombs and on murals at Pompeii. Their reproduction, which are published are not convincing. The pre-Columbian presence of the crop in the Old World still remains non-proven.

Pineapples spread rapidly throughout the tropics after discovery of the New World. They were used to provision ships, and were probably a useful anti-scorbutic. The crowns can bear longer period of desiccation and provide viable planting material. This fact must have favoured its early establishment in the Old World. They spread by two routes: (i) across the Pacific with the Spanish to the Philippines and (ii) with the Portuguese from Brazil to Madagascar and India.

8. Further Developments: The most important cultivar of pineapple is Cayenne also called Kew or Sarawak and the proportion of Cayenne growth has increased as rapidly as planting material was made available (Collins, 1960). Collins has traced the early history of this cultivar back to five plants sent by Perrottet from French Guiana to France sometime before 1840, when European were interested in 'pineries' and pineapples, The plants were multiplied vegetatively, spread to England and thence to Australia, Florida, Jamaica and Hawaii. Pineapple agriculture has a very narrow genetic base. However, numerous mutations have built up asexual generations and Cayenne now consists of a number of heterozygous clones.

The pineapple is improved by simple selection of mutant clones within cultivars and by hybridization between cultivars followed by selection of the highly heterozygous progeny. Most breeding programmes start with Cayenne and paid attention to flesh colour vitamin-C content, uniformity of ripening, disease resistance, etc. Cayenne is highly susceptible to mealy bug wilt. Some other cultivars and the wild species are resistant. Cabezona is a natural triploid cultivar. Tetraploid clones are also produced artificially, but bear smaller fruits than the diploids and lower sugar content. Both triploid and tetraploid pineapples remain self-incompatible and seedless.

9. Epilogue: In breeding programmes, the wild species of *Ananas* and *Pseudananas* have been used. *Ananas ananassotdes* contributed distinctive flavour, disease resistance and ability to grow in cool, moist conditions. *Ananas bracteatus* is also a source of disease resistance. Pineapple breeding is, however, a slow process. An Fl plant four years to produce fruit and, in the case of an interspecific cross, four backcross generations are generally necessary to produce a plant suitable for commercial use. Thus a new cultivar may be developed 25 years after the initial cross. It is then be multiplied and distributed. Advances in methods of vegetative propagation enable about 25 daughter plants to be obtained from a single parent. The use of the wild species in breeding programmes will continue. It would also seem worthwhile to collect and conserve pineapple clones from the less accessible parts of South America. Such collections could be used as necessary to broaden the genetic base of the commercial cultivars.

7.24 POMEGRANATE

Punica granatum L. (Punicaceae)

Resume: *Pomegranate is native of Iran and is cultivated mostly in tropical subtropical regions. The fruits are mainly used for dessert purpose being sweet-acidic in taste and of exquisite quality. It grows well in a variety of soil types and in a wide range of climatic conditions. The fruits keep well for a long time. Because of versatile adaptability and hardy nature, it is widespread. Chromosome number is n=9.*

1. Introduction: Pomegranate is one of the favourite table fruits of tropical and subtropical regions and grown commercially being sweet-acidic taste. It is native of Iran and widely cultivated in the Mediterranean countries and also in China, Japan, India and earstwhile Russia. It has a versatile adaptability to wide range of climatic conditions.

2. Plant Characteristics: A large shrub or small tree with dark grey bark and thorny branches; leaves opposite, ovate or elliptic-oblong; flowers orange-red, showy borne solitary or in 3-flowered cymes; berries generally orange to red containing numerous, arillate seeds.

3. Parts Used: Fruits (Seeds).

4. Chemical Constituents: Protein (1.6%), fat (0.1%), carbohydrate (14.5%), minerals (calcium, phosphorous, magnesium, iron), oxalic acid, thiamine, riboflavin, nicotinic acid and vitamin-C (Anonymous, 1952).

5. Uses And Applications: The fruits are mostly used for dessert purposes. The fresh fruit is of exquisite quality. Some processed products are prepared from seeds *e.g.* bottled juice, syrups and jelly. The fruit juice is fermented for the production of wine, citric acid and sodium citrate for medicinal purpose. The juice is heated to a thick, viscous consistency by adding sugar. This product keeps well and used like tomato sauce or ketchup.

6. Karyotaxonomic Evidence: In *Punica granatum* (Pomegranate), the haploid chromosome number is found to be n=9. The chromosome variation is found to be very little (Raman *et al.,* 1971). B-chromosomes have been recorded in some cultivars. They are smaller than A-chromosomes and do not pair with the latter (Masoud, 2007).

7. History: The genus *Punica* contains two species *viz., P. granatum* and *P. protopunica*. The former is being cultivated mostly in tropical and subtropical regions, whereas the latter grows wild in the Socotra island. The cultivated pomegranate is divided into two subspecies *viz.,* (i) Chlorocarpa which is found in Transcaucacus region and (ii) Porphyrocarpa growing in Central Asia. Several cultivars have been identified. The cultivar Schahvar bears fruits with thin rind and the Robab with thick rind. These are grown in Iran. In Uzbekistan, 46 cultivars are evaluated, of which eight cultivars are recognized promising. In India, several types of pomegranates are cultivated. These are distinguished on the characteristics such as shape and colour of fruit, thickness of rind, colour of aril and taste. Of these, some cultivars *viz.,* Alandi, Dholka, Kabul, Kandhari Muskat Red, Paper shell, Spanish Ruby and Ganesh are recommended for commercial cultivation based on yield and physico-chemical characters. Some cultivars produce seedless fruits (*i.e.* with soft seeds) *e.g.* Jyothi, Dholka, Bessein (*cf.* Patil and Karale, 1996).

8. Further Developments: The two factors *viz.,* heterozygosity and cross-pollination, bring about the variation in the seedlings and this naturally induced variation or natural hybrid is used for selecting new strains. The 'Ganesh' is one such example with prolific yield, medium-sized fruits with soft seeds, better pulp and sweet juice with agreeable flavor. Clonal selections from 'Ganesh' is also reported which are more promising considering the fruit quality (*cf.* Patil and Karale, 1996).

The success in any breeding programme depends on the extent of genetic variability present in the plant populations. The Muskat cultivar has large variation in fruit characters which offers a good scope for improvement of pomegranate by resorting to seedling selection and subsequently propagating them clonally. Four promising seedlings among Muskat cultivar are superior to 'Ganesh' in respect of fruit weight, grain weight and other quality components. Likewise, some Russian cultivars had deep colour of aril. This character is introduced in 'Ganesh'.

9. Epilogue: The pomegranate is not very particular about soil requirement and can be grown on diverse types of soils. Also, it has a versatile adaptability to wide range of climatic conditions. It can grow in the plains and also on the hills up to an elevation of about 1850 meters. It grows best in semi-arid climate where cool winter and hot summer prevail. It is winter hardy and drought-resistant. It can be raised from seedlings. The common method adapted is but vegetative propagation by rooting hardwood cuttings. The pomegranate fruits keep well for a long time.

Thus the versatile adaptability, hardy nature, low maintenance cost, seedy and high yield, fine table value and characteristics mentioned above are all responsible for its wide spread.

7.25 RASPBERRIES AND BLACKBERRIES

Rubus Spp. (Rosaceae)

Resume: *Raspberries and blackberries belong to the genus Rubus. They are favourite garden fruits in Europe and also North America since many centuries. The former is both an American and European fruits, whereas the latter is almost entirely an American fruit. Presently, these are cultivated in both Old and New worlds. They are most commonly used in industrial products in the form of jam, flavourings, freezed, canned or fresh. Various species of the genus Rubus have basic chromosome number ×=7. They are diploids as well as polyploids.*

1. Introduction: Raspberries and blackberries have been favourite garden fruits in Europe and North America for several centuries and they have now become important commercial crops, supplying some 100,000 tons of fruit annually. for jam making, canning, freezing, yoghurt and flavourings. Commercial red raspberry production is particularly concentrated in eastern Scotland, eastern Europe and western North America, notably Oregon, Washington and British Columbia, while black raspberry production is limited to America. Commercial blackberry production has been neglected in Europe but forms a major part of the fruit-growing industry of Oregon, USA.

Both Raspberries and blackberries are favourite fruits in Europe and America. The blackberry is almost entirely an American fruit. The raspberries are both American and European fruits. The European species has been cultivated from great antiquity and was highly esteemed by the Greeks and Romans. At present, it is produced on large scale in the parts of the former USSR. In India, different species are grown in the Himlayan region, Sikkim, Western Ghat, Karnataka, Kumaon, Khasi Hills and Manipur.

2. Plant Characteristics:

(i) Raspberries: Smaller shrubs, reaching 2m height, generally with vigorous, erect, bushy habit, small prickled; leaves pinnate, 3-7 serrated leaflets, fluppy beneath; flowers white, in clusters of usually 10 flowers; aggregate fruits separating from receptacle when ripe and leave a cavity on one side.

(ii) Blackberries: Erect, decumbent or creeping shrubs, usually armed with prickles and thorns, erect 'canes' die down to the ground after every few years and are renewed from the rootstalks; leaves compound with five ovate leaflets, toothed, sessile, bright, white hairy beneath; flowers white or pink, borne in clusters at the end of branches; black velvety fruits aggregated, consisting of numerous ovaries of the flower ripened into small druplets.

3. Parts Used: Fruits.

4. Chemical Constituents:

(i) Raspberries: (*Rubus idaeus* Linn.): Protein (1%), fat (1%), sugars (5. 6%), minerals (calcium, potassium, sodium, magnesium, iron, copper) carotene, riboflavin, thiamine, nicotinic acid, pantothemic acid, vitamin B6, and C, biotin, folic acid, ascorbic acid.

(ii) Blackberries (*Rubus fruticosus* L.): Protein (3%) fat (1%), carbohydrate (3%), vitamin A, C, Minerals (Ca, Fe).

5. Uses And Applications:

(i) Raspberries: Fruits are used fresh or cooked, and are utilized for jellies, jams, vinegar and as a flavouring agent. Large quantities are canned and frozen.

(ii) Blackberries: Fruits are used fresh and for jams, jellies, cordials, preserves and canning.

6. Karyotaxonomic Evidence: Three sections of the genus *Rubus* inclusive plants yielding fruits are: (i) daeobatus (including raspberries) in which the ripe fruits detach from the receptacle, (ii) Eubatus (including blackberries) in which the drupelets of the ripe fruits attached to the receptacle and both separate from the calyx and (iii) Cylactis, including the herbaceous north-circumpolar or alpine species. A fourth group, Anoplobatus (including the flowering raspberries), has also been employed in breeding. The basic chromosome number is 7.

Cultivated raspberries are originated from three diploid species or subspecies *viz.,* (i) *Rubus idaeus* or *R. idaeus vulgatus*, the red raspberry of northern Europe and Asia (ii) *R. strigosus* or *R. idaeus strigosus*, the red raspberry of North America, and (iii) *R. occidentalis* the black raspberry, also native to North America, but generally with a more southerly distribution than *R. strigosus*. Wild forms of *R. idaeus* and *R. strigosus* are outbreedcrs. Fertilization is governed by a multi-allelic oppositional type of gamatophytic incompatibility system (Keep, 1968). These are rendered thereby highly heterozygous and exhibit considerable inbreeding depression. Domestication involved growing in clonal mono-culture and therefore demanded the selection of self-compatible form.

Three groups of blackberries are brought under domesticated (Clausen *et al.*, 1945) are: (i) the European blackberries extend over the whole of Europe and much of Asia. They contain large number of polyploid forms, mostly tetraploid. These can be conveniently regarded as a single population characterized by morphological diversity and ecological differentiation. The only surviving primary diploids are *R. tomentosus* and *R. ulmifolius*, (ii) there are the eastern American erect blackberries and trailing dewberries. These contain large number of polyploid forms and also five primary diploids *viz., R. allegheniensis, R. argusus, R. setosus, R. cuneifolius and R. trivialis*, (iii) the western American trailing blackberries are geographically separated by the prairies and rockies. They have much higher chromosome numbers, the most common being 56 and 84. There is no trace of lower ploidy levels. The area shows a considerable diversity of some other *Rubus* sections. It is likely that the early progenitors of this blackberry group combined genomes from several of these. They have probably the most complex origin of all the blackberries.

Most diploid blackberries have proved self-incompatibles. Polyploidy appeared to have led to loss of this system, Species of the European and eastern American groups are thought either facultative or obligate apomicts. Recent evidence raised doubts on this fact. European blackberries have a versatile reproductive system in which all the gametes are formed by meiosis. Some egg cells may be fertilized and give sexual offspring. Others are, however, 'diploidized'. In *R. caesius*, this occurs by fusion of two haploid nuclei produced by division of the haploid egg nucleus. There are several unusual features of this process, which begins only after a pollen tube has entered a synergid nucleus (Gerlach, 1965). The breeding system clearly represents a successful restriction of full sexuality and is not the evolutionary dead-end associated with apomixis. In contrast, the western American blackberries responded to polyploidy. The resultant loss of incompatibility established dioecy and maintained a predominantly sexual breeding system.

7. History: Domestication, during 16th century in Europe involved wild forms of raspberries with fewer, stronger canes, stronger fruiting branches and large fruits. However, distinct cultivars were not recorded until late in the 18th century. About this time, cultivars of European origin were first grown in America, where their superior fruits won them preference over the indigenous kinds, which did not appear in cultivation until later. Cultivated forms of black raspberry appeared later about 1830 but were preceded by their natural hybrids with the European red raspberry, known as purples or purple-cane hybrids. A major event in the evolution of the raspberry occurred with the advent of natural hybrids between the European and American red raspberries. These combined the fruit qualities of European kinds with the hardiness and heat and drought tolerance of the American ones. Two famous ones were Cuthbert discovered in a New York garden about 1865, and (ii) Preussen discovered in a German garden about 1915. These later figured prominently in controlled breeding.

Selection for large-fruited variants also led to the selection of autotetraploid raspberries. They were all autumn-fruiting kinds and mostly originated particularly in France. Their selection clearly depended on the occurrence of high temperatures which maximize fruit-set in these sub-fertile genotypes. Their success was but short-lived. Autotetraploidy played a role in the evolution of the Siberian raspberry *R. sachalittelesis*. This is probably an autopolyploid derivative of *R. strigosus*. The eastern American blackberry flora was mostly diploid until the time of settlement but a much longer time scale operated in Europe. The culture of blackberries was stared late in Europe, possibly because of the abundance there of hedgerow fruit. It began in the USA between 1840 and 1860, 18 cultivars being listed by 1867. Among these was the Evergreen or Cutleaf blackberry (*R. laciniotus*), was introduced from central Europe about 1850. Himalaya Giant (*R. procerus*) was added from southern Europe about 1880. Both of these escaped from cultivation. They became naturalized in western America and hybridized with the indigenous flora. Both are highly successful commercial blackberries in Europe and America. A thornless mutant of *R. laciniatus* found wild about 1930. It has become America's highest yielding cultivar.

The western blackberries were popular with early settlers. The dioecious habit precluded domestication though their natural hybrids with other forms were successful. The most successful of these was the hexaploid Loganberry. Judge Logan discovered it in his California garden in about 1881. Its origin is thought due to crossing between the octoploid Californian blackberry (*R. vitifolius*) and an unreduced gamete of the European raspberry.

8. Further Developments: The evergreen blackberry (*R. laciniatus*) and the Loganberry can be cited as examples of chance discoveries made nearly a century ago. What began by accident was pursued and improved upon by intent. Further crosses between red and black raspberries were made. Considerable crossing was executed between the European and American forms of red raspberry. Rapid degeneration of successful cultivars through virus infection then prompted breeding for resistance to the aphid vectors of these diseases. Effective major-gene controlled resistances to the main European and American aphids were noted. Notable advance was the development of raspberries adapted to the southern states of the USA by using Idaeobatus species like *R. parvifolius, R. bifiorus* and *R. kuntzeanus* to contribute genes for resistance to leaf diseases, drought and tolerance of heat.

Breeding with the western American blackberries (Waldo, 1968) combined their excellent flavour with a perfect-flowered and more productive growth habit. Both natural hybrids and hybrids obtained from controlled crosses involving the introduced forms of *R. laciniatus, R. procerus* and

Loganberry have been used to achieve this. Spinelessness was a main objective of all blackberry breeders. All of the many spine-free mutants found are chimeral and confined to the outer layer (I) of the meristem. These are inaccessible for breeding, but the diploid *R. rusticanus*. var. *inermis* (a spine-free form of *R. ulmifolius*) was valuable source. It was crossed in England with tetraploid *R. thyrsiger*, an unreduced gamete functioned and successful spinefree cultivars from this source have been developed in the USA.

The western American blackberries have early ripening quality, but the only early European cultivar is Bedford Giant, which is a blackberry-raspberry hybrid and probably owes its earliness to raspberry germplasm.

9. Epilogue: Evolution in the wild natural state showed the wide range of germplasm combinations possible in *Rubus*. The breeders found that most gene transfers can be achieved without specialized techniques. Breeding was aimed to bring together the very considerable range of available germplasm. In the short-term, the objectives are to breed raspberries and blackberries which meet the specialized requirements of processing capable of being machine harvested. In case of raspberries this means combining the fruit size of *R. idaeus*, the firm fruit texture of *R. occidentalis*, the bright fruit appearance of *R. strigosus* and the disease and pest resistance of several species. In case of blackberries it means combining the erect growth habit of *R. argutus*, the flavour and early ripening of the western American blackberries and the spinelessness of *R. rusticanus*. Other objectives include transferring genes for specific yield components. It is possible to reduce maintenance costs by restructuring the raspberry into a self-supporting plant which preferably bears all its crop on its first year's canes (Knight and Keep, 1966). Wild forms of *R. arcticus* and *R. stellatus* and *R. chamaemorus* are popular in Scandinavia, especially for making liquors. These are domesticated and other improvements are being sought by crossing them with *R. idaeus* (Larsson, 1969).

7.26 SAPOTA

Manilkara zapota (L.) P. van Royan

[(Syn.) *Achras sapota* L.; L. zapata L.] (Sapotaceae)

Resume: *Sapota is a native of Mexico and Central America. It is basically a tropical fruit plant. It is used mainly as dessert fruit being delicious, pulpy and sweet. It is a hardy tree and grows well in a variety of soil but with tropical weather. It has fair diversity of cultivars and can be grown by seeds and vegetative method, the latter method offers true type. It responds well to irrigation, although it can tolerate drought conditions to some extent. Chromosome number is 2n=26.*

1. Introduction: Sapota, when ripe, is a sweet-smelling delicious fruit. The Sapota or Sapodilla is a native of Mexico and Central America. It was grown in the West Indies long back, and Oviedo, who was there during 1513 to 1525, considered it as the best of fruits. The Spaniards introduced it in the Philippines in the early days and thence it spread westwords to Malaysia and other countries (Purseglove, 1968). It's time for introduction in India is unknown (Singh *et al.,* 1963) but its cultivation was taken up for the first time in Maharashtra (a state of India) in 1898 in Gholwad village (Cheema *et al.,* 1954). It is now cultivated widely throughout tropical region.

2. Plant Characteristics: An evergreen, sturdy, much-branched tree with brown, longitudinally fissured bark; leaves elliptic to lanceolate, acute to acuminate and coriaceous; flowers white, solitary and axillary; berries globose, ovoid or ellipsoid, rusty brown containing few, oblong, compressed, smooth, shining seeds.

3. Parts Used: Berries.

4. Chemical Constituents: Ripe fruit: Protein (0.7%), carbohydrate (21.4%), fat (1.1%), minerals (calcium, phosphorous, potassium, magnesium, zinc, iron), ascorbic acid, thiamin, riboflavin, niacin, vitamin B, C, riboflavin, niacin, pantothenic acid.

5. Uses And Applications: Ripe berries (pulp) are delicious and sweet and eaten as dessert. They are good source of sugar. The pulp is also made into sherbet and halwa (Singh, *et al.,* 1963). The mature fruits are used for making mixed jams. They are valuable source for the manufacture of industrial glucose, pectin and natural fruit jellies. They are also canned as sliced.

6. Karyotaxonomic Evidence: Peiris (2014) has revealed chromosome number 2n=26. Genetic improvement programmes involving hybridization to improve fruit yields and characteristics have been carried out (Sambamurty and Ramalingam, 1954).

7. History: There are several cultivars of Sapota. In India, they are grouped under four types on the basis of nature of branches and colour of foliage *e.g.* branches in whorls or irregular, deep green or light green leaves, shape and size of leaves, fruit size, fruit colour, smell of pulp, etc. Generally, a good table sapota have a few seeds with melting, sweet pulp. Thick-skinned, hard-fleshed cultivars with sandy texture are considered inferior. Several cultivars have been developed in India *e.g.* Kalipatti, Chaatri, Dhola, Diwani, Long, Bhuri, Jinger, Vanjet, Pla, Oval, Kirthabharti, Dwarapudi, cricket ball and some others.

In Florida, USA, cultivars Prolific and Brown Sugar have been found to be good producers. There are also other cultivars *viz.,* Modello and Russel but do not fruit heavily or regularly. There are few more cultivars but have been proved unfruitful in some areas and hence not recommended (*cf.* Sulladmath and Reddy, 1996).

8. Further Developments: Sapota is propagated by seeds and vegetative methods. In earlier days, seedlings were used for planting but they had some disadantages *e.g.* slow growth, very long pre-bearing age of 8-10 years, huge size or height or not being true to type, etc. (Cheema *et al.,* 1954). The vegetative method is advantageous *e.g.* being true to type, earliness in bearing, dwarf or easily manageable tree size, favourable rootstock, etc.

Information on floral biology, fruit set growth and development are important to achieve desired yield. Sapota starts bearing small crops from 2nd or 3rd year of planting but economical yield is obtained from 7th year onwards. There are generally two main seasons of flowering and harvesting. Intercrops can be taken by short duration fruits like banana, papaya or vegetables.

Nine microsatellite markers, consisting of seven novel markers and two markers identified in a related species, were used to characterize the genetic diversity and population genetic structure in three populations of *M. zapota* collected from Gautemala and Belize (Thompson *et al.,* 2015). Four populations of *Manilkara zapota* were analysed from Mexico using four polymorphic RAPD markers (Heaton *et al.,* 1999). These are not matched by genetic divergence.

9. Epilogue: Sapota is a hardy tree and can be grown on a wide range of soils. The most ideal soils are deep alluvium, sandy loams and medium black soils. Being basically a tropical species, it can be grown from sea level upto 1200 m. It prefers a warm and moist weather and grows in dry and humid areas. It is affected by insect pests and diseases. Necessary plants protection should be taken up in time. The period of storage depends on the respiration, relative humidity, temperature, enzymic activity and CO_2 content. Under ordinary conditions, fruits keep well for 7-8 days from picking. Fruit ripen in about five days after harvest at room temperature. The ripening period depends upon the cultivar type, stage of maturity and temperature.

7.27 STRAWBERRY

Fragaria ananassa Duchesne (Rosaceae)

Resume: *Majority of cultivated forms are derived from Fragaria chiloensis which is a native of Western America from Chile northward to the mountains of Mexico. It is favourite fruit of temperate regions of the world. Primarily, it is a dessert fruit and also used as canned, frozen or preserved, as flavouring a great in drinks and confectionary. Species of the genus Fragaria exhibit basic chromosome number ×=7. They are either diploids or polyploids.*

1. Introduction: It is one of the important small fruits. It is favorite in temperate countries. It has been grown in Europe since the 14th century and in America since colonial days. In US, it is grown commercially since 1860. *Fragaria virginiana* is a native of eastern North America and was grown by the early settlers. They took it to England where it is cultivated since the 17th century. The majority of cultivated forms are derived from *F. chiloensis* which is a native of Western America from Chile northward to the mountains of Mexico.

2. Plant Characteristics: A low perennial herb with a very short stem; leaves basal, 3-foliate; flowers white, in clusters; fruit an aggregate fruit, consisting of several small dry achenes embedded on the surface a large fleshy receptacle.

3. Parts Used: Fruit.

4. Chemical Constituents: Fruits: Sugars (as invert sugar), pentosan, pectin, citric acid, malic acid.

5. Uses And Applications: Primarily, strawberry is a dessert fruit. They are also canned, frozen and used in preserves and jams. They are also used as a flavouring agent especially for drinks and confectionery.

6. Karyotaxonomic Evidence: The various species of *Fragaria* form a polyploidy series from diploid to octoploid, with a basic chromosome number $\times$=7. *Fragaria vesca* is the commonest wild diploid and distributed in north temperate Europe, Asia and America. It occurs also in North Africa and South America. Other diploids are confined to Asia, although *F. viridis* occurs widely in Europe and Asia. The three known tetraploids occur in eastern Asia only, and the only hexaploid species, *F. moschata* is distributed in western Asia and Europe.

Octopoid *Fragaria ananassa* is (2n=8x=56) is orignated from crosses between the American octoploids *F. chiloensis* and. *F. inrginiana*. *Fragaria chiloensis* occurs discontinuously along the Pacific coast, commonly on sand dunes, from the Aleutian Islands to California, and then in Chile. *Fragaria virginiana* flourishes in open woodland and hill meadows in North America, from the east coast to the Rocky Mountains and from New Mexico to Alaska. It is rare near the Pacific coast where some natural hybrids with *F. chiloensis* have been reported. Natural introgression may have added to the diversity of both species, especially in *F. oirginiana* of which some variants have been described as separate species. A third octoploid species, *F. iturupensis*, is reported from the Kuril Islands, northeast of Japan (Staudt, 1973).

There has been little differentiation in the homology of chromosome sets in any of the species. Each of the natural polyploids appears to be an autopolyploid and this prevents any simple analysis of evolutionary relationships. The separate distributions of the tetraploid, hexaploid and octoploid species indicated that each ploidy group originated independently. Two possibilities are by unreduced and by doubled unreduced gametes, both of which have been shown to occur (Ellis, 1962; Bringhurst and Senanayake, 1966). However, the finding of *F. iturupensis* (Staudt, 1973) extends the range of the octoploids towards that of tetraploid *F. orientalis* on mainland Asia.

Several diploids, including *F. vesca*, are monoecious, self-compatible and mostly inbreeding. Three diploid species are self-incompatible. Most of the polyploids are entirely or predominantly dioecious. Hermaphrodite forms also occur and have been selected in cultivation. These are self-compatible, but cross pollination is either essential or advantageous for full fruit set.

7. History: *Fragaria* species yield palatable fruits. It is probable that they were collected in the wild. Strawberries (probably *F. vesca*) were tended in gardens by the Romans in 200 BC. This species was planted in England and France during the 14th and more extensively in the 15th and 16th centuries, when some ever-bearing forms were used. These strawberries were grown for ornament and medicine as well as for food. Many improved alpine varieties of *F. vesca* were selected in the 19th century, especially in France. Some are still grown on a small scale (Darrow, 1966; Wilhem and Sagen, 1974).

The European species were also used, *F. viridis* on a small scale and *F. moschata* quite extensively. The latter called the musk or hautbois strawberry. It was established in cultivation by the 16th century. Distinct varieties were selected in England and France. The names Black, Apricot and Raspberry suggest that considerable variation existed. This species is grown on a small scale for its unique flavor and aroma.

American octoploids were used before Europeans settled in America. North American Indians used fruit of *F. virginiana* to flavour bread and beverages. They planted as well as gathered. *Fragaria chiloensis* was cultivated in Chile before the Spaniards reached. Chilean Indians had different names for wild and cultivated forrns. The fruits were used fresh and in wine and also as raisins after drying in the sun. The Spaniards, impressed by the large firm fruits of three colours (red, yellow and white), introduced the cultivated *F. chiloensis* (called frutilla into Peru and Ecuador) where some varieties are still grown.

Seeds of *F. virginiana* were taken to Europe by 1556, but the first record of *F. virginiana* in cultivation dates from 1624. Some of the first introductions did not grow well. Further introductions were made and self-compatible hermaphrodite forms selected. Varieties, known as the Scarlet strawberry, became popular for fresh fruit and conserves during the 18th century. At least 30 varieties were used by 1820. Most of these were later replaced by *F. ananassa*, quite quickly in Europe and more slowly in America. The large fruits of the cultivated Chilean strawberry are reported during the 16th century. The first introduction to Europe was but in 1714. Possibly influenced by the vegetative propagation used in Chile, Frezier brought plants rather than seeds and five plant survived the journey to France. Some of these could have been of the same clone. The plants were given to different gardens and it seems likely that only one or two of them formed the basis of cultivated *F. chiloensis* in Europe and later became the source of *F. ananassa*. All five plants were female. They mostly remained sterile. Fruits were obtained locally in Brittany. By 1740 it was realized that satisfactory fruiting could obtained if *F. moschata* and/or *F. virginiana* are grown near to female *F. chiloensis* plants. A system of planting in alternate rows hence developed. The strawberry named as Plougastel became popular in Paris and London. This Chilean clone was grown until the end of the 19th century.

The two American species stimulated interest in new strawberries but the crossing that produced the first plants of the new cultivated strawberry, *F. ananassa*, was accidental. The first plants were obtained and selected in a botanic garden instead by the farmers growing the successful variety of *F. chiloensis*. It seems likely that interest in finding a new strawberry was maintained where the female *F. chiloensis* fruited rarely and never produced large fruits like those reported from Chile.

The new Pineapple or Pine strawberry was first described by Miller in London in 1759. Miller obtained the plant from a garden near Amsterdam where it had first appeared about 1750. Duchesne obtained plants from London and observed that they had characters of both *F. chiloensis* and *F. virginiana*. The cross produced viable seeds and progeny.

Similar new strawberries were already grown in other gardens about the same time. These may have originated separately from other crosses between one of the few clones of *F. chiloensis* and different forms of *F.virginiana*. Since *F. moschata* was probably growing in the same gardens, and *F.chiloensis* × *F.moschata* hybrids can be vigorous and partly fertile. It has since been repeatedly confirmed that *F.chiloensis* and *F.virginiana* cross readily and that some of the progeny are vigorous, fertile and similar to *F.ananassa*.

8. Further Developments: Introductions of *F. chiloensis* and *F. virginiana*, in earlier times, must have been highly heterozygous. Several varieties of the Pineapple strawberry (*F. ananassa*) were selected before 1800. These were grown along with varieties of *F. virginiana* and their success stimulated further breeding especially in England and then in France and America. Many of the early selections were from progenies of *F. ananassa*, but the collections of *F. virginiana*, the original Chilean *F. chiloensis* and some wild *F. chiloensis* from California were also employed.

The early breeding programmes exploited narrow genetic base although not as narrow as is sometimes thought. American breeders used the native octoploids, especially *F. virginiana*. An early stimulus to extend the genetic base was the occurrence of plants which fruited for a longer time. A perpetual flowering segregant was selected in France in 1866 and a number of ever-bearing variety was produced. American breeders were impressed by the availability of fruit of *F. chiloensis* in California throughout most of the year. Other collections of both progenitor species have been used in breeding. It is realized that both species have extensive distributions and corresponding ranges of genetic variation.

Useful genetic variation was obtained from the wild octoploids. For example, *F. ovalis* (*F. virginiana*) was used in the breeding of American varieties which are resistant to drought and low temperatures and both *F. chiloensis* and *F. virginiana* have provided resistance to pests and diseases such as Vertuillium wilt, red stele (Phytophthora fragariae) and aphids. In some cases, it has been shown that such useful characters can be separated from unwanted characters by three or four backcrosses (Scott *et al.*, 1972). Several characteristics make *F. ananassa* suitable for this sort of interspecific transfer. A recurrent problem in strawberry breeding from the early years onwards has been the occurrence of partial sterility. It seems to be associated with a recurrence of dioecy. It may also be a consequence of interspecific hybrid ongm.

9. Epilogue: The principal objectives in breeding are high fruit characters and resistance to pests and diseases. There is increasing interest also in suitability for new methods of propagation, cultivation, harvesting and processing. Characters such as high yield from first year plants, the restricted production of runners and uniformity of fruit size and quality are of increasing significance. The extension of fruiting time generates problems of pollination and fruit set. The need for enhanced self-pollination or flowers adapted to different pollinators is warranted.

There is still much unused genetic variation in the natural octoploids, both in the wild species and in *F. ananassa*. Some genetic variation has probably been lost from the cultivated strawberries during the last century. This may be due to the requirements of commercial production necessitated other priorities.

The genetic base of strawberry breeding may be further extended by the use of other species. The variation may be in diploids, hexploids or octoploids. A series of synthetic octoploids of diverse origin has been produced and crossed with natural octoploids. Many of the hybrids are sufficiently fertile to be used in breeding (Jones, 1966). Since most species combinations can be obtained, either directly \ or indirectly, it now appears probable that any useful genetic variation in the genus can be made available for breeding.

Many Fragaria × Potentilla crosses stimulate seed development, but viable hybrids have been obtained with only four *Potentilla* species. However, since much of the work has been on a small scale, without the use of advanced techniques for overcoming crossing barriers, and since few species of *Potentilla* have been tried, the potential may be considerable. (Ellis, 1962; Asker, 1971; Bringhurst and Barrientos, 1973). These cross with octoploid strawberries and such *Potentilla* crosses may supply useful characters.

References:

Airy Shaw, H.K. (1966) J.C. Willis' A Dictionary of the Flowering Plants And Ferns. (7th Ed.) University Press, Cambridge.

Aliyu, O.M. and J.A. Alwopetu (2007) Chromsome studies in Cashew (*Anacardium occidentale* (L.) *African Journal of Biotechnology* 6(2):131-136.

Anonymous (1948-1976) The Wealth of India: A Dictionary of Indian Raw Materials and Industrial Products. Vol.1-11: CSIR, New Delhi, India.

Anonymous (1952) The Wealth of India: Raw Materials. Vol. III. DE CSIR, New Delhi, India.

Anonymous (1966) Annual Report, Agriculture University Department of Horticulture, Punjab Agriculture University, Ludhiana, India.

Anonymous (1987) All India Co-ordinated Research Project on Arid Zone Fruits. Technical Report, Document No.20.

Antonios, P., Jeroen van der K., Elina Chrysanthou, Susan McCallulm, Julie Graham and Robert D. Hancock (2018) Photosynthetic limitation as a factor influencing yield in highbush blueberries (Vaccinium corymbosum) grown in a northern European environment. *Journal of Experimental Botany* 69(12):3069-3080.

Asker, S. (1971) Some viewpoints on Fragaria x Potentilla intergeneric hybridisation. Hereditas, 67:181-190.

Atkinson, R.G., Cipriani, G., Whittaker, D.J. and R.C. Gardner (1997) The allopolyploid origin of Kiwifruit, Actinidia deliciosa (Actinidiaceae). *Plant Systematics And Evolution* 205 (1-2):111-124.

Badillo, V.M. (1971) Monografia de la jamilia Caricaceae. Maracay, Venezuela.

Bailey, L.H. (1949) Manual of Cultivated Plants. Macmillan Co., New York, USA.

Bajpai, P.N. and H.S. Shukla (1996) Aonla. *In:* Fruits: Tropical And Subtropical (Ed. T.K. Bose and S.K. Mitra) Naya Prakash, Calcutta, India, pp.757-767.

Bajpai, P.N. and O.P. Chaturvedi (1990) Jamun. *In:* Fruits: Tropical And Subtropical (Ed. Bose, T.K. and S.K.Mitra). Naya Prakash, Calcutta, India. pp.747-756.

Bally, Ian S.E. (2006) Mangifera indica (Mango). Species Profiles for Pacific Island Agroforestry. www.traditionaltree.org Ver. 3.1 (April):1-24.

Baradi, T.A. El. (1968) Date growing: A review. *Trop. Abstr.* 23:473-479.

Beal, J.M. (1937) Cytological studies in the genus Phoenix. *Bot. Gaz.* 99:400-407.

Bergh, B.O. (1975) Avocados. *In:* J.Janick and J.N.Moore (Eds), Advances in fruit breeding. Lafayette, Indiana, USA, pp.541-567.

Bergh, B.O., Scora, R.W. and W.B. Storey (1973) A comparison of leaf tempenes in Persea subgenus Persea. *Bot.Gaz.* 134:130-134.

Bharad, S.G., Kulwal, P.L. and S.A. Bagul (2009) Genetic diversity study in A.quamosa by morphological, biochemical and RAPD markers. *Acta Hort.* 839:615-623.

Brewbaker, J. Land D. D. Gorrez (1967) Genetics of self-incompatibility in the monocot genera, Ananas (pineapple) and Gasteria. *Amer. J. Bot.* 54:611-616.

Bringhurst, R.S. and F. Barrientos (1973) Fertile Fragaria chiloensis x Potentilla glandulosa amphiploids. *Genetics* 74:530.

Bringhurst, R.S. and Y.D.A. Senanayake (1966) The evolutionary significance of natural Fragaria chiloensis X F. Vesca hybrids resulting from unreduced gametes. *Amer. J. Bot.* 53:1000-1006.

Bunyard, E.A. (1921) A revision of the red currants. *J. Pomol.* 2:38-55.

Burkill, I.H. (1935) A Dictionary of The Economic Products of The Malay Peninsula. Vol. I. Crown Agents of the Colonies, London, UK.

Cameron, J.W. and H.B. Frost (1968) Genetics, breeding, and nucellar embryony. *In:* Reuther, Batchelor, and Webber (eds.), The citrus industry, Berkeley 2:325-371 (Revised).

Carpenter, J.B. (1979) Date Growers's Institute Report. 54:25-33.

Carpenter, J.B. and C.L. Ream (1976) Date Growers' Institute Report 46:11-14.

Chakma, S.P., Radhid A.S.M. Harunor, Roy, S. and M. Islam (2014) Effect of NPK on the yield of Dragon fruit [H.costaricensis (F.A. (Weber) Britton & Rose)] in Chittagang Hill tracts. *American-Eurasian. J. Agric. & Environ. Sci.* 14(6):521-526.

Cheema, G.S., Bhat, S.S. and K.C.Naik (1954) Commercial Fruits of India. Macmillan & Co., London, UK.

Chevalier, A. (1952) Recherches sur les Phoenix africaines. *Rev. Int. Bot. Appl. Agric. Trop.* 32:205-233.

Chiu, A. and A.B. Kimball (2003) Topical vitamins, minerals and botanical ingredients as modulators of environmental and chronological skin damage. *Br. J. Dermatol.* 149:681-691.

Clausen, J, K. eck, D. D. and W. M. Hiesey (1945) Experimental studies on the nature of species. 11. Plant evolution through amphiploidy and autoploidy with examples from the Madiinae, Carnegie Inst, Wash. Pub. 564.

Collins, J.L. (1960) The Pineapple: Botany, Cultivation And Utilization, L.Hill. London, UK.

Condit, I.J. (1928, 1933, 1964) Cytological and morphological studies in the genus Ficus. I, II, III. Univ. *Calif. Publ. Bot.* 1:233-44; 17:61-74.

Condit, I.J. (1955) Fig varieties: a monograph. *Hilgardia* 23:323-538.

Condit, I.J. (1969) Ficus: the exotic species. Berkeley, Calif., USA.

Darrow, G.M. (1966) The Strawberry. History, Breeding And Physiology. Rinehart and Winston, New York, USA.

De Candolle (1904) Origin of Cultivated Plants. Kegan Paul, London, UK.

De Candolle, A. (1886) Origin of Cultivated Plants (Translated from 2nd Ed.in French, 1859. Hafner, New York, USA.

De Candolle, A.P. (1825) Prodrumus Systematis Naturalis, Paris, France.

De Costa, C. (1578) Tratodo des Drogase Medicines das Indian Orientals (1964) Ed. Junta De investigacoes do Ultramar, Lisbon, Portugal.

De Oviedo Valdes, Gonzalo Fernadez [1535 (1851-1855)] La Historia general de las Indias (Complete work published in 1851-1855 and was Edited by J.A. de los Rios, for the Spanish Academy of History, Seville).

Deverson, Tony and Kennedy, Graeme (2005) The New Zealand Oxford Dictionary. Oxford University Press Oxford, UK.

Devi, P.S., Thangam, S.M., Desai, A.R. and P.G.Adsule (2002) Studies on variability in physico-chemical characteristics of different jamun (Syzygium cumini Skeels) accessions from Goa. *Indian J. Hort.* 58(2):153-156.

Ellis, J.R. (1962) Fragaria-Potentilla intergeneric hybridisation and evolution in Fragaria. *Proc. Linn. Soc. Lond.* 173:99-106.

Esen, A. and R.K. Soost (1973) Seed development in Citrus with special reference to 2x x 4x crosses. *Amer. F. Bot.* 60:448-462.

Farriol, M. *et al.* (1994) Effect of vitamin-C and vitamin-E analog on aged tibroblast. *Revista Espanola de Fisiologia* 50(4):253-288.

Firminger, T.A. (1947) Firminger's Manual of Gardening For India. Thacker Spink Co.Ltd., Calcutta, India.

Fosberg, F.R. (1960) Introgression in Artocarpus (Moraceae) in Micronesia. *Brittonia* 12:101-113

Frost, H.B. and R.K. Soost (1968) Seed reproduction: development of gametes and embryos. *In:* Reuther, Batchelor, and Webber (Eds), The citrus industry, Berkeley 2:290-324 (Revised).

Gerlach, D. (1965) Fertilisation and autogamy in Rubus caesius. *Biologisches Zentralblatt* 84:611-633.

Ghosh, S.P. (1996) Citrus. *In:* Fruits: Tropical And Subtropical (Ed. Bose, T.K. and S.K. Mitra) Naya Prakash, Calcutta, India. (Rev. Ed.) pp.63-131.

Hamilton, R. A. and T. Izuno (1967) A revised concept of sex inheritance in *Carica papaya*. *Agron. trop.* 17:401-402.

Hatton, R.G. (1920) Black currant varieties. A method of classification. *J. Pomol. I,* 65-80:145-154.

Hayes, W.B. (1957) Fruit Growing In India. Kitabistan, Allahbad, India.

Hayes, W.B. (1970) Fruit Growing In India. Kitabistan, Allahabad, India.

Hazarika, T.K. (2014) Underexploited and lesser known fruits of Mizoram. *In:* Future Crops, Vol.2 (Ed.Prof.K.V.Peter), Daya Publishing House, New Delhi, India pp.85-102.

Heaton, H.J., Whitkus, R. and A. Gomez-Pompa (1939). Extreme ecological and phenotypic differences in tropical tree Chicozapote [*Manilkara zapota* (L.) *P.Royen*] *Mol.Ecol.* 8:627-632.

Helbaeck, H. (1959) Domestication of food plants in the old world. *Science*, N.Y. 130-365.

Hodgson, R.W. (1961) Taxonomy and nomenclature in Citrus. *In:* Price (Ed.), Intnl Org. Citrus Virologists, Proc. 2nd ConI, Gainesville, Florida, 1-7.

Hofmeyr, J. D. J. (1967) Some genetic and breeding aspeets cf Carica papaya. *Agron, trop.* 17:345-351.

Horovitz, S. and Jiminez, H. (1967) Cruzarnientos interespecificos e intergenericos en Caricaceas y sus implicaciones fitotecnias, *Agron. trop.* 17:353-359.

Hsiao-Han Lin, Pei-Shan Tsai, Su-Chen Fang and Jen-Fang Liu (2011) Effect of Kiwifruit consumption of sleep quality in adults with sleep problems. *Asia Pacific Journal of Clinical Nutrition* 20(2):169-174.

https://en.wikipedia.org/wiki/kumquat

https://hort.purdue.edu/newcrop/morton/kumquat.html.

Hunt, B. W. (1911, 1912) Fig breeding. *Georgia Univ. Bull.* 11:146-148; 12:106-110.

Jahemski, W. F. (1973) Large vineyard discovered in ancient Pompeii. *Science*, N. Y., 180:821-830.

Janczewski, E. de (1907) Monograph of the currants, Ribes. *Mem. Soc. Phys. Hist. nat. Geneve.* 35:199-517.

Jarrett, F.M. (1959) Studies in Artocarpus and allied genera. III a revision of Artocarpus subgenus Artocarpus. *J. Arnold Arb.*, 40:113-368.

Jelenkovic, G. and H. P. Olmo (1969) Cytogenetics of Vitis. *V. Allotetraploids* of *V. vinifera* x *V. rotundifolia. Vitis* 8:265-279.

Jindal, P.C. (1996) Grapes. *In:* Fruit: Tropical And Subtropical (Ed. Bose, T.K. and S.K. Mitra) Naya Prakash, Calcutta, India. Rev.Ed. pp.186-251.

Jin-Hu Wu (2012) Manipulation of ploidy for Kiwifruit breeding and the study of Actinidia genomics. Proc.7th on the Vitro Culture and Horticultural Breeding (Ed. D.Geelen). *Acta Hort.* 961 (ISHS): 539-546.

Johannes Von Cuba (1485) Herbarius (Ortus sanitatis). Peter schoffer the Elder (German publisher). Germany.

Jones, J. K. (1966) Evolution and breeding potential in Strawberries. *Sci. Hort.* 18: 121-130.

Keep, E. (1975) Currants and gooseberries. *In:* (Ed.J.Janick and J. N. Moore) Advances in fruit breeding, Purdue Univ. Press, pp:197-268.

Keep, Elizabeth (1968) Incornpatibility in Rubus with special reference to R. idaeus L. *Can. J. Genet. Cytol.* 10:253-262.

Kester, D.E. (1969) Pollen effects on Chilling requirements of almond and almond-peach hybrid seeds. *J. Am. Hortic.Sci.* 94:318-321.

Khader Abdul, Chellappan K., Chatopadhyayay, P.K. and O.N.A. Pillai (1996) Banana. *In:* Fruits: Tropical And Subtropical (Ed. Bose, T.K. and S.K.Mitra) Rev. Ed. pp.132-185.

Khoshoo, T.N. and N. Singh (1963) Cytology of North-East Indian trees: *Ziziphus Jujuba* and *Z.rotundifolia. Silvae Genetica* 12:141-180.

Knight, R.L. (1963) Abstract bibliography of fruit breeding and genetics to 1960: Malus and Pyrus. Tech. Commun. Commonw. Bur. Hart. Plantr. Crops, 29, pp. 535.

Knight, R.L. (1969) Abstract bibliography of fruit breeding and genetics to 1965; *Prunus*. Tech. Commun. Commonto. Bur. Hort. Plantn Crops, 31, pp. 649.

Knight, R.L. and E. Keep (1958) Abstract bibliography of fruit breeding and genetics to 1955. Rubus and Ribes a survey. *Tech. Commun. Commonw. Bur. Hart, Plantn Crops*, 25:254.

Knight, R.L. and Keep, Elizabeth (1966) Breeding new soft fruits. *In:*Fruit, present and future, London (Royal Horticultural Society) 2:98-111.

Krug, C.A. (1943) Chromosome numbers in the subfamily Aurantioideae with special reference to the genus Citrus. *Bot. Gaz.* 48:602-611.

Kumquat

Kumquat_Wikipedia.

Kumquats Information, Recipes and Facts - Specialty Produce. com/produce/ kunquats_906.php.

Lai, R., Wolley, D.J. and G.S. Lawes (1989) Fruit growth in Kiwifruit (Actinidia deliciosa): Patterns of assimilate distribution between laterals. *Scientia Horticulturae* 40(1):43-52.

Lamb, J.G.D. (1973) High blueberries: A new fruit for Europe. *J.Roy.Kort.Soc.* 98:401-404.

Larsson, E.G.K. (1969) Experimental taxonomy as a base for breeding northern Rubi. *Hereditas* 63:283-351, 283-351.

Lavie, P. (1970) Contribution a fitude caryosystimatique des Vitacies. These, Faculte des Sciences de Monrpellier, 1:pp. 213.

LeBlanc, Cheryl M. and Donald J. Leopold (1992) Demography and age structure of a central New York Shrub-Carr 94 years after fire. Bulletin of The Torrey botanical Club. 119(1):50-54.

Levadoux, L. (1946) Etude de la fleur et de la sexualite chez la vigne. Ann. Ecole Nat. Agr. Montpellier, 27, pp, 89.

Lichtenzveig, J., A660, S., Nard, A., Tel-Zux, N. and Y. Mizrahi (2000) Cytology and mating systems in the climbing cacti Hylocereus and Selenicereus. *Am. J. Bot.* 87(7):1058.106.5

Liu Y. *et al.* (2017) Rapid radiations of both Kiwifruit hybrid licences and their parents shed light on a two-layer mode of species diversification. *New Phytol.* 215(2):877-890.

Loureigo, J. de (1796) Flora Cochinchinensis. Ulysipone, Lisbon, Portugal.

Louvet, J. and G. Toutain (1973) Recherches sur lcs Fusarioses, VIII. Ann. *Phytopathol* 5:35-52.

Macmillam, H.G. (1914) A Handbook of Tropical Gardening And Planting, Macmillan Co., New York, USA.

Mann, Charles C. (2005) 1491: New Revelations of the Americas Before Colmbus. New York Publishing Group, New York, USA.

Masoud Sheidal (2007) B-chromosome variability in pomegranate (*Punica granatum* L.) Cultivars. *Caryologia* 60(3):251-256.

Mawe and John Abercrombie Thomas (1778) Dictionary of Gardening And Botany. Printed for G. Robinson and T. Cadell, London, UK.

Mcpherson, H.G., Richardson, A.C., Snelgar, W.P., Patterson, K.J. and M.B.Currie (2001) Flower quality and fruit size in Kiwifruit (*Actinidia deliciosa*). *New Zealand Journey of Crop AndHorticultural Science* 29:93-101.

Merrill, E.D. (1954) The Botany of Cook's Voyages. Vol.14. No.5/6. Chronica Botanica Co. Waltham. Mass. (Ambarella).

Miranda, M., T. Motomura, F. Ikeda, T. ohgawaro, W. Saito, T. Endo, OM.Omura and T.Moriguchi (1997) Somatic hybrids obtained by fustion between Ponicirus trifoliate (2x) and Fortunell hindsii (4X) protoplasts. *Plant Cell Reports* 16(6):401-405.

Misra, R.S. and P.N. Bajpai (1973) Bloosom bud differentiation in Jamun (Syzygium cumini Skeels). *Indian J.Hort.* 28:273-277.

Misra, R.S. and P.N. Bajpai (1975) Studies on flower biology of jamun [Syzygium cumini (L.) Skeels]. Indian Journ.*Horticulture* 32:15-24.

Mitra, S.K. and T.K. Bose (1996) Guava. *In:* Fruits: Tropical And Subtropical (Ed. Bose and Mitra) Rev. Ed. pp.280-303. Naya Prakash, Calcutta, India.

Morton, J. (1987) Indan Jujube. *In:* Fruits of Warm Climates. pp. 272-275. and http://ww.hort.purdue.edu/newcrop/morton/Indian_jujube.html.

Morton, J. (1987) Kiwifruit. In: Fruits of Warm Climates. Julia F. Morton, Miami, Fl. pp.293-300.

Morton, J.F. (1963) The Jambolan (Syzygium cumini Skeels) its food, medicinal, ornamental and other uses. *Proc. Florida State Hort. Soc.* 76:328-338.

Moshfeghi Nazli, Maddavi, O., Shahhosseini, F., Malekifar S. and S.K. Taghizadeh (2013) Introducing a new natural product from Dragon Fruit into the market. *IJRRAS* 15(2):269-272.

Mowry, H., Toy, L.R. and H.S. Wolfe (1941) Miscellaneous tropical and subtropical florida fruits. Agricultural Extension Service, Gainesville. *Florida Bulletin* 109:11-21.

Mukherjee, S.K. (1972) Origin of mango (Mangifera indica). *Econ. Bot.* 26:260-266.

Mukherjee, S.K., Singh, R. N., Majumder, P. K. and D. K. Sharma (1968) Present position regarding breeding of mango.in India. *Euphytica* 17:462-467.

Munier, P. (1974) be problerne de l'origine du palmier dattier et l'atlantide. *Fruits d'Outre-Mer* 29:235-240.

Muthukrishnan C.R. and I. Irulappan (1996) Papaya. *In:* Fruits: Tropical And Subtropical (Ed. Bose, T.K. and S.K.Mitra) Rev. Ed. Naya Prakash, Calcutta, India.

Muthukrishnan, C.R. and J.M.M.MD. Abdul Khader (1996) Avocado. *In:* Fruits: Tropical And Subtropical. (Ed. Bose, T.K. and S.K.Mitra) Rev.Ed. pp. 547-565. Naya Prakash, Calcutta, India.

Naik, K.C. (1963) South Indian Fruits And Their Culture. P. Vavadachary And Co., Madras, Tamil Nadu, India.

Nakasone, H.Y. and R.E. Paull (1998) "Annonas" *In:* Tropical Fruits (Ed.Naksone, H.Y. and R.E.Paull) CAB International. London, UK pp.45-75.

Nambiar, M.C., Baskara Rao, E.V.V. and P.K. Thankamma Pillai (1996) Cashew. *In:* Fruits: Tropical and Subtropical (Ed.Bose, T.K. and S.K.Mitra) Rev. Ed. pp. 386-419. Naya Prakash, Calcutta, India.

Negrul, A.M. (1938) Evolution of cultivated forms of L grapes. *C.R. Acad. Sci.,* U.S.S.R. 18:585-588.

Nishiura, M. (1964) Citrus breeding and bud selection in Japan. *Proc. Florida State Hort. Soc.* 77:79-83.

Nixon, R.W. (1966) Growing dates in the United States. U.S.D.A. agric. Inform. Bull., 207, pp. 50 (Revised).

Nixon, R.W. and J.B. Carpenter (1978) Agriculture Information Bulletin, 207, USDA pp.63.

Nixon, R.W. and J.R. Furr (1965) Problems and progress in date breeding, Rep. Date Growers' Inst. 42:2-5.

O'Connor, B.A. (1959) Exotic plant pests and diseases: A handbook of plant pests and diseases to be excluded from spreading within the area of the South Pacific Commission. S.P.C., Noumea, New Caledonia.

Ochse, J.J., Saule, M.J. Jr. Dijkman, M.J. and C. Wehlberg (1961) Tropical And Subtropical Agriculture. Macmillan, New York, USA.

Ohri, D. (2015) How small and constrained is the genome size of angiosperm woody specie. *Silvae Genetica* 64(1-2):20-32.

Olmo, H.P. (1976) Grapes. *In:* Evolution of Crop plants. (Ed. Simmonds, N.W.) Longman, London, UK. pp.294-298.

Olmo, H.P. and C. Baris (1973) Obtention de raisins de table apyrenes, Symp. Intern. Raisins de Table, Limasol (Chypre) July 16-21.

Pareek, O.P. (1996) Date Palm. *In:* Fruits: Tropical And Subtropical (Ed.Bose, T.K. and S.K.Mitra) Naya Prakash, Calcutta, India. Rev. Ed. pp:667-689.

Patil, A.V. and A.R. Karale (1996) Pomegranate. *In:* Fruits: Tropical And Subtropical (Ed. Bose T.K. and S.K.Mitra). Naya Prakash, Calcutta, India. pp.616-637.

Patino, V.M. (1963) Varietal variations on pollen size and fertility in pineapple (Ananas comosus (L.) Merr. Imprenta Departmental. California, Columbia.

Peiris KHS (2014) Sapodilla. Manilkara zopata L. *In:* Underutilised Fruit Trees (Ed. Puspakumara, D.K.N.G., Gunasena, H.P.M. and V.P.Singh) World Agroforestry Centre, South Asia Office, New Delhi, India pp.183-224.

Percy-Lancaster, S. and T.K. Bose (1965) Studies on hybrids of Syzygium. *Indian J.Hort.* 22(1) 87-88.

Perez *et al.* (2005) Wound healing properties of Hylocereus undatus on diabetic rats. *Phytother. Res.* 19:665-668.

Pieringer, A.P., Edwards, G.H. and R.W. Wolford (1964) The identification of Citrus species and varieties by instrumental analysis of Citrus leaf oils. *Proc. Amer. Soc. Hort, Sci.* 84:204-212.

Popenoe, W. (1920) Manual of Tropical And Subtropical Fruits. Macmillan Co., New York, USA.

Popeoe, W. (1920) Manual of Tropical And Subtropical Fruits. Macmillan, New York, USA.

Prabhuraj, H.S. (2002) Correlation and path coefficient analysis in jamun (Syzygium cumini) trees. *My Forest* 38(2):177-182.

Production-226-progress Towards The Development of A Mechanical harvester for Wild Blueberries. https://extension.umaine.edu/blueberries/factsheets/production/progress-towards -the-development-of-a-mechanical-harvester....

Purseglove, J. W. (1968) Tropical crops. Dicotyledons. Longman, London, UK. 2:377-386.

Purseglove, J.W. (1974) Tropical Crops Dicotyledons. Longman, London, UK.

Raman, V.S., G. Manimekalal and S.R. Sreerangaswamy (1971) Chromosome behavior at meiosis in *Punica granatum* L. *Cytologia* 36:400-404.

Rathore, D.S. (1996) Custard apples. *In:* Fruits: Tropical And Subtropical (Ed. Bose, T.K. and S.K. Mitra) Rev. Ed. pp. 449-468. Naya Prakash, Calcutta. India.

Rehder, A. (1954) Manual of cultivated trees and shrubs hardy in North America. New York, 2nd Ed. Macmillan Co., USA.

Richardson Annette C. *et al.* (2011) Fruit development of the diploid kiwifruit, Actinidia chinensis 'Hort 16A'. BMC Plant Biology 11:182. http://www.biomedicentral.com/ 1471-2229/11/182.

Richardson, D.L.; Hamilton, K.S. and D.J.Hutchinson (1965) Notes on bananas 1. Natural edible tetraploids. *Trop Agriculture* 42:125-137.

Roy, B. and S.S. Visweswariya (1951) Cytogenetics of mango and banana. Maharashtra Association for the Cultivation of Science, Poona, (M.S.), India.

Ruehle, G.D. (1963) The Florida avocado industry. Univ. Florida agric. Exp. Sta. Bull., 602, pp.102.

Rumphius, G.E. (1962) Herbarium amboinenge, Uitgevery en Drukkerij Hollandis, N.V.Barn.

Sambamurty, K. and V. Ramalingam (1954) A note on hybridization in the sapota (Achras zapota L.). *Indian Jour. Hort.* 11:57-60.

Sauer, C.O. (1952) Agricultural Origins And Dispersals, New York, USA.

Schmid, R. (1972) A resolution of the Eugenia-Syzygium controversy (Myrtaceae). *Amer. J. Bot.* 59(4):423-436.

Schroeder, C.A. (1968) Prehistoric avocados in California. Calif. *Avocado Soc. Yrbk.*, 52:29-34.

Scott, D. H., Draper, A.D. and L.W. Greeley (1972) Interspecific hybridisation in octoploid strawberries. *Hurl. Sci.* 7:382-384.

Sen, S.K. (1996) Pineapple. *In:* Fruits: Tropical And Subtropical (Ed. Bose, T.K. and S.K.Mitra) Rev. Ed. pp.252-279. Naya Prakash, Calcutta, India.

Shanmugavelu, K.G. (1967) A note on air laying of Eugenia (Eugenia jambolana Lam.) and Neem (Azadirachta indica A.Juss.) *South Indian Horticulture* 54:70-71.

Shanmugavelu, K.G. (1970) Effect of gibberellic acid on seed germination and development of seedlings of some tree plant species. *Madras Agricultural Journal* 57(6):311-314.

Sharma, D.K. and R.N. Singh (1970) Investigation on self-compatibility in Mangifera indica. Symposium on 'Mango And Mango Culture'. ISHS, The Hague, The Netherlands.

Sharma, S.R. (2008) Effect of vitamin-C on collagen biosynthesis and degree of birefringence in polarization sensitive optical coherence tomography (PS-OCT). *African Journal of Biotechnology* 9(12):2049-2054.

Sharma, V.P. and Y.N. Kore (1996) Ber. *In:* Fruits: Tropical And Subtrophical. (Ed. Bose T.K. and S.K. Mitra) Rev. Ed. Naya Prakash, Calcutta, India.

Singh, L.B. (1969) Mango (Mangifera indica). *In:* (Ed. Ferwerda, F.P. and Wit. F.), Outlines of perennial crop breeding in the tropics. Wageningen, pp.309-327.

Singh, R. (1969) Fruits. National Book Trust, New Delhi, India.

Singh, S.K., Krishnamurty, S. and S.L.Katyal (1967) Fruit Culture In India. ICAR, New Delhi, India. pp.255-259.

Singh, Sham, Krinamurthy, S. and S.L. Katyal (1967) Fruit Culture In India. ICAR, New Delhi, India.

Singh, Sham, Krishnamurti, S. and S.L. Katyal (1963) Fruit Culture of India. ICDR, New Delhi, India.

Smith, C.E. (1969) Additional notes on pre-Conquest avocados in Mexico. *Econ. Bot.*, 23:135-140.

Snyder, E. (1937) USDA Year Book of Agriculture. pp.263-268.

Soost, R.K. (1964) Self-incompatibility in Citrus grandis. *Proc. Amer. Soc, Hort, Sci.*, 84:137-140.

Staudt, G. (1973) Fragaria iturüpenisis, eine neue Erdbeerart aus Ostasien. *Willdenowia* 7:101-104.

Storey, W.B. (1953) Genetics of the papaya. *J. Hered.* 44:70-78.

Storey, W.B. (1958) Modifications of sex expression in papaya. *Hort. Ado.* 2:49-60.

Storey, W.B. (1967) Theory of the derivations of the uni-sexual flowers of Caricaceae. *Agron. trop.* 17:273-321.

Storey, W.B. (1969) Pistillate papaya flower: a morphological anomaly. *Science* 163:401-405.

Stout, A.B. (1936) Seedlessness in grapes. N. Y. *Agric. Exp. Sta. tech. Bull.*, 238, pp. 68.

Sulladmath, U.V. and M.A. Narayana Reddy (1996) Sapota. *In:* Fruits Tropical and Subtropical (Ed. Bose, T.K. and S.K. Mitra), Naya Prakash, Calcutta, India. pp. 565-591.

Swingle, W.T. (1908) Breeding new and superior figs. Calif. State Fr. Growers Conv. Rep. 34:187.

Swingle, W.T. (revised by Reece, P.C.) (1967) The botany of Citrus and its wild relatives. *In:* (Ed. Reuther, Webber, and Batchelor) The citrus industry, Berkeley 1:190-430 (Rev. Edn.).

Takami, K., Matsumura, A., Yahala, M., Imayama, T., Kunitake, H. and H. Komatsu (2004) Production of intergeneric somatic hybrids between round kumquat (Fortunella japonica Swingle) and 'Morita naval' orange (Citrus sinensis osbeck) *Plant Cell Reports* 23:39-45.

Tanaka, T. (1954) Species problem ill Citrus. Japanese Society for the Promotion of Science. Tokyo, Japan.

Tenore, G.C. *et al.* (2012) Nutraceutical potential and antioxidant benefits of red pitaya (Hylocereus polyrhizus) extracts. *Journal of Functional Foods* 4(1):129-136.

Thaper, A.R. (1958) Farm Bull. No. 42, Indian Council of Agricultural Research, New Delhi, India.

Thompson, K.M., Culley, T.M., Zymberger, A.M. and D.L Lentz (2015) Genetic variation and structure in the neotropical tree M.Zapota (L) Royen (Sapotacee).

Tolkowski, Y. (1938) Hespierides: A History of the Culture and use of the Citrus Fruits. John Bale Sons & Curnow, London, UK.

Tydeman, H.M. (1962) Rootstocks. *In:* Handbuch der Pflanzenziichtung, 6, Berlin, 2nd Ed., 547-572.

Uttal, L.J. (1987) The genus Vaccinium L. (Ericaceae) in Virginia. *Castanea* 52:246.

Vacciniium caesarense. PLANTS Profile. United States Department of Agriculture. Retrived September 13. 2010. eol.org/pages/484405/details#cite_note_PLANTS_1.

Vakili, N.G. (1967) The experimental formation of polyploidy and its effect in the genus Musa. *Amer. J. Bot.* 54:24-36.

Vander Kloet, S.P. (1980) The taxonomy of the highbush blueberry, Vccinium corymbosum. *Canad. J. Bot.* 58:187-1201.

Vander Kloet, S.P. (1988) The genus Vaccinium in North America. Publication 1828, Ottawa: Research Branch, Agriculture Canada, pp.201.

Vander Kloet, S.P. and P.J. Austin-Smith (1986) Energetics, Patterns and timing of seed dispersal in Vaccinium section Cyanococcus. American Mildland. *Naturalist* 115:386-396.

Vavilov, N.I. (1951) The origin, variation, immunity and breeding of cultivated plants. *Chronica Bot.* 13:1-364.

Venkatraman, L. (1963) Fruit Culture in India (Ed. Sham Singh, S. Krishnamurty and S.L. Katyal) ICAR, New Delhi, India.

Vijaykumar, N. and D. Subramanian (1985) Cytotaxonomical studies in South Indian Myrtaceae. *Cytologia* 50:513-520.

Waldo, G.F. (1968) Blackberry breeding involving native Pacific coast parentage. Fruit Var. *Hort, Dig.* 22:3-7.

Walter, S. Judd (2000) The PLANTS Database database (version 5.1.1). National Plant Data Center, NRCS, USDA. Baton Rouge, LA 70874-4490 USA. http://plants.usda.gov.vaccinium corymbosum.

Walter, S.Judd (2010) Ericaceae of North America Update Database (Version 2010) (Updated By The Flora of North America Expertise Network, in connection with an update for USDA PLANTS (2007-2010) Vaccium Corymbosum.

Ward, C. and D. Courtney (2013) Kiwifruit: Taking its place in the global fruit bowl. *Advances In Food And Nutrition Research* 68:1-14.

Watanabe, K., Takahashi, B. and K. Shirat (1990) Chromosome numbers in Kiwifruit (*Actinidia deliciosa*) and related species. *Journal of the Japanese Society for Horticultural Science* 58(4):835-840.

Watt, G. (1889) Dictionary of Economic Products of India. Govt. of India & Periodical Experts, Shahadar, Delhi, India.

Watt, G. (1892) A Dictionary of the Economic Products. Cosmos Publications, India (Reptr. Ed.).

Watt, G. (1908) The Commercial Products of India. John Murray, London, UK.

Webber, H.J. (1967) History and development of the citrus industry. *In:* (Ed. Reuther, Webber and Batchelor) The citrus industry, Berkeley 1:1-39 (Revised Reuther, W. and Lawton, H.)

Wilcox, A.N. (1962) The apple: I Systematics. Handbuch der Pflanzenzuchtung, 6, Berlin, 2nd Ed., pp. 637-645.

Wilhelm, S. and J.E. Sagen (1974) A history of the strawberry from ancient gardens to modern markets. *J.Food.Sci.* 35:448-452.

Xiaoyong XU Jihong Liu and Xiuxin Dent (2005) FCM, SSR and CAPS analysis of intergeneric somatic FCM, SSR ad CAPS analysis of intergeneric somatic hybrid plants between Changshou Kumquat and Dancy tangerine. *Bot.Bull.Acad.Sin.* 46:93-98.

Yasuda, K., M. Yaheta and H.Kunitake (2016) Phytogeny and classification of Kumquats (Fortunella Spp.) inferred from CMA karyotype composition. *The Horticultural Journal* 85(2):115-121.

Young, T.W. and R.B. Ledin (1954) Mango breeding. *Proc. Fla. Hart. Soc.* 67:241.

Zylka, D. (1970) Die Vertoendung von Wildarten der Gauung Primus in der Sortenziichtung lind als Unterlage. Wiesbaden.

8

Vegetables

INTRODUCTION

Owing to a greater appreciation of their food value, vegetable growing is one of the major enterprises of horticulture and became very popular. People consume more fresh vegetables today than ever before. They are an important part of the diet of millions of people worldwide, more so in the tropics. Vegetables provide variety, flavor and zest to the dies, making meals more appetizing. Their importance has increased because they produce a maximum quantity of food for the area planted and they also grow quickly.

Several vegetable crops are of local importance. Some are of world-wide importance. The phenomenal growth of the vegetable industry since World War II has been due to greatly improved production, handling and transporting techniques. Now, a wide variety of fresh and processed vegetables from distant markets are available in all the great cities at all times of the years. Processed vegetables in the form of frozen, canned, dehydrated and pickled products are available throughout.

In a popular sense, the term vegetable applies to those plants or plant parts that are usually consumed with the main course of a meal and are commonly salted and boiled or used for dessert and salads. Thus, tomatoes, pepper, okra, eggplant, cucumber and squash, although botanically fruits are traditionally regarded vegetables. Rhubarb, botanically petiole, is commonly served as a dessert. Squash is used as a fruit in pies.

The food value especially of leafy and fruit vegetables is low because of large amount of water. Root crops, however, contain large amounts of carbohydrates but are poor in protein and oils, and rank next to cereals as a source carbohydrate food. Grain legumes are different from other vegetable article of human diet, being rich sources of oil and proteins as well.

Source of Vitamins:

Carotene, a precursor of vitamin A, is abundant in several vegetables *e.g.* carrots, sweet potatoes, spinach, lettuce, amaranths, kale, turnip green, muskmelon and watermelon. It is also found in raw tomatoes, cabbage and green peppers. Vitamin-C is abundant in many vegetables *e.g.* kale, Brusse sprouts, broccoli, lettuce, tomatoes, pepper, potatoes, cabbage, greens, cucumber, bitter gourd and onion, especially when they are uncooked. Thiamine, niacin and riboflavin occur in succulent vegetables and legumes. The latter contain five times as much riboflavin and ten times as much thiamine as cereals. Green vegetables, *e.g.* collard green, kale and spinach have more vitamin E than iceberg head lettuce. The quantity of vitamin E in leafy vegetables increases with their greenness.

Source of Minerals:

Vegetables are important sources of minerals that are needed for proper growth and development. Potato, sweet potato and onion contain appreciable amounts of phosphorous. Calcium is present in beans, spinach, onions, lettuce, tomatoes and varieties of the cabbage family *e.g.* head cabbage, cauliflower, broccoli, kale, and collard greens. Soybean, spinach, peas, chillies, radish, garlic, beans and tomato are god sources of iron. Vegetables like okra, onion and asparagus provide iodine. Usually, legumes contain more iron, calcium, and potassium salts. Magnesium, calcium and potassium act as neutralizers of acids which are produced in the course of digestion of some foods.

Sources of Roughage:

Many vegetables have a high percentage of refuse (indigestible cellulose material). It has no direct nutritional value but the presence of this material in the human digestive system is of considerable value. The spongy mass helps to satisfy the appetite and also aids in moving food through the alimentary canal by helping the muscular action of the intestine. It thus avoids the sufferer from constipation.

Fresh vegetables are naturally low in fat, salt and sugar. This makes the vegetables an excellent food choice. Vegetables provide, apart from minerals and vitamins, energy and fibres. Some vegetable contain higher levels of carbohydrate, for example, roots and tubers like potatoes, yams, taro, etc. They are obviously higher in energy. Other vegetables are non-starchy. They tend to have a higher water content and lower in energy. They are often richer in vitamins and minerals. Some vegetables contain phytonutrients which impart bright colours, for example, tomatoes, beets, carrots, etc. Such vegetable sources are said to promote health and reduce risk of disease. Vegetables contain fibre, some have more than others. Fibre keeps the digestive system healthy. It helps to keep a healthy body weight and to lower cholesterol levels.

(I) Leafy Vegetables

Under this category, for the sake of convenience, many unrelated plants whose stems or foliage are consumed as vegetables are included. Shoots, bamboos, asparagus, Chinese bean sprouts and palm shoots are only a few examples of juvenile stems which are eaten in different regions of the world. Dark green vegetables such as kale, collard greens, spinach and New Zealand spinach are important leafy vegetables. They contain higher amounts of proteins, vitamins and minerals than other leafy vegetables. They also provide the roughage and are a rich sources of iron, potassium, calcium, vitamins A and C, riboflavin and folacin. The amount of vitamins A and E in leafy vegetables increases with their greenness. Celery and rhubarb are two important examples where the part consumed is the leaf petiole. Different varieties of the cabbage family such as cabbage (leafy bud), brussels sprouts (axillary buds), broccoli and cauliflower (inflorescence), and globe artichoke of the Asteraceae (immature heads together with the fleshy bases of the involucral bracts and the thickened receptacle). It is to be noted that leafy vegetables have little energy value and contain a large amount of water.

Production statistics for most of these vegetables are not available as they are consumed locally as a fresh produce. There is little international commerce in leafy vegetable as they are very perishable. In technologically advanced countries (especially North America), a sizeable and very specialized lettuce industry has grown up. Celery, too, may be marketed widely. Many commercial vegetables, *e.g.* broccoli, cauliflower, brussels sprouts and spinach tend to lose their form when preserved by canning but preservation of perishable vegetables by freezing has become an important business. The vegetables must be kept frozen until sold and consumed. Of late, leaf protein concentrate is being employed as an animal feed but because of its green colour, however, it is unpalatable for human consumption.

Green leafy vegetables are recognized as one of relatively inexpensive sources of so many important nutrients. They are more ideal for weight management as they usually contain low fat content, high dietary fibre and rich in folic acid, vitamin-C, minerals like potassium and magnesium. They are also major source of iron and calcium. Salad greens provide a whole range of important nutrients and phytochemicals to keep us healthy. They contain green pigment *viz.*, chlorophyll which one of key nutrients. It benefits us in accelerating wound healing to detoxify liver, to boost immune system and help treat anemia.

8.1 CABBAGES

Brassica oleracea L. (Brassicaceae)

Resume: *Brassica oleracea is polymorphic and its various parts are modified into storage organs. The ancient Germans, Saxtons and Celts were the first to grow cabbage in northern Europe. Its cultivation is very old, at least since 2500 BC. The Sea Cabbage or the Colewort is thought its ancestor which is weedy species of the seacoasts of south-western Europe and the Mediterranean. B. oleracea is a diploid with 2n=18 chromosomes. Various modified parts viz., stem, leaves, sprouts, inflorescence, etc. are consumed.*

1. Introduction: *Brassica oleracea* L. is polymorphic and provide mankind important herbage vegetables. Its various parts have been modified into storage organs which are utilized by man *e.g.* (i) numerous overlapping leaves surrounding the terminal bud (cabbage), (ii) enlarged axillary buds (Brussels sprout), (iii) Swollen bulb-like stem (Kohlrabi), (iv) the inflorescence and flower buds thickened and fleshy (Cauliflower, broccoli).

The wild ancestor, the Colewort (Sea cabbage) a stout weedy perennial of the seacoasts of Great Britain and South Eastern Europe is still growing there. By selection or mutation, the great variety of cultivated forms have arisen. The ancient Germans, Saxons and Celts were the first to grow cabbage in Northern Europe. Its cultivation is very old, at least since 2500 BC. Other varieties *viz.,* Cauliflower, broccoli, true cabbage, etc. were known to the Greeks and Romans. The Cabbage was held in high esteem by the Greeks and the Egyptians are said to have worshipped it.

2. Plant Characteristics: The important features of various forms are: (i) Cabbage: The stem is so short that the great mass of thick overlapping leaves tends to form a head (ii) Cauliflower and Broccoli: A short stem is produced with an underdeveloped inflorescence. The whole inflorescence forms a large head of abortive flowers of thick hypertrophied branches. (iii) Kohlrabi: The short stem is transformed into a juicy mass of edible tissue, which stands out of the ground. It is large, spherical and turnip-like. (iv) Brussels sprouts: The axillary buds on the erect stem develop little heads, instead of forming branches. (v) Kale: It resembles cabbage but instead of producing compact heads, the short and thick stalk bears crinkly blue-green leaves.

3. Parts Used: They are modified vegetative parts as described under plant characteristics.

4. Chemical Constituents:

(i) Cabbage: Carbohydrate (6.3%), protein (1.8%), fat (0.1%), minerals (calcium, phosphorous, iron), vitamin A, B and C.

(ii) Cauliflower or Broccoli: Protein (5.3%), fat (0.4%), carbohydrates (5.3%), minerals (calcium, phosphorous, iron), vitamin A, B and C.

(iii) Brussels Sprouts: Protein (4.7%), carbohydrate (9.2%), vitamin A and C, riboflavin, niacin, minerals (calcium, iron), phosphorous.

(iv) Kohlrabi, Knol-kuhl: Protein (1.1%), fat (0.2%), carbohydrate (5.9%), minerals (calcium, phosphorous, iron), vitamin A and C.

(v) Kale: Protein (3. 9%), fat (0. 6%), carbohydrate (6. 0%), minerals (calcium phosphorous, iron), vitamins A, B, B2, C.

5. Uses And Applications: (i) Cabbage: It is consumed raw, as salad, or cooked. It is also used after steaming. Red cabbage used for pickling. In Russia, it is used in the form of soups and stuffing cake. (ii) Cauliflower: It is consumed as vegetable. (iii) Knolknol: The swollen stem is eaten as vegetable. Leaves are used as vegetable. (iv) Kale: The young leaves and tender shoots are eaten as vegetable. (v) Brusels Sprouts: The sprouts are cooked like cabbage and eaten. (vi) Kohlrabi: It is cooked as vegetable and eaten. (vii) Broccoli: It is usually eaten cooked, but also in soups.

6. Karyotaxonomic Evidence: *Brassica oleracea* is a diploid with 2n=2x=18. A basic chromosome number for the genus *Brassica* has been interpreted as x=5, making *B. oleracea* a modified amphidiploid from a cross between two primitive 5-chromosome species with subsequent loss of one pair of chromosomes. Other authors have different view. They suggested that x=6 and that the constitution of the haploid set of *B. oleracea* is AABBCCDEF (Yarnell, 1956). The occurrence of up to two bivaents in kale haploids (Thompson, 1956) lends some support to the latter hypothesis. Six different chromosomes types were recognised by Robbelen (1960) based on (i) total length, (ii) symmetry of arms and (iii) on the shape of the heterochromatic centromere region in the three *Brassica genomes*. In each genome, different basic chromosomes had been duplicated and the haploid constitution of *B. oleracea* was given as A BB CC D EE F.

The wild cabbage is native to the coasts of north western Europe and also to the Mediterranean. It is but doubtful whether the man cultivated varieties of *B. oleracea* evolved exclusively from the wild cabbage, *B. oleracea* var. *oleracea*. Many other wild diploid relatives may have contributed. These wild relatives can be different species and include *B. cretica, B. insularis* and *B. rupestris* (Yarnell, 1956). Crosses between them, however, are interfertile and so they have, along with B. oleracea, been referred to one cytodeme (Harberd, 1972). Some of these wild forms exhibit characteristics quite unlike typical *B. oleracea e.g.* (i) the short, flat siliqua of *B. macrocarpa* and (ii) the corymbose inflorescence of *B. cretica* may have resulted in geographical isolation of populations on the sea clifts of Europe, the fixation of distinct characters, combined with the absence of mating berries. Perhaps, the whole complex should be classified as one species, *B. oleracea*, in which morphological variability is not restricted to the cultivated types.

7. History: The different cultivated types evolved early in the Mediterranean area and /or in Asia Minor. However, the development of cultivars from wild species took place in several different areas of the Mediterranean *e.g.* the eastern Mediterranean was the evolutionary centre for the cauliflower (Helm, 1963). Cultivation of kales probably occurred first several thousand years ago. The Greeks grew them as early as 600 BC. Many other types, including a heading cabbage and possibly kohlrabi, were described by writers in ancient Rome. These types would have been highly self-incompatible and almost completely cross-fertilized. The maintenance of distinct types must have been difficult, as all cultivar groups of *B. oleracea* are completely cross-fertile.

All species of the genus *Brassica* contain glucosinolates which, in crushed leaves, are broken down by the enzyme myrosinase to give bitter tasting and goitrogenic substances. Selection, at an early stage in domestication, must have been for plants which were less bitter tasting than usual. Estimations of these substances showed that the stem, petiole and lamina of wild cabbage can contain four times the quantity present in organs of cultigens eaten by man (Josefsson, 1967). The present Mediterranean wild species are annuals. They grow in the warm and moist winter. As the early cultivars spread to northwestern Europe, biennial types must have been selected with greater winter-hardiness and a considerable cold vernalization requirement for flowering. This change over probably took place from out-crossing with native biennial or wild cabbages. Less fibrous, thicker stems and more succulent storage organs were other early selection criteria by the early man.

Leafy unbranched kales and branching thin-stemmed kales were the earliest cultivated Brassicas. These *B. oleracea* var. *acephala* types gave rise to the forage kales. An open heading cabbage. 'Tritianon', with a head 30 cm in diameter was although described by Pliny, Helm (1963) considered that the real hearting cabbage, *B. oleracea* var. *capitata*, evolved in Germany. Both red and white cabbages were known about AD. 1150 in Germany and in England by the 14th century. Three types of Savoy cabbage, varying from plants with loose rosettes of crinkled leaves to others with well-formed hearts but only slightly crinkled leaves, were documented in a German herbal in 1543. Savoy cabbage probably originated in Italy and spread to France and Germany in the 16th and 17th centuries.

Pliny described "a Brassica in which the stem is thin just above the roots, but swells out in the region that bears the leaves, which are few and slender" a description which he considered must refer to the kohlrabi. This vegetable was described in the 16th century herbals. Gerarde stated that it was grown in Italy, Spain and Germany (Helm, 1963). Boswell (1949) opined that Pliny's description applied to the cauliflower and indicated that the kohlrabi developed in northern. Europe in the 15th century. However, Helm (*loc. cit.)* said that the cauliflower, in its present form, was unknown before the early Middle Ages. Three different cultivars of 'Chou de Syrie' were described in a work by an Arab botanist in Spain in the 12th century, but it is not known whether the cauliflower was grown in Spain then. It was only described in 1576 as *Brassica florida botrytis* and in 1583 as *Brassica cauliflora*. It was probably introduced into Italy by the Genoese from the Levant or Cyprus about 1490. It was grown for seed around the Gulf of Naples. At the beginning of the 17th century, it reached Germany, France and England.

Sprouting broccoli, *B. oleracea* var. *italica*, came from the Levant, Cyprus or Crete to Italy. It was not mentioned until 1660 and was referred to as 'sprout 'colli-flower' or 'Italian Asparagus' in Miller's Gardener's

Dictionary of 1724. As all flower buds open in sprouting broccoli, sprouting broccoli might be expected to have preceded the cauliflower in the evolution of the species. It is to be noted that the flower buds in the cauliflower are fleshy, deformed and sterile. It is not known whether sprouting broccoli did not exist until the 17th century or whether it was not marked out from the cauliflower before that time.

The Brussels sprout, *B. oleracea* var. *gemmifera*, in its present form, appeared only in about 1750 in Belgium as a 'sport', but by 1800 had reached France and England. From 1820, it was known as 'Chou de Bruxelles'. Records from 1231 and 1472 in Belgium refer to 'un cent de sproq', but may have referred to *B. capitata polycephalos*, illustrated by Dalechamps in 1587. This had a large number of sprouts but lacked the characteristic terminal head (Helm, 1963).

Marrow-stem kale, *B. oleracea* var. *acephala*, is recorded with certainty only from the Vendee region of France in the early 19th century and its origin is not known. From France, it spread early to Germany and Denmark. It was introduced into England about 1900. It is grown in New Zealand under its French name, 'chou moellier'.

8. Further Developments: Brussels sprouts, cabbages and cauliflowers were originally grown only in gardens. They have become market garden crops later. The acreage of kale for autumn and winter cattle grazing reached a peak in 1960 in England and Wales. Since the late 19th century, cultivars of Brussels sprouts and cauliflowers were bred for marketing at different times of the year. Breeding for improved quality, in white curds and round heads in winter cauliflowers and small, firm Brussels sprouts, became important to meet marketing and processing requirements. Resistance to *Fusarium* wilt of cabbage in America saught serious attention.

Breeders achieved greater uniformity by more intense selection using mass selection methods. Cultivars were frequently developed from very few parental lines. This was probably because of the tendency to self-compatibility found in Brussels sprouts, cauliflowers and kohlrabi, in comparison with agricultural kales, which are morphologically less specialized and have received comparatively little attention from the breeders. Thompson and Taylor (1966) found, in this sporophytically controlled incompatibility system, that at least two of the three same pollen-recessive S-alleles were found in all varieties of *B. oleracea* examined, But their frequency was much higher and that the number of S-alleles high in the dominance series was lower in these partially self-compatible vegetable cultivars than in a commercial stock of marrow-stem kale. Self-compatibility was determined by genes outside the S-allele system. It occurred only in the absence of S-alleles high in the dominance series (Thompson and Taylor, 1972).

The only fully self-compatible group in *B. oleracea* are the summer cauliflowers Watt (1965) found little if any inbreeding depression on selfing in this crop, in contrast to the more self-incompatible autumn and winter cauliflowers. The change over from outbreeding to inbreeding appear to have occurred within about 100 years, since the middle of the 19th century. The evolution of self-compatibility in the summer cauliflower may find parallel in the breeding of the self-compatible winter cauliflower, Cambridge Dwarf April, which was obtained by rigorous selection for 30 years. It followed a cross between two plants of the winter cauliflower, St George, which is one of the most self-incompatible of its kind (Watt, 1963).

9. Epilogue: Because of the high costs and increasing mechanisation, growing emphasis is now being put on breeding varieties which can be harvested mechanically in one operation.

There have been attempts to meet the demand for uniformity combined with vigour by the production of F1 hybrids from uniform, inbred lines, mainly by exploiting knowledge of the self-incompatibility system. Unacceptable numbers of selfed or sibbed plants have occurred in some Brussels sprout hybrids. This was due to the partially self-compatible and the pollinating honey bees prefer to visit flowers of one inbred rather than the flowers of both lines indiscriminately. Inbreds that are homozygous for S-alleles high in the dominance series should be more self-incompatible and the self-incompatibility should be less influenced by environmental factors. The production of three way crosses and double-cross hybrids, based on single-crosses between morphologically similar but genetically unrelated inbred lines, have been tried, as well as synthetic varieties.

8.2 FENUGREEK

Trigonella foenum-graecum L. (Leguminosae-Papilionatae)

Resume: *Fenugreek is a native of India, Persia. Greece, Italy, Spain and Mesopotamia. De Candolle (1886) mentioned it only of Indian origin. It is widely used as food, spice and medicine since ancient times. Its leaves and seeds are useful parts. Somatic chromosome number have been found to be 14, 16, 30 and 46 along with B-chromosomes. Processed dietary products are being used in recent time. It has also wide medicinal applicability.*

1. Introduction: Fenugreek (Methi) finds place widely in food spices and medicine. About 260 species of the genus *Trigonella* occur worldwide (Acharya, 2006). De Candolle (1886) mentioned it to be of Indian origin. It is also thought a native of Mesopotamia desert, Persia, Greece, Italy and Spain, besides India (Linnaeus, 1753; Fazli and Hardman, 1968). Its leaves were employed to yield Egyptian incense 'Kuphi', a holy smoke used in fumigation and embalming rites (Rosengarten, 1969). The plant details are available, dating back to 16th century, in Kolozsvar Herbarium (Petropoulos, 2003). Some miscellaneous and medicinal uses, especially to expel placenta after childbirth, have been reported from 16th and 17th century (Hidvegi, 1984; Howard,. 1987).

2. Plant Characteristics: Erect annuals, 10-15 cm tall, glabrous, younger parts hairy; leaves 3-foliate, leaflets obovate or oblanceolate, nearly glabrous; flowers creamy yellow or whitish, borne terminally, solitary or few in short raceme; pods flat, beaked, many-seeded, seeds yellowish-brown, subquadrate-oblong, glabrous, smooth.

3. Parts Used: Leaves, seeds.

4. Chemical Constituents : (i) Seeds: Carbobydrates (58%), protein (23%), fat (6.4%), minerals (calcium, iron, magnesium manganese, phosphorous, potassium, sodium, zinc), thiamine, triboflavin, niacin, vitamins B6, Vitamin-C, folate B9 (*cf.* Fenugreek-Wikipedia). (ii) Leaves: Protein, vitamin-C, niacin, potassium, alkaloids, lysine, L-tryptophan, steroidal saponins, falvonoids, phenols (Brar *et al.*, 2013).

5. Uses And Applications: Fresh tender leaves and shoots are eaten as curried vegetable since ancient period in India and Egypt. It is also used as food additive. As a spice, fenugreek seeds add nutritive value and flavour to foods. In the Middle East and Far East, it is a popular leaf vegetable where meatless diets are customary. Dried leaf powder is used for garnishing and flavoring of foods in Indian cuisine. In Egypt and Ethiopia, it is a popular ingredient of bread. It is known as 'Bulba' and 'Abish' in the Arabian countries and Ethiopia respectively. The seeds, raw or boiled, are consumed with honey in Greece. In the USA, seeds are used in chutneys and various spice blends. It is specially used in mango pickle in India. It is used in recipes like 'Hearty Vegetable', 'Bean Soup' and 'Fenugreek Beef Stew', etc. Some processed products are available in the markets in the form of fixed oil, volatile oil, oleoresin and methi leaves. It is also useful as medicine. In tablet form in the UK, Mexico and Central America, it is used to make the birth control a success. It is beneficial in diabetes. In Java, it is used in hair tonic preparations and also as a cosmetic. Its oil is important for hair growth and hence hair-tonics are manufactured. A wide range of human afflictions are benefitted by its applications in various forms. It is also fed to cows to increase the flow of milk (Patil and Dhale, 2013).

6. Karyotaxonomic Evidence: The somatic chromosome numbers for the genus *Trigonella* has been reported as 2n=14, 16, 30 and 46 along with B-chromosomes (Martin *et al.*, 2011; Vaidya *et al.*, 2013). Genome size of Fenugreek is around 685 Mbp (Young *et al.*, 2003; Vaidya *et al.*, *loc. cit.*). Although, fenugreek is useful widely in food, spices and medicine, its whole genome sequencing is not investigated yet. Only transcriptone sequencing data is available. Particularly, targeting transcriptone sequencing of Fenugreek has been studied in order to identify the genes of diosgenin pathway (Vaidya *et al.*, *loc.cit.*).

7. History: It is thought to have been brought into cultivation in the Near East. Zohary *et al.*, (2012) pointed out uncertainty which wild strain of the genus *Trigonella* gave rise to domesticated Fenurgreek. Charred fenugreek seeds have been recovered from Tell Halal, Iraq (carbon dated to 4000 BC.), and Bronze Age levels of Lachish and desiccated seeds from the tomb of Tukankhamen (Zohari *et al.*, *loc. cit.*). Cato the Elder listed fenugreek with clover and vetch as crop grown to feed cattle (Fenugreek-Wikipedia). The Romans flavoured wine with fenugreek (Curry, 2010). The specific name of fenugreek 'foenum-graecum' means 'Greek hay' indicating its use as a forage crop in the past. Presently, India is the leading fenugreek producing country in the world (Anonymous, 1996).

8. Further Developments: In India, some improved varieties of fenugreek have been released *e.g.* Rajendra Kanti, LamSel 1, Pusa Early Bunching, Pusa Earlier, Bangalore-Local, Local Bobes, T8, HM-46, IC-74, and many others depending upon the needs of agricultural zones in India (Edison, 1995; Komal *et al.*, 1987; Parodo and Karwasra (1975); Singh *et al.*, 1994, Singh and Singh, 1974). Similar releases of different improved varieties of fenugreek in various countries *viz.*, Egypt, Italy, Greece, France, England are on record (*cf.* Chaudhary *et al.*, 2018).

Fenugreek is a rich source of saponins of which diosgenin is the most important secondary metabolite. It is used as raw precursor for the industrial, large scale synthesis of steroidal drugs and hormones. It also exhibited anticancer and anti-aging activities, beside cardioprotective and contraceptive properties (*cf.* Chaudhary *et al.*, 2018).

The fenugreek seed protein content ranging from the highest 28.55% to 24.7%. These variations have attribute to different cultural practices, soil and environmental conditions. This also the case of fat (Sulieman, 1995).

Yadav and Sehgal (1997) found that fresh leaves contain ascorbic acid and beta-carotene. They noted that ascorbic acid was reduced in sun and oven-dried leaves. They suggested that for better retention of nutrients in leaves, they should be stored in refrigeration, dried in oven, blanched for a short period of time (05 minutes) and should be cooked in pressure cooker. They also showed that there is no change in the calcium and zinc content of the processed leaves.

Balch (2003) suggested that fenugreek seeds have powerful antioxidant property that has beneficial effect on liver and pancreas. Since antioxidant properties have been linked to health benefits of natural products. Such properties are studied with germinated seeds which are observed to be more beneficial than dried seeds. The geminated seeds increase the bioavailability of different constituents of fenugreek.

Addition of fenugreek fiber to refined flour helps to fortify with a balance of soluble and insoluble fiber. Flour fortified with 8-10% fenugreek fiber has been used to prepare bakery foods *e.g.* pizza, bread, muffins and cakes with acceptable sensory properties (Srinivasan, 2005).

Fenugreek seed flour and fenugreek leaf powder have been used for the development of extruded snacks. A mixture of about 1.78% seed flour and 0.66% leaf powder with base material was found to have high preference levels for parameters of physical, functional and colour, and could be extruded with acceptable quality characteristics (Wani and Kumar, 2016).

9. Epilogue: Fenugreek is found to possess significant phytoconstituents such as alkaloid, amino compound, ketone compound, vitamin compound, steroid and pyran compound (Kumaravel, *et al.*, 2017). Fenugreek have been used in various traditional systems of medicines. The presence of various bioactive compounds justifies its use against various human diseases. Fenugreek is thus a plant of phyto-pharmaceutical importance, besides its wide role in diets. Processed products derived from fenugreek may be promoted from commercial point of view. It grows in a wide range of habitats and can be easily cultivated by seeds. It is fast growing and can be grown in different seasons. Thus it can be available throughout a year. Its roots bear bacterial nodules and thereby improves the fertility of soil. It thus benefits by its cultivation also.

8.3 LETTUCE

Lactuca sativa L. (Asteraceae)

Resume: *It is thought to be derived from Lactuca serriola L., a ruderal in southern Europe, North Africa and Western Asia. It is depicted on Egyptian tombs of about 4500 BC and known to be cultivated by the Greeks and Romans. It has chromosome number 2n-18. Previously, it was grown in home gardens, now it is a valuable commercial crop and world's most popular leafy salad crop.*

1. Introduction: World's most popular and important salad crop is the lettuce. It is said to be descended from the wild lettuce *L. serriola* L., a ruderal of southern Europe, North Africa and Western Asia. It is a herbage vegetable of great antiquity. It was depicted on Egyptian tombs of about 4500 BC. It is also known to have cultivated by the Greeks and Romans. It also appeared on the royal table of Persian kings as long as 550 BC. By the 7th century, its use spread to China. The Mors developed many new cultivars which are still grown. Spanish settlers carried lettuce to the Western Hemisphere. Formerly, it was cultivated only in the home gardens. Now, it is grown as a valuable commercial crop. It is grown for fresh market.

Cultivated lettuce spread to Greece and Rome and throughout the Mediterranean from Egypt (Lindavist, 1960). The first written records of lettuce cultivation are credited to Herodotus who mentioned that about 550 BC. a cos-like lettuce was eaten at the Persian Court (de Vries, 1997). Lettuce was brought to the New World by Christopher Columbus. Cultivated lettuce was introduced in China between 600 and 900 AD (de Vries, 1997).

2. Plant Characteristics: An annual or biannual herb, with shortened fleshy stem 10-50 cm tall; a rosette of smooth, radical spirally arranged leaves; yellow flowers in a homogenous capitulum, characteristics vary depending upon the cultivar types.

3. Parts Used: Leaves

4. Chemical Constituents: Protein (1. 36%), carbohydrate (2. 79%), fat (0. 15%), minerals (calcium, phosphorous, sodium, magnesium, potassium, manganese, zinc, copper), vitamin (A, B6, E, K, folate, niaein, riboflavin, thiamin, pantothemic acid) (Bahadur and Singh, 2011).

5. Uses And Applications: It is mainly used as a salad and in sandwitches, hamburgers, tacos and other dishes. The leaves may be boiled like spinach and eaten as vegetable. In China and Egypt, stems are used either as cooked, raw, pickled, dried or as a sauce.

6. Karyotaxonomic Evidence: *Lactuca* is a large genus of the Asteraceae with more than 100 species, chiefly indigenous to north temperate regions. The genus contains species with 8, 9 and 17 pairs of chromosomes (Babcock *et al.,* 1937). *Lactuca sativa* have nine pairs of chromosomes and are interfertile with each other. The others are *L. serriola, L. virosa* and *L. saligna.* All four are native to the Mediterranean basin. These species appear to be isolated within the genus by genetic barriers from other nine-chromosome species. Extensive tests revealed strong incompatibility barriers between the cultivated *Lactuca* species and the 17-chromosome species.

7. History: Lettuce is an old and major salad crop. Evidently, it was first cultivated about 4500 BC. (Lindqvist, 1960). Leaves are planted on the walls of Egyptian tombs. This suggests that it was a common crop, widely known and appreciated at the time. The leaves appear to represent those of the romaine or cos type. Keimer (1924) suggested that the early Egyptians first cultivated lettuce as a seed crop for the edible oil extracted from the seeds.

It spread quickly throughout the Mediterranean basin at an early date. Numerous references to the crop in Greek and Roman literature are cited indicating that it was a popular and extensively used vegetable at the apogee of these civilizations (Hedrick, 1919). It probably spread with the Roman legions to France, England and the rest of Europe. Lettuce cultivation quickly spread in the New World after 1492. It was reported to be abundant in Haiti by 1565 and cultivated in Brazil in 1647. As early as 1806, an American seeds-man listed 16 cultivars in his catalogue (Hedrick, 1919).

The origin of cultivated lettuce is uncertain. The sativa-serriola complex seems to be large, polymorphic and capable of free interchange of genes with little, if any, reduction in fertility. Through selection, *L. sativa* may have been originated directly from *L. serriola*, because nearly all the variations in cultivated lettuce are observed in *L. serriola*, except the extreme forms of head formation and shape of involucre. Lindqvist (1960) suggests three possibilities *viz.*, (i) both species might have originated from hybrid populations that diverged into two groups, *L. sativa* cultivated by man, and *L. serriola* adapted to man-made waste habitats, (ii) the ancestors of *L. sativa* might have been hybrids between *L. serriola* and a third species; (iii) *L. serriola* might be a product of hybridization between *L. sativa* and some other species.

The early human selection must have on non-shattering seedheads, absence of early flowering, non-spininess, decrease in latex content and increase in seed size, as well as on the hearting character. These are the major features separating cultivated from wild forms. Later programmes used to emphasize resistance to bolting and added resistance to disease, particularly lettuce mosaic and down mildew.

8. Further Developments: It is considerably variable crop and variation is mostly restricted to the vegetative parts, which are the economic parts of the plant. There is much variation in leaf length, shape, colour, texture, size and hearting type. Perhaps, the observed variation results from the origin and early development of the species in the Mediterranean, an area of many overrunning civilizations. These cultures may have exerted some selection pressure on the species. Early cultivars mostly appear to have been pointed, narrow-leaved, and non-heading. Cultivar development in Europe and, later, in the United States emphasized the head form, first of the butterhead, then of the crisphead type. European cultivars are mostly butterheads, but the crispheads predominated in the United States at an early date. The first definite evidence for existence of head lettuce was in 1543 (Helm, 1954).

Butterhead cultivars have been developed for winter culture in the Mediterranean area and for summer culture in northern Europe. Butterhead and loose-leaf types are used for glasshouse production in France, Holland and Great Britain. Crisp-head types are very popular and hence their production in France, Spain and North Africa has increased.

Plant breeding programmes emphasized the development of salad types for market gardens around urban areas and for home gardens in US. Early in this century, crisphead lettuce became dominant because it could be grown on a large scale in the irrigated vastness of the western United States and shipped under ice to the cities of the East and mid-West. These programmes emphasized size, weight and the ability to withstand the rigours of long-distance shipment in a cold, wet environment. At the turn of the century, the cultivar New York became predominant. The appearance of a disease *viz.*, brown blight led to the development, beginning in 1926, of a large group of resistant cultivars. Continued emphasis on colour, size, weight and resistance to bolting and diseases led to the development of the Great Lakes cultivars, first appearing in 1941, which exceeded the Imperial group in many of these characteristics.

The Imperial and Great Lakes cultivars were susceptible to a new race of downy mildew (*Bremia lactucae*). It appeared in 1932. The resistant cultivars Valverde and Calmar have been developed in response to this new race. Calmar and its derivatives are the latest group of cultivars to achieve dominance in the western United States. They are resistant to the races of downy mildew prevalent in the region.

Lettuce breeding programmes exploited wild species of *Lactuca*. Collections of *L. serriola* have been used in a few crosses. This species was the source of the downy mildew-resistance gene incorporated into Calmar and Valverde. The cultivar Vanguard is the only known derivative of the difficult cross between cultivated lettuce and *L. virosa*, This wild species seems to have contributed genes for the dark-green colour and excellent leaf texture that characterize Vanguard and its derivatives.

Breeding methods used with lettuce are the standard ones for a self-fertilizing, polymorphic species. Some backcrossing were practised. There is a little evidence for heterosis. Even if hybrid vigour were shown, it could not, for technical reasons, be easily exploited.

9. Epilogue: Considerable changes in the direction of breeding programmes are clear. In cultivar usage, popularity of crisphead lettuce is markedly increased in western Europe and Japan, This trend continued and have two consequences. Breeding programmes with crisphead lettuce increased emphasizing those characteristics that accompanied the rise of the crisphead industry in the United States. There will be continued emphasis on disease resistance and increased emphasis on insect resistance. In the vast production areas of the western United States, growers anticipated mechanical harvesting and post-harvest handling. Mechanization create two breeding needs: (a) the exploitation of such characters as head shape, head size, head density and resistance to rough treatment, all of which will facilitate machine-handling; and (b) uniform harvest density. In northern Europe, the development of crisphead seems likely as well as of butterheads and leaf types adapted to year-round lettuce production, particularly in greenhouses in winter.

8.4 SPINACH

Spinacea oleracea L. (Chenopodiaceae)

Resume: *Spinach is a naïve to the erstwhile Persia. It is a diploid with 2=12 chromosomes, however, triploid and tetraploid cultivars are also known. It* *is a dioecious and have separate male and female plants. Leaves are rich in phytonutrients and hence used in various food recipes and exclusively as vegetable. It benefits in many health benefits.*

1. Introduction: Spinach (*Spinacea oleracea* L.) is considered to have originated in ancient Persia (modern Iran and adjacent countries). In China, it is known as 'Persian Vegetable'. The earliest available record of the spinach plant was recorded in Chinese, stating it was introduced into China via Nepal, probably in 647 AD (Victor, 1949). In AD 827, the Saracens introduced spinach to Sicily (Rolland, 2006). The first written evidence of spinach in the Mediterranean region was recorded in the 10th century works *viz.* (i) the medical work by Al-Razi (known as Rhazes in the West) and two agricultural treatises (one by Ibn Wahshiyah and another by Qustus al-Rumi). It become popular in Spain in the latter part of the 12th century (Ibn and Yahya, 2014). Generally, three types of spinach are recognised *viz.*, (i) Savoy Spinach (dark green curly leaves), (ii) Flat-leafed spinach (smooth leaves), (iii) Semi-Savoy spinach (hybrid between other two types).

2. Plant Characteristics : Annual dioecious (separate male female plants), stem branched or unbranched; leaves in a basal rosette, leaf blade, light green, simple triangular-ovate to hastate, slightly succulent, margin entire or lobed, leaves on stem lanceolate; flowers unisexual, male flowers in leafless spikes or panicles with greenish parienth; female flowers in leaf axils, clustered; utricles with spiny appandages or smooth, ovoid or subglobose.

3. Useful Parts: Leaves.

4. Chemical Constituents: Carbohydrates (36%) , protein (29%), fat (0.4%), minerals (calcium, iron, magnesium, manganes, phosphorus, potassium, sodium, zinc), vitamin-A, thiamine (B1), riboflavin (B2), niacin (B3), vitamin B6, folate (B9), Vitamin-C, vitamin-E, vitamin-K.

5. Uses And Applications: Spinach leaves are cooked and used as vegetable. It is useful to flavour soups, sauces, pasta, and rice dishes. It is employed in baked foods such as muffins, pancakes or breads. It is rich in vitamin-K which is excellent for bone health. Its vitamin-A content is important for healthy skin and hair. It reduces aging of skin and well-known for promoting good eye health. Being in rich dietary fiber and anti-inflammatory flavonoids, spinach is also beneficial for digestive system. It reduces irritation in the digestive tract and allows for regular bowel movements and better adsorption of food.

6. Karyotaxonomic Evidence: Spinach (*S.oleracea*) is diploid with 2n=12 chromosomes. Triploid and tetraploid cultivars also known Sex in spinach is controlled by a series of alleles Y, X^m and X. Females are characterized by XX, male by XY and X^mY, whie X^mX and X^mX^m result in monoecious plants (Frankel and Galun, 1977). Most cultivars have three distinct morphs *viz.*, male, female and monoceious. The latter may be predominantly male or female or intermediary, according to the flower ratio, but in secondary sex characters and some other growth parameters they are significantly different from both males and females (Vitale and Freeman, 1985).

7. History: Spinach (*Spinacea oleracea* L.), a leafy vegetable (Chenopodiaceae) is closely related to *Spinacea tetranda* and also through its ancestor. Other allied species are *S.spinosa*, *S.inermis* and *S.turkstanica*. The genus is native of South-West Asia. The arabs probably brought it to Spain in the 14th century. It then spread from Spain to other countries by the 16th century (George, 1985). Various species of the genus *Spinacea* can be deciphered using morphological characteristics such as seed (prickly or smooth), leaf texture (crinkled or smooth), colour, shape and position of leaves and petiole length. Girenk (1968) emphasizes the distinctness of *S.oleracea* from its two diploid relatives *viz.*, *S.tetrandra* and *S.turkstanica* and described primitive forms of *S.spinacea* from north-western India and Nepal.

8. Further Developments: The growth of *Spinacea oleracea* L. is investigated by Saurer and Possingham (1970) from approximately 1cm long to its adult size. Cell division in leaves continued until the leaves reach from one-third to one-half full size. Cell division within the lamina of the leaves is found not uniform. It ceased at an early stage of development in the leaf tip region and continued for a extended period at the base.

Yang *et al.* (2004) used spinach leaf protein digests to know antihypertensive properties. They obtainied leaf protein containing approximately 50% rubisco (ribulose biphosphate carboxylase/oxygenase) from fresh spinach leaf using a simple extraction method. Pepsin and pepsin-pancreatin digests of spinach leaf protein have protent angiotensin-I converting enzyme inhibitory properties with IC50 values of 56 and 12 μg/mL respectively. Both digests of leaf protein have antihypertensive effects after oral administration to spontaneously hypertensive rats (SHR) with minimum effective doses of 0.25 and 0.5g/kg respectively. The maximum antihypertensive effect for the pepsin digest was observed 4h after oral administration, while for the pepsin-pancreatin digest, the maximum effect was observed 2h after oral administration. Undigested spinach leaf protein did not exert any significant antihypertensive effect after oral administration to SHR at doses of 0.5 and 19/kg. The pepsin digest of leaf protein is inferred useful in treatment of hypertension. Yao and Giesman used the spinach (Cultivar American Long Standing) to remove malathion (insecticide). They sprayed malathion on spinach crop. They washed the leaf sample thrice and the clean spinach was steam blanched for three minutes. Spinach kept in plastic bags was processed for quick freezing at -22 °F. Spinach was then kept in the cold room (40 °F). It was again washed thrice and then cooked. The unit operations involved trimming, sorting and soaking, plain washing and detergent washing in warm water (100 °F), steam blanching, followed by cooling and quick freezing. Home preparation methods included trimming, and sorting, washing in cold water and then cooking in boiling water for 6 or 7 minutes.

They concluded that: (a) in field conditions, rapid decomposition of malathion deposits was observed. A faster rate of decomposition occurred on the first day of after application and the decomposition slowed considerably afterwards (b) washing procedures aided removal of malathion residues in raw products. Washing with addition of detergent greatly improved the removal of residues (c) blanching and cooking reduced or completely removed the trace amount of residues remaining after washing (d) the initial concentration of residues in the spinach crop does not significantly affect the percentage of residues removed (e) this study indicated that if by some accidental spinach was obtained with a malathion residue, the residue could be significantly reduced. In most cases, a single washing reduced the malathion residue below the FDA tolerance. This means that the processor has a means for reducing residues and assuring the consumer of a wholesome food supply.

Robert *et al.* (2009) investigated the role played by lutein and zeaxanthin present in spinach in eye and skin health. They concluded that lutein and zeaxanthin may provide significant protection against the potential damage caused by light striking the retina. In the eye, these have been shown to filter high-energy. Wavelength of visible light and act as anti oxidants to protect against the formation of reactive oxygen species and subsequent free radicals. Studies have also provided evidence of significant efficacy of lutein and zeaxanthin present in skin against light-induced skin damage, especially the ultraviolet wavelengths.

9. Epilogue: Spinach is mainly cultivated for its foliage. Its leaves are rich in protein, aminoacids, minerals and vitamins. It is hence used in cuisines worldwide. It can be employed in various recipes to enrich their nutritional contents and works well in their preparation. It is easily cultivable on different types of soil and environments.

(II) Root Vegetables

Root crops (earth vegetables) are all the forms in which food is stored in the swollen underground parts such as true roots or modified stems like tubers, corms, and bulbs. All these structures are well adapted to storage because of their protected position. They are also the world's chief sources of energy. Root crops (potatoes, sweet potatoes, yams and cassava) provide about 8% of the total human energy intake compared to 20% for cereal grain. Most root crops contain about 25% starch or other carbohydrates. They have, however, a much greater calorie value per unit area of land than cereals.

Since the ancient times, primitive people have always made use of native roots, bulbs and tubers. Although not closely related, these plants have much in common. The majority are of tropical origin, although the potato comes from the highlands whereas the others are lowland species. They are all propagated by vegetative means rather than by seeds, Many of them have practically lost the ability to reproduce by means of seeds, probably because they have been propagated vegetatively for longer past. Archaeological evidence for most of them is scanty as they grow in locations where plant material is less likely to be preserved than in drier regions. They contain much carbohydrates but all are most deficient of proteins. In general, root crops thrive best in a relatively cool climate and in a deep friable soil and have similar cultural requirements. They need little care after planting and still giving good yields. Some of them can be left in the soil without undergoing deterioration until they are to be consumed. Once harvested, they have a poor keeping quality. The handling, preservation and transportation of fleshy root crops is generally less satisfactory.

Major root crops *e.g.* the potato, sweet potato and cassava are the staple food in the diet of many people, others like yams and taro are equally important, mostly in the wet tropics. Root crops typically consumed as fresh vegetables. They are often consumed locally. Bulk shipping is more costly than that of grains and pulses. Many root crops, little known elsewhere are regularly marketed locally *e.g.* Oca (*Oxalis tuberosa* Molina), ullucu or melloco (*Ullucus tuberosus* Caldas), anu (*Tropaeolum tuberosum* Ruiz & Pav.), arracacha (*Arracacia xanthorrhiza* Bancr.), yam bean (*Pachyrrhizus erosus* (L.) Urban), maca (*Lepidium meyenii* Walpers) are certain sedges (*Cyperus*) and aroids.

8.5 AROIDS

Alocasia, Colocasia, Cyrtosperma, Xanthosoma Spp. (Araceae)

Resume: *The edible aroids belong to the araceous genera viz., Colocasia, Alocasia, Cyrtosperma and Xanthosoma. Except Xanthosoma, which is tropical American, other three genera originated in Old World. Xanthosoma is characterized by chromosome number 2x=26, whereas other three have 2=28, 26 and 3x=42. All these aroids are grown for their edible corms, leaves and petioles and consumed only after cooking (boiled, roasted, baked or toasted). They are rich in carbohydrates.*

1. Introduction: In many parts of the tropics, aroids are important food crops for native people. They are grown, for their edible corms, cormels, leaves and petioles. They rarely enter into commerce except in local markets. Most of this production undoubtedly resulted from *Colocasia* and *Xanthosoma* which are used extensively in the Pacific and Caribbean islands and in West Africa. *Alocasia* and *Cyrtosperma* are important subsistence crops in some

Pacific islands. *Cyrtosperma* is the main starch crop of coral atolls. *Colocasia* and *Cyrtosperma* have special importance as crops for difficult lands, for they can produce large yields under flooded or swampy conditions. *Alocasia, Colocasia* and *Xanthosoma* are mainly grown as upland rainfed crops. Few processed products are made from them, athough *Colocasia* flour has been marketed successfully in Hawaii and India. Their food value equals to that of most root crops. Most aroids contain calcium oxalate raphides in their corms and leaves. These irritate the mouth and throat. Some varieties are less acrid than others. There is hardly any breeding programmes for aroid development in past.

2. Plant Characteristics:

(i) *Alocasia:* A stout large, suberect rootstock, stout thick stem, 1-2 m tall; leaves large broad-sagittate, triangular-ovate, with conspicuous midrib; inflorescence spadix, spathe 15-30 cm long, pale yellowish-green.

(ii) *Colocasia:* A cormatous plant growing into 2-3 m in height; leaves spirally arranged, long petioled and shield-shaped, leaf blades being thick, green, glabrous and nearly rounded; spadix inflorescence containg female flowers at base and separated from the upper male flowers by a sterile region, subtended by spathe.

(iii) *Cyrtosperma:* Plants 3-5m high, robust; rhizome cylindrical-conical, 15-20 kg; leaves 6-8, 1-2 m long, coriaceous, olive-green; inflorescence spadix; fruits globose, orange-scarlet berries.

(iv) *Xanthosoma:* Differs from Colocasia in having sagittate leaves and the terminal appendage absent in the spadix, plant height about 7-8 ft, tuberous.

3. Parts Used: Corms (underground rootstock), leaves.

4. Chemical Constituents:

(i) *Alocasia:* Carbohydrate (16. 9%), protein (0. 6%), fat (0. 1%), calcium, iron phosphorus, riblaflavin, thaminage.

(ii) *Colocasia:* Corm carbohydrates (13. 30%), protein (1. 5-3. 0%), vitamin-C, thiamine, riboflavin and niacin, calcium, phosphrous.

(iii) *Cyrtosperma:* Carbohydrate (1. 8%), protein (0. 5%), fat (0. 2%), calcium, sodium, potassium, magnesium, zinc.

(iv) *Xanthosoma:* Starch (17-34. 5%), protein (1. 3-3. 7%), calcium, thiamine, riboflavin, niacin, ascorbic acid.

5. Uses And Applications:

(i) *Alocasia: (i) A. indica:* Stem and rhizome are boiled and then eaten. Its flour is nutritious food. *(ii) A. macrorrhiza:* Stem is sliced, washed and boiled for before consumption.

(ii) *Colocasia:* The corm *C. esculenta* is boiled and then consumed. It has a sweet nutty flavor. The leaves are used as vegetable and source of vitamin A and C. They are bleached before use. Flour and chips are consumed prepared from it. Cakes are prepared from it. It is also baked or toasted. Flour is used for soups, gravies, gruels and puddings. Taro-lactin and Taro-malt are good for infants.

(iii) *Cyrtosperma:* Starch obtained from rhizome is used for various culinary and industrial purposes.

(iv) *Xanthosoma:* The starchy tubers are nutritious and eaten roasted or boiled. It is cooked in coconut milk and then consumed. It is also sun dried and then canned.

6. Karyotaxonomic Evidence: Edible aroid genera are included in two subfamilies *viz.,* (i) Lasiodeae (*Cyrtosperma* and *Amorphophallus*) and (ii) Colocasoideae (*Alocasia, Colocasia,* and *Xanthosoma*).

Only a few aroids can be considered as economic plants, although some species are harvested for subsistence or emergency food. . As is common with vegetatively propagated ancient crops, the taxonomic position of many species is questioned, Major species alongwith their vernacular name, chromosome number and origin are:

I.	**Asian Origin:**		
i.	*Alocasia indica* (considered by some to be synonymous with *A. macrorrhiza*)	—	(2x=28)
ii.	*Alocasia macrorrhiza*	Known as Giant taro, taamu, ape	(2x=26, 2x=28)
iii.	*Colocasia esculenta* (some authors refer to *C. antiquorum* but Hill, 1939 considered *C. esculenta* to be *polymorphic*)	Known as Taro, eddo, old cocoyam, keladi, arvi, dasheen (W. Indies)	(2x=28, 3x=42)
iv.	*C. esculenta* var. *globulifera*	Known as Dasheen (Pacific and Asia)	(3x=42)
v.	*Cyrtosperma chamissonis* (sometimes called *C. edule* or *C. merkusii*)	Known as Giant swamp taro	--
II.	**South American origin:**		
i.	*Xanthosoma atrovirens*	Known as Tanier, yautia, cocoyam	(2x=26)
ii.	*Xanthosoma sagittifolium*	Known as Tanier, yautia, cocoyam	(2x=26)

iii.	*Xanthosoma violaceum*	Known as Tanier, yautia, cocoyam	(2x=26)

Great confusion exists concerning the taxonomy of edible *Xanthosoma*. Other edible species often include: *X. brasiliense, X. belophyllum, X. caracu, X. jacquini* and *X. mafaffa* (Coursey, 1968; Purseglove, 1972) to be the 'new' cocoyam of West Africa. It may be that, with the exception of *X. brasiliense*, which is grown for its edible leaves, edible *Xanthosoma* should be considered as one polymorphic species (namely *X. sagittifolium*) as most species distinctions are made on vegetative characteristics. It appears that there are probably four major ones: *Alocasia macrorrhiza, Colocasia esculenta, Cyrtosperma chamissonis* and *Xansthosoma sagittifolium*.

7. History: Development of aroids is due to vegetative mutation of chance seed setting and their selection by prehistoric mankind. All varieties known today were selected and propagated by subsistence farmers of Africa, Asia, Oceania and Latin America. The aroids probably merged in swamps or in high rainfall forest areas. Some anthropologists believe that *Colocasia* may have been the first irrigated crop. The ancient 'rice' terraces of Asia were originally constructed for *Colocasia*. Wild aroids were collected for food by ancient man. He observed that non or little acrid types existed and that cast-off plant parts *e.g.* small corms, corm fragments and tops grew readily in waste heaps, in moist camp sites and along streams. Then, the next step of planning and cultivating the plants can be understood. An estimate of 4000 to 7000 years ago seems reasonable period for this fact.

Colocasia, Alocasia, and *Cyrtosperma* originated in the Old World. *Alocasia* was probably native to Ceylon (Shri Lanka) or India. Thence it spread eastwards to Oceania to become a crop in parts of Micronesia, Polynesia and Melanesia. *Colocasia* originated in Indo-Malaya (Plunknett *et al.,* 1970). Ethnobotanical evidence warrants India as the place of origin (Spier, 1951). It spread eastwards into Southeast Asia, eastern Asia, including China and Japan, and the Pacific Islands. It was noted as an important food crop in China about 100 BC. (Plunknett *et al.,* 1970). It moved westwards with megalithic people to Arabia, the eastern Mediterranean and Egypt where it was known for over 2,500 years. From there it was spread, perhaps more than 2,000 years ago along the coast to East Africa and across the continent to West Africa by voyagers, explorers and traders. *Colocasia* is known as 'old coco yam' in West Africa. This suggested that it was present there before the introduction of 'new' cocoyam, *Xanthosoma sagittifolium*. *Colocasia* was carried to the Caribbean and tropical America on slave ships as a source of food.

Cyrtosperma originated in southeast Asia, perhaps in Indonesia. It spread throughout the Malayan Archipelago and eastwards to the Pacific and become a minor crop in Melanesia and parts of Polynesia, but a major food in Micronesia.

A tropical American plant *viz., Xanthosoma* was developed by Amerindian people of the continent and the Caribbean Islands. During the slave-trading era, *Xanthosoma* was, taken to Africa where it became known as 'new' cocoyam. Today in West Africa, *Xanthosoma* is more important than *Colocasia*, ranking third after cassava and the yam. During the middle and late 19th and early 20th centuries, *Xanthosoma* was spread throughout the Pacific and into Asia. It achieved minor crop status there because of its resistance to disease and pests (Massal and Barrau, 1955).

8. Further Developments: There has been no role in the development cultivars of aroids by the plant breeding programmes. Cultivars in use are the products of indigenous agriculturists. It has been suggested that attempts be made to hybridize Xanthosoma and Colocasia in order to transfer disease resistance to Colocasia. Some decline because of diseases such as *Phytophthora colocasiae* and certain viruses has been experienced in *Colocasia*.

Colocasia breeding has been attempted in Japan and Hawaii (Plucknett *et al.,* 1970). Seedlings from natural and cross-pollination studies were grown and observed in Japan and Hawaii. No new varieties resulted from this work.

Colocasia and *Xanthosoma* are worth crops for improvement through breeding. Studies of flowering and chance seed-set in them could be fruitful in providing a basic for plant breeding. Many differences are noted between cultivars w. r. t. corm colour, corm quality, disease susceptibility, ability to grow under varying moisture regimes, including flooding; soil or water salinity levels, and in edibility of corms or cormels.

9. Epilogue: Aroids are often considered either static or declining crops. This may hold true for *Alocasia* and *Cyrtosperma* and perhaps, *Colocasia* but not for *Xanthosoma. Xanthosoma* is increasing in some Pacific Islands, West Africa and elsewhere. The future for the aroids depends largely on four factors *viz.,* disease and pest resistance of cultivars, food yield and quality, their role in more efficient land use and their importance as emergency, reserve or famine crops.

Colocasia and *Xanthosoma* seem to be most suitable for wider use. Some Colocasia cultivars are resistant to flooding and salinity and are not easily destroyed by storms. The starch is easily digestible. The corm proteins are non-allergenic. It can, therefore, be made into specialty foods for allergic infants, hospital patients and persons with gastric disorders. Disadvantages, of some cultivars, include: (i) acridity of corms and leaves, (ii) susceptibility to certain fungal and virus diseases, (iii) and a relatively long crop life of 9-15 months. It is a subsistence crop under shifting cultivation and a commercial crop under irrigated or rainfed conditions.

Xanthosoma is more drought resistant and somewhat more shade tolerant than *Colocasia*. It is a successful intercrop or shade crop and as well as an important crop in shifting cultivation. Its main advantages are remarkable resistance to pests and diseases and high yields under difficult conditions are remarkable. *Xanthosoma* and *Colocasia* are complementary crops since they are adapted to different agricultural regimes.

Alocasia has several disadvantages *e.g.* some cultivars are very acrid, corms are often coarse, crop life is quite long (2-4 years), and it grows in upland areas suited to upland *Colocasia* or *Xanthosoma.* The crop is but exceptionally hardy, requires little care and yields are high.

Cyrtosperma fills a special niche in the array of food plants. It is advantageous because of extraordinary hardiness, low maintenance requirement, ability to grow in stagnant, often brackish swamps, and potential as an emergency or famine food. Long crop duration, as much as four years or more, of course, is a disadvantage.

Concerted efforts are needed to collect, preserve and study aroid cultivars around the world. Many clones are being lost each year. Those surviving cultivars represent a legacy of uncounted centuries of activities of subsistence farmers. They saw special virtues in these forms and preserved them. Some of these could be used selecting desirable traits to help solve our food problems.

8.6 CARROT

Daucus carota L. (Apiaceae)

Resume: *Most wild forms of carrot are denizen of Europe and southwestern Asia. Carrot is derived from the wild carrot or 'Queen Anne's lace (D. carota L. subsp. carota) which occurs naturally in Europe, Asia and also in Africa. The Greeks and Romans at first used them for medicinal purposes. It has been brought under cultivation recently before 2000 years. It is a cool weather crop and has 2n=18 chromosomes. The cultivated D. carota intercrosses freely with wild forms. It is consumed raw or cooked. It is a well-known source of carotene (precursor of vitamin-A).*

1. Introduction: It has been under cultivation for over 2000 years. The Greeks and Romans previously used it for medicinal purpose. In recent times, it became a widespread human food and cultivated as popular vegetable worldwide. It reached Europe early in the Christian era. It was introduced in Virginia in 1609 and New England in 1629. The Indians carried it over to the rest of America. It is new grown all over the world. It is a well-known source of carotene, precursor of vitamin-A, following industrial synthesis of carotene. In recent times, cultivation of carrot exclusively for this purpose has declined.

2. Plant Characteristics: An erect, biannual plant producing a thick conical tap root; radical whorled leaves, long-petioled and pinnately compound; inflorescence a terminal compound umbel containing small, white flowers.

3. Parts Used: Tap-root.

4. Chemical Constituents: Carotene.

5. Uses And Applications: Carrots are consumed raw or cooked into vegetable. They are used to flavor stews, curries, soups, etc., besides its use as salad and in sweetmeats. Juice extracted from it is also used as a drink. They are pickled and used to prepare jam. Some varieties of it are used for the preparation of a beverage, called 'Kanji' in India. It is supposed to be a good appetizer. Carrots are canned or dehydrated. Carotene concentrate is used to colour butter and other foods.

6. Karyotaxonomic Evidence: Cultivated carrots and the wild ones are known from Europe. They all belong to the species *Daucus carota* witn 2n=2x=18. Most wild *Daucus* forms inhabit southwestern Asia and the Mediterranean, a few in Africa, Australia and America. All wild forms investigated in Asia, Asia Minor, Japan and the USA have the same chromosome number as their European relatives. Polyploidy and structural changes in the chromosomes seem to have played no role in the divercification of the species (Whitaker, 1949).

The commonest wild from *viz., Daucus carota* spp. *carota* is found in Europe and southwestern Asia. The ssp. *maximus* is noticed from the Mediterranean and eastwards to Iran while *maritimus, commutates, hispanicus* and *gummifer* flourish in the western part of the Mediterranean. Likewise, *fontanesii* and *bocconet* grow in the central part, and *major* in the Balkans.

The ssp. *carota* found in Afghanistan, is rather variable in many morphological characters (Mackevic, 1929). The roots differ in degree of fleshiness, ramification and colour. Some are white and some coloured depending upon anthocyanin content. The cultivated *D. carota* freely intercroses with ssp. *carota* and also probably with all or most of the other wild forms. It is an outbreeder. Incompatibility has not been observed. Selfing of individual flowers or umbels is naturally prevented by protandry.

7. History: The western carotene carrot is a derivative of the eastern anthocyanin carrot. The latter has probably been derived from anthocyanin-containing forms of *D. carota* ssp. *carota* such as those found in Afghanistan. Forms with smoother, fleshier less forked and better coloured roots appear to have been selected to provide the earliest cultivars. Their root may have been shades of purple or violet, from light to almost black, and their shapes generally long-conic. Afghanistan is the centre of diversity of the anthocyanin carrot. This country is inferred as the primary centre of origin (Mackevic, 1929).

The cultivated anthocyanin carrot, in most cases accompanied by a yellow mutant devoid of anthocyanin, spread westwards and eastwards. They can be traced in Asia Minor in the 10th and 11th centuries, in Arab occupied Spain in the 12th century, in continental northwestern Europe in the 14th century and in England in the 15th century. Their introduction to China was in the 14th century and to Japan early in the 17th century.

The carrots were all purple or yellow and of long-conic shape. They were growing northwestern Europe before and during the 16th century. The purple type was at first the more popular because of its milder taste. However, when cooked, the root exuded anthocyanin which gave soups and sauces a brownish purple colour. The yellow type, therefore, became more preferred. In The Netherlands, about 1600, a more orange coloured type from the yellow was selected. The orange colour was gradually intensified and, early in the 17th century, the Long Orange established. From this variety, the finer Horn carrots were selected by 1763 differentiated into three orange varieties which differed in earliness and size *viz.,* (i) the Late Half Long (the biggest) (ii) the Early Half Long and (iii) Early Scarlet Horn (the smallest) All varieties of the western carotene carrot are developed from these four closely related varieties (Banga, 1963) either by simple selection or by intercrossing different types. There are also white carrots which probably originated as mutants of the yellow. They had a limited popularity but are of little importance.

There has been significant change in photoperiodic reaction. When plants adapted to subtropical latitudes are grown in the northern summer, they respond to long days by bolting before the root has thickened properly. Long-day tolerance must have been a consequence of prolonged rogueing of bolters. Simultaneously, selection must have favoured types with increased sensitivity to vernalizing temperatures, since after storage in a cool place during winter, such types flower and seed more readily.

8. Further Developments: The methods previously employed were mass-selection or combined mass-pedigree selection. The outbreeding character of the crop and consequent inbreeding depression forbade the use of inbred lines and the development of a high degree of uniformity. Uniformity needed mechanical harvest and handling. This encouraged the adoption of the F_1 hybrid approach, based upon male-sterile lines. By the use of a suitable cytoplasm and one or more genes, it has proved possible to facilitate production and maintenance of hybrids. The use of hybrids was started in USA. They were not dominant but their share slowly increased. In Japan, some hybrids are in commercial production. In Europe, they are in an advanced experimental stage. The objectives of carrot breeding was largely concerned with yield, shape, colour, earliness, tap-root ratio, resistance to bolting and quality characteristics.

Yielding capacity depends on factors such as root length, root shape, dry matter content, resistance to unfavourable conditions and pathogens. A cylindrical root weighs more than a conical one of the same length. The latter is but less readily affected by unfavourable growth conditions. The popular varieties are conical. A fairly high dry matter content is preferred as this favours good storage and transport characters.

Earliness can be achieved either by smaller size or higher growth rate or both. In the series: Amsterdam Forcing-Nantes-Berlikum, the first variety is the smallest and the earliest. On the other hand, the orange (but annual) North African variety, Muscade, is larger but nevertheless earlier than Amsterdam Forcing because of its higher growth rate.

The length of this phase is longer in types with large tops and strong taproots, shorter in those with small tops and weak taproots. This led to the evolution of huge-top varieties in continental climates and of small-top varieties in more maritime environments. High carotene types with two or three times the ordinary carotene content were bred in the UK between 1940 and 1945 to minimize night blindness among aviators. As carotene is now synthetically manufactured, breeding programmes are aimed at an even colouring throughout the whole root in both core and flesh and tenderness.

9. Epilogue: The genetic base of modern carrots is narrow. Eastern genetic material could be used to widen the base. There may be opportunities to use annual material to improve earliness under short day-lengths. Resistance breeding, already begun in some programmes, will surely be extended.

8.7 ONION AND ALLIES

Allium Spp. (Liliaceae)

Resume: *Except chives, cultivated species of the genus Allium are believed to have come from the eastern Mediteranean region. Onion and garlic especially formed part of the daily diet of the labourers who constructed the pyramids of the Pharohs. Onion are cultivated at least 4800 years ago as they their pictures are carved on Egyptian monuments. They were eaten during pre-biblical period. Old world Allium species have basic chromosome number x=8, whereas North American species have x=7. Their bulbs or leaves are eaten as vegetable, spice or in various culinary uses.*

1. Introduction: The onion is believed to have originated in Southern Asia or the Mediterranon region. It was cultivated at least 4800 years ago as evidenced by its pictures carved on Egyptian monuments. Onion and garlic were given to the workmen who built the great pyramids. Onions were employed in religious and funerary offerings and also found in mummies. They were consumed by the Israelites before the time of their exodus from Egypt and are recorded in the Bible, in which it is mentioned that the Israelites complained of their hardships because Moses did not bring onions for food. Druids used them in their ceremonies about 2000 years ago. The American Indians were unknown about onions until introduction by the Spaniards and Portuguese. In Medieval Europe, garlic was extensively used to camouflage the smell and flavor of salted meat and fish.

2. Plant Characteristics:

(i) Onion: A biennial crop, storing food in the bulb, root system shallow and fibrous; leaf consists of two main parts *viz.*, a sheating leaf base and a hollow, linear, cylindrical or flattened blade; bulb consists of a short plate-like stem surrounded by several concentric layers of fleshy leaf bases; small, white flowers borne in umbellate head; fruits globular capsules.

(ii) Garlic: Resembles the onion except flattened solid leaf blades and a composite or compound bulb consisting of several small densely crowded, angular, tunicate bulblets (cloves).

(iii) Leek: A robust perennial with rather thick flat broad leaves and small bulbs.

(iv) Chives: Hollow and cylindrical leaves with very small clustered bulbs.

(v) Shallots: Cylindrical hollow leaves with large clustered bulbs; leaves are short and awl-shaped.

3. Parts Used: Bulbs (when present) and leaves.

4. Chemical Constituents:

(i) Onion: Protein (1. 2%), fats (0. 1%), carbohydrates (11. 6%), minerals (calcium, phosphorous, iron), vitamin A, B and C.

(ii) Garlic: Protein (6. 3%), fat (0. 1%), carbohydrate (29. 0%), minerals (calcium, phosphorous, iron), vitamin-C.

(iii) Leek: Protein (1. 8%), fat (0. 1%), carbohydrates (17. 2%), minerals (calcium, phosphorous, iron), vitamins A, B and C.

(iv) Chieves: Protein (0. 2%), fat (0. 2%), carbohydrates (0. 5%), vitamin A, K, C, minerals (Ca, K, Mn, P).

(v) Shallots: Protein (1%), carbohydrates (1%), vitamin A, C, B6, folate, minerals (Ca, Fe, Mg, P, Na).

5. Uses And Applications:

(i) Onion: Mature and immature bulbs are eaten raw as a salad. They are used as fries, boiled and roasted. Onions are extensively used for flavouring soups, ketchups, canned meat products, sausages, sauce, stews, curries and other vegetable dishes. They are used in the form of pickled and chutneys. Processed products like dehydrated onion, in vinegar or brine are now available.

(ii) Garlic: It is generally used as a flavouring agent rather that a vegetable, especially in soups, stews, pickles and salads. Processed products like dehydrated garlic in a pulverised or granular state are available for home and industrial uses.

(iii) Leek: The bases of leaves are mild flavoured and edible. They are often blanched like asparagus and also used in seasoning.

(iv) Chieves: The young leaves and bulbs are used as condiment and also for seasoning.

(v) Shallots: Bulbs are used for pickling.

6. Karyotaxonomic Evidence: Most of the North American species of *Allum* have a basic x=7, whereas the Old World Alliums are predominantly x=8 (Traub, 1968). The domesticated Alliums, except chives, come from the Near East or from central or eastern Asia. They all have basic chromosome number x=8. The common onion, *A. cepa* (and also shallot, multiplier and potato onion) has 2n=16 and is known only in cultivation, with its primary gene centre being Afghanistan. The closest related species (*A. vavilovii, A. oschaninii, A. pskemense* and *A. galanthum*) are wild diploids (2n=16) of central Asia. *Allium fistulosum* 2n=16, the Japanese bunching or Welsh onion, is not known in the wild but the related *A. altaicum* (2n=16) is a wild plant of northern Mongolia and adjacent Siberia. The chive, *A. schoenoprasum*, with 2n=16, 24, 32, is a variable and widespread species from the Arctic to far southwards in the mountains of North America, Europe and Asia. Garlic, *A. sativum*, known only in cultivation is diploid (2n=16). The diploid *A. longicuspis* is thought endemic to central Asia, to be its wild ancestor. The extremely variable *A. ampeloprasum* (2n=16, 24, 32, 40, 48) grows as a wild plant from southern Europe and north Africa through the Middle East to western and southern USSR. Two cultivated forms of *A. ampeloprasum*, garden leek (syn. *A. porrum*) and the similar kurrat (syn, *A. kurrat*), are tetraploid with 2n=32, whereas third form, great-headed garlic, is hexaploid. Rakkyo, *A. chinense*, a denizen of central and eastern China and cultivated throughout eastern Asia, is 2x, 3x and 4x. The Chinese chive, *A. tuberosum*, a wild and cultivated plant throughout eastern Asia, is 2x and 4x.

7. History: The wild *A. cepa* collected by 19th century botanists in western Asia was more likely one of the related species that range from northern Iran and Afghanistan, northwards into the central Asiatic region of the USSR (Wendelbo, 1971). According to Vavilov (1926), the primary centre of origin of onion and garlic is central Asia and of the leek the Near East and Mediterranean. The Near East is a secondary centre for onion and the Mediterranean a secondary centre for garlic and the large-size types of onion.

References to onions, garlic and leek as food, medicine or religious objects can be traced back to the 1st Egyptian Dynasty (3200 BC.) and to the biblical account of the exodus of the Israelites from Egypt (1500 BC.). Introduction into Egypt must have occurred much earlier. The use of onions and garlic in medicine in India in the 6th century BC. can be inferred from later Indian writings. By the time of the Greek and Roman authors, from Hippocrates (430 BC.) and Theophrastus (322 BC.) to Pliny (AD. 79), several onion cultivars were named and described as long or round; white, yellow or red; mild or pungent.

Cultivars evolved more and diversified w. r. t. shape, colour, flavour, keeping quality and critical adaptations to cultivation in new climates and environments. The modern onion is self-compatible. It is an outbreeder and is subject to inbreeding depression. It appears to have retained the breeding system of its early ancestors and evolved independently of contacts with other species for several thousand years. Experimental hybrids with related wild species (*A. galanthum, A. oschaninii, A. pskemellse*) are highly sterile.

In China and Japan since prehistoric times, *A. fistulosum* has been the main garden onion. It is used for edible tops or long leaf bases. It is not bulbous, It developed an array of cultivars in Southwestern Asia and Europe. The centre of origin of *A. fistulosum* is considered as central and western China. *Allium fistulosum* is an outcrosser and is self-compatible. It crosses with *A. cepa* but the hybrids are highly sterile.

Leek and kurrat, known to the ancient civilizations of the Near East, closely resemble wild *A. ampeloprasum*. They differ mainly in a lesser tendency to form bulbs. Leek was known in Europe in the Middle Ages and is still popular there. Many varieties are selected for long, white, edible leaf-bases and green tops, winter hardiness, and resistance to bolting. Only the green leaves of kurrat, grown in Egypt and other countries of the Near East, are consumed. Leek and kurrat are interfertile. The hexaploid *A. ampeloprasum*, great-headed garlic, is vegetatively propagated by its large cloves or smaller ground bulblets. It flowers but sets few or no seeds.

Garlic is an ancient crop of central Asian origin, known only in cultivation. It is mentioned in ancient Chinese, Egyptian and Greek writings. The early domestication of garlic took quite a different turn from that of the largely seed-propagated leek and onion. It became exclusively vegetatively propagated by cloves and, in those cultivars that still bolt, by inflorescence bulbils. Garlic thus presents problem as to the origin of the many cultivars which differ in maturity, bulb and clove size, and number, scale colour, bolting, scape height, number and size of inflorescence, bulbils and presence or absence of flowers. We do not know how much variation was selected while *A. sativum* or its ancestors were still sexual and when garlic became vegetatively propagated, viviparous and sterile. The presumed wild ancestor, *A. longicuspis*, is itself viviparous and seedless.

Cultivated chives are said to have a Mediterranean origin. They are not recorded earlier than 16th century Europe. They have remained unchanged much from the polymorphic, wide-ranging wild *A. schoenoprasum*. Probably that may have been brought under cultivation many times from wild populations. Both outbreeders and inbreeders are known (Ved Brat, 1965). Chinese chives, *A. tuberosum*, and Rakkyo, *A. chinense* were domesticated since ancient times in the Orient. Chinese chives is probably little changed from the wild. It is an abundant seed producer. Rakkyo is vegetatively propagated by bulb multiplication. Cultivated forms are nearly all tetraploid, sparsely-flowering and seed-sterile. Flower structure indicates that both Rakkyo and Chinese chives are outbreeders.

8. Further Developments: Onions, garlic and leek are vegetables in Europe in the Middle Ages. They were mentioned by Chaucer (about 1340) and written about by many of the herbalists. *Allium cepa* was likely to have introduced in the West Indies by Columbus (1494) and reintroduced to the New World many times. It was cultivated in Massachusetts in 1629. Vilmorin listed 60 varieties in 1883. Further developments took place by mass or individual selection at the diploid level. Involvement with other species has been rare. The relatively unimportant top onion, *A. cepa* var. *viviparum*, may have derived through some hybridization between *A. cepa* and *A. fistulosum*. An amphidiploid *A. cepa* × *fistulosum* ('Beltsville Bunching') was introduced as a green bunching onion superior to *A. fistulosum*. 'Delta Giant' shallot is a backcross of shallot (*A. cepa*) with a *fistulosum* shallot amphidiploid.

Many programs were started to improve horticultural qualities and develop cultivars resistant to pests and diseases (Jones and Mann, 1963). Allium fistulosum was found to be resistant or immune to pink root, downy mildew, smut, yellow dwarf and thrips. However, these resistances remain unexploited in onion breeding because of hybrid sterility and limited genetic recombination, Efforts to use *A. galanthum, A. pskemense* and *A. oschaninii* have met similar difficulties (McCollum, 1974).

9. Epilogue: The methods of simultaneously transferring cytoplasmic male sterility and developing inbred parent lines by backcrossing, progeny testing and selfing will continue hybrid combinations. The efficiency of this process can be improved as more biometrical genetic studies of economic characters and disease resistances are undertaken. Three-way and double-cross hybrids are advised to improve seed yields. Selection of cultivars was for specific purposes w. r. t. mild flavour for the fresh market, pungency and high solids content for the dehydration industry and single centre bulbs for processing as french-fried onion rings. Knowledge of the features *e.g.* volatile flavour, odour and lachrymatory components and of other biochemical constituents can contribute increasingly to quality breeding. Restrictions on the use of chemicals to control pests and diseases will continue the pressure on breeders for cultivars. Resistance to some of them has been found within locally adapted *A. cepa* material. Breeders may have little choice but to introduce unadapted foreign germplasm into their programmes. The resultant broadening of the genetic base could be beneficial because many US programmes have become rather narrowly based, especially since the beginnings of hybrid onion development. Even less use is made of several potentially valuable but difficult-to-use related species. *Allium fistulosum, A.*

galanthum, A. oschaninii and *A. pskemense* cross with *A. cepa* but the hybrids are highly sterile. Accessions in living collections in the USA under the name *A. vavilovii* are interfertile with *A. cepa* but their identities and origins need checking. *Allium farctum* (2n=16), described by Wendelbo (1971) from West Pakistan and Afghanistan, is a new species for Section Cepa. Its crossability with *A. cepa* can be determined. There is one report that *A. roylei* (2n=16) from the western Himalayas can be crossed with *A. galanthum*.

Little change is foreseen in the breeding of the other Alliums. Genetic male sterility discovered in Japan could be used for hybrid seed production of *A. fistulosum* by rogueing of fertiles and vegetative propagation of the sterile clones. Such use of genetic male sterile leek has been proposed by Schweisguth (1970).

8.8 POTATO

Solanum tuberosum L. (Solanaceae)

> ***Resume:*** *Potato or Irish Potato is a native of the highlands of Peru and Bolivia. It was grown as staple food by the Incas and represented on pottery designs recovered during archaeological excavations in the graves as early as 800-1500 AD. It was unknown to the Old World before the 16th century. Potato is the only staple food derived from tubers, other are all grains. It is widely used throughout the world in various forms for human consumption. It is particularly rich in starch.*

1. Introduction: Potato is a native of the highlands of Peru and Bolivia, probably the high plateau neighbouring the lake Titicaca. It was grown as staple food by the Incas. It is represented in designs on early pottery recovered during archaeological excavations of prehistoric graves dating as early as 800-1500 AD. Each character of potato is represented on these potteries. This is an evidence of importance of the potato in the lives of early Andean people. Ever since, it had dominated the Andean region of South America and still continues to be so.

It was unknown to the Old World before the 16th century. It was also unknown in North and Central America at the time of discovery of America by Columbus. The potato was introduced into Europe soon after 1580 by the Spaniards probably from the northern Andes. It spread over entire Europe by the end of the 17th century. The potato was introduced independently to England in 1586 and it was Sir Walter Raleigh who introduced it into Ireland from England. Sometimes potato is called 'Irish Potato', this should not mean that Ireland is its native. In India, during 17th century, it was introduced by the Portuguese explorers. Its acceptance in Europe was not immediate. It was thought to be an aphrodisiac. There was also violent opposition to potato in Europe because of its relationship to the poisonous mightshade family, the

Solanaceae. It achieved recognition as a crop of commercial importance in the latter half of the 18th century. To compel its plantation, royal edicts were issued in Germany and Sweden in 1744 and 1764 respectively.

2. Plant Characteristics: A branched, erect annual, reaching 2 to 3 ft in height; leaves pinnately compound; fine fibrous roots and stolons form tubers at their tips; flowers white, yellow or purple in terminal clusters; fruits spherical berry.

3. Parts Used: Tubers.

4. Chemical Constituents: Carbohydrates (10-30%) (mainly starch and little sugar), protein (1-3%), fibre (2-3%), fats (0.1%), vitamin-C, minerals (potassium, phosphorous, iron, magnesium).

5. Uses And Applications: Tubers are consumed as a fresh vegetable in various forms *e.g.* boiled, steamed, fried, baked or roasted. They are used after processing *e.g.* chips, crisps, dehydrated mashed potatoes, potato flour, frozen French-fries and canned potatoes. A Russian alcoholic drink, called Vodka, is prepared by fermenting cooked potatoes. Potato starch is used for sizing cloth and paper.

6. Karyotaxonomic Evidence: The tuber-bearing Solanums are classified in Subgenus Pachystemonum, section Tuberarium, subsection Hyperbasarthrum (Correl, 1962). About 170 species are included, of which fewer than ten are relevant to the evolution of the crop. The basic chromosome number is x = 12. Among the wild forms rather less than half occur in north and central America. They include diploids, allotetraploids and allohexaploids (2n=24, 48, 72). These have marginal bearing on cultivar history as a few having been used in potato breeding. The majority of species are South American and most of these are diploids. It is from a few of these diploids (2n=24) that the cultivars derived one of the rare tetraploids (*Solanum acaule*).

Amongst the wild potatoes is the contrast between very variable, outbred diploids and the less variable inbred polyploids. The former have a gametophytic S-allele incompatibility system and are highly intolerant of inbreeding. The polyploids are self-compatible and often self-pollinated. The polyploids may have been selected at the limits of geographical and altitudinal range because of reason of their potential for inbreeding.

7. History: Potato tubers of wild species are bitter and contain steroidal alkaloids. The first step in the selection at the gathering stage of clones that were less bitter vis-à-vis less toxic than usual. The emergence of alkaloid-free diploids which could safely be consumed was the first step occurred is unknown. The general historical-archaeological evidence suggests the period 2000-5000 BC., concurrently with the domestication of the *llama* (Jensen and Kautz, 1974). The wild species concerned belong to series Tuberosa *viz., S. brevicaule, S. leptophyes, S. canasense, S. soukupii* and *S. parsipilum*, etc.

Related wild species hybridize easily with each other and with cultivated diploids. Several wild forms could have been derived before or after the primary domestication (Ugent, 1970). The cultivated diploids developed a great deal of new variability in foliage and tuber characters. Although they have been classified into a number of species, they are all readily intercrossed. They can be treated simply as 'cultivated diploids'. They retain the self-incompatibility and outbreeding habits of their putative ancestors (Dodds, 1965).

The region of domestication may be considered where in the wild as well as cultivated diploids are still growing. The region may be the high plateau of Bolivia Peru, in the general region of Lake Titicaca. The diploids are still there but have been largely superseded by cultivated tetraploids and a few triploids. The tetraploids are referred to as *Solanum andigena, S. tuberosum* subsp. *andigena* or Group Andigena. They are autotetraploids, though segmental allopolyploidy is favoured by some workers (Howard, 1970). They exhibit great range of variability of habits, leaf characters, and tuber shapes, colours and sizes as their diploid ancestors with which they share common S-alleles (Dodds, 1965). Andigena potatoes are self-compatible but intolerant of inbreeding and often rather infertile.

8. Further Developments: The first recorded European contact with the potato was in 1537 in the Magdalena valley. The Spanish invaders became familiar with the crop and it was probably about 1570 that a Spanish ship first introduced potatoes to Europe. From Spain, potatoes were widely spread round Europe by the end of the century. There were few initial introductions. They were all Andigena types. Probably the first introduction to Britain in about 1590 was independent of the earlier Spanish introduction. The first potatoes to reach North America came from Europe about 1621. A source in the northern Andes for the first introduction to Europe appears very likely. (Salman, 1949; Hawkes, 1967: Simmonds, 1969).

The potatoes of the central Andes were adapted to the prevailing short days of those latitudes. They developed tuber very late or not at all in the long days of a north temperate summer. Andean potatoes were therefore ill-adapted to Europe. It was nearly 200 years before the crop began to have any significant agricultural impact in its new home. By the late 18th century, clones adapted to long days. This was accomplished by selection for earliness and size of crop among seedlings raised from open-pollinated berries.

In South America the Chilean tetraploids, a local offshoot of the Andigena Group, also adapted to long days. A Chilean variety, however, was probably important in a later phase of potato breeding.

The 19*th* century in Europe and North America witnessed rapid and sustained progress in potato breeding. The products (*S. tuberosum* subsp. *tuberosum*, Group Tuberosum) differed from their Andigena ancestors in day-length response and smaller haulms and fewer flowers, larger leaves, shorter stolons, fewer, larger and smoother tubers, less pigmentation and enhanced disease resistance. Potato blight (due to *Phytophthora infestans*) was a significant episode from the 1840s. Besides the tremendous social and historical impact of the disease it imposed a major new selective factor on the potatoes of the time. Some resistance became available in the 1870s (Salaman, 1949).

The modern potatoes emerged around the end of the 19th century. They were all founded genetically upon the original Andigena introductions supplemented by later accessions. The first interspecific crosses in potatoes were tried about 1850. Wild species were not used in potato breeding. The procedure adapted was to cross cultivars by species having desirable characters and backcross to cultivars. By this means breeding stocks have become somewhat introgressed by: *S. demissumi* (race-specific hypersensitivity to blight); *S. stoloniferum, S. chacoense, S. acaule* (resistances to viruses); *S. multidissectum, S. spegazzinii, S. kurtzianum; S. oplocense* and *S. vernei* (resistances to eelworms). In the late 1960s, only four varieties out of 30 bore any genetic contribution from wild species (all from *S. demissumi*) and the leading varieties were all pure Tuberosum. The average age of the leading varieties was over 50 years (Howard, 1970).

The objectives of disease resistance dominated breeding programme. Of the diseases; blight received the largest share of the effort by the use of R (hypersensitivity) genes from *S. demissum*. General failure ensued because the fungus rapidly evolved new specificities. Even multiple R-genes proved to be ineffective. By the late 1950s, it was shown that the use of polygenic 'field-resistance' was feasible and most potato breeding programmes now ignore R-genes.

9. Epilogue: Potato breeding has almost entirely been directed towards the needs of temperate countries. The last few decades seen a trend towards widening still further the range of adaption of the crop by extending it, first, to low latitudes at middle elevations and second, perhaps, even to the lowland tropics (Simmonds, 1971). Basic materials have so far mostly been Tuberosum stocks of north temperate origin. The breeding potential of Andigena and Neo-tuberosum in these circumstances have been considerable.

It is advisable always to consider the potential of diploids and dihaploids (2n=24). The latter have been produced in quantity from tetraploids pollinated by diploids (Peloquin *et al.,* 1965; Howard, 1970). They represent, in effect, a sample of female gametes. A autotetraploid outbreeder, are weak and infertile but crosses with cultivated diploids can yield remarkably vigorous and productive seedlings. Dihaploids, will have their uses in synthesizing special parents.

8.9 RADISH

Raphanus sativus L. (Brassicaceae)

Resume: *The area of maximum diversity from the eastern Mediterranean to the Caspain is the likely area of origin of the cultivars of radish. It was grown by the ancient Egyptians, Assyrians, Greeks and Romans. It is a popular root vegetable. Apart from its fleshy root, leaves and pods are also used as vegetable. It has chromosome number 2n=18 and intercrossed with its four types and related wild species.*

1. Introduction: There are four different types of radish depending upon characteristics such as weather in which it grows, economic significance, utility of part, development and size of radish. It was grown by the ancient Egyptians, Assyrians, Greeks and Romans. (i) The small short-season type of radish is a cool-weather plant. It is a popular garden vegetable. Its culture is of commercial interest, especially early production under glass or large scale production in the field. (ii) The larger radish has a wider range of temperature adaptation. Early varieties grow well in cool weather. Summer varieties can stand, or even need, relatively high temperatures. The larger type is an important vegetable in Japan, Korea, China, India and other eastern countries. It is be eaten raw or cooked or preserved by storage, pickling, canning or drying. The fleshy root, the leaves and the young seed pods provide vegetables. In Europe, half a century ago, the larger radish was used to be eaten more than nowadays. It is but popular in parts of southern Germany. (iii) Mougri-radish do not develop fleshy roots. It is grown in southeast Asia especially for its leaves and young seed pods. The pods may become 20-100 cm long. (iv) Fodder-radish also bears little or no fleshy root. It is also grown for its foliage which is used as fodder or green manure. Its rapid growth has given it a limited but promising popularity in western Europe during the last quarter century.

2. Plant Characteristics: An annual or biennial bristly herb of about 20-100 cm height; leaves in radical rosette, lyrate and pinnatifid; flowers small, white to lilac, cruciform, borne on inflorescence axis; fruit fleshy siliqua of about 3-7 cm length.

3. Parts Used: Root, leaves, green pods.

4. Chemical Constituents: Pectin, protopectin, volatile isothiocynates (mustard oil glycosides).

5. Uses And Applications: The fleshy roots are used as salad. The leaves are cooked as vegetable. The green, tender fruits are consumed in India as a vegetable. In Japan and China, the root is generally pickled in brine. In France, radish is eaten with butter.

6. Karyotaxonomic Evidence: The aforesaid four types of radish all belong to the single species *Raphanus sativus* with 2n=2x=18. Botanically, they are known as varieties *radieula, niger, mougri* arid *oleifera* respectively. Recent botanical nomenclature can be equated with the above-mentioned varieties respectively as (i) *R. sativus* var. *radicula*, (ii) *R. sativus* var. *longipinnata*, (iii) *R. sativus* var. *caudatus* and (iv) *R. sativus* var. *esculentus*. All four types intercross freely among each other and also with related wild species. The most important wild species are: (i) *Raphanus raphanistrum* (n=9). Europe eastwards to the Volga and Mediterranean eastwards to the Caspian, (ii) *R. maritimus* and (iii) *R. landra*, which is considered to be an inland form of martimus (n=9). Mediterranean eastwards to the Caspian and, in Europe, along the coasts of France, Belgium, Holland and Great Britain; and *R. rostratus* from Greece eastwards to the Caspian.

All three wild species mentioned are variable. No one of the wild species can be claimed as the source of the cultivated radish. Possibly, all three contributed to its development. The area of maximum diversity from the eastern Mediterranean to the Caspian, is the likely area of origin of the cultivars. Werth (1937) has but argued for a more easterly source. Crossing of *Raphanus sativus* with *Brassica species* and *Sinapis arvensis* warrant a fairly close relationship between the three genera, However, there is no evidence that such crossing has played a role in the evolution of the crop.

7. History: In Egypt (about 2000 BC.), inscriptions on the inner walls of pyramids tell us that the *niger* type of radish was an important food. It has been inferred that it was already used so by about 2700 BC. (Becker, 1962). It is thought to have spread to China about 500 BC. and to Japan about AD. 700 (Sirks, 1957). This suggests that radish was well known in the eastern Mediterranean at least two millennia ago. It was known in China later.

Its variability decreases from Europe to China and from there to Japan (Sinskaja, 1928). It is doubtless that Europe adopted much of the Mediterranean material. Some area east of the Mediterranean was the probable source of the crop. The radicula type must be much younger than the niger type. A long white form of it first appeared in Europe at the end of the 16*th* century. Its origin is unknown. Its origin from niger type seems reasonable (Helm, 1957). In the 18*th* century globular forms were developed. The white form preceded the red one. There was great variation in shape and colour. There were long, half-long, globular, pear-shaped, even flattened roots coloured yellow, black, white and red.

8. Further Developments: Radishes are normally self-incompatible and entomophilous. In the sporophytic incompatibility system the papillae on the stigma surface form the barrier to pollen tube penetration. It is possible to get normal seed set by selfing. Bud pollination is another, and so far more generally used way of doing the same (Roggen and van Dijk, 1973; Dickinson and Lewis, 1973).

Male sterility has been found in Japanese radishes. Ogura (1968) described a type controlled by the interaction of a recessive gene, ms, and S-cytoplasm (Bonnet, 1970). Hybrid varieties can be developed using either incompatibility or male-sterility for the control of crossing. Japanese breeders breed hybrids for the production of niger types.

An important breeding objective was adaptation to growing seasons of different length. Selection of varieties suitable for early culture in spring has thus led to small size, high growth rate and an annual life cycle. Both *radicula* types and the earliest *niger* types are annual. They do not need vernalizing temperatures. They are not good for production in full summer. However, selection for summer, autumn and winter radishes has tended to produce greater sensitivity to vernalizing temperatures and a biennial character. Roots stored during the winter in a cool place flower the following season only when they have had enough cold. (Tsukamoto and Konishi, 1959; Banga and van Bennekom, 1962).

Raising the growth rate to promote earliness may, however, increase sensitivity to 'pithiness'. It revealed first as dry, white patches in the inner tissues (Hagiya, 1952-59). Many pithiness-resistant varieties have been introduced, the first being A. R. Zwaan's Cherry Belle, which has spread all over the world. Lines resistant to *Fusarium oxysporum* f. conglutinans race 2 and *Albugo candida* have been bred by Pound and coworkers (Williams and Pound, 1963) in the USA and lines resistant to virus disease by Shimiru and coworkers (Shimiru *et al.,* 1963) in Japan. Radishes, as noted above, are all diploid. Experimental autopolyploids have been made, especially in Japan and East Germany.

9. Epilogue: Most breeding works were aimed at further adaptation to different growth conditions, at improved resistance to pests and diseases and at improved marketing characteristics. Breeding methods were, until recently, rather simple but the production of F_1 hybrids opened new perspectives. The success of fodder-radish in western Europe as a result of recent German breeding work suggested that this form may have greater potential.

8.10 TURNIP AND RELATIVES

Brassica campestris L. (Brassicaceae)

Resume: *Turnip is a native of Central and Southern Europe. It has been under cultivation in Europe since 4000 years or more. It still grows wild in Europe and Siberia. Turnip and Swede are forms of Brassica compestris and have chromosome number 2n=20. Their swollen fleshy roots are edible, which are rounded or vary in size, shape and colour. These are consumed usually as vegetable and rarely otherwise, particularly in Europe and North America.*

1. Introduction: Turnips (*Brassica rapa* L.) is a native of central and southern Europe. In Europe, it was been cultivated for 4000 years or more and still grows wild in Europe and Siberia. It reached Mexico in 1586, Viginia in 1610, and England in 1628. It's another closely related species is Swede [*B. napobrassica* (L.) Mill.]. It is of recent origin, possibly in the 17th century. It is believed to have originated accidently from a cross between cabbage and turnip, probably in Bohemia. It is extensively grown in Europe and to some extent in North America.

2. Plant Characteristics:

(i) Turnip: A herbaceous, rough biennial with an enlarged root; swollen turnip root consists of hypocotyle, round, flattened or cylindrical flesh generally white; leaves hairy; flowers cruciferous.

(ii) Swede: Differentiated from the turnip by the smooth, thick and non-hairy leaves; and presence of thickened, fleshy, elongated neck, flesh generally yellow, and less commonly white.

3. Parts Used: Roots.

4. Chemical Constituents: Protein (0.5%), fat (0.2%), carbohydrate (7.6%), minerals (calcium, phosphorous, iron, vitamin-A, B and C).

5. Uses And Applications:

(i) Turnip: It is used as a cooked vegetable and is also pickled. It is eaten raw as well or in the forms soups.

(ii) Swede: The roots are eaten boiled and buttered.

6. Karyotaxonomic Evidence: *Brassica campestris* (2n=2x=20) is polymorphic. The group has been classified into a number of subspecies on the grounds of complete inter-fertility (Olsson, 1954). Few genes appear to separate some of the subspecies, particularly the oriental forms.

The wild-type, ssp. *eu-campestris*, is a slender rooted, branching annual. Truly wild *B. campestris* probably still exists today. It is a common weed both in Europe (wild rape) and in North America (field mustard).

The ssp. *oleifera*, turnip-rape, seems closest to the wild-type morphologically and also probably phylogenetically. There are summer and winter varieties generally equated to as var. *annua* and var. *biennis* respectively. The true turnip *viz.,* ssp. rapifera is biennial. The useful part is the storage organ (botanically swollen hypocotyl). The stubble-turnip (Dutch turnip) is a separate class containing both lyrate and strap-leaved forms. There are also quick growing, early maturing varieties used as vegetables.

The ssp. *chinensis* (Pak-choi or Chinese mustard) is a leafy annual of which the young shoots, more or less blanched, are an important vegetable in China. Selection has favoured some extreme forms with greatly enlarged leaf petioles and only a small fringe of lamina. Ssp. *pekinensis* (Pe-tsai or Chinese cabbate) forms distinct heads of leaves. It is used as a salad vegetable in the Far East and elsewhere. Ssp. *narinosa* is a compact form with small, puckered leaves. It is used as a salad vegetable in China. The ssp. *nipposillica* forms rosettes of very numerous leaves. There is a strap-leaved form and variety with highly dissected leaves (var. *laciniata*). Both are used for greens or as pickled vegetables in the Orient. The ssp. *dichotoma* (Toria or Indian rape) is an annual oil-seed form, as is ssp. *trilocularis*, yellow-seeded Sarson. Indian mustard oil is generally obtained from varying mixtures of Toria and Sarson, together with Rai or wild turnip (*B. tournefortii*) another 2n=20 chromosome species It is not inter-fertile with *B. campestris* (Olsson, 1954).

Brassica campestris crosses readily with *B. napus* (2n=4x=38), when the latter is used as female parent. Both spontaneous and artificial amphidiploids (*B. napocampestris*, 2n=58) have been produced (Frandsen and Winge, 1932; McNaughton, 1973; Olsson, 1963) involving bulb-forming, leafy and oil-seed forms of the parental species. These are not proved agriculturally useful. However, there are possibilities of introgression of desirable characters from one species into another by way of such amphidiploids.

Hybridization with *B. oleracea* is difficult. It may be but facilitated by embryo culture (Harberd, 1969). Hybrids are highly sterile but fertility is restored in the amphidiploid *B. napus*. Artificial forms of *B. napus* have been used as a 'bridge' by Japanese breeders for the introgression of disease resistance from *B. oleracea* (cabbage) into *B. campestris* (Chinese cabbage).

Brassica campestris (AA) hybridizes with great difficulty with *Raphanus sativus* (radish, RR, 2n=18). Amphidiploids (AARR, 2n=38) have been reported by Japanese workers (Hosoda, 1947). They are rather infertile.

7. History: Two main centres of origin are indicated: (i) The Mediterranean area is thought to be the primary centre of European forms. (ii) Eastern Afghanistan and the . adjoining portion of Pakistan is considered to be another primary centre, with Asia Minor, Transcaucasus and Iran as secondary centres (Sinskaia, 1928).

When studied on comparative morphological ground, it seems likely that subsp. *oleifera* is the basic cultivated form nearest to the wild type. Vavilov (1926) suggested its origin from *campestris* weeds of older seed crops such as flax. The time and place of domestication are unknown. Old Araband Hebrew names for the crops suggest that it may be somewhere in southwestern Asia in pre-Classical times. According to Appelqvist and Ohlson (1972), rape seed was not cultivated by the Romans and was used only by people who had neither the olive nor the poppy. They also remarked that Sanskrit records show that sarson has been used in India since 2000-1500 BC. This suggests multiple domestication of annual oil-seed forms from the Mediterranean to India perhaps about two millenia BC.

In Europe, cultivation of oil-seed turnip-rape is thought to have started in 13th century. It was important as a source of lamp oil. It was replaced then by petroleum products (Appelqvist and Ohlson, 1972). Presumably, biennial oil-seed forms were the source of the turnips. The botanical, historical and philological evidences point to an origin in the cooler parts of Europe. There are also old Anglo-Saxon, Welsh and Slav names for the crop. The Romans were familiar with turnips in northern France and were probably introduced by them into Britain. In Europe, therefore, the turnip probably long antedates the oil-seed use of the crop.

The stubble-turnip has been selected for very rapid early growth from late sowing. It is a recent development, originating probably, during the great turnip era in Europe especially 15th to 18th centuries. Stubble-turnips are an autumn crop which were usually grown in the (rye) stubbles. There were, however, many variations in their culture. Forms resembling modern Dutch cultivars were being grown in Britain in the early 19th century.

There is no evidence that *B. campestris* was anciently cultivated in China. Rape seed reached Japan from China via Korea as recently as 200 years ago (Appelqvist and Ohlson, 1972). The oriental subspecies are thought to have evolved in China, very probably from oil-seed forms, by selection for leafiness. The ssp. *nipposinica* developed in the direction of increased leaf numbers. There are increase in leaf size and apical bud development in *chinensis* and *pekinensis* leading to head-forming characteristics. These changes were paralleled by forms of *B. oleracea* in Europe.

8. Further Developments: In Britain, in the early 19th century, various forms of turnips, together with swedes (*B. napus*), with which they cross readily, were reported to have been commonly multiplied side by side. As a result, it was not possible to obtain true-breeding strains. Since then, but before modern plant breeding, some progress was probably achieved by mass selection or by mild inbreeding. This was coupled with the realization of the need for adequate isolation distances during multiplication.

Heterosis for dry matter yield of turnips has been demonstrated (Wit, 1966). *Brassica campestris* is an out-breeder with a sporophyric incompatibility system based on S-alleles. The production of F_1 hybrid varieties would, therefore, be a possibility, In Holland, difficulty of maintaining inbred lines, due to Increased susceptibility to disease and reduced winter hardiness, has been a deterrent to the commercial production of hybrid stubble-turnips. Synthetic varieties ever, outyielded the best commercial cultivars (Wit, 1966). Hybrid varieties of the leafy oriental salad vegetables have been available in Japan for some years. All cultivars listed in seed catalogues are hybrids. Most of them have resulted from utilization of the incompatibility system.

Tetraploid stubble-turnips are commercially available. They are produced by Dutch breeders. They reported to yield dry matter yields marginally superior to the diploid strains from which they are derived. One such cultivar, Sirius, produced in Sweden by inter-crossing tetraploid strains from various sources, has given dry matter yields up to 14% higher than the best commercial control. This gain is probably attributable more to heterosis than to polyploidy *per se*.

Progress in breeding oil-seed turnip-rape were mainly due to improved analytical techniques and screening methods and to an ever increasing knowledge of the chemical contents the seed and their genetic control. Strains with very low erucic acid content have been produced, particularly in Canada. Breeding and improved cultural methods have jointly resulted in increases of 40% in seed yields over a 25-year period (Downey, 1971).

9. Epilogue: It is possible to combine the leafiness of ssp. *pekinensis* and *nipposinica* with the hardiness of turnip and of the biennial form of turnip-rape. *Plasmodiophora* resistance must be kept in view in such hybrids because the oriental forms are generally susceptible.

Yellow-seeded strains of oil-seed rape are proved to have a higher oil content with more protein and less fibre than brown-seeded strains of the same genetic make-up. Removal of undesirable glucosinolates in rapeseed meal is another breeding objective (Downey, 1971).

The breeding of hybrid varieties is not an attractive possibility. Hybrids between different cultivars of oil-seed turniprape, produced in Sweden, have given yields only equal to the best commercial material. However, inter-subspecific hybrids between *oleifera* and *chinensis, dichotoma, narinosa* and *pekinensis* showed yields superior to the better parent. This suggests that future work should not be confined to ssp. *oleifera* alone but should consider the whole species (Olsson, 1955).

(III) Fruit And Seed Vegetables

Under this category are included vegetables that are true fruits. These are mostly used as cooked vegetables. They are largely composed of water and on the whole are inferior sources of proteins, fats and carbohydrates. However, they do provide some vitamins and minerals. They add variety and flavour to our diet. Among the vegetables that have contributed significantly to man's subsistence are the vine crops (cucurbits), the solanaceous crops, lady's finger (okra), grain legumes and sweet corn. The family Cucurbitaceae and Solanaceae are especially important in providing fruit vegetables, These vegetable fruits are almost entirely grown as herbaceous annuals, although many among them may be perennials in the tropics. The members of both families grow rapidly under tropical conditions but are treated as summer annuals in temperate countries. The solanaceous vegetable crops are erect or sprawling herbs and those of Cucurbitaceae are trailing vines.

Fruits are emphasized as vegetable sources rather than the fruits that are tasty and enjoyed desserts. Technically or botanically they are 'fruits' formed after ripening of ovary. Vegetable fruits are those traditionally find place in culinary tradition. They are used to make savory dishes and typically considered vegetables. Thus fruits are served for both purposes, as dessert and vegetable. Thus edible plant part that contain seeds and surrounding pulpy tissue and has a sweet or dark taste generally consumed as breakfast, lunch side-dishes, sweet-dishes, in jellies and jams, or deserts. Thus fruit vegetable is a culinary grouping.

Some fruits are used as vegetables but not entirely. Only their seeds constitute or used for culinary purpose, for example, pulses. Pulses are rich in protein and thus necessary also for health. They used either fresh (green) or dried. Some of them are even used after sprouting. Thus the fruit walls are not edible component. These (pulses) can be used even after their season is over. These can be stored for future. Thus they are not perishable immediately as other fruit vegetables.

8.11 CLUSTER BEAN

Cyamopsis tetragonoloba (L.) Taub. (Leguminosae: Papilionatae)

Resume: *It probably came from Africa and domesticated in dry region of Western Asia. The African species C. senegalensis is considered its immediate acestor. Chromosome number is n=7 is reported for it. Its tender pods yield a vegetable 'Guar Gum' is obtained from seeds. It has wide applications in various industries.*

1. Introduction: It is also called 'Guar'. The species probably came originally from Africa. It was probably domesticated in dry region of Western Asia. Arab traders are thought to have introduced it to Asia via southern India. The African *Cyamopsis senegalensis* seems to be immediately ancestral to guar, which is not anywhere found in the wild state (Hymowitz, 1972). It was introduced into USA originally for soil improvement, being a leguminous crop, but is now grown as a cash crop in Texas and Okhahoma.

2. Plant Characteristics: Robust, bushy annuals, growing upto 3m tall, stem angular and round; leaves alternate, 3-foliate, leaflets ovate, serrate; flowers pinkish-white, borne in dense axillary racemes; pods compressed, linear, ridged, erect and clustered, 4-10 cm long, seeds oval, grey or black.

3. Parts Used: Seeds, pods.

4. Chemical Constituents:

(i) Pods: Protein (3. 7%), carbohydrate (9. 9%), fat (0. 2%), minerals (calcium, phosphorous, iron), carotene, vitamin A&C. (Anonymous, 1948-1976).

(ii) Seeds Protein (6. 63%), carbohydrates (40. 73%), (Anonymous, 1948-1976)

5. Uses And Applications: The tender green pods are used as vegetable. They are also used after drying and salting. Mature dry seeds constitute a valuable feed for cattle. It is extensively used in India, Pakistan, Indonesia and other Eastern countries as a forage crop and green manure. Guar Gum (not a true gum) is obtained from endosperm of its seeds. It is very mucilaginous and has high viscosity nearly 5 to 8 times the thickening power of ordinary starch. This high viscosity is on account of mannogalacton (a complx polysaccharide) or galactomanon (guar gum). Guar flour is, therefore, used chiefly as a stabilizer for sizing textiles, paints and paper products, as a emulsifier in salad dressings; as a flocculent and filter-aid in the mining industry. It is also employed in cosmetics and drugs. It is valuable for backing postage stamps. It is used on large scale in detergents. Actually, the extra whitening and brightening of the detergents is mainly due to the carboxymethyl cellulose content present in it.

6. Karyotaxonomic Evidence: Chromosome number has been recorded 2n=2x=14 for the species. As stated earlier, the African *C. senegalensis* appears to be immediate ancestral to *C. tetragonoloba*.

7. History: In India, two cultivars of guar are generally recognized, the giant and the dwarf. The former produce large pods and seeds. It is also thought probably indigenous to India but is not found in wild in India.

8. Further Developments: Two varieties are recognized in northern states of India *viz.,* (i) giant, (ii) dwarf. In the erstwhile Bombay state, three forms have been identified *viz.,* var. Pardeshi, Sotia-guar and Deshi. The former two are used as pod (green) vegetable and the last one mainly for dry seeds.

9. Epilogue: It is drought resistant and hardy crop species. It is adapted to arid and semi-arid regions and can grow in areas with average rainfall 30-40 cm. It is grown as a pure or mixed crop. It adopts well on deep alluvial soils and sandy loams. It matures in three to three and half months. It is usually cultivated as a mixed crop.

8.12 CUCURBITS

Cucumis, Citrullus, Cucurbita, Lagenaria Spp. (Cucurbitaceae)

Resume: *Cultivated species of cucurbitaceous genera viz., Cucumis, Citrullus, Cucurbita and Lageneria have been domesticated in both, New and Old worlds. Strictly speaking they are native of India, South Africa, South and Central America and Mexico. Chromosome number have been reported n=11, 12 and 20 for them. Archaeological remains in pre-historic times before 7000 to 10000 BC have been recovered from Mexico and Peru. Their fruits are eaten as salad or cooked in various ways.*

1. Introduction: In the New as well as Old World, in prehistoric times, many species of cucurbits were domesticated. Some of them constituted the mainstay foods of the then aboriginal societies. For example, species of *Cucurbita* were the main constituents of the food of American Indians. The seeds of some species also were reputed for consumption. The pumpkins and squashes of the New World are recorded in archaeological deposits in Mexico dating back to 7000 BC. They formed staple food alongwith maize and beans. After the discovery of America, these were carried to Europe and Asia. Many species of cucurbits in the Old world *e.g. Cucumis, Citrullus, Luffa* and *Momordica* played similar role in prehistoric times. The bottle ground (*Lagenaris siceria*) is considered native of Africa. However, the archaeological remains of fruits have been discovered in Peru dated to before 10,000 BC and in both Mexico and Thailand dated to about 7000 BC. Possible pre-Columbian contacts between the New and Old worlds are thereby demonstrated.

2. Plant Characteristics: In general, cucurbits characterized by shallow, extensive root system and angled stem with bicollateral vascular bundles; leaves simple, alternate, long-petiolate, palmately-lobed; flowers generally large and yellow occurring singly mostly on monoecious plants as well as on dioecious and andromonoecious forms; fruits large, called 'pepo', normally indehiscent with hard skin (rind).

3. Parts Used: Fruits (pulp, seeds).

4. Chemical Constituents:

(i) *Cucumis sativus:* Protein (1%), carbohydrate (3%), fats (0. 5%), minerals (potassium, calcium, magnesium, manganese, phosphorus, zinc), vitamins A, C, E, K, niacin, riboflavin, thiamin, periodoxine, folates and pantothenic acid.

(ii) *Cucumis melo:* Protein (1. 5%), carbohydrate (6. 5%), fats (1%), minerals (calcium, copper, iron, magnesium, manganese, zinc), vitamins A, C, E, K, folates, niacin, pantothenic acid, pyridoxine, riboflavin, thiamin).

(iii) *Citrullus lanatus:* Protein (1%), carbohydrates (6%), fats (0. 5%), minerals (calcium, copper, iron, magnesium, manganese, zinc), vitamins A, C, E, folates, niacin, pantothenic acid, pyridoxine and thiamin.

(iv) *Lagenaria siceraria:* Protein (1%), carbohydrates (2. 5%), fats (0. 5%), minerals (sodium, potassium, calcium, copper, iron, magnesium, manganese, phosphorus, selenium, zinc), vitamins A, C, folates, niacin, pantothenic acid, pyridoxine, riboflavin and thiamin.

(v) *Cucurbita pepo:* Protein (54%), carbohydrates (8%), fats (164%), minerals (sodium, potassium, calcium, copper, iron, magnesium, manganese, phosphorus, selenium, zinc), vitamins A, C, folates, niacin, pantothenic acid, pyridoxine, riboflavin and thiamin.

(vi) *Cucurbita maxima:* Protein, carbohydrates, fats, minerals (Ca, Fe, Mg, P, K, Na, Zn, Mn, Cu), vitamins A, B6, B12, C, riboflavin, thiamin, niacin, pantothenic acid, and folates.

(vii) *Cucurbita moschata:* Protein (2%), carbohydrates (9%), fats (0. 5%), minerals (calcium, copper, iron, magnesium, manganese, phosphorus, selenium, zinc), vitamins A, C, E, K, folates, niacin, pantothenic acid, pyridoxine, riboflavin and thiamin.

5. Uses And Applications:

(i) *Cucumis:* Fruits of *C. sativus* Linn. are eaten raw, pickled or cooked. Fruits of C. melo Linn. are also eaten similarly.

(ii) *Citrullus:* Fruits of *C. lanatus* are consumed as a dessert.

(iii) *Lagenaria:* Fruits of *L. siceracia* (Mol.) Standl. are consumed as vegetable.

(iv) *Cucurbita:* Fruits of *C. pepo* are used as vegetable and for making ketchups, as also the fruits of *C. maxima* Duch.

6. Karyotaxonomic Evidence: The Cucurbitaceae are difficult cytologically and only recently have good methods been devised (Varghese, 1973). The large space requirements of individual plants and partly because crops are mainly grown in small patches are some obstacles to know cytogenetics of Cucurbits. Thus, units are small and they have, therefore, generated little research support. The major cucurbits have been revealed cytologically. Their common names, haploid chromosome numbers and nativity is provided in Table-I.

Table 1: Cucurbits, their chromosome and nativity.

	Plant Name	Common Name	Chromosomes & Nativity	
(i)	*Cucnmis sativus*	(Cucumber)	n=7	India
(ii)	*Cucumis melo*	(Muskmelon, cantaloup, etc.)	n=12	Africa, India
(iii)	*Citrullus lanatus*	(Watermelon)	n=11	South Africa
(iv)	*Lagenaria siceraria*	(White-flowered gourd)	n=11	Africa, the Americas
(v)	*Cucurbita pepo*	(Summer squash, pumpkin, marrow)	n=20	North America north of Mexico City
(vi)	*Cucurbita mixta*	(Winter squash)	n=20	Mexico, Central America
(vii)	*Cucurbita maxima*	(Winter squash)	n=20	Mexico, Central America

(viii)	*Cucurbita ficifolia*	(Winter squash, pumpkin)	n=20	Northern South America, Central America
(ix)	*Cucurbita ficifolia*	(Fig-leaf gourd)	n=20	Mexico, Central America, northern South America

(a) The cucumber (*Cucumis sativus*) is the only species in the genus with 2n=2x=14 and native to India. Other species of *Cucumis* have x=12 and are mostly African, primarily from southern Africa. Some 55 loci have been identified. These include genes governing fertility, sex expression, fruit type, fruit flavor, fruit set and disease resistance. The genetics of sex expression in cucumber has been clearly explored. This information is exploited in the commercial production of hybrid seed. Another interesting group of genes of wide occurrence in the cucurbits are those responsible for producing terpene compounds, cucurbitacins, that produce bitter flavours in foliage and fruits. A mutant lacking cucurbitacins has made it possible for breeders to produce non-bitter cucumbers. The presence of cucurbitacins but makes the plants resistant to some insects though more attractive to others. Another interesting mutant, useful in the production of greenhouse cucumbers, enables the fruit to develop parthenocarpically.

(b) Several cultivars of mustkmelon (*Cucumis melo*) were investigated cytologically. All have 2n=2x=24 with regular meiosis and pollen fertility. Breeding work was aimed at improvements in the dessert quality of the fruit and in resistance to diseases and insect pests. Studies have also been conducted on the inheritance of plant habit, sex expression, male sterility, fruit skin colour, flesh colour and fruit smoothness.

Resistance to powdery mildew involves two genes, a dominant (Pm1) for resistance to race 1 of the pathogen and a partial dominant (Pm2) for resistance to race 2. Two modifiers, in addition to Pm2, are responsible for extreme resistance. Cultivars are developed incorporating some of these genes (Bohn and Whitaker, 1964).

(c) For watermelon (*Citrullus lanatus*) plant breeders have collected much useful information about the inheritance of characters *e.g.* leaf shape, plant habit, sex expression, male sterility, fruit colour, fruit surface, fruit shape, seed size, seed coat colour, fruit bitterness, disease and insect resistance. Species of *Citrullus* have 2n=2x=22 and *C. lanatus* is highly pollen fertile. Orton (1911) opined that resistance to *Fusarium* wilt is determined genetically. He also proved that resistance could be transferred from the wild form to edible cultivars. Later, many cultivars resistant to *Fusarium* have been developed. The inheritance of resistance, however, remains unknown. A single dominant gene for resistance to at least two races of anthracnose (*Colletotrichum lagenarium*) has been reported. The seed coat colour depends upon a system of three major genes and a specific modifier active in combination with the black seed coat phenotype. This three-gene system accounts for most of the seed coat colour phenotypes from black to tan, red and white (Poole *et al.*, 1941). For seed length, digenic inheritance is reported. The short and long genotypes are recessive to medium (Poole *et al.*, 1941; Prem Nath and Dutta, 1973).

The mutant occurring in watermelon *viz.*, male sterility (msg) is associated with a glabrous condition. Seed production in the original mutant was sparse but has been increased to an acceptable level by selection. Glabrous plants occasionally produce a little pollen late in the season.

The commercial production of seedless watermelons was an attractive objective since 1939 when seedless polyploids were first produced (Kihara and Nishiyama, 1951). Andrus *et al.*, (1971) showed that triploids are feasible, have superior consumer qualities. In fact, they are much superior to diploids. They almost constitute a new product.

(d) A bottle gourd (*Lagenaria siceraria*) is revealed 2n=2x=22. The work of Pathak and Singh (1950) showed that fruit colour was determined on a monogenic basis, as were several fruit shapes. Time of fruit maturity and seed size were polygenic. Tyagi (1973) demonstrated heterosis for the number of pistillate flowers, number of fruits per plant, weight of fruits and number of seeds per fruit. Varghese (1973) suggested that modern forms were secondary polyploids derived from an ancestor with a basic x=5; speculatively x=5+5+1=11.

(e) Five species of *Cucurbita*, are domesticated cenospecies and are genetically isolated from each other. The wild species contain many ecospecies. They are separated geographically but are genetically cross-compatible. They belong to five natural groups (Bemis *et al.*, 1970). All species have 2n=2x=40 and are suggested that they are polyploids based on x=10. The centre of diversity of the genus is in the tropics near the Mexico-Guatemala border. The wild species are variable and they range from mesophytic to xerophytic and from annuals to perennials. The domesticated species are: (i) *C. pepo* includes both summer and winter type squash; cultivars of summer squash are mostly brachytic plants; has some cold resistance (ii) *C. moschata* high quality warmer squash; vine type plants; tolerant of the squash vine borer. (iii) *C. maxima* high quality, soft skinned winter squash with vigorous vines and, sometimes, extremely large fruit. (iv) *C. mixta* similar to *C. moschata* but largely restricted to Mexico; used both as summer and winter squash; seeds are roasted as a confection; fruits have a large, hard, corky peduncle. (v) *C. ficifolia* restricted to the highland, tropical regions of Mexico, Central America and northern parts of South America.

Since each of the cultivated species has desirable horticultural qualities, there has been interest in interspecific crosses. Whitaker and Davis (1962) listed the results of hydridization studies and only three out of the 58 showed some degree of fertility in the F_2. Amphidiploids of *C. moschata* × *C. maxima* have been made. Fertility was enhanced but seed production was not economical (Pearson *et al.,* 1951). A number of wild species have some degree of compatibility with the cultigens and may have certain desirable horticultural characters. It is sometimes possible to transfer these characters from wild to cultivated species. Some wild × cultivated amphidiploids have been made. The entire genome of the wild species but remains intact. Its undesirable characters cannot be omitted.

Curtis (1939) studied heterosis in summer squash (*C. pepo*). He proposed a method of producing F_1 hybrids. Hybrid seed is produced commercially by field emasculation of brachytic plants as females. This method is economically feasible since Cucurbita flowers are large and male buds can be removed some days before anthesis.

Genic male sterility is revealed in *C. pepo, C. maxima* and *C. moschata*. Populations are maintained by pollinating male sterile (recessive) plants with heterozygous male fertiles but the expense of field-roguing the male fertile segregates has prevented production of hybrid seed by this method.

7. History:

1. The cucumber is thought to be indigenous to India. It has been not found in the wild state in India. A tremendous array of fruit sizes, shapes, skin colours, skin ornamentations and vegetative characters are found in India and adjacent countries. Also a small, bitter cucumber, with sparse, stiff spines, *C. hardwickii*, a close relative of the cultivated kinds, grows wild in the foothills of the Himalayas. These observations suggest that *C. hardwickii* is either a feral form or, more likely, a progenitor of the cultivated cucumber. De Candolle (1882) thought that the cucumber had been cultivated for over 3,000 years in India.
2. The muskmelon is thought to be indigenous to Africa. It is probable that truly wild forms of *C. melo* are found only in eastern tropical Africa south of the Sahara. Wild forms occurring in India are more likely to be feral escapes derived from local cultivars. *Cucumis*, except for the cucumber, is an African genus with x=12. The muskmelon was evidently a latecomer to man's list of crops. Once domesticated, however, it exploded into several cultivars in India, which may thus be regarded as a secondary centre. Cultivars of *C. melo* were rapidly dispersed throughout Europe and, at an early date, into the Americas.
3. The watermelon (*Citrullus lanatus*) is reported to be feral in the warm parts of the globe but it is truly native only in any sandy areas of South

Africa, particularly in the Kalahari Desert. It occurs in two biochemical forms, in this area one having bitter fruits, containing cucurbitacin (E-glucoside), and the other devoid of this constituent. Forms with nonbitter fruits are a source of food and water for the Bushmen of the Kalahari. There are some forms of watermelon (called citrons) which have fruits with firm, hard, greenish flesh and large seeds. Citrons should be regarded as cultivars of the watermelon, not as progenitors.

4. The white-flowered gourd (*Lagenaria siceraria*), is most likely indigenous to the tropical lowlands of Africa south of the equator. The archaeological records indicate that it was present in the western hemisphere as early as 7000 BC., probably as a cultivated plant. Schweinfurth (1884) has identified remains of *L. siceraria* from Egyptian tombs dated at about 3300-3500 BC. Thus the white-flowered gourd has a long history of cultivation in both hemispheres documented by well-preserved remains of fruits, seeds, peduncles and even intact fruit. The widespread . occurrence of gourd remains in many archaeological sites of the western hemisphere is perhaps a result of the importance of the crop in the day-to-day affairs of primitive societies. The shells were used as containers for liquids, for food storage, as eating utensils and for musical instruments, etc.

5. Cucurbita (the squashes, pumpkins and gourds) was associated with mankind in the Americas for at least 10,000 years (Whitaker and Cutler, 1971). The five cultivated species *viz.,; C. pepo, C. moschata, C. maxima, C. mixta* and *C. ficifolia* were selected by American Indians long before the Discovery of New World in 1492. They were staple items of their diet. The advanced civilizations of the Americas (Aztec, Inca, Maya) prior to the European influx, were founded upon a food complex of squash alongwith maize and beans.

Wild species of *Cucurbita* closely allied, to the cultivated forms in Mexico and Gautemala *viz., C.lundelliana* and *C.martinezii* which are to some extent cross-compatible with the cultivars. They are vigorous plants bearing small fruits with tough rinds, coarse, bitter, stringy flesh and abundant seeds.

8. Further Developments: Attempts in recent times with cucumber has stressed disease resistance, downy mildew, powdery mildew mainly to scab and cucumber mosaic virus. Sources of resistance to scab have been found and incorporated into commercial cultivars. Tolerance to downy and powdery mildews are utilized but cultivars with satisfactory resistance or tolerance to cucumber mosaic virus have not yet been developed.

The muskmelon (*C. melo*) includes fruits popularly known as cantaloups, honey dews, casabas etc. Muskmelons are grown throughout the warmer parts of the world but are of peculiar economic importance in the states of Arizona, California, New Mexico and Texas. In these areas, muskmelons are grown on large acreages for distant markets. Effort has been to develop cultivars with superior consumer qualities such as improved flesh colour and texture, increased soluble solids and fruits. These features withstand severe shipping stresses, so as to arrive at distant markets in good condition.

The principal objectives of the plant breeder in *Cucurbita* has been to improve the quantity and quality of the fruit flesh of the winter types and the earliness and productivity of the fruit of the summer types. The primary breeding method has been to make selections within interspecific crosses.

Breeding work with watermelon has been directed towards producing cultivars with good consumer qualities, tough rinds for long distance transport and multiple disease resistances. The cultivar, Charleston Gray, is a good example of this effort.

9. Epilogue: Cucurbits are interesting group of fruit yielding plants, morphologically and from horticulture and economic point of view. (1) Cucumber production are directed mainly with two issues: (i) the genetic manipulation of sex expression designed to find inexpensive means of producing hybrid seed; and (ii) breeding of a plant-type adapted to 'once over' harvest. Breeders are also attempting to develop cultivars with short internodes and large number of fruits that can be harvested mechanically and are suitable for processing. (2) A search for biological means of controlling insect pests of the crop has revealed collections within *C. melo* tolerant of the melon aphid (*Aphis gossypii*). This character is controlled by a single dominant gene (Bohn *et al.,* 1973). (3) Efforts are made to breed watermelons, *Citrullus lanatus*, that bear smaller fruits and are more prolific than those currently available; small fruits are easier to handle than large ones. They are thus less possibility to damage. (4) There has been a remarkable revival of interest in the white flowered gourds for use as ornamentals and in handicraft. They are made into bird-feeders, planters, musical instruments, animal figures, ornaments, etc. This activity constitutes a small but thriving trade for growers and craftsmen appears likely to continue, as more people with leisure time adopt the hobby. (5). In *Cucurbita*, F_1 hybrids have been shown to be more desirable than open pollinated cultivars. The production of vine-type F_1 hybrids is not economic. However, ethylene producing compounds, when sprayed on seedlings of normally monoecious *Cucurbita* plants, induce a gynecious condition which persists for at least ten flowers after treatment. By judicious management of the chemical, the production of hybrid seed is possible (Robinson, *et al.,* 1971). One of the most exciting developments in the utilization of *Cucurbita* is the possible domestication of the wild xerophytic species. These species are perennial and produce abundant fruit with a less supply of water. The seeds contain 30-35 per cent high quality vegetable oil and 30-35 per cent protein. If these projections prove reliable, new sources of oil and protein of great potential for arid countries would be a good objective.

8.13 EGGPLANTS (BRINJAL)

Solanum melongena L. (Solanaceae)

Resume: *It is believed to be native of India and have domesticated in northeastern India where wild forms till grow. It reached Europe in the 15th century and China before 1500 years ago. It belong to the genus Solanum and has chromosomes 2n=24 as in other related species. It is now cultivated throughout the tropics and subtropics. It is an important article of diet consumed in a great variety of ways. It contains fair amount of carbohydrates, protein and minerals.*

1. Introduction: The eggplant (Brinjal or Aubergine) is thought to have been domesticated in northeastern India. Wild forms are still found in India. About 1500 year ago, its seeds were carried to China where small-fruited types were developed. The early Arabian and Persian traders introduced eggplant from India to the eastern and southern shore of the Mediterranean during the early the middle ages. The Portuguese took it to Brazil. It is now widespread in the tropics and subtropics, particularly in southern Europe and southern United States.

2. Plant Characteristics: A spiny, much-branched perennial reaching to 0. 5 to 1. 5 m in height; leaves simple, alternate, ovate to oblong-ovate with sinuately lobed margin; flowers purplish and borne singly or in few-flowered cymes; fruits bears glossy, smooth, pendant berry of variable colour, size and shape.

3. Parts Used: Berry.

4. Chemical Constituents: Fruit: Carbohydrates (6%), protein (1. 0%), fats, minerals (calcium, phosphorous, iron) and vitamin-B.

5. Uses And Applications: It is consumed in a variety of ways. In India, it is made into 'Bharta' a preparation by roasting, mashing and seasoning with salt, onion, chilies, tomato, coriander leaves and fatty oil. The fruit is consumed in curries. In Japan and other countries of the Orient, it is pickled.

6. Karyotaxonomic Evidence: *Solanum* is a large genus. Among the 22 Indian species, there is a group of five related ones *viz., melongena, coagulans, xanthocarpum, indicum* and *maccant*. They are all diploids with 2n=2x=24. Occurrence of polyploidy within *S. melongena* also has been reported. *Solanum melongena* produces fertile hybrids in crosses with *S. incanum*. The cross *S. xanthocarpum* × *S. melongena* is successful when the former is used as the pistillate parent. However, the resulting seeds are inviable. There are variable reports regarding the compatibility of *S. indicum* to *S. melongena* and *S. xanthocarpum*. It appears that *S. melongena* is more closely related to *S. incanum* than to any other species. *Solanum khasianum* crosses with *S. melongena* only when the latter is used as the female parent. The fruits formed by the hybrid have shrunken seeds. Wild *S. melongena* is an armed perennial shrub with bitter fruits. It occurs in India (Bhaduri, 1951). There have been suggestions of an African origin of the crop (Sampson, 1936). There are spiny African Solanums but there is no evidence that *S. melongena* is native there.

7. History: The crop is old in India. This is clear from an exhaustive list of old names. No date for domestication can be, however, suggested. Wild forms are spiny and bitter. It appears clear that objective of early selection must have been against these characters and in favour of large fruit size. Based on distribution and variability, Vavilov (1928) regarded the crop as of Indian origin. He added that some secondary variability had also developed in China. It has been known in China since the 4th century BC.

The crop is extremely variable in India. The species has perfect flowers and is self-compatible and an inbreeder. The amount of natural cross-pollination varies with the variety and environment.

The eggplant was taken to Africa by Arab and Persian travellers before the middle ages in Europe. First records in Europe were in the 15th century (Hedrick, 1919) but the crop did not become generally known there until the 17th century. The early European name, eggplant, suggests that first introductions were small fruited; an early Italian name, melazana, meaning mad apple, later became melongene and Linnaeus's Latin, melongena (Heiser, 1969). The French name, aubergine, comes from the Arabic by way of the Spanish berenjena.

8. Further Developments: Breeding programmes of eggplant have been carried on in India and Japan, USA and Europe. The principal methods used for improvement of the crop are selection from inbred lines and intervarietal crosses. The occurrence of hybrid vigour in eggplants was recorded as early as 1892 by Munson in the USA and in 1926 in Japan by Nagai and Kida. Hybrid eggplants are now commonly used in many countries, especially in Japan. The male-sterile lines have been detected, but F_1 hybrid seed is still produced by emasculation and hand pollination.

9. Epilogue: Objective of breeding for disease resistance will still continue to be important, especially in connection with *Phomopsis, Verticillium* and bacterial wilts and little leaf disease. Breeding for resistance to insects like shoot and fruit borer (*Leycinodes orbonalis*), jassids, aphids and root knot nematodes should also receive proper attention. Several related and wild species might offer opportunities for interspecific transfer of disease, insect and nematode resistance. The crosses have so far proved difficult. The *Solanum* spp., *nigrum, khasianum, sisymbrifolium, integrifolium* and *gilo*, are possibilities. There are also prospects of improving protein and vitamin content.

8.14 FIELD BEAN

Vicia faba Linn. (Leguminosae: Papilionatae)

Resume: *The centre of origin of field bean is thought to be the Near East. It was introduced to agriculture in the late Neolithic period. There have been finds from Egypt from 1800 BC. and numerous references from classical Greek and Roman literature. The crop is diploid with 2n=12 chromosomes. It bears easily threshed pods and large seeds with high protein content.*

1. Introduction: It is also popular as Broad, Horse, Scotch or Windsor bean. It is said to be the only bean known to Europe before the time of Columbus. It was under cultivation since prehistoric times and originated probably in Algeria or South Western Asia. In the Old World exclusively, more than 1000 varieties are grown. Presently, China is the largest grower of field bean. Egypt, Ethiopia, Morocco and Italy are in descending order of production.

2. Plant Characteristics: A hearty, erect annuals, reaching 2 to 4 ft in height; leaves pinnate, leaflet 1-3 pairs, elliptic to oblong; flowers white but blotted purple, in racemes ±40 cm long; pods leathery, large, broad; chick, seeds large, brown, green, purplish or black, much compressed.

3. Parts Used: Pods, seeds, foliage.

4. Chemical Constituents: Fresh beans have protein, fat, carbohydrate, fibre, minerals (calcium, phosphorous, magnesium, sodium, potassium, etc.) carotene, thiamine, niacin, vitamin-C, etc.

5. Uses And Applications: Young pods and seeds are consumed as vegetable. Seeds are aphrodisiac. Decoction of leafy shoots is used as a diuretic. It is good forage crop for domestic animals. Field bean is also used as a cover crop, in rotation crop and for silage. Seeds are good feed for racing pigeons.

6. Karyotaxonomic Evidence: *Vicia faba* is a diploid with 2n=2x=12 and the chromosomes are characterized by their large size, high DNA content and one long chromosome. However, variability in karyotype, including translocations is evidenced. Other *Vicia* species have 2n=10, 12, 14 and none can be hybridized with *V. faba* (Schafer, 1973). Wild ancestor of *V. faba* is unknown. It has been suggested that some other species of *Vicia e.g. V. narbonsis, V. galilaea* had a common ancestor with *v. faba* (Zohary and Hopf, 1975). There is no direct evidence of this and chromosome numbers, size and DNA content suggest that any evolutionary relationship must have been in the very remote past.

Intraspecific classification based mainly on seed size has been proposed (Muratova, 1931). This recognized subsp. *faba* (vars. *faba, equina minor*) and subsp. *paucijuga*. However, Hanelt (1972b) considered *paucijuga* to be only a geographical race of subsp. *minor* and recognized subsp. *faba* (vars *faba*, *equina*) and subsp. minor var. minor. Cubero (1974) distinguished only the four varieties *viz., minor, equine, faba* and *paucijuga*. Sub-varieties have been although named fine systematic subdivision serves little purpose in this partially allogamous species. There is considerable variation within populations and no sterility barriers between sub-species.

7. History: The bean is unlikely to have been among the first crops to be cultivated. Archaeological evidence indicated that *V. faba* was introduced to agriculture in the late neolithic period (Schultze-Motel, 1972). There have been no prehistoric finds east of a line running close to the coast from Israel to Turkey and Greece (Hanelt *et al.,* 1972). Cubero (1974) therefore supposed the centre of origin to be in the Near East, with the species radiating out in four directions: (i) to Europe, (ii) along the North African coast to Spain, (iii) long the Nile to Ethiopia, and (iv) from Mesopotamia to India. Secondary centres of diversity may have later become established in Afghanistan and Ethiopia.

The culture of *V. faba* was fairly well established in Europe, including Britain by the Iron Age. There have been finds from 1800 BC. in Egypt and there are numerous references from classical Greek and Roman literature. The species was not grown to any extent in China before AD. 1200. Hanelt (1972a) thought that it was then brought there at the beginning of the silk trade and, by the 16th century. Considerable cultivation was started there. Chinese *V. faba* are of the large-seeded 'major' type (Muratova, 1931) and large-seeded types are not known to have existed anywhere until about AD. 500. Thus the 'minor' subspecies is presumably the more primitive. According to Cubero (1973), it has the least genetic diversity, so that the greater evolutionary potential now lies in the 'major' and 'equina' groups. The characteristic *viz.,* indehiscent pod is recessive to shattering. The latter is hence thought to be the more primitive condition but non-shattering types have been favoured by selection in the more arid regions. Within this zone, 'minor' can be seen to have accumulated at the eastern ends. *viz.,* Ethiopia and India and 'major' at the western end. 'Major' varieties have moved to Russia, China and Japan in recent times. In central and northern Europe all three 'major, 'equina' and 'minor' are now cultivated. They have pods mostly capable of shattering. *Vicia faba* is generally thought to have reached Mexico and South America in the hands of the Spaniards, Most of the central and South American *V. faba* are of the 'major' type.

The main features of the early history of the bean appear to be of adjustment of length of life-cycle, plant habit and pod dehiscence to habitat. These changes were developed hand in hand by local increase in seed size, by local adjustment of the breeding system towards more or less inbreeding and by the development of varied seed colours.

8. Further Developments: In recent times, much selection have taken place *e.g.* for large seededness. Winter hardiness was probably selected independently in 'equina' and 'minor' types. Darwin (1858) noted reduced pod set in a cage from which bees were excluded. Sirks (1931) analysed genetically a large number of variants. Muratova (1931) carried out systematic and geographical studies.

Variability has been induced and investigated through mutations and cytological work at Svalof, Sweden through the study of collections and classification at Gaterslesben, Germany after 1966. Through the study of genotype relationships at Cordoba, Spain, some of the variability has been exploited in achieving small increases in yield, earlier maturity, resistance to a few diseases and in the breeding of seed types suited to special purposes.

Other breeding methods which have been used include mass selection and progeny-test selection under open pollination but control of pollination in enclosures is being increasingly employed, particularly for: (1) the production of inbreds (especially for the evaluation of lines in regard to characters of low heritability and the subsequent composition of synthetics); and (ii) control over recurrent selection in populations. Current varieties are all either mass-selected populations or synthetics.

9. Epilogue: The objectives, in temperate regions, appear to be changing towards stabilizing yield rather than improving maximum yield. Uncertainty of pollination, which may be overcome by autofertility, is an exclusive limiting factor. The necessity to reduce flower loss may demand a trend towards more determinate plant structure, such as that which exists in certain mutants and in some of the Asian, African and Mediterranean populations. Determinate growth may provide a more stable basis of yield though it would be less responsive to good seasons than the tall, indeterminate habit which has evolved in the long seasons of central and north-western Europe. There is likely to be more emphasis on breeding for adaptation to new areas of cultivation and for improving intrinsic characters *e.g.* as protein content and quality, for the reduction of toxic factors, and for suitability for industrial processing. There is some variability to exploit in the above factors and in resistance to pests, disease and other hazards. The prospects for bean improvement would be greatly enhanced by successful hybridization with other species of *Vicia*.

8.15 OKRA

Abelmoschus esculentus (L.) Moench. (Malvaceae)

Resume: *It is believed to be native of tropical Africa. It still grow there in wild. Its seeds reached to Arabia 1300 years ago and thence to Spain. It reached the New World after 1700. It is now grown in the tropics and subtropics and warm temperature countries. Variable chromosome number* *from n=33, 36 and 59 to 72 have been reported for it. Some consider it an amphidiploid. Tender fruits are favoured as fresh vegetable.*

1. Introduction: Okra or Lady's Fingers is considered to be a native of tropical parts of Africa. It still grows wild along the River Nile in Egypt and Ethiopia. Its seeds probably were taken from Ethiopia to Arabia before 1300 years and thence introduced to Spain across northern Africa. French colonist took Okra to the New World after 1700. It is now under cultivation throughout in the tropical and subtropical countries, besides warmer temperate areas.

2. Plant Characteristics: An erect, robust annual, reaching upto 2. 0 m in height; leaves large, alternate, cordate and lobed, the upper ones being usually more deeply lobed; flowers yellow, large, borne singly in leaf axils on peduncles; capsular fruits 10-30 cm long, ridged and dehiscing longitudinally when dry.

3. Parts Used: Fruits

4. Chemical Constituents: Fruits: Carbohydrate (9.7%), protein (2.2%), fibre (1.0%), fats (0.2%), ash (0.9%), minerals (especially iodine), vitamin A and B.

5. Uses And Applications: The green, fresh and tender fruits are consumed as vegetable. They are processed by canning, freezing or preserving in brine. The young fruits, being mucilaginous, are used to thicken soups, sauces and stews. Roasted okra seeds are used as a substitute for coffee. An edible oil obtained from ripe seeds.

6. Karyotaxonomic Evidence: The valid botanical name is *Abelmoschus esculentus*. It was designated earlier as *Hibiscus esculentus* L. under the section *Abelmoschus* of *Hibiscus*, established by Linnaeus (1737). The generic name, *Abelmoschus*, proposed by Medikus, was upheld by Hochreutiner (Joshi *et al.,* 1974) on the basis of the constant distinguishing feature of caducous calyx. Index Kewensis lists over 30 species of *Abelmoschus* in the Old World, 04 in the New World and 04 in Australia.

Variable chromosome numbers (n=36; about 33; 59 to 72) have been reported for *A. esculentus* (Joshi *et al.,* 1974). The same authors reported on the cytotaxonomy of *A. esculentus* in India in which they repeatedly found n=65. They also reported on its hybrids with *A. tuberculatus* (n=29), a new species recently described from India. In their opinion, *A. esculentus* is an amphidiploid comprising two genomes: one with 29 chromosomes (T1), homologous with that of the related wild species, *A. tuberculatus* (T), and the other with 36 chromosomes (Y). The latter genome of *A. esculentus* exhibited very little homology with the M genome of *A. moschatus* (n=36) and greater, but still very incomplete homology with the *F. genome* of *A. ficulneus* (also n=36). To know evolutionary origin of *A. esculentus*, it is necessary to determine the present-day geographical distribution of *Abelmoschus* taxa that might have contributed the 36-chromosome genome to it and also the relationships of material (of Ceylon and Japanese origins) for which low chromosome numbers (n=33, 36) are reported.

Kuwada (1957) also studied the cytology of the hybrid between *A. esculentus* and *A. tuberculatus* and its amphidiploid, but reported different chromosome numbers for the two species. Joshi *et al.,* (1974) reported on meiosis in the sterile hybrids of *A. esculentus* (n=65, T'+Y) with *A. pungens* (n=69) and *A. tetraphyllus* (n=69). Kuwada (1957) reported on the cytology of the sterile hybrid, *A. esculentus* (n=62) × *A. manihot* (n=34), and its amphidiploid.

Singh *et al.,* (1975) at the Indian Agricultural Research Institute, New Delhi, studied a number of cultivars from West Africa (where the crop is known as gumbo). They identified one accession from Ghana as being immunes. Hybrids of this cultivar with Indian okra (n=65) were only partially fertile; those between the Ghanaian accession and *A. tetraphyllus* (n=59) were completely sterile.

Okra is thus probably not a single species but a polytypic complex which exhibits both high polyploidy and hybridity and the parental wild species are yet undetermined. Even basic chromosome numbers are unknown; thus the haploid numbers n=29, 36 must themselves be polyploid so Indian okra, with n=65, though proximately tetraploid, could well be basically 16-ploid.

7. History: The cultivated okra is native of Old World. According to Vavilov, it was probably domesticated in the Ethiopian region and according to Murdock in West Africa (Joshi *et al.,* 1974). There is a late 19*th* century record of its occurrence in the wild state on the White Nile in the Sudan (Singh *et al.,* 1975). It was apparently introduced to tropical and subtropical America during the early post-Columbian period. The accepted view that okra is African should not, however, be taken as established. The cytotaxonomy of *Abelmoschus* is not that an Asian origin of whole or part of the cultigen does not appear impossible. It is to be remembered that Indian okra bears genomes (T', Y) that show a varying degree of homology with the genomes of two Indian wild species (T, F) and that African-Indian hybrids show some sterility. The crop could perhaps yet turn out to be polyphyletic.

8. Further Developments: Result of breeding on okra are available from India (Singh *et al.,* 1975) and the USA (Boswell and Reed, 1962). Smooth-fruited varieties have priority over varieties whose fruits are prickly even in the green tender stage. The latter turn smooth and consumable on cooking. Processing industries generally prefer shorter fruits of manageable size. Local preferences, however, differ also for five-angled, many-angled or round-fruited types. Young green fruits are rich in vitamin-C but content of this vitamin falls as the fruits grow in age and length, though still green and tender. Pusa Sawani variety of okra in India gives fruits up to 18 cm long, tender and stringless. Until recently, this variety was reputed for its freedom from the symptoms of the yellow-vein-mosaic virus disease in the field. A dual purpose, ornamental-vegetable type, with red fruits, has been reported from southern India (Anonymous, 1966).

9. Epilogue: Breeder of okra pay attention to breeding for resistance to diseases and pests and to early fruiting. Many new sources of resistance to yellow-vein-mosaic have been searched out by Prem Nath (1970). Sources of tolerance or resistance to root knot mematodes are noted (Singh *et al.*, 1975) also *Fusarium* wilt by Grover and Singh (1970) and Corley (1965); and to *Verticillium* wilt by Silveira *et al.*, (1970). Mutation experiments for induction of new variability have been reported by Nandpuri *et al.*, (1971) and Kuwada (1970). Mutagens used were gamma and X-rays, respectively. In-depth studies on various aspects and objectives are still desirable.

8.16 TOMATO

Lycopersicon esculentum Mill.

[(Syn.) *L.esculentum* (L.) Mill.] Karsten (Solanaceae)

Resume: *It is native of Peru and Ecuador. It is domesticated in Central Mexico and spread in pre-Columbian times. It was simply grown as ornamental and was called 'Love Apple'. It became widespread and gained popularity due to colour, flavor and being rich in vitamin-C. The Spanish explorers introduced it in 1523 in Europe. It is a common ingredient of various culinary preparations. It was, at first, little used being related to poisonous deadly nightshade family. Its wild form viz., L. esculentum var. cerasiforme (Dun.) Alef. is still found in Mexico and Central and South America. Chromosome of all species of the genus have 2n=24 and morphologically similar.*

1. Introduction: Tomato is a native of Peru and Ecuador. Its name 'Xtomatl' by Mayans suggests its familiarity in Mayan culture. It had spread in pre-Columbian times to Mexico, where it was domesticated. Thence, it was introduced by the Spanish explorers in 1523 into Europe. At first, it was little used for food as it was considered poisonous because of its relationship with the deadly nightshade family. It was grown only for ornamental purpose and was called 'Love Apple' and referred as an 'object of affection'. It was so referred also in France and hence called 'Pomme d'amour'. Tomato was consumed only in Italy in earlier times. Colonel R. G. Johnson of USA announced that at noon on 20th September 1820, he would publically eat tomato to prove its edibility. This event of 'Sure suicidal nature' was witnessed by thousands of people. He ate tomato and noting wrong happened. Onwards, people started eating tomato with taste. Now, it ranks next to potatoes in importance as food. From Europe, it was carried across the Pacific into southeast Asia before 1650.

2. Plant Characteristics: A short-lived weak-stemmed, branched perennial; leaves spirally arranged, uneven inparipinnate compound; flowers yellow, borne in clusters; fruits fleshy berries, juicy, shiny, bright orange, yellow or reddish when ripe.

3. Parts Used: Fruits (Green or ripe).

4. Chemical Constituents: Soluble sugars, organic acids (citric acids, malic acids), minerals, vitamin-A, B1, B2, besides C, thiamin, niacin.

5. Uses And Applications: Tomatoes are eaten raw or cooked, and are also preserved. They are used in the preparation of tomato soups, pickles (when green), ketchups, sauces, etc. It is a commonly used ingredient of salads, being valued for distinctive flavor, pleasing acidic taste and colour.

6. Karyotaxonomic Evidence: The genus *Lycopersicon* contains 8 to 10 closely related species, tomato proper is one of them. Most of the species are short-lived perennial herbs. All species are native to western South America; the wild form of *L. esculentum*, var. *cerasiforme*, (Dun.) Alef. is also found in Mexico, Central America and other parts of South America, as well as the subtropics of the Old World. It is weedy and its true native range is obscured by its aggressive colonizing tendencies. The chromosomes of all species are alike morphologically and in number (2n=2x=24). The cytology of interspecific F_1 hybrids yields little or no evidence of structural difference. *Lycopersicon esculentum* and its near relatives are self-fertile. The former is outcrossed to a considerable extent in its native region in other subtropical areas. Elsewhere, it is almost completely self-pollinating (Rick, 1950). In contrast, the other species are self-incompatible and consequently entirely outcrossed. *Lycopersicon esculentum* can be hybridized with, and thereby receive genes from, all other species, with varying degrees of difficulty. Numerous desired characters have thereby been transmitted to acceptable cultivars.

7. History: In the New World at the time of conquests, the tomato was already a well-developed cultigen. It was then reached from Europe to many areas of the world. Despite earlier wide acceptance of Peru as the centre of domestication, the bulk of the historical, linguistic, archaeological and ethnobotanical information favours Mexico, especially the Vera Cruz-Puebla area, as the source of the cultivated tomatoes that were transported to the Old World. Mexico was also thought most likely centre of domestication in spite of the undisputed distribution of the genus in South America. These conclusions are based on fragmentary or negative evidence. The archaeological data does not positively favour any particular region. However, representations of the tomato in Peruvian artefacts is considered significant of certain pre-Columbian cultures, for depicting their cultigens on pottery. The poor preservation qualities of tomato plant parts obstructs progress in detecting remains of either fossil or archaeological nature.

Jenkins (1948) stresses the great morphological variability in the cultivated tomatoes of Mexico. This level of variation can be matched by that of Central America (Lehmann and Schwanitz, 1965) and coastal Peru. Genetic variability *per se* is not very diagnostic in determining centres of origin. Far more critical would be genetic distance as determined by the degree of genetic differentiation between modern cultivated tomatoes and their counterparts in putative areas of origin. Unfortunately the data are not critical, although they do reveal that the tomatoes introduced by the Spanish explorers represent only a minute portion of the germplasm available in the same and closely related species. Distribution of alleles at the Ge (Gamete eliminator; Rick, 1971) and Prx (peroxidase allozymes; Rick *et al.,* 1974) loci, the currently available European and US cultivars are exceedingly homogeneous. The few deviations from the 'normal' genotype appear in tiny enclaves, which might represent either *de novo* appearance by mutation or ancestry in western South America. In these assays, the greatest overall similarity in allele frequency is with cultivars and var. *cerasiforme* from Mexico and Central America. The differing alleles found in the Ecuadorean and Peruvian cultivars are also common to var. *cerasiforme* and *L. pimpinellifolium* from their respective regions.

Immediate ancestor of tomato was probably var. *cerasiforme*, the common, weedy, wild counterpart of the same species (Jenkins, 1948). The latter bears greater genetic resemblance to the cultivated tomato than the only other likely candidate *L. pimpinellifolium*, which is probably a by-product rather than a member of the stem-line of the crop. Although these taxa differ considerably from each other in size characteristics, the magnitude of genetic difference does not appear to be great. Thus, many breeding experiments have succeeded in restoring the large fruit size of *lycopercium* (*esculentum*) cultivars after a cross with *L. pimpineliifolium*. This was followed by only three generations of backcrossing. Selection has demonstrated that it is possible by successive steps of induced mutation and selection to increase fruit size in *pimpinellifolium* to a level comparable with that of smaller fruited cultivars (Stubbe, l971).

8. Further Developments: Before 1920's, tomato improvement depended largely upon selection of chance variants that originated as a result of mutation, spontaneous outcrossing or recombination of pre-existing genetic variation. Such selection certainly succeeded in developing cultivars adapted differing environments as the long warm season in Italy, the short season in outdoor culture in northern Europe, greenhouse production in Britain, the mesophytic conditions of the eastern and midwestern USA and the warm, arid districts of the western USA.

Rapid progress in tomato breeding and utilization has come about during the last century. Modern methods resulted in the breeding and introduction of hundreds of cultivars adapted to widely differing conditions and intended for diverse purposes as fresh vegetables and for various forms of preservation. The accelerated rate of varietal introduction has led to rapid replacement of cultivars held supremacy for few decades.

Single hybridization followed by pedigree selection is highly efficient for achieving desired combinations of parental characteristics. Certain dominant characters can be combined and heterosis exploited in F_1 hybrid cultivars. The following principal achievements were attained: (a) Increased yields by way of larger fruit size and increased fruit number. The ability of flowers to set fruit has been improved and, in certain types, a more concentrated fruit set achieved. (b) Improved fruit quality: shape, texture, colour, flavour; uniformity in all characteristics. (c) Plant habit modified to facilitate cultural and harvest operations. The most significant change has been the discovery and exploitation of determinate restricted growth conditioned by the s gene. (d) Improved handling and storage durability (but sometimes to the detriment of culinary quality). (e) Pest resistance. The list of accomplishments includes resistance to various species of insects, viral and fungal parasites, and nematodes.

Floral structure changed so as to improve fruit setting ability. Intentionally or otherwise, the position of the stigma has been markedly changed. The cultivars from Latin America tend to have well exserted stigmas. If such types are cultivated in foreign areas lacking the appropriate pollinating insects, fruit setting is greatly diminished. When the tomato was first introduced to Europe, an intense selection would have been expected for less exserted stigmas. In glasshouse culture, the selection would have been even stronger because insects are excluded as well as wind-induced vibrations, which can aid in pollination. The stigma of most cultivars became fixed at the mouth of the anther tube. During the past decades, stigma position was shifted again, this time to an even lower level, well within the anther tube (Rick and Dempsey, 1969). This change has been incorporated in the leading California cultivars which further improved self-pollination and consequent fruit set and practically eliminated outcrossing.

Perfection of completely mechanized harvest has had a strong impact on the evolution of the crop. Although applied primarily to the crop grown for processing, the method adapted for harvest for the fresh fruit market. NBPGR, New Delhi evaluated several germplasm accessions from diverse sources for various agro-botanical and economic characters. About 188 exotic accessions were evaluated also at NBPGK, Hyderabad w. r. t. plant growth, flower and inflorescence, fruit colour, seed shape, etc. (*cf.* Swamy and Sadashiva, 2007).

Tomato F_1 hybrids were investigated and numerous F1 cultivars have been introduced to the trade. Advantages of such hybrids are increased earliness, improved quality and yield, besides combinations of desirable dominant parental alleles. Capable of setting fruit under adverse conditions, F_1 cultivars tend to be popular with home gardeners. Their use in commercial plantings depends on economic considerations. The extent of F_1 plantings therefore varies according to country and distict. Progress in tomato breeding depends on the inheritance of desired characters, linkage relationships and cytogenetics of interspecific hybrids. Some 250 genes have been allocated to their respective chromosomes. Chromosome arm contents have been well established by cytogenetic identification of centromere positions (Rick, 1974).

9. Epilogue: Desired characters, not otherwise attainable, are sought in induced mutants and other species. Large resources of germplasm are available. Efforts are concentrated on the analysis of flavour constituents, on their genetic control and on the utilization of such knowledge to improve quality. Total yield may receive less attention, in as much as the yield potential of existing cultivars is already very high. The emphasis on mechanized harvest will undoubtedly have a strong impact on the direction of breeding efforts. The objectives sought are compact plant habit, concentrated fruit set, durability of fruits and retention of fruits on trusses to reduce losses from shattering. The appearance of new pests and new strains of existing ones will require constant surveillance and breeding resistance thereto. Resistance is badly needed to certain diseases, notably bacterial wilt.

8.17 LABLAB BEAN

Lablab purpureus (L.) Sweet

[(Syn.) *Dolichos purpureus* L.)] (Leguminosae-Papilionatae)

Resume: *It is native to India. It grows wild and has several cultivars in India. Its archaeological remains from Harappan (ca 2100-1900 BC.), Chalcolithic-Inamgaon and Dharmabad (Maharashtra: India) (ca. 1700-1450 BC. are on record. Probably, it was under cultivation during Harappan period. Chromosome number n=11,12 are reported for it. Tender pods and seeds are widely as vegetable.*

1. Introduction: It is also known as Hyacinth Bean, Seim Bean or Indian Bean. It is native to India, where it grows wild. Now, it is also grown in South and Central America, the East and West Indies.

2. Plant Characteristics: Climber or bushy perennials; leaves alternate, 3-foliate, stipulate, leaflets ovate, entire; flowers purple-violet or white, in erect, axillary long-stalked, stiff racemes; pods 5-15 cm in length, flattened, seeds of different sizes and colour, seeds coat ranging from white to reddish-brown to black with or without speckling.

3. Parts Used: Seeds, pods, foliage.

4. Chemical Constituents: Protein (24.9%), fats (0.8%), carbohydrates (60.1%), fibre (1.4%), ash (3.2%).

5. Uses And Applications: Tender seeds and pods are widely used as vegetable. The dried and ripe seeds are eaten as a split pulse. In India, sprouted seeds are boiled and mashed into a paste which is then fried with spices. Foliage provides hay, silage and green manure. It is also grown as a cover crop. Both husk and broken pieces of pulse (dal) are used as a cattle feed. Green plants are used as fodder for horses and other cattle

6. Karyotaxonomic Evidence: It has chromosome number 2n=2x=22, 24. Basic chromosome number x=11 and 12, is curiously shared with the related genera *viz., Phaseolus* and *Pueraria*. Several cultivar forms are known including scramblers and a tall climbing type. In India, two subspecies with numerous cultivated races are recognized: *viz.,* (i) subsp. *bengalensis* (Jacq.) Verdc., a short-lived perennial twining herb with longer pods, and (ii) subsp. *purpureus*, a long-lived, semi-erect, unpleasant bushy perennial with relatively short pods.

7. History: It is called 'Khalva' in vedic literature *viz.,* Yajurveda and Atharveda and other ancient Sanskrit literature *viz.,* Upanishada. It is also known as 'Nisav' in ancient Sanskrit literature *viz.,* Patanjal's Mahabhasya. It is reported in archaeological remains from Harappan (ca. 2100-1900 BC.), Chalcolithic-Inamgaon and Dhaimabad, Maharashtra (ca. 1700-1450 BC.) and Megalithic-Naikund and Bhagimohari (Nagpur), Maharashtra (ca. 500 BC-400 AD.). Its cultivation as green vegetable since Harappan period is an indication of its domestication (Singh, 2008). Two varieties are recognised *viz.,* (i) *D. lablab* var. *typicus* Praini and *D. lablab* var. *lignosus* Prain. These are not but readily recognised (Synonym is used here). The former is grown as a garden crop, while the latter as field crop. The former is called 'Lablab Bean' and the latter as 'Field Bean' (Anonymous, 1948-1976).

8. Further Developments: The so called 'Lablab Bean' is valued for pods and the 'Field Bean' for the seeds.

9. Epilogue: It is grown as a dry-land crop especially in regions with low rainfall. It is a hardy and drought resistant and grows well on poor soils. In India and Myanmar, it is often cultivated on exposed sandy river banks when monsoon floods recede. In India, it is intersown with other crops. *e.g.* Ragi (*Eleusine coracana*) and also in home gardens.

References:

Acharya, S.N., Thomas, J.E. and S.K. Basu (2006) Fenugreek: An 'Old World' crop for the 'New World'. *Biodiversity* 7(3-4):27-30.

Andrus, C.F., Seshadri, V.S. and P.C. Grirnball (1971) Production of seedless watermelons. *USDA Tech. Bull.* 1425:12.

Anonymous (1948-1976) Wealth of India: A Dictionary of Raw And Industrial Materials. CSIR, New Delhi, India.

Anonymous (1966) Red bhendi - a promising ornamental-cumeconomical type for kitchen garden. *Madras agric. F.* 53:297.

Anonymous (1996) Ayurvedic Pharmacopoeia. Ministry of Health And Family Welfare. Govt. of India, New Delhi, India. pp.1:43.

Appelqvist, L.A. and R. Ohlson (Ed.) (1972) Rapeseed: Cultivation, Composition, Processing And Utilization. Elsevier, Amsterdam, Netherlands.

Babcock, E.B., Stebbins, G.L. and G.A. Jenkins (1937) Chromosomes and phylogeny in some genera of the Crepidineae. *Cytologia, Fujiiluh* 1:188-210.

Bahadur, A. Anand and K.P. Singh (2011) Lettuce. *In:* Future Crops. Vol. I. (Ed. Prof. K.V. Peter) Daya Publishing House, New Delhi, India. pp. 79-91.

Balch, P.A. (2003) Prescription of Dietary Wellness. Avery Pengun Group, New York, USA.

Banga, O. and J.W. De Bruijn (1968) Effect of temperature on the balance between protein synthesis and carotenogenesis in the roots of carrot. *Euphytica* 17:168-172.

Banga, O. and Van J.L Bennekom (1962) Breeding radish for winter production under glass. *Euphytica* 11:311-326.

Becker, G. (1962) Rettich und Radies (Raphanus sativus) Handbuch der Pfianzenziichtung 6:23-78.

Bemis, W.P., Rhodes, A. M., Whitaker, T.W. and S.G. Carmer (1970) Numerical taxonomy applied to Cucurbita relationships. *Amer. J. Bot.* 57:404-412.

Bhaduri, P.N. (1951) Inter-relationship of non-tuberiferous species of Solanum with some consideration on the origin of brinjal. Indian). *Genet. PI. Br.* 2:75-86.

Bohn, G.W. and T.W. Whitaker (1964) Genetics of resistance to powdery mildew race two in muskmelons. *Phytopath.* 54:587-591.

Bohn, G.W., Kishaba, A.N., Principe, J.A. and H.H. Toba (1973) Tolerance to melon aphid in Cucumis melo. *J. Amer. Soc. hart. Sci.* 98:37-40.

Bonnet, A. (1970) Comportement d'une sterilire male cytoplasrnique d'origine japonaise dans les varietes europeenne de radis, Eucarpia (hort. Sect. Rep.), CRNA, Versailles pp. 83-88.

Boswel, V.R. (1949) Our vegetable travellers. *Natl. Geogr. Mag.* 96:145-217.

Boswell, V.R. and L.B. Reed (1962) Okra Culture. Leafl. US Dep. Agric. 499:498.

Brar, J.K., Raj Rai, D., Singh, A. and N. Kaur (2013) Biochemical and physiological changes in Fenugreek (Trigonella foenum-graecum L.) leaves during storage under modified atmosphere pckiging. *J.Food.Sci.Technol.* 50(4):696-704.

Chaudhary, S., Chaudhary, P.S., Chikara, S.K., Sharma, M.C. and M.Iriti (2018) Review of Fenugreek (Trigonella foenum-graecum L.) and its important secondary metabolite diosgenin. *Nat. Bot. Horti Agroba* 46(1):22-31.

Corley, W.L. (1965) Some preliminary evaluations of okra plant introductions. Georgia *Agric. Exp. Stat. Bull.* 145, pp. 16.

Correll, D.S. (1962) The potato and its wild relatives. Renner, Texas, USA.

Coursey, D.G. (1968) The edible aroids. *World Crops.* 20:25-30.

Cubero, J.I. (1973) Evolutionary trends in Vicia faba. *Theoret. appl. Genet.* 43:59-65.

Cubero, J.I. (1974) On the evolution of Vitia faba. *Theoret. appl. Genet.* 45:47-51.

Curry, A. (2010) A 9000-Year Love Affair. National Geograhic 231(2):46.

Curtis, L.C. (1939) Heterosis in summer squash (Cucurbita pepo) and the possibilities of producing F 1 hybrid seed for commercial planting. *Amer. Soc. hort, Sci. Proc.* 37:827-828.

Darwin, C. (1858) Gardeners' Chronicle. Annals of Natural History, 3rd Series pp. 459-465.

De Candolle, A. (1886) Origin of cultivated plants. (Translated from 2nd Ed. in French, 1959) Hafner, New York, USA.

De Candolle, A. (1886) Origin of Cultivated Plants. English Translation of the 2nd Ed. 1967. Hafner New York, USA.

De Vries (1997) Origin and domestication of Lactuca Sativa L. *Genet. Resou. Crop. Evol.* 44:165-174.

Dickinson, H.G. and D.Lewis (1973) Cytochemical and ultrastructural differences between interspecific compatible and incompatible pollinations in Raphanus. *Proc. R. Soc., London* (B)183:21-38.

Dodds, K.S. (1965) The history and relationships of cultivated potatoes. In J.B. Hutchinson (Ed.), Essays in crop plant evolution, Cambridge Univ. Press, Cambridge, UK. pp. 123-141.

Downey, R.K. (1971) Agricultural and genetic potentials of cruciferous oilseed crops. *J. Am. Oil Chon. Soc.* 48:718-722.

Fazli, F.R.Y. and R. Hardman (1968) The spice fenugreek (Trigonella foenum-gracecum L.). Its commercial varieties of seeds as a source of diosgen. *Tropical Science* 10:66-78.

Fenugreek-Wikipedia

https://en.wikipedia.org./wiki/Fenugreek.

Frandscn, H.N. and B. Winge (1932) Brassica napocampestris, a new constant amphidiploid species hybrid. *Hereditas* 16:212-218.

Frankel, R. and E.Gulun (1977) Pollination Mechanisms, Reproduction And Plant Breeding. Springer, Berlin, Heidelberg, New York.

George, R.A.T. (1985) Vegetable Seed Production. Longman, London, UK and New York, USA.

Girenko, M.M. (1968) Classification of Spinach (In Russian).Trud Priklas. *Bot. Genet, Seleckc.* 40:28-36.

Grover, R.K. and G. Singh (1970) Pathology of Okra (Abelmoschus esculentus) caused by fusarium oxysporum *f. vasinfectum*, its host range and histopathology. *Ind. J. Agric. Sci.* 40:989-996.

Hagiya, K. (1952-59) Physiological studies on the occurrence of pithy tissue in root crops. *F. hort, Ass. Japan*, 21:81-86, 165-173; 26:111-120, 121-125; 27:68-77; 28:109-114.

Hanelt, P. (1972a) Zur Geschichte des Anbaues von Vicia faba und ihre Gliederung. *Klilturpji* 20:209-223.

Hanelt, P. (l972b) Die infraspezifische Variabilitat von Vitia faba und ihre Gliederung. *Kulturpfl.* 20:75-128.

Hanelt, P., Schafer, H. and J. von Schultze-Motel (1972) Die Stellung von Vida [aba in der Gattung Vicia und Betrachtungen zur Entstehung dieser Kulturart. *Kulturpfl.* 20:263-275.

Harberd, D.J. (1969) A simple effective embryo culture technique for Brassica. *Euphytica* 18:425-429.

Harberd, D.J. (1972) A contribution to the cytotaxonomy of Brassica (Cruciferae) and its allies. *Bot. J. Linn. Soc.* 65:1-23.

Hawkes, J.G. (1967) The history of the potato. *F. Roy. hort. Soc.* 92:207-224, 249-262, 288-302, 364-365.

Hedrick, U.P. (1919) Sturtevant's notes on edible plants. N.Y. Dep. *Agrie. Ann. Rep.* 27:212-685.

Heiser, C.B. (1969) Nightshades: The paradoxical plants. W. H. Freeman and Co., San Francisco, USA.

Helm, J. (1954) Lactuca sativa in morphologisch-syste matischer Sicht. *Kulturpjlanze* 2:72-129.

Helm, J. (1957) Ueber den typus der Art Raphenus sativus, deren Gliederung und Synonymic. *Kulturpfianze* 5:41-54.

Helm, J. (1963) Morphologisch-taxonomische Gliedcrung der Kultursippen von Brassica oleracea. *Kulturpfianze.* 11:92-210.

Hidvegi, M., El-Kady, A., Lasztity, R., Bekes, F. and S.L. Simon (1984) Contribution to the nutritional chracterisstion of fenugreek (Trigonella Foenum-graecum L.) *Acta Alimentaria* 13(4):315-324.

Hosoda, T. (1947) On the fertility of Raphanus-Brassica and Brassica-Raphanus obtained by colchicine treatment. *Jap. J. Genet.* 22:52-53.

Howard, H.W. (1970) The Genetics of the potato-Solanum tuberosum. Logos Press Ltd. London, UK.

Howard, M. (1987) Traditional Folk Remedies: A Comprehensive Herbal. Century Hutchinson Ltd., London, UK.

Hymowitz, T. (1972) The transdomestication concept as applied to guar. *Econ. Bot.* 26:49-60.

Ibn al-Awwam and Yahya Ibn Muhammad "23.8". Kitab al-Filahah (Retrived July 30, 2014).

Jenkins, J.A. (1948) The origin of the cultivated tomato. *Econ. Bot.* 2:379-292.

Jensen, P.M. and R.R. Kautz (1974) Preceramic transhumance and Andean food production. *Econ. Bot.* 28:43-55.

Jones, H.A. and L.K. Mann (1963) Onions and their allies. Interscience, London and New York, pp. 286.

Josefsson, E. (1967) Distribution of thioglucosides in different parts of Brassica plants. *Phytochem.* 6:1617-1627.

Joshi, A.B., Gadwal, V. R. and M. W. Hardas (1974) Okra. *In:* (Ed. Hutchinson, J. B.), Evolutionary studies ill world crops. Diversity and change in the Indian sub-continent. Cambridge pp. 99-105.

Keimer, L. (1924) Die Gartenpfianze im alien A'gypten. Aamburg and Berlin, Germany.

Kihara, H. and I. Nishiyama (1951) An application of sterility of autotriploid to the breeding of seedless watermelons. *Seiken Ziko* 3:5-15.

Kuckuck, H. and G. Kobabe (1962) Kuchenzwiebel, Allium cepa. Handbuch der Pflanzenzuchtung, 2nd Ed. 6:270-312.

Kumaravel, S., Muthukumaran, P. and K. Shanmugapria (2017) Chemical composition of Trigonella foenumgrae-cum through gas chromatography mass spectrometry analysis. *Journal of Medicinal Plant Studies* 5(3):1-3.

Kuwada, H. (1957) Cross compatibility in the reciprocal crosses between amphiploid and its parents (Abebnosclms esculentus and A. manihot) and the characters and meiotic division in their hybrids obtained among them. *Jap. J. Breed.* 7:103-111.

Kuwada, H. (1970) X-ray induced mutations in okra (Abelmoschus esculemus)]. *Tech. Bull. Fac. Agric. Kagawa Univ.* 21:2-8.

Lehmann, C.O. and F. Schwanitz (1965) Ein Beitrag zur Kenntnis der Formenmannigfaltigkeit der Kulturtornaten (L. esculentum). *Kulturpfl.* 8:545-585.

Lindqvist, K. (1960) On the origin of cultivated lettuce. *Hereditas* 46:319-350.

Linnaeus (1737) Genera Plantarum. Critica Botanica, Leideen, Netherlands.

Linnaeus, C. (1753) Species Plantarum. Silvius. Stockholm.

Mackevic, V.I. (1929) The carrot of Afghanistan. *Bull. appl. Bot., Genet. Pl. Br.* 20:517-557.

Massal, E. and J. Barrau (1955) Pacific subsistence crops taros. South Pacif. *Comm. quart. Bull.* 5(2):17-21.

McCollum, G.D. (1974) Chromosome behavior and sterility of hybrids between the common onion, Allium cepa; and the related wild A. oscharinii. *Euphytica* 23:699-709.

McNaughton, I.H. (1973) Brassica napocampestns (2n=58). I. Synthesis, cytology, fertility and general considerations. *Euphtica* 22:301-309

Munson, W.M. (1892) Notes on eggplants. Maine agric. Expt. Sta. ann. Rep. 1892:76-89.

Muratova, V. (1931) Common beans (Vicia faba). Bull. appl. Bot. Genet. Plant Breed., Suppl. 50 pp. 285.

Nagai, K. and S. Kida (1926) Experiments on hybridization of various strains of Solanum melongena. *Jap. J. Genet.* 4:10-30.

Nandpuri, K.S., Sandhu, K.S. and K.S. Randhawa (1971) Effect of irradiation on variability in okra (Abelmoschus esculentuss *J. Res., Punjab Agric. Univ.* 8:183-188.

Ogura Hiroshi (1968) Studies on the male sterility in Japanese Radish with special reference to the utilization of the sterility towards the practical raising of hybrid seeds. *In:* Memories of The Faculty of Agriculture, Kagoshima University 6:39-78.

Olsson, G. (1954) Crosses within the campestris group of the genus Brassica. *Hereditas* 40:398-418.

Olsson, G. (1955) [Heterosis in spring turnip rape]. Sver. Utsddesfii: Tulskr. 65:215-219 (Swedish).

Olsson, G. (1963) Induced polyploids in Brassica. *In:* (Ed. Akcrbcrg and Hagberg), Recan Plant Breeding Research, Svalof 1946-1961. New York and London.

Orton, W.A. (1911) The development of disease resistant varieties of plants. IV Calif. Int. Genetic Paris, C. R. *et al.,* Rapp. 247-65.

Pathak, G.N. and S.N. Singh (1950) Genetical studies in Lagenaria leucantha (L. vulgaris) *Ind. J. Cell. and PI. Breed* 10:28-35.

Patil, D.A. and D.A. Dhale (2013) Spices And Condiments: Origin, History And Applications. Daya Publishing House, New Delhi, India.

Pearson, O.H., Hopp, R. and G. W. Bohn (1951) Notes on species crosses in Cucurbita. *Amer. Soc. hart. Sci. Proc.* 57:310-322.

Peloquin, S.J., Hougas, R.W. and A.C. Gabert (1965) Haploidy as a new approach to the cytogenetics and breeding of Solanum tuberosum. *In:* (Ed. Riley and Lewis), Chromosome manipulations and plant genetics. Edinburgh and London.

Petropoulos, G.S. (1973) Agronomic, genetic and chemical studies of Trigonella foenum-graccum L. Ph.D. Thesis, Bath University, England, UK.

Placknett, D.L., de la Pena, R.S. and F. Obrero (1970) Taro (Colocasia esculenta). *Field Crop Abstr.* 23:413-426.

Poole, C.F.; Grimball, P.C. and D.R. Porter (1941) Inheritance of seed characters in watermelons. *J. agric. Res.* 63:433-456.

Prem Nath (1970) Problem oriented breeding projects in vegetable crops. SABRAO Netosl. *Mishima* 2:125-134.

Prem Nath and O.P. Dutra (1973) Inheritance of some seed characters in Citrullus lanatus. *Ind. J. Hort.* 30:388-390.

Purseglove, J.W. (1972) Tropical Crops. Monocotyledons 1. Longman, London, UK. and New York, USA. pp. 58-74.

Rick, C.M. (1950) Pollination relations of Lycopersicon esculentum in native and foreign regions. *Evolution* 4:110-122.

Rick, C.M. (1971) The tomato Ge locus: linkage relations and geographic distribution of alleles. *Genetics* 67:75-85.

Rick, C.M. (1974) The tomato. *In:* (Ed.) King, R.C., Handbook of genetics, New York, USA. pp. 247-280.

Rick, C.M. and W.H. Dempsey (1969) Position of the stigma in relation to fruit setting of the tomato. *Bal. Gaz.* 130:180-186.

Rick, C.M., Zobel, R.W. and J.F. Fobes (1974) Four peroxidase loci in red-fruited tomato species: Genetics and geographic distribution. *Proc. Natl. Acad. Sci.* USA 71:835-839.

Robbelen, G. (1960) Beitrage zur analyse des Brassica genorns, Chromosome, II: 205-228.

Robert, R.L., Green, J. and B.Lewis (2009) Lutain and Zeaxanthin in eye and skin health. *Clin.Dermatol.* 27(2):195-201.

Robinson, R.W., Whitaker, T. \to and G. W. Bohn (1971) Promotion of pistillate flowering in Cucurbita by 2-chloroethylphosphonic acid. *Euphytica* 19:180-183.

Roggen, H.P. J. R. and A.J. Van Dijk (1973) Electric aided and bud pollination. *Euphytica* 22:260-263.

Rolland, Jacques Land Sharma, Carol (2006) The Food Encyclepedia. Robert Rose, Toronto, Canada. pp.335-338.

Salaman, R.N. (1949) The history and social influence of the potato Cambridge Univ. Press, Cambridge, UK.

Sampson, H.C. (1936) Cultivated crop plants of the British Empire and the Anglo-Egyptian Sudan. Bull . and the misc. *Inf. Roy. bot Gdn Kew add. Ser.* 12:159.

Saurer, W. and J.V. Possingham (1970) Studies on the growth of Spinach leaves (Spinacea oleracea). *Journal of experimental Botany* 21(1):151-158.

Schafer, H.I. (1973) Zur Taxonomie der Vicia narbonensisfiruppe. *Kulturpfl.* 21:211-73.

Schultze-Motel J. Von (1972) Die archaologischen Reste der Ackerbohne Vitia faba und die Genase cler Art. *Kulturpl.* 19:321-58.

Schweinfurth, G. (1884) Further discoveries in the flora of ancient Egypt. *Nature* 29:312-315.

Schweisguth, B. (1980) Etudes preliminaries a Pamelioration du oireau A. porrum. Proposition d'une method d'amelioration. Ann. Amelior. Plantes 20:215-231.

Shimiru, S., Kanazawa, K., Kono, H. and Y. Yokota (1963) Studies on breeding radish for resistance to virus diseases. *Plant Br. Abstr.* 34:30-50.

Silveira, A.P. Dn, Cruz, B.P.B., Bernardi, J.B. and V Silveira, S.G.P. On (1970) Response of some varieties of Hibiscus esculentus to Verticillium wilt. *Biologico* 36:63-68.

Simmonds, N.W. (1969) Prospects of potato improvement. *Ann Rep. Scott. Pl. Br. Sta.* 48:18-38.

Simmonds, N.W. (1971) The potential of potatoes in the tropics. *Trop. Agriculture Trin.* 48:291-299.

Singh, Ajay (2008) Plants In Ancient India Civilizations. Agam Kala Prakashan, New Delhi, India, pp. 152.

Singh, H.B., Swarup, V. and B. Singh (1975) Three decades of vegetable research in India. ICAR, Tech. Bull., New Delhi, India.

Sinskaja, E.N. (1928) The oleiferous plants and root crops of the family Cruciferae. *Bull. appl. Bot. Genet. Pl. Br.* 19(3):1-648.

Sirks, M.J. (1931) Beitrage zu einer Genotypischen Analyse der Ackerbohne Vicia faba. *Genetica*, 13:209-631.

Sirks, M.J. (1957) Japanese genetica, *Genen en Phaenen* 2:2-10.

Spier, R.F.G. (1951) Some notes on the origin of taro. Southwestern F. *Anthropot* 7:69-76.

Srinivasan, K. (2005) Role of Spices beyond food flavoring neutroceuticals with multiple health effects. *Food Rev. Int.* 21:167-188.

Stubbe, H. (1971) Weitere evolutionsgenetische Unterschungen in der Gattung Lycopersicon. *Biol. Zbl.* 90:545-559.

Sulieman, A. Elhadi (1995) Fenugreek Supplemented Backed Products: Quality Aspects. M.Sc. Thesis, University of Gezira, Wad Meadni, Sudan.

Swamy, K.R.M. and A.T. Sadashiva (2007) Tropical vegetable crops. *In:*Biodiversity In horticultural crops, Vol. I. (Ed. Prof. K. V. Peter and Dr. 2 Abraham). Daya Publishing house, Delhi. India, pp. 109-183.

Thompson, K.F. (1956) Production of haploid plants of yarrow-stem kale. *Nature, Lond.* 178:748.

Thompson, K.F. and J.P. Taylor (1966) The break-down of self-incompatibility in cultivars of Brassica oleracea. *Heredity* 21:345-362.

Thompson, K.F. and J.P. Taylor (1972) Self-compatibility in kale. *Heredity* 27:459-471.

Traub, H. P. (1968) The subgenera, sections and subsections of Allium. *Plant Lift* 24: 147-163.

Tsukamoto, Y. and K. Konishi (1959) Studies on the vernalisation of growing plants of radish. *In:* Mem. Res. Inst. Food Sci., Kyoto Univ. 18:41-47.

Tyagi, I.D. (1973) Heterosis in bottle gourd. *Ind. J. Hort.* 30:394-399.

Ugent, D. (1970) The potato, Science, N. Y. 170:1161-1166.

Vavilov, N.I. (1926) Studies on the origin of cultivated plants. Bulletin of Applied Botany And Plant Breeding. 26:1-248 (In English) Leningrad.

Vavilov, N.I. (1928). Geographical centres of our cultivated plants. *Proc. V Intnl Congr. Genet.* New York, USA. 342-369.

Vavilov, V.N.L. (1926). Studies in the origin of cultivated plants. Trudy Byuro Prikl. Bot. 16:139-248.

Ved Brat, S. (1965) Genetic systems in Allium. III. Meiosis and breeding systems. *Heredity* 20:325-339.

Verghese, Benny M. (1973) Studies on the cytology and evolution of South Indian Cucurbit aceae. Ph. D. Thesis, Kerala Univ., India.

Victor R. Boswell (1949) Garden Peas and Spinach from the Middle East. Reprint of 'Our vegetable Travelers', *National Geographc Magazine* 96:2.

Vitale, J.J. and D.C. Freeman (1985) Secondary sex characteristics in Spinacea oleraceae L.:Quantitative evidence for the existence of at least three sexual morths. *Amer J. Bot.*72:1061-1066.

Wani, S.A. and P. Kumar (2016) Fenugreek enriched extrude product: Optimization of ingredients using response surface methodology. *Int. Food. Res. J.* 23(1):18-25.

Watt, L.E. (1963) Investigations into the breeding system of cauliflowers Brassica oleracea var. botrytis. L Studies of self-incompatibility. *Euphytica* 12:323-340.

Watt, L.E. (1965) Investigations into the breeding system of cauliflowers Brassica oleracea var. botrytis II. Adaptation of the system to inbreeding. *Euphytica* 14: 67-77.

Wendelbo, P. (1971) Alliaceae. *In:* K.H. Rechinger (Ed.), Flora Iranica, Graz, Austria, 76 pp, 100.

Werth, E. (1937) Abstammung und Heimat des Rettichs. *Angew. Bot.* 19:194-205.

Whitaker, T.W. (1949) A note on the cytology and systematic relationship of the carrot. *Prot. Amer. Soc. hort. Sci.* 53:305-308.

Whitaker, T.W. and G.N. Davis (1962) Cucurbits. Botany, cultivation and utilization, Leonard Hill, London, UK.

Whitaker, T.W. and H.C. Cutler (1971) Pre-historic cucurbits from the Valley of Oaxaca. *Icon. Bot.* 25:123-127.

Williams, P.H. and G.S. Pound (1963) Nature and inheritance of resistance to Albugo candida in radish. *Phytopath.* 53:1150-1154.

Wit, F. (1966) The use of inbred lines in turnip breeding. *Qualitas Pl., Hater. veg.* 13:305-310.

Yadav, S. and S. Sehgal (1997) Effect of home processing and storage on ascorbic acid and beta-carotin content of bathua (Chemopodium album) and fenugreek (Trigonella foenum-graceum) leaves. *Plant Foods for Human Nutrition* 50:239-247.

Yang *et al.* (2004) Antihypertensive properties of Spinach leaf protein digests. *J.Agric Food.Chem.* 52(8):2223-2225.

Yao, B.C. and J.R. Geisman (1972) Removal of malathion from spinach by commercial and home preparation procedures. Reserch Circular 188. Oho Agricultural Research And Development Center, Wooster, Oho. pp.1-22.

Yarrnell, S.H. (1956) Cytogenetics of the vegetable crops II. Crucifers. *Bot. Reo.* 22:81-166.

Zohary Danial, Hopf Maria and Weiss Ehud (2012) Domestication of Plants In the Old World: The origin And Spread of Domesticated Plants In South-East Asia, Europe And The Mediterraneal Basin (4th Ed.) Oxford University Press, UK.

Zohary, D. and M. Hopf (1973) Domestication of pulses in the Old World. *Science*, N. Y. 182:887-94.

9

Oils and Fats

INTRODUCTION

Since prehistoric times, oils, fats and waxes have been in use. Probably the fats initially used by man were of animal origin such as tallow, obtained from cattle and sheep, and grease from pigs. They were separated from the other tissues by 'rendering', a process involving heating either in the presence or absence of water. The art of recovering vegetable oils from the oil-bearing materials with the help of primitive mills was known to the ancient Chinese and Hindus much earlier than in any other parts of the world. The old Chinese oil presses were driven by manpower. The primitive Indian extractor known as the ghani', 'chekku' or 'kulhu' was driven by bullocks. Even today, the ghanis are still being used in remote rural areas of India and operates on the principle of a pestle rotating in a mortar, generally made of wood.

In ancient time, the Egyptians and Phoenicians used vegetable oils for food and for anointing their bodies. Egyptian chariot wheels were lubricated with greases consisting of fats and lime as early as 1400 BC. From the Egyptians and Phoenicians, the knowledge of the use of fats and oils spread to the Hebrews and thence to the Greeks. Homer mentioned oil as an aid to weaving. Pliny made difference between hard and soft soaps. These oils and fats are put to endless uses in modern time.

Vegetable oils and fats (fixed or expressed oils) have stimulated little exploitation and exploration because of their universality, abundance and interchangeability or substitution. Their demand has, however, increased continuously with times with their use as a raw material in industrial activities. Our dependence on fatty oils can be expected to increase further because of constant depletion of non-renewable mineral resources.

2. Chemical Composition: Oils and fats contain only carbon, hydrogen and oxygen in a different proportion with very low percentage of oxygen comparatively. Vegetable oils and fats are triglycerides of complex organic fatty acids. Naturally occurring fats are usually mixtures of triglycerides of various fatty acids. The fatty acids are of two principal kinds *viz.,* (i) saturated and (ii) unsaturated. The common saturated fatty acids in fats are palmitic and stearic, the former being the most abundant in vegetable fats. The common unsaturated fatty acids are oleic, linoleic and linolenic acids.

Oils, fats and waxes are very similar in their composition. Oils are liquid at ordinary room temperature, fats are solid or nearly solid. In general, the triglycerides of unsaturated fatty acids are liquid and glycerides of saturated fatty acids with 12 or more carbon atoms are solid at ordinary temperatures. Coconut oil, palm oil, palm kernel oil, cocoa butter and shea butter are examples of fats. Coconut oil is in liquid state in the tropical climate and solid in temperate regions.

On the other hand, waxes are fatty acid esters of monohydroxy alcohols. They occur mostly as protective coverings on the surfaces of the leaves and stems. They greatly minimize the loss of water through transpiration. Waxes are very rarely found within living plant cells. The carnauba and candelilla waxes are amongst the most valuable to mankind.

3. Comparison: Vegetable oils differ from volatile or essential oils (ethereal oils) in a number of important respects: (i) they do not volatilize at ordinary room temperatures, (ii) they cannot be distilled without being decomposed, (iii) they leave a permanent greasy stain on paper, (iv) being glycerides, they form soaps with alkali (v) they lack strong taste and odour of essential oils and may become rancid on long air exposure.

4. Classification: Depending upon the ability to absorb oxygen from the atmosphere, vegetable oils are generally classified into three main groups *viz.*, (i) non-drying oils, (ii) semi-drying oils, and (iii) drying oils.

(i) Non-drying oils: At normal temperatures, they are in liquid state. They are not capable of forming elastic films even after long exposure to air since they do not react with atmospheric oxygen. They are largely glycerides of saturated acids and oleic acid, with little or no linoleic and linolenic acids. The iodine number is less than 100. They never undergo oxidation to form a film, hence, are of no use in the paint, varnish or lacquer industry. They are useful in the manufacture of soaps, as lubricants and as food. These oils are found notably in plants of tropical regions. Groundnut, palm, olive, castor, rape and almond oils are some examples of this class.

(ii) Semi-drying oils: These are intermediate between the drying and non-drying oils. They are characteristic in having large amounts of linoleic and saturated acids but no linolenic acid. The semi-drying oils absorb atmospheric oxygen slowly, producing only a soft film after prolonged exposures to air. They do not form a tough elastic film as is the case in drying oils. The iodine number is between 100 and 130.

(iii) Drying oils: They are fairly rich in glycerides of the unsaturated fatty acids, especially linoleic and linolenic with few oleic compounds. They readily absorb oxygen on exposure to air and form a tough, elastic but resistant film. They are, therefore, used as solvents for pigments in the paint and varnish industries. Drying oils have a high iodine. number, more than 130. Temperate plants. notably linseed, soyabean, tung, safflower and hempseed are important types included under this class.

5. Location of Oils: Vegetable oils and fats occur in the form of small insoluble droplets within the plant cells. They are usually found in seeds, most commonly in the endosperm and cotyledons. Fats occur almost exclusively in the embryo in case of most cereals. Olive and palm oils are two important examples that are obtained from the fleshy pericarp of the fruit. Oils and fats are also extracted from the roots, stems and foliage, but less frequently.

6. Utilization of Vegetable Oils and Fats: Vegetable oils and fats are used, as food, for industrial and pharmaceutical purposes. Fats and lipids contribute to daily diet considerably: (i) They are highly concentrated reserve of energy as they contain very little oxygen. All fats have a high caloric value. One gramme of fat supplies about nine calories, whereas a gramme of carbohydrate or protein produces about four. (ii) fats and other lipids have a role in feeling of satisfaction and delay the onset of hunger as fats retard the rate at which food leaves the stomach. (iii) fats add flavour and variety to foods and make meals palatable and satisfying. (iv) fats act as solvents for the fat soluble vitamins such as vitamins A, D, E, K and pro-vitamin A.

Olive, groundnut, soybean, corn and cottonseed are used as cooking oils and in oleomargarine, which consists of about 80% refined and hydrogenated oil, some 14-16 % cent cultured whole or skimmed milk together with small amounts of emulsifying agents *e.g.* lecithen and monoglycerides, salt, vitamin A and benzoate of soda.

Now-a-days, oils and fats are utilized in the manufacture of non-edible products *e.g.* paints, varnishes, lacquers, linoleum and oilcloth, soaps, detergents, candles, plastics, synthetic fibres, printer's ink, artificial leather, lubricants, polishes and cosmetics. Fixed oils and fats *e.g.* castor and croton oils are used pharmaceutically.

9.1 COCONUT

Cocos nucifera L. (Arecaceae)

Resume: *It is thought to be a native of the Pacific coast of tropical America. The Indo-Pacific region is also interpreted as its place of nativity. It is regarded one of nature's greatest gift to mankind. No truly wild coconut species survive today. It has chromosome number x=16. It is widely used for edible purposes, besides many uses in industrial products.*

1. Introduction: Coconut palm is regarded as one of nature's greatest gift to mankind since practically all its parts are useful in various ways. The place of its origin have been much debated and is a subject of many discussions and speculations. It was believed to be a native of the Pacific coast of tropical America. It's nativity is also discussed taking into consideration the Indo-Pacific region (Vavilov, 1951; Beccari, 1917). The plant is now extensively grown in coastal and deltaic regions of tropical and subtropical countries.

2. Plant Characteristics: A tall, handsome tree reaching a height of 15-30 m, main stem (trunk) is unbranched with a swollen base surrounded by a mass of adventitious roots; crowned by 20-30 large, paripinnate leaves, leaves about 6. 0 m long; male and female flowers borne on a large inflorescence (spadix) subtended by a spathe; mature fruit fibrous, ovoid drupe, differentiated into exocarp, mesocarp and endocarp.

3. Parts Used: Kernels.

4. Chemical Constituents: Coconut oil: Lauric acid (44.51%), myristic (13.1-18.5%), palmitic (7.5-10.5%), caproic (0.2-0.5%), caprylic (5.4-9.5%), capric (4.5-9.7%), stearic (1.0-3.2%), arachidic (0-1.5%), oleic (5.0-8.2%) and linoleic (1.0-2.6%).

5. Uses And Applications: The fresh kernels are consumed raw and used in the preparation of sweets, puddings and chutneys. Coconut milk obtained from tender fruits is a very delicious and refreshing drink. Coconut oil is edible. It is also used in the manufacture of margarine, lubricants, cosmetics (face creams), shampoos, rubber, synthetic resins and hydraulic brake fluids. Its stearin fraction is used in the manufacture of candles

6. Karyotaxonomic Evidence: The genus is monospecific. It has 2n=2x=32 chromosomes. No truly wild coconuts are known and nothing is known of cytogenetic relationships with South American relatives. The species is very variable. It grows throughout the inter-tropical zone. Distinctive local populations have often been given local names. Such names are merely useful provisional labels. There are more so-called varieties in southeast Asia than in Africa or South America. The distinction is made between tall (or common) and dwarf types. They are referred to as 'typica' and 'nana' respectively by some authors. Talls tend to be mainly outbreeding and dwarfs usually inbreeding and smaller fruited. Niu Leka dwarf palms, found in some Pacific islands, resemble talls in characteristics other than height. Dwarfness is, probably, due to a single gene of heterozygous penetrance, so that heterozygotes are semi-tall.

7. History: The nativity of coconut has been much debated and disputed. Some writers have argued in favour of a South American origin, others seek for an origin in Polynesia or Asia. Fremond *et al.,* (1966) discussed the main reasons for considering a southeast Asian origin the most probable. Variability of local names and uses are greater in this region than elsewhere (Wester, 1920; Tammes and Whitehead, 1969).

The spread of coconuts from the main centre of diversity in southeast Asia to other areas can be explained. Places remote from the main dispersal route would have few seeds arriving and less variable resulting populations than places nearer to the main route. Variability would also be low in areas in which initial establishment or subsequent growth was difficult because of unfavourable environments. Large land masses and high elevations would form barriers to dispersal.

The Central and South America is at the ends of both eastward and westward migrations from southeast Asia. The main points are: (i) palms on the Pacific coasts of tropical America are distinct from those in the Caribbean and on the Caribbean/Atlantic coasts (Whitehead, 1965, 1966a, 1968a; Harries, 1971), (ii) it seems likely that spread across the Pacific was from west to east, (iii) spread across the Atlantic was from east to west.

Let us focus the first point of possibility. Caribbean islands and the Atlantic/Caribbean coasts have Talls similar to the Jamaica Tall, whereas on the Pacific coasts, the Panama Tall (previously called San Blas) occurs. There are exceptions to the general pattern *e.g.* in Nicaragua, where Jamaica Tall type palms are found on both coasts, and in Venezuela and Surinam where both types are found (Harries, 1971).

Secondly, there is little barrier to spread through the Pacific from New Guinea to French Polynesia and from there to western tropical America, though the distance between the last two is considerable. This factor would reduce the number of seed-nuts likely to be transported. Spread might have been by drifting. It might have been also assisted by man. The coconuts on Pacific islands more resemble the Panama Tall than the Jamaica Tall in major characteristics (Whitehead, 1966a). Populations in Fiji, Tonga, Western Samoa and New Guinea are more variable than those in French Polynesia. This suggests that there was more frequent spread to these areas or that the environment was more suitable for the crop.

There also appears a third possibility. Coconuts in the Caribbean islands and on the eastern coasts of tropical America broadly resemble those of West Africa (Fremond *et al.*, 1966; Whitehead, 1965, 1966a). They are basically similar to those of East Africa, the Seychelles, Mozambique, Sri Lanka and some varieties in India. Palms bordering the Atlantic are later introduction. Although the Spaniards found coconuts on the western coast of Panama, there were none on either side of the Atlantic at the time of Columbus (Purseglove, 1963; Cook, 1910). It, therefore, appears plausible to assume that Atlantic-American palms arrived relatively recently from southeast Asia via Africa. Portuguese shipments to Brazil and Spanish introductions to Puerto Rico would have been possible in the 15th century. The spread of coconuts to East Africa from southeast Asia perhaps via the Seychelles, presents no serious difficulty. The number of seeds initially introduced into West Africa was probably small. The most possible route would have been overland. Fremond *et al.*, (1966) remarked for the genetic poverty of West American populations.

The Jamaica Tall and the Panama Tall coconuts appear distinct geographical races at opposite ends of the range, the tropical American mainland forming a barrier that has, until recently, prevented the merging of the two populations. The relatively slow germination of the Jamaica Tall supports the broad pattern of dispersal described (Whitehead, 1965). Diseases *e.g.* 'Lethal Yellowing' and 'Cadang-Cadang' and pests *e.g.* 'Rhinoceros beetle' have exercised significant selection in some countries. Similar consideration is also for peculiar local environments *e.g.* coralline soils, drought and high rainfall (Fremond *et al.,* 1966). Hurricanes have influenced coconut evolution, initially reducing variability but sometimes necessitating seed introduction from neighbouring areas.

Relatively isolated islands in the South Pacific, *e.g.* Rennell, Rotuma and the Wallis Islands, have a high proportion of large-fruited palms. These probably originated from fruits selected by fishermen for large nut size. In the New Hebrides and the Solomons, fruit size is small but nut numbers are high. This sort of situation might be the result of initial selection for high nut numbers when plantations were established. Dwarf palms have been preserved and propagated in many areas since they are ornamental, early bearing and easily harvested. They are often uniform, having originated from very few seeds. In Jamaica, two palms from seeds introduced in 1940 had over 15,090 descendants by 1966 (Whitehead, 1966b).

8. Further Developments: Outstanding parents, high yielding and good combining ability, were identified but most of the seed released to growers has simply selected from healthy palms which appeared to be high yielding. Systematic breeding was begun only recently. Some basic research was done, notably in Indonesia in 1930-42 and in India and Sri Lanka. Attempts were made to assemble collections but numbers of seeds successfully introduced were usually few and fairly unrepresentative.

Coconut breeders always need attention on dependence on seed production, small number of nuts per bunch and per pollination, long generation time and the large amount of land needed, shortage of information on coconut botany (including variability) and breeding techniques, cost of prospecting and collecting on any worthwhile scale. The high priority for objectives in coconut breeding can be: (i) resistance to diseases and, where possible, pests, (ii) early bearing, (iii) high yield of copra, (iv) varietal adaptability. The emphasis depends on the local conditions.

Geethalakshmi *et al.* (2005) studied genetic diversity among coconut (*Cocos nucifera* L.)genotypes using isozymes leaves of 30 cultivars. Ten enzyme systems viz., alcohol dehydrogenese, glutamate exalocetate transaminase, glucose-6, phosphate dehydrogenase, peroxidase, esterase, alpha-amylase, phosphorylase, malate dehydrogenase, super oxide dismutase and aid phosphatase were studied using discontinuous polyacrylamide get electrophoresis. Twenty loci and 40 alleles were observed, of which, 14 loci were polymorphic. Seven loci were heterozygous. Null alleles were observed to ACP, ADH and MDH. Greater heterozygosity was observed for glucose-6 phosphate dehydrogenase and least for super oxide dismutase. Among the cultiars, Nudora tall and Calangute Tall showed maximum heterozygosity while Kulasekharam Green Dwarf showed least heterozygosity.

Triura *et al.* (2018) studied the performance of indigenous and exotic coconut genotypes for yield and nut quality. Wide variations on vegetative parameters were observed within genotypes. Among the 19 genotypes, the maximum vegetative characters *viz.*, palm height (15.70m) in Java Tall, trunk girth (1.26 m each) in FMS and Andoman Giant, number of functional leaves (38.25) in Ceylon Tall petiole length and length of leaflet bearing portion (4.16) were recorded in Seychelles Tall. The maximum number of bunches per palm per year and nut yield per palm per year were recorded by Tiptur Tall (10.13 and 67.25). The maximum nut characters *viz.*, whole nut weight (1999.509), dehusked nut weight (1323.009), kernel weight (510.00g) and copra content (390.50g) were recorded in Andaman Giant. However, the maximum nut husk weight was recorded by San Roman (677.50g) compared to lowest record by Arasampatti. Tall (187.099). The number of nuts were maximum in genotype Seychells Tall. But genotype Andaman Giant performed better than other genotypes if quantitative and qualitative characters considered.

9. Epilogue: Coconuts are outbreeders. Heterosis has been demonstrated in a wide range of inter-origin crosses which is partly a reflection of the narrow genetic base of most local strains. Since clonal propagation on a commercial scale is inconceivable with present technology, improvement is likely to depend upon the systematic identification of parental lines of good combining ability. The product may be either improved populations or 'hybrids'. Fairly rough control of hybridity would be economically acceptable, especially if hybrids could be identified in the nursery. In the longer term, the outcome is likely to be extensive recombination between local ecotypes that were previously isolated and themselves too narrowly based genetically to provide an effective source of improvement.

The Malayan Dwarf has local acceptance. It has been significant in providing a valuable supplement to traditional stocks. It has performed well on its own and yielded productive hybrids. It has also proved a source of resistance to Lethal Yellowing disease in Jamaica where most of its crosses show resistance (Whitehead, 1966b, 1968b). There will be continuing need for exploration, collection, maintenance and botanical analysis of local populations, for genetic analysis of economic characters and disease resistance and for further search for disease resistance coupled with the development of rapid screening methods.

9.2 FLAX AND LINSEED

Linum usitatissimum L. (Linaceae)

Resume: *Linseed is one of the earliest plants to have been recorded for extraction of fibres in the Mediterranean region and Southern Asia. It attained prominence as a dual-purpose crop, as a source of both fibre and oil. It was cultivated for 1000 or more years BC. in Egypt and the Middle East. The Egyptians used the linen to wrap their mummies and linseed oil was used in embalming them. The genus is very variable in chromosome numbers ranging from x=8 to 16, Linun usitatissimum has n=15 chromosomes. The oil is used for edible purposes and in the manufacture of industrial products.*

1. Introduction: The history and development of oil-yielding and fibre-yielding plants has gone hand in hand in the past. Linseed has attained prominence as a dual purpose crop. *Linum usitatissimum* is an annual grown for its fibre (fibre flax), or for its seed (oil flax, seed flax, linseed), or for both (dual purpose flax). It is a national flower of Belarus.

Flax, possibly *L. angustifolium,* is associated with the earliest records of civilization in the deposits of the Swiss Lake Dwellers, and *L. usitatissimum* was cultivated for a thousand or more years BC. in Egypt and the Middle East. Linen was worn in ancient Egypt and the Egyptians used it to wrap their mummies. Linseed oil was used in embalming. The early Greeks and Romans also cultivated flax for fibre and seed. The Romans probably stimulated its culture in Britain where they later established a linen industry. The plant (or its products) is mentioned in the Bible. References reveal the associated vocabulary of the time, *e.g. linteum domi retum* (homespun linen), *negotiatio lintea* (linen trade). 'Flax' and 'flaxen' are from the Anglo-Saxon *fleas.* Elsewhere, fibre flax has been grown from the earliest times in Europ, Russia and Central Asia. Linseed oil was used in ancient rituals in India.

2. Plant Characteristics: An annual plant with a slender stem; leaves alternate, small ovate or lanceolate; flowers sky-blue or white, borne in loose terminal racemes or open cymes; fruits indehiscent, globular capsules.

3. Parts Used: Seeds.

4. Chemical Constituents: Seed oil: Stearic (6-16%), palmitic (6-16%), oleic (13-36%), linoleic (10-25%), and taces of myristic and arachidic acids.

5. Uses And Applications: Raw cold pressed oil is used for edible purposes in India. Linseed oil is used chiefly in making paints, varnishes, oil-cloth, linoleum, soft soap, and printing ink.

6. Karyotaxonomic Evidence: The genus *Linum* contains nearly 200 species spread over the temperate and warm temperate zones of the northern hemispheres. They are abundant in Europe and Asia but with about 50 species in America. Species are inbreeding or outbreeding annual or perennial herbs or half-shrubs. The chromosome numers in the genus are remarkably variable (x=8, 9, 10, 12, 14, 15, 16) but there are evident nodes at 2n=2x=18 and 2n= 2x=30. Flax belongs to the second, 2n=30, group. Bivalents have been found in at least one haploid (n=15) plant. Hybrids have been obtained between several species of the 2n=18 group and between several of the 30 group, many showing one or more translocations (Gill and Yermanos, 1967). The *L. pratense* group (2n=18) in North America is close to the *L. perenne* group (2n=18) of Eurasia, It is thought to have entered via Siberia and Alaska. The higher numbered groups (2n=30, 36) were probably separate introductions to America. Many species have been assigned more than one chromosome number *e.g. L. usitatissimum*, 2n=30, 32.

Cultivated flax is one of a well-defined group of North African and Eurasian-species with 2n=30 chromosomes: *L. africanum, L. angustifolium, L. corymbiferum, L. decumbens, L. neruvsum* and *L. pallescens*. They produce fertile hybrids (Gill and Yerrnanos, 1967) and are possibly variants of a single species. *Linum usitatissimum* is the nearest rest to the highly variable *L. angustifolium* (2n=30, 32), which is strongly branching and tillering perennial or biennial with dehiscent capsules. It is common in southern Europe and western Asia, less frequent in western Europe, with which it is easily crossed and differs from it cytologically by one translocation. The flowers of cultivated flax are adapted for self-fertilization, although cases of substantial cross-fertilization have been recorded. *L. angustifolium* cross-fertilizes more readily and is said to outcross naturally with *L. usitattssimum*.

7. History: Based on a study of more than a thousand indigenous types in Europe and Asia, Vavilov (1951) defined two geographical areas, one in southwestern Asia (India, Afghanistan, Turkestan) containing seed types, and the other with fibre types in the Mediterranean countries (from Spain and Algeria to Greece and Egypt), with a transitional region (Asia Minor, Caucasus, Black Sea and Caspian Sea area) containing dual purpose forms. The great diversity of forms found in India suggested that flax originated here and spread northwards and westwards. The tall unbranched fibre types have a shorter vegetative period than the shorter more branched seed types. The indigenous fibre types of the USSR could have arisen by natural selection from the southwest Asian seed types as they spread northwards. The Mediterranean fibres may derived from the dual purpose forms in the transitional area or, possibly, from another centre of diversity in Abyssinia; or they could have spread from Russia via Finland and eastern European countries. Others have suggested that fibre types originated in the Baltic region and became shorter, more branched and with larger seeds as they spread southwards. Indian linseeds may have been derived from Central Asian, Egyptian and Abyssinian forms. Any southerly migrations that occurred were subsequent to the spread northwards at an earlier period.

In a nutshell, the origin of cultivated flax is uncertain. The most likely progenitor is the widely distributed, biennial or perennial. *L. angustifolium* but other species of the group could have contributed. Domestication involved selection for annual habit and non-shattering capsules and more efficient self-fertilization. The place of origin is also unknown. There could have been several centres now obscured by subsequent migrations. But procumbent linseeds probably appeared first, then the erect, followed by dual-purpose and finally fibre types.

8. Further Developments: Seed was taken to the New World in the 17th century by the early Colonists. Cultivation migrated westwards during the next 200 years as new settlements arose because, apart from supplying fibre for hand spinning, it acted as a good first crop on newly broken ground and wilt disease reduced the acreage sown on old cultivated land. Fibre flax cultivation then gradually declined to a negligible level in the United States. It was replaced by linseed which later became an important crop. Large quantities of linseed are cultivated in Argentina and in Canada where fibre flax cultivation also continued simultaneously.

Flax remained the most important plant fibre among western nations almost up to the beginning of the 20th century when cotton took its place. In Britain, fibre flax acreage declined from a peak in the 1860s, when large quantities were grown in Scotland and Northern Ireland. Both linseed and fibre crops showed temporary increases in acreage under the stimulus of the Second World War. A successful programme of fibre flax cultivation was speedily initiated in New Zealand (Hadfield, 1953).

Many cultivar groups are identified in the USSR, from the flaxes of the Baltic region (Riga, Pskov flax), tall, unbranched, fine quality fibre flaxes. These contributed to the development of many of the world's best fibre flaxes, to the drought and frost-resistant linseeds of Central Asia. At one time Russia was practically the sole exporter of fibre flax seed.

Flax is grown in India for its seed. There are two main linseed types, both considerably different from the Russian: (i) the one grown in the river alluvium of the Ganges and the Indus with reduced taproot and numerous secondary roots near the soil surface and (ii) the other (Peninsular type) grown further south, with deeply penetrating taproot and laterals at a greater depth. Genetically, the difference could be small, because an Indo-Ganges type has been derived by irradiating Peninsular type plants.

After 1900, flax breeding began to develop. At first by 1930, several countries had produced improved lines. The Stormont (Northern Ireland) varieties, and the Liral series bred by the Linen Industry Research Association at Lamberg, near Belfast. Flax is an inbreeder and, naturally, pedigree pure line breeding methods were adopted. Early studies of the inheritance of wilt disease reaction in the USA resulted in the cultivation of wilt resistant linseeds. Flor (1955) in Canada provided evidence for a gene-far-gene relationship between virulence of flax rust and resistance in the host plant. He later developed flax varieties for identifying rust genotypes.

Apart from disease resistance, other characters which received attention are: seed size, yield, oil content; iodine number, height, uniform ripening and drought, cold and herbicide resistance. In fibre flax, fine quality and length of fibre have to be balanced against lodging resistance. The 20th century witnessed the replacement of local populations by improved cultivars.

Chandrawati *et al.* (2014) carried out AFLP fingerprinting of 45 Indian genotypes of linseed (*Linum usitatissimum* L.) to determine genetic relationship among them. Sixteen primer combinations produced 1142 fragments with 1129 as polymorphic and 13 as monomorphic fragments. The PIC value varied from 0.19 to 0.31 and genetic similarity from 0.16 to 0.57 with an avegage of 0.25 ± 0.05. The maximum genetic similarities (57%) were detected between genotypes Him Alsi-1 and Him Alsi-2, followed by G541 and LC-54. The genotypes R-552, Himani, RKY-14, Meera, Indira Alsi-32 and Suyog were found to be more divergent genotypes. The NJ clustering grouped all the 45 genotypes into three major clusters. In general, the genotypes of cluster III had high oil content and those of cluster I had low oil content. At the population level, within population variance was much higher than between populations variance.

Arpna Kukmari *et al.* (2018) reported genetic diversity in a collection of 28 genotypes of flax (*Linum usitatissimum* L.) including nine exotic ones. They used ramdamly amplified polymorphic DNA (RAPD) and inter-simple sequence repeat markers (ISSR). The characterized genotypes showed high level of genetic variation with respect to oil and fibre contents of genotypes. Altogether 27 markers amplified 130 fragments with an average of 4.8 fragment and mean polymorphism information content and marker index values of 0.325 and 1.90 respectively. Jacards coefficient based dendrogram grouped all genotypes into three groups and separated genotypes based on their oil and fibre contents. STRUCTURE analysis showed two genetic stocks for all the analysed genotypes and recorded higher admixture in indigenous germplasm. Ovarall, the characterized genotypes were highly diverse genetically and can be used in future breeding and improvement programmes of this crop. Genotypes *viz.* KL-257, KL-236, Surbhi, Bahgsu and Himani can be used as diverse parents in breeding programmes for generating improved lines in context to oil content.

9. Epilogue: There is sufficient reserve of intraspecific genetic variation among the innumerable forms of *L. usitatissimum* to continue to supply breeders with material for crop improvement. Irradiation has also provided mutations of potential economic value such as larger seeds, longer straw, increased fibre content, disease resistance and high seed yield (Bari, 1971). *Linum usitatissimum* can be crossed easily with the (2n=30) species listed above. These fertile hybrids could be an important source of genetic material, including genes for disease resistance, and they are also the most promising for further studies on the origin of cultivated flax. Other crop improvement possibilities exist among the derivatives of crosses with autotetraploids (2n=4x=36) of species with 2n=18 chromosomes.

9.3 GROUNDNUT

Arachis hypogaea L. (Leguminosae- Papilionatae)

Resume: *Arachis is a South American genus and has a centre of diversity in the Mato Groso of Brazil. It is an allotetraploid possessing 2n=4x=40 chromosomes. The earliest archaeological records are from Peru dating 2000-3000 BC. It is geocarpic and grown chiefly for its vegetable oil but the protein fraction presents a favourable amino acid profile. It is the second largest source of vegetable oil.*

1. Introduction: It is native of South America, particularly Brazil. It then spread to Peru, Argentina, Ghana and then to Jamaica, Cuba and West Indies. It was unknown to the Old World before the discovery of America. In the 16th century, the Portuguese brought them from Brazil to West Africa and the Spaniards took these across the Pacific to China, Japan, Malaysia, India and Malagasy Republic. According to Ninan (1989) and Gregory *et al.,* (1951) groundnut is introduced into India during first half of the 16th century. It is introduced from Africa to Western India or from pacific islands of China to India. The commercial development has come about since the Civil War. As many as 20 kinds of peanuts are grown selecting habit and pod size.

2. Plant Characteristics: A low-growing annual 1.5 to 3.0 ft in height, much branched; leaves compound with two pairs of opposite leaflets; flowers yellow, borne on lower nodes, singly or 2 to 4 together, chasmogamous and cleistogamous, fruit elongated oblong, indehiscent pods with 1 to 3 seeds, hull (shell) fibrous.

3. Parts Used: Seed, foliage.

4. Chemical Constituents: Nuts contain protein (26%) (Arachin and conarachin important ones), oil (45%), phosphrous, vitamins (thiamine, riboflavin, niacin), fatty acids oleic (56%), linoleite (25%), palmatic (6-12%) and little stearic, arachidic and higher saturated acids.

5. Uses And Applications: The seeds (peanuts) are used for roasting or salting, in candy, and for preparation of peanut butter. Peanut oil is important food oil. The oilcake is used as fodder. Its protein is used for manufacturing ardil (a synthetic fibre). Its oil, on hydrogenation, is made into vegetable ghee. The kernels are used in various foods and confectionery. Some products *e.g.* groundnut milk, peanut ice-cream, peanut massage oil for infantile paralysis, lubricants, etc. are obtained from it. Hulls are used as filler for fertilizers or ground into meal for insulation blocks. The foliage constitute a good green or dry fodder for cattle.

6. Karyotaxonomic Evidence: *Arachis* is a South American genus. Most are diploids with 2n=2x=20. Seven sections are recognized by Gregory *et al.,* (1973) and six by Krapovickas (1973), of which one section Arachis (Axonomorphae) is of our interest. The entire genus has a centre of diversity in the Mato Groso of Brazil. Species in areas of high rainfall are perennial. Those growing in semi-arid areas are annuals. Section Arachis contains annual and perennial diploids and two allotetraploids with 2n=4x=40. Of these, *A. monticola*, from northwestern Argentina, is an annual that crosses freely with cultivated *A. hypogaea*. It is supposed to be the current wild descendant of the amphidiploid species ancestral to the cultigen.

Section Arachis possessed a pair of small 'A' chromosomes similar to that reported by Husted (1936) in *A. hypogaea*. This distinctive pair was lacking in the erectoid form *A. paraguariensis* and he suggested that the ancestral form of *A. hypogaea* could have arisen by hybridization between two diploid species having these distinct karyotypes.

The somatic chromosomes of the hybrid *A. hypogaea* × *A. helodes* showed that only a single 'A' chromosome occurs (Smartt and Gregory, 1967). The implication is that *A. helodes* (perennial, sect. Arachis) does not possess a pair of , A' chromosomes. Since both *A. helodes* and the undescribed annual species belong to the section Arachis, the erectoid *A. paraguariensis* is not required to account for the 'A' chromosome constitution of *A. hypogaea*. In fact, an interspecific hybrid between such annual and perennial species of sect. Arachis is a distinct possibility. Such a hybrid would have a karyotype not unlike that of *A. hypogaea-monticola*. However, an experimental re-creation of *A. monticola-hypogaea* is a very real possibility and would be of considerable interest.

Thus, the groundnut is an allotetraploid thought derived directly from a wild allotetraploid (Krapovickas and Rigoni, 1957). That the wild diploid ancestors of this tetraploid were probably fairly related. This is suggested by the observations of occasional meiotic multivalents in *A. hypogaea* and also suggestive of segmental allotetraploidy (Husted, 1936; Gregory *et al.,* 1951; Smartt, 1964). All members of *Arachis* are geocarpic and, presumably all, like the cultivars, are highly inbred. Wild species can sometimes be intercrossed but the hybrids are usually more or less sterile. (Smartt and Gregory, 1967; Gregory and Gregory, 1967).

7. History: The crop was domesticated well to the southeast and must be presumed to have had a much longer history. It was, in fact, domesticated by the predecessors of modern Arawak-speaking people. The earliest archaeological records are, however, from Peru, 2000-3000 BC. (Krapovickas, 1968; Hammons, 1973).

A. hypogaea has more robust, non-fragile pods and shorter carpophores than wild *A. monticola*. These changes, which would be disadvantageous in the wild, were brought about by early human selection. In subsequent spread and diversification, a great deal of variability in plant habit and seed characters have accumulated.

Krapovickas (1968, 1973) has distinguished five major geographical groups of cultivated groundnuts in South America. These appear representing secondary centres of diversity that have developed from a primary centre of domestication in southern Bolivia and northern Argentina. The longstanding notion of the Brazilian origin of *A. hypogaea* (Chevalier, 1933) does not receive support from current data.

The differences between the geographical groups in relation to established intraspecific classification can be briefed as follows (Krapovickas, 1968, 1973). *Arachis hypogaea* comprises two subspecies and four varieties, the former differing in branching habit.

	Name & Taxon	Type	Characteristics
A.	subsp. *hypogaea*		No floral axes on main axis; pairs of vegetative and floral axes alternating along lateral branches.
	1. var. *lypogaea*	Virginia	Less airy, with short branches
	2. var. *hirsuta*	Peruvian Runner	More hairy, branches long.
B.	subsp. *fastigiata*		Floral axes on the main axis; continuous runs of one-many floral axes along lateral branches.
	1. var. *fastigiata*	Valencia	Branching little.
	2. var. *vulgaris*	Spanish	Branching more.

These four basic varietal types are to some extent characteristic of the secondary centres of diversity. Centre-I contains mostly B (fastigiata) types and is thus the centre of Valencia Spanish groundnuts. Centre-II is also mostly B. Centre-III contains a unique form of AI. Centre-IV is mostly Al but also shows some Bl types and gives evidence of inter-subspecific hybridization. Centre V contains AI, A2 and a special fastigiate form not found in Centres-I and II. Centre VI may be regarded as tertiary.

Some slight genetic isolation between the subspecies, presumably a reflection of long separation in the secondary centre is evidenced (Krapovickas, 1973; Wynne, 1974). Groundnuts were generally distributed, by the time of Columbus, through South America, the Caribbean and Mexico and were repeatedly recorded (Krapovickas, 1968; Hammons, 1973). There is no evidence that the crop moved out of the western hemisphere before this period.

It probably reached eastern Asia by way of Mexico and the Philippines, in the hands of Spanish travellers; Africa from Brazil by Portuguese contacts about 1500 and, later, India via Africa. Varietal types and diversity present in Asia and Africa broadly agree with these views (Gibbons *et al.,* 1972). First introductions to temperate North America were probably made in the 17th century from Africa (Higgins, 1951), although the crop was already established in the Caribbean, Central America and Mexico at that time (Hammons, 1973). By the middle-to-late 19th century, groundnuts had assumed considerable agricultural importance in the USA, India and West Africa.

8. Further Developments: Early in the 20th century, improvement of the crop by individual on plant and line selection began (Gregory *et al.,* 1951). The groundnuts of A1 are nearly strict inbreeders; those of B1 and B2 somewhat less so. Experimental crossing on any scale is difficult. Earlier efforts at improvement by plant and line selection were useful. These were but inevitably limited in scope. Hybridization, followed by pedigree selection of pure lines, was not fully adopted, even in the USA, until very recent years (Gregory *et al.,* 1951; Norden, 1973).

Borkar and Dhoranguttikar (2014) assessed groundnut (*Arachis hypogaea* L.) genotypes for physiological traits. There evaluated 22 genotypes during summer, 2011. They made observations plant height, number of branches, leaves, leaf area, LAI, dry matter production and its distribution in component parts of plant, photosynthetic rate, transpiration rate, stomatal conductance, water use efficiency, stomatal frequency, chlorophyll content, protein and oil content and yield, contributing characters. The genotypes significantly differed for seed yield. The highest dry pod yield was recorded by the genotypes *viz.,* ICG-8420, ICG-8473 and ICG-8506 due to significant favourable yield contributing characters such as number of pods per plant, pod yield(g) and kernel yield (g) per plant, shelling percentage and harvest index. Based on morpho-physiological traits and biochemical parameters genotypes *viz.,* ICG-8420, ICG-8473, ICG-8506, ICG-0845, ICG-8316 and ICG-8525 were considered as promising genotypes for future breeding programs for yield improvement and protein and oil content.

Souina *et al.* (2016) studied genotype x environment interaction and kernel yield-stability of groundnut (*Arachis hypogaea* L.). They was kernel yield of 13 varieties for revealing genotype by environment interaction (GEI) in Cameron, using stability parameters, additive main effects and multiplicative interaction analysis (AMMKI). Combined analysis of variance showed significant differences between genotypes, locations and GEI, suggesting differential response of varieties across testing locations, and the need for stability analysis. Differences between genotypes, environments and GEI accounted respectively for 68.91%, 16.00% and 15.08% of the total variation. Highest kernel yield of 2105, 1900, 1882, 1845 and 1841 kg ha^{-1} was recovered from genotypes *viz.,* Daurou, Bhani, Gobo-55-432 Ad-

Manipenta and JL 28 respectively. Genotypes NW-Red Esimbi and Quest-A2 gave the lowest kernel yield of 454 and 1284 kg ha^{-1}. Respectively. Two IPCA of AMMI were significant and captured the largest portion of variation of the total GEI. Stability analysis identified Dourou as the last lines due to its stability and high yield, while Gobo-55-437 and Blanc were specifically adated to Atoa and Gobo locations.

9. Epilogue: The peanut is one with great breeding potential. There is much genetic variability, both in cultivars and wild species. These helped utilization as a new source of adaptation and of resistance to pathogens. Several hundred interspecific hybrids of *Arachis* and many of their amphidiploid derivatives already exist. These amphidiploids combined among themselves and with *A. hypogaea* promise the release of a great deal of new variability.

9.4 MUSTARDS

Brassica Spp. (Brassicaceae)

Resume: *Mustards belong to the genus Brassica, which is native of north temperate parts of the Old World, particularly the Mediterranean region. It is frequently mentioned in the Bible and in the extensive literature for Greek and Roman times onwards. The cultivated species are either amphidiploids or allotetraploids, basic chromosome number being x=8 chromosomes. The oils and seeds obtained from different species are used for edible purposes and also as a spice.*

1. Introduction: The genus *Brassica* contains over 150 species. Some of them are useful multipurposely. They are used as herbage vegetable, spice and oils. Some species are used for dual purpose. However, these purposes differ country to country. They are mostly native to the north temperate parts of the Old World, especially the Mediterrnean region. These species of *Brassica* are generally referred by the term 'mustard' which is believed to be derived from the use of seeds as condiment. The sweet 'must' of old wine was mixed with crushed seeds to form a paste, called 'hot must' or 'mustum ardens', and hence mustard. It is one of the oldest recorded spices in India. It is mentioned in ancient Sanskrit literature dating back to about 3000 BC (Mehra, 1968). It is frequently mentioned in the Bible and also in the extensive literature from Greek and Roman times onwards (Rosengarten, 1969). Four species of *Brassica* are commonly involved in literature and even modern uses *viz.,* (i) *Brassica alba* (L.) Rabenh. (Syn. *Synapsis alba*): White Mustard, (ii) *B. nigra* (L.) Koch. : Black or True Mustard, (iii) *B. juncea* (L.) Czern & Coss. Brown or Indian Mstard, (iv) *B. carinata:* Ethiopian Mustard. Taxonomists agree that considerable confusion exists as to the identification and nomenclature of species within the genus *Brassica.*

2. Plant Characteristics: In general, mustard plants slender, erect, branches, annuals reaching 0. 6 to 1. 5 m in height; leaves lyrate and auricled; flowers yellow, typically cruciferous, borne in corymbose racemes; fruits siliqua containing small, spherical, yellow-brown to black in colour.

3. Parts Used: Seeds.

4. Chemical Constituents: Seeds: protein (20%), erucic acid (40-50%), oleic acid (47%), linoleic acid (47%), ad palmitic, stearic and linoceric acids in very small quantities, also a glucoside (Sinalbin, mucilage, a fixed oil).

5. Uses And Applications: (i) White Mustard: It is used as a condiment in food stuffs and pickles. The seeds yield fatty oil. (ii) Black mustard: Ground black mustard seeds are used as a condiment in pickles. The seed oil is used for cooking purposes. (iii) Indian (Brown) mustard: The seeds are used as a condiment in pickles. The seeds oil is used for salads, pickles and cooking purposes. (iv) Ethiopian mustard: Cake after oil extraction is used to feed animals. Young leaves are used as vegetable, oil may be used as biofuel.

6. Karyotaxonomic Evidence: The common nomenclature of mustards remains somewhat confused. 'White' mustard, *B. alba* (syn. *Synapsis alba*) is known as 'Yellow', *B. hirta* in North America; the names 'Brown', 'Black' and 'Indian' are to a degree interchangeable, and there are cultivars of 'Brown' with yellow seed coats.

Brassica nigra (x=8) is one of the three basic diploid cultivated *Brassica* species which form amphidiploids described by Japanese workers, principally U (1935). The other two *Brassica* mustards are allotetraploids. These amphidiploids have been synthesized by Frandsen (1943) and others. Considerable additional evidence on cytotaxonomic relationships in *Brassica* has recently been reported by Harberd (1972) and Vaughan *et al.,* (1963). They have explored serological and protein analysis methods in further elucidation. Studies by Vaugnan *et al.,* (1963) based on chemical and morphological variation in *B. juncea* led to the identification of three secondary centres of diversity, that x=10 Brassicas other than *B. campestris* (*e.g. B. japonica, B. pekinensis, B. trilocularis*) could have been parents of *B. juncea*. However, these three species are sometimes treated as subspecies of *campestris*. Number of chromosomes with x=8 is recorded other than *B. nigra*.

7. History: *Sinapis* alba is a wind pollinated species with a sporophytic incompatibility system as in various horticultural Brassicas and kales. Much material, however, seems to sib-compatible. The centre of origin is thought to be eastern Mediterranean. Wild forms occur around most of the Mediterranean littoral, especially in the Aegean. They are low growing, much branched, short-season dehiscent plants, often with brown or black seeds of the form recorded as *melanosperma*. One feature of wild material is the presence of a single seed in the siliqua beak in addition to 1-3 in each of the two locules. As such a seed is unable to germinate until the beak has rotted away. It has been suggested that this is a natural survival mechanism. The feature has virtually disappeared from cropped land-races, perhaps by unconscious human selection. In bred cultivars, the seed number per locule may rise as high as seven. Seed weight per pod is a more constant factor than seed number.

Brassica nigra probably originated in the Asia Minor Iran area. It was so early used as a commercial spice. It became widespread in Europe, Africa, Asia, India and the Far East. Early forms are of short season, spreading, semi-erect growth, up to one metre tall. Selection of more erect, taller material has evidently long been practised though no known land-race or cultivar.

The primary centre of origin of *B. juncea* is considered to be Central Asia-Himalayas, with migration to three secondary centres in India, China and the Caucasus. The strong evidence of diversity in several aspects reported by Vaughan *et al.,* (1963) could suggests independent hybridization of *B. nigra* with local n=10 species (or subspecies) in those areas. Yellow seeded forms arose in the Orient. It is also thought that *B. carinata* arose from *B. nigra* by crossing with a *B. oleracea* form in north-eastern Africa. Selection of both species for use as vegetables must have occurred over long periods of time in many areas, particularly of *juncea* in south-east Asia and China.

8. Further Developments: In many European countries from the Middle Ages, land-race forms of *B. alba* and *B. nigra* were cultivated and acquired joint origin/trade nomenclature as, *e.g.* 'English', 'Danish', 'Dutch', 'Hungarian' White, 'Sicilian', 'Bari', or 'Russian Brown'. *Brassica juncea* was generally known as 'Indian' and *B. carinata* 'Ethiopian'.

Land-race forms in *B. alba* from northern Europe tend to be of much taller, fleshier growth than those from the Danube Basin, and these more southerly forms have mainly been used in breeding for modern cultivation. In Sweden, a series ('Primex', 'Secu', 'Trico' successively) was successfully produced at Svalof in the 1940-50s, Primex was the first notable crop cultivar to arise from an X-irridation programme. Interest in tetraploid forms declined after some breeding in Sweden and Germany. The named German varieties arising from work at Giessen and Gross-Lusewitz have been bred particularly for use as green manure crops. The major centre of breeding of *B. alba* for crop yield, habit, full mechanization and for the many parameters of seed quality for condiment usage was Norwich, England where 'Bixley' and 'Kirby' have been successfully introduced. Cultivation in Denmark is mainly of English, German or Swedish cultivars or seedsmens' land-races.

Breeding or selection work has been very little on *B. nigra*. As stated earlier, *B. nigra* was everywhere abandoned in favour of the more suitable *B. juncea*. Along history of breeding *B. juncea* as an oil-seed is available in India, with many recorded types, such as 'Manipuri-rai', 'Gohna-sarson' and more modern pure lines such as 'RT -11', 'RL-9', etc. Similarly, in Japan and

China, established vegetable and oil-seed varieties with a multiplicity of local names are on record. Japanese workers (Aoba, 1972) have elucidated the inheritance of plant and seed characters. Attempts to develop *B. juncea* as an oilseed at Svalof ('Fiskeby-senap') were made in the 1940s. Breeding for condiment usage, usually under the stimulus of manufacturers, has gone on during the last few decades at Giessen, Germany ('Primus'), Douai, France ('Burgonde' and 'Ekla'), Lethbridge and Saskatoon, Canada ('Lethbridge 22A' and imminent further introductions) and Norwich, England (Trowse', 'Stoke' and 'Newton'). Lethbridge 22A, Stoke and Newton are all yellow-seeded as it is observed that this characteristic, with associated higher kernel content, aids extraction during milling. 'Ekla', 'Stoke' and 'Newton' are of greatly enhanced pungency. The work at Norwich originated from very extensive world collections showing a great variability of both morphological and chemical characteristics, enabling close definition of breeding objectives for habit, yield and quality. All breeding of *B. juncea* is by means of pure lines. Self-pollinating is noted exclusively in this species.

9. Epilogue: Studies on genetics of *B. juncea* were continued in India and Japan and the breeding of this species for oil-seed usage progressed several centres in India, Pakistan and Russia. For condiment, the predominant centres were Norwich and Saskatoon. At the latter centre there was also interest in industrial usage of\strains with varied fixed oil fatty acid content. *Brassica alba* also received attention as a potential protein crop.

Besides the possible exploitation of heterosis in *S. alba*, breeding methods are unlikely to change. Polyploidy is disadvantageous and irradiation as an aid to producing mutants proved too drastic a method. Very large collections of both *B. alba* and *B. juncea*, easily maintained by the long-lived seed, serve as substantial gene pools and wild material of all four mustard species widely distributed in several continents.

9.5 OIL PALM

Elaeis guineensis Jack. (Arecaceae)

Resume: *It is of African origin. It spread throughout the tropics in both hemispheres. It is perennial source of palm oil which continues for 60-70 years of its age. It has basic chromosome number x=16. Numerous vernacular names for oil palm in West Africa lends linguistic support to its prominent role played in village economy. Apart from its extensive use for edible purposes, it is widely used in various industrial products. Moreover, the oil contains a large quality of carotene, particularly β-carotene, the precursor of vitamin-A.*

1. Introduction: Palm oil obtained from the nuts of the African oil palm *viz., E. guineensis*. It is a native of western Africa but has spread all through the tropics of both hemispheres. Malaysia and Indonesia are the largest palm oil producing countries of the world. The oil palm is very productive tree. It bears fruits at the age of 5 to 6 years, reaches full bearing at 15, and continues until 60 to 70 years of age.

2. Plant Characteristics: A handsome erect tree reaching a height 20 m or more, trunk studded with persistent leaf-bases; spirally arranged crown of large leaves, leaves 3 to 5 m in length with 100 to 160 pairs of leaflets; male and female flowers borne in separate inflorescence in a compound spike or spadix; fruits borne in large bunches, fruit being sessile, ovoid drupe of nearly 4×2 cm size.

3. Parts Used: Fruits.

4. Chemical Constituents:

(i) Palm oil: It is obtained from the fleshy pulp of the fruit. This oil contains: (a) β-carotene (precursor of vitamin-A) (b) Fatty acid composition of palm oil: myristic acid (1. 0-2. 5%), palmitic acid (32-43%), stearic acid (1-9%), oleic acid (40-43%), and linoleic acid (2-11%).

(ii) Palm kernal oil: It is derived from the kernel or endosperm. Its composition is: caprylic acid (3. 4%), capric acid (3-7%), lauric acid (46-52%), myristic acid (14-17%), palmitic acid (6. 5-9%), oleic acid (13-19%) and linoletic acid (0. 5%-2. 0%).

5. Uses And Applications:

(i) Palm oil: It is extensively used for edible purposes. It is used in the manufacture of soaps, candles, lubricants and in tin plating industry. It can be used as a source of vitamin-A.

(ii) Palm kernal oil: Its stearin fraction is used in confectionery and olein fraction in biscuits and cakes. The oil is used in hard water soaps and in the manufacture of glycerin, shampoos and candles. Its better grades are used in the manufacture of margarine.

6. Karyotaxonomic Evidence: The genus *Elaeis* contains three species: (i) *E. guineensis*, endemic to West Africa, the major economic species and generally referred to simply as 'Oil Palm' (ii) *E. oleifera* (formerly *Corozo oleifera* or *E. melanococca*), endemic to the northern Amazon basin, extending into Central America as far north as Costa Rica (iii) The relatively little known *E. odora* (formerly *Barcella odora*) reported to occur in various places in the Amazon region.

Although the centres of origin are on different continents, close relationship between *E. guinneensis* and *E. oleifera* is suggested by Hardon and Tan (1969) because the species hybridize readily and produce fertile off-spring. Systematic relationships suggest that the genus must have originated in the New World; but how and when it became established in West Africa is unknown. *Elaeis guineensis* and *E. oleifera* differ in several morphological characteristics and are readily separable. This suggests that isolation must have occurred some considerable time ago. This is supported by the presence of fossil pollen similar to that of *Elaeis* in Miocene strata in the Niger Delta (Zeven, 1967). The chromosome number is 2n=2x=32.

7. History: From the 15th century onwards, early Portuguese, Dutch and English seafarers mention palm oil (and palm, wine) from the coast of West Africa. Several vernacular names for the oil palm in various West African languages lends linguistic support to the important role the oil has played in the village economy for many centuries (Zeven, 1967, 1972). The natural habitat is banks of rivers, lakes, swamps and other places too wet for rain forest. Most authors consider it to be originally a species of the transition zone between savanna and rain forest, but now more widely spread. The palm cannot compete itself with rain forest species. Its present extensive distribution throughout the rain forest belt of West Africa, from Cape Verde down to Angola, is possibly to be due to human activity which afforded proper habitats in forest clearings around temporary settlements, Most oil palm populations fall into this semi-wild or 'camp-follower' category (Zeven, 1972). Main centres of such populations are noted in the Ivory Coast, eastern Nigeria and in the Congo basin extending into Angola. The palms so grown are little, if at all, removed from the wild type.

Elaeis oleifera has not been used to the same extent in South and Central America. Low human population density and a more migratory pattern of existence of the South American Indians in the rain forest zone did not provide the same conditions as in West Africa for the palms to become established in sizeable groves. The oil has minor uses for lamps and as an edible oil in the Amazon basin and in Colombia.

8. Further Developments: The international market for vegetable oils grew fast during the latter part of the 16th century. Palm and kernel oils became extensively used in stearic candles, margarine and soap. Industrial use was found in the tin-plate industry and palm oil was even used as a lubricant. To have a secured and reliable supply of these oils, Sir William Lever obtained extensive concessions, first in Sierra Leone and later (1911) in the Congo. Oil mills were erected. This stimulated more efficient exploitation of existing palm groves and the planting of selected material. During the same period, plantation culture was started in Sumatra, followed a few years later by Malaysia. And thus a new industrial crop was born. The evolution of the crop truly started from this period onward.

The plantation industry in southeast Asia was established on material originating from four specimens in the Botanical Garden of Bogar, Java. They were obtained as seedlings in 1848 from Mauritius where oil palms must have been introduced from Africa as ornamental plant at an earlier date. Descendants of the original Bogar palms were distributed throughout the Indonesian archipelago in small trial plantings and as ornamentals.

Extensive rows of avenue palms, lining roads on tobacco estates in Deli on Sumatra, furnished the seeds for the first estates in that area around 1911. Material of this origin is generally referred to as the Deli Dura. It immediately proved to be high yielding and with good fruit characters.

In the early 1920s, selection programmes were started employing, at first, by simple mass selection and followed then by controlled crosses to produce bi-parental families. Both total yield of fruit bunches and fruit characters were taken into consideration in selection. During the same period from 1922, oil palm breeding was started in the Congo (now Zaire) by INEAC (Institut National pour I'Etude Agronomique du Congo Beige) at Yangarnbi (Pichel, 1956). The starting material in Sumatra and Malaysia consisted of a type with thick-shelled fruit (*E. guineensis* var. *dura*). In Zaire, the programme started from open pollinated bunches of palms with thin-shelled fruits (*E. guineensis* var. *tenera*) collected at Eala and Yawenda. The Tenera form of the palm was considered superior because fruits with a thin endocarp (shell) had correspondingly more oil bearing mesocarp per fruit.

Tenera × Tenera crosses were found to segregate in 1:2:1 ratio of Dura, Tenera and a third type (Pisifera), characterized by the absence of a shell in the fruits, which however generally abort at an early stage of development (Beinaert and Vanderweijen, 1941). This discovery of single-gene inheritance of shell thickness allowed the production of pure Tenera (sh+, sh-) families by crossing Dura (sh+, sh+) (female) with Pisifera (sh-, sh-) (male). Thus the heterozygote is the agriculturally desirable phenotype. Breeding programmes were, changed after the Second World War from family and individual palm selection to selection of Dura and Pisifera lines which, in combination, would give high-yielding Tenera families.

The programme of INEAC in Zaire, was based on a diallele between six Teneras thereby high-yielding combinations were reproduced by crossing Duras derived from selfing one Tenera parent with Pisiferas of a selfing of the other Tenera parent. In Malaysia and Sumatra, Pisiferas were derived from Tenera crosses of new introductions and by introgressing Teneras into Deli Dura and backcrossing these to Teneras and Pisiferas of unrelated new introductions. In the Ivory Coast, IRHO (Institut de Recherche des Huiles et Oleagineux) extended its breeding programme in the 1950s based on a large exchange programme involving crosses between materials of various origins: Deli Dura (Malaysia and Sumatra), Yangambi Tenera (Zaire), La Me Tenera (Ivory Coast) and Pobe Tenera (Dahomey). The highest yields were obtained by crossing Deli Dura wit Pisifera of Yangambi and La Me origin (Meunier and Gascon, 1972). This led to a breeding programme based on reciprocal recurrent selection. The same system was followed by the Nigerian Institute for Oil Palm Research (NIFOR) but using material of various Nigerian origins (Sparnaaij, 1969).

9. Epilogue: Oil palm breeding have been essentially concerned with the relative importance of general combining ability and specific combining ability. In Malaysia, emphasis was placed on the creation of new, genetically more variable populations to allow efficient utilization of the former. In the Ivory Coast, mild inbreeding within existing populations is tolerated and commercial seed production is based on inter-population (inter-origin) crosses. The expectation, is that this will allow exploitation of both former and latter.

In Malaysia, Indonesia, Ivory Coast, most of the commercial planting material was based on crosses between Deli Dura and Pisifera mainly of Yangambi origin. In cross-pollinating species, a considerable part of the commercial planting material in southeast Asia, parts of West Africa and in South America goes back to four botanical specimens in Java and only a few palms in Zaire.

The narrowness of the genetic base of present breeding populations was generally realized and there was increasing interest in the collection of wild material. Collecting expeditions have been carried out by INEAC in Zaire and by NIFOR in Nigeria (Hartley, 1967). The IRHO (Meunier, 1969) and MARDI (Malaysian Research and Development Institute) were involved in extensive collecting programmes. In Malaysia, considerable attention was given to crop-physiological analysis. Interspecific hybrids have been produced between *E. guineensis* and *E. oleifera* (Hardon and Tan, 1969). Of prime interest was the low height increment of the hybrids, a character inherited from the oleifera parent. Hybrid bunch yields are promising but oil content of the mesocarp is lower than in *E. guineensis* though oil quality is good.

9.6 OLIVE

Olea europaea L. (Oleaceae)

Resume: *It is ancient and characteristic plant of the Mediterranean region. Cultivated olives are all diploid with 2n=40 chromosome number. Although, it is ancient plant, it has little changed. It is being cultivated since biblical times. It is largely used for culinary purposes, besides in the manufacturer of high class toilet preparations, cosmetics and pharmaceuticals.*

1. Introduction: Olive trees have been under cultivation since biblical times. They played an important role in the lives of the Israelites and neighboring peoples. The tree is native of western Asia and is presently grown extensively in the countries bordering the Mediterranean sea. It is a source of olive oil from the fruits. This tree is one of the most characteristic plants of the Mediterranean basin. It is ancient and of immense historical importance as the principal source of edible oil for the people of the Mediterranean area for a period of millenia. Though ancient, it has changed but little and is now in decline, a decline which can be arrested only by adaptation of oleiculture to modern conditions.

2. Plant Characteristics: A narrow leaved xerophytic, large, evergreen tree reaching to a height of 15-18. 5m; leaves linear, small, margin entire, opposite; flowers off-white, in racemose panicles, flowers being unisexual; fruit a one-seeded drupe, globular, oblong or crescent-shaped, green at first and then changes colour from blue to purple to red to black, mesocarp or pulp oily, containing elongated brown seed.

3. Parts Used: Fruits.

4. Chemical Constituents: Olive oil: Oleic acid (65.86%), palmitic acid (7-20%), linoleic acid (5-15%), myristic acid (0-1%) and stearic acid (0.3%).

5. Uses And Applications: Olive oil is used for edible purposes, in cooking, salad oils and food presentation *e.g.* sardine canning. It is also used in the manufacture of high class toilet preparations and cosmetics. It is used in wool combing.

6. Karyotaxonomic Evidence: The Olive is only member of the genus that has survived in the area. It is one of the relicts of the tropical mid-Tertiary flora of the Mediterranean. Two botanical varieties are usually recognized *viz.*, (i) cultivated *sativa* and (ii) wild *oleaster*. The latter, in the light of evolutionary views outlined below, must be regarded as escaped rather than truly wild or in any sense ancestral to the cultivars.

Olea includes about 35 species. They are distributed widely from Africa to New Zealand. Three species deserve special mention in view of the evolution of the crop: (i) there is *O. laperrinii*, with leaves silvery-white beneath, native to the Saharan mountains (Benichow, 1962); related forms are found in the Macaronesian archipelago (Canary Islands and Azores) and North Africa. Apparently related forms occur by the Red Sea and Persian Gulf. (ii) there is *O. africana* (=*chrysophylla*=*verrucosa*) which is African, from far southern Africa to the Sudan-Egypt border, (Verdoorn, 1956). It has narrow leaves which are golden beneath. And (ii) *O. ferruginea* (=*cuspidata*) is native from Iran to the Himalayas. Cultivated olives are nearly all diploid with 2n=2x=46. Occasional triploids and tetraploids are reported and also one case of polysomaty (2n=55). It is to be noted that the large fruited cultivars that have been examined are diploid.

7. History: The parents of the cultivated olive are unknown. They are generally inferred to have been two species, one a form with narrow leaves golden beneath and the other (the 'proto-Olea' of Ciferri), the contributor of the oily pulp character. The former could have been *O. africana* or *O. ferruginea* mentioned above. The latter *viz.*, 'proto-Olea' is thought to have become extinct. The crop probably originated as a hybrid swarm in the mountains of the eastern Mediterranean: the Taurus, Amanus and Lebanese mountains, as far as High Galilee. A hybrid origin is indicated by the great diversity of kernel types found by M. Dunand at Byblos (on the coast of Syria) and dated to the 4th millenium BC. Some were more or less identical to those of *O. africana*, others were with thin shells, elongated and sometimes curved.

The crop moved westward from the eastern Mediterranean generating a secondary centre of diversity in the Aegean area and what might be regarded as a tertiary one further west still, in the area southern Italy-Tunisia. The Aegean centre is characterized by large-fruited types which are found only as impurities in Near Eastern populations. The Tunisian populations in the tertiary centre may have been influenced by local wild forms. The crop was known to the Romans by 600 BC.

The history of the crop has been influenced: (i) by the longevity of the tree, (ii) by its capacity for regeneration by suckering from the rootstock and (iii) by the low rate of vegetative multiplication that has been possible. It is doubtless that chance seedlings have generated a supply of new clones over the millennia. There is evidence, nevertheless, that the rate of clonal turnover is very low. Some Tunisian clones are thought to have survived from Roman times. The general pattern is of cultivars composed of mixtures of clones, fairly homogeneous and with recognizable dominant types, sometimes widespread but sometimes also broken up into a diverse mosaic in diverse environments.

8. Further Developments: This pattern of clonal mixture has persisted from ancient times down to the present. Even the expensive plantings of the late 19th and early 20th centuries lacked a means of rapid vegetative multiplication. They thus merely reduced variability without establishing outstanding clones.

Classification of olive clones have been attempted on in a number of places for a good many years but are far from satisfaction. (Cantarelli, 1962; Scaramuzzi, 1950). It is unfortunate that the very promising studies by Ciferri (1950 a,b) in Italy, using the concept of 'proles', were never followed up.

There have been many attempts by horticulturists and nurserymen to select clones for yield, oil content and resistance *e.g. Cycloconium oleagiuum* (leaf spot). Selected clones have been distributed but only in one trial in France. They have been field-tested before distribution. It seems very likely, a *priori*, that superior clones exist and could be isolated. However, the practical problems of propagation and time-scale are severe. However, mist-propagation of mature leafy cuttings is widely used and clonal selection proceeded (Almeida, 1963). The earliest attempt of Olive breeding was begun 40 years ago by Ortego-Nieto at Jaen (southern Spain) using open-pollinated seeds. Many seedlings were killed by frost. The survivors showed curious juvenile characters and some high oil contents. They were not superior to local cultivars. Controlled crossing has yet to be exploited. There is some evidence of an erratic self-incompatibility.

'Bragelia *et al.* (2017) proposed a simplified approach for olive (*Olea europaea* L.) genotyping and a cultivar traceability. They used the Tubulin Based Polymorphism (TBP) method to genotype olive cultivars of different origin and to produce short-size cultivar-specific molecular probes. Both the first and second intron of the members of the olive beta-tubulin gene family were exploited as sources of DNA polymorphism. Compared with the data obtained with the use of a set of 11 SSR markers selected from an *Olea europaea* L. database, TBP is shown to provide similar information about the polymorphic content of the olive genomes, releasing, at the same time, a simple and discriminator DNA barcode specific for any of the analysed cultivars. Such as barcode is the source for the preparation of variety specific molecular probes.

Begnard *et al.* (2001) searched for genetic relationships in the olive (*Olea europaea* L.) cultivars. They sampled 102 olive RAPD profiles from all around the Mediterranean Basin. Twenty four clusters of RAPD profiles were shown in the dendrogram based on the Word's minimum variance algorithm using chi-square distances. Factorial discriminant analysis showed that RAPD profiles were correlated with the use of the fruits and the country or region of origin of the cultivars. This suggested that cultivar selection has a occurred in different genetic pools and in different areas. Mitochondrial DNA RFLP analyses were also performed. These mitotypes supported the conclusion also that multilocal live selection has occurred. This prediction for the use of cultivars will help olive growers to choose new foreign cultivars for testing the before an eventual introduction if they are well adopted to local conditions.

9. Epilogue: Attempts to classify olive clones and to isolate and propagate superior selections should be continuous. There is general awareness of the need to test selections. Meanwhile, the needs of mechanical harvesting by shaking become apparent, *viz.,* easily detached fruits of synchronous maturity. The temptation is to try to adapt established plantings for the purpose.

9.7 SAFFLOWER

Carthamus tinctorius L. (Asteraceae)

Resume: *It is believed to have originated in Southern Asia. It is cultivated for centuries in India, the Middle East and East Africa. The earliest evidence of safflower as a cultivated plant is available from the archaeological record in Egypt dating about 1600 BC. Previously, it was used as a source of dye but abandoned later by production of synthetic dyestuff. Now, it is grown for oil source. Cultivated safflower has chromosome number 2n=24. It is used for edible purposes and for some industrial products.*

1. Introduction: It is thought to have originated in Southern Asia and has been under cultivation for centuries in India, the Middle East and East Africa. Primarily, it was used as a source of orange dye. But only recently, it has also become a prominent source of an unsaturated oil. The said dye was later abandoned because of its impermanent nature. It was naturally forced to abandon as the synthetic dyestuff were available. Safflower is now grown primarily as a oil seed crop and has become popular in India, China, Egypt, Italy, Spain, Australia and United States.

2. Plant Characteristics: An erect, much-branched herbaceous annual, reaching 0.5 to 1.5 m in height; leaves spinosely serrate, oblong-ovate to lanceolate; inflorescence a dense globular terminal capitulum surrounded by spiny bractsl, flowers yellow to orange-red; fruits one-seeded achenes, shining, 4-angled and nearly oblong.

3. Parts Used: Achenes.

4. Chemical Constituents: Oil: Fatty acids (myristic, palmitic, stearic, arachidic, oleic, linoleic; glucerides (myristo-oleo-linolein, myristo-di-linolein, and few others).

5. Uses And Applications: The oil is used for making soups, paints and varnishes. It is also used for edible purposes and as an illuminant.

6. Karyotaxonomic Evidence: Safflower which is under cultivation belongs to a group of closely related diploids (2n=2x=24). These extends from central Turkey, Lebanon and Israel in the west to northwestern India in the east. Two successful weedy species are *viz., C. flavescens* (Turkey, Syria and Lebanon primarily) and *C. oxyacantha* (Iran and Iraq to northwestern India). The former is self-incompatible and the latter mixed self-compatible and incompatible. *Carthamus palaestinus* (desert areas from Israel to western Iraq) is self-compatible. *Carthamus gypsicolus* is found from the Caspian Sea to the Aral Sea and *C. curdicus* is reported only from western Iran. Flowers of wild species exhibit various shades of yellow and all have yellow pollen (Imrie and Knowles, 1970).

A closely related group of species with 2n=2x=20 occupies the Middle East from Libya to Iran. All these have flowers that range from blue through to white and all have white pollen. *C. nitidus* (2n=2x=24) crosses with cultivated safflower. The F1 shows low meiotic pairing and is sterile. In appearance, *C. nitidus resembles C. leucocaulos* (2n=2x=20) and both are self-compatible. *Carthamus divaricatus*, a self-incompatible diploid with x=11 is found only in Libya. It resembles x=10 species in general appearance but has flower colours that range from white through yellow to blue and it has yellow pollen.

There exists three polyploids. All are self-compatible. *Carthamus lanatus* (2n=4x=44) is presumed to be a product of a cross of (x=10) × (x=12) type. It was not been possible to synthesize the species from crosses of presumed parents. *Carthamus turkestanicus* (2n=64) occurs from Turkey to the easternmost range of the genus and into Ethiopia. *Carthamus baeticus* (also 2n=64) is found from Turkey westwards to Spain. Polyploidy has extended the boundaries of the genus in Europe, Africa and central and eastern Asia. Moreover; the most successful emigrants have been the polyploids, with *C. lanatus* and/or *C. baeticus* reported to grow in Australia, California and Chile.

7. History: The earliest evidence of safflower as a cultivated plant is available from the archaeological record in Egypt. There, about 1600 BC., safflower was grown for its flowers which showed the same variability in colour as may be found presently in that country. The reddish orange florets were sewn sideways on narrow strips of papyrus or cloth to form long garlands that were wrapped about the necks and bodies of mummies. Then, or at some later date, it was discovered that the orange and red flowers could serve as a source of dye to colour cloth. This indicates that man became interested in safflower when he first found plants with red or orange flowers.

Safflower had become one of the two most important plant sources of dye-stuffs by the early 19th century. Its importance declined, along with that of indigo, when synthetic aniline dyes were developed. The dried flowers still sold in the bazaars of cities from Cairo to Tehran as colouring for foods. The pigment a chalcone derivative is carthamone.

Possibly, safflower seed was used as a source of oil during the Roman period in Egypt and perhaps earlier. The date of its first culture for oil in India is not known, but it was probably also very early. It was not tested as a potential oil crop in any other areas until the later part of the 15th century.

It appears that the safflower was first domesticated in the Near East as a dye plant and then very widely used for this purpose in the old world subtropics. At an early stage, it was locally converted into an oil-seed. In cultivation, over a extensive area and for a longer time, considerable diversity is developed. There is evidence of incipient genetic differentiation.

Centres of diversity with pre-dominant types in each are: (i) Far East: late, spiny, tall, red-flowered, (ii) India-Pakistan: early, very spiny, short, orange-flowered; (iii) Middle East: late, spineless, tall, red-flowered, (iv) Egypt: variable, usually large-headed, (v) Sudan: early, very spiny, yellow-flowered, (vi) Ethiopia: late, very spiny, tall, red-flowered, (vii) Europe (Mediterranean): variable (Knowles, 1969b).

There seems to have been introgression from the *C. oxyacantha* of northern Pakistan into the domestic safflower of that area; and from the safflower of the Sudan centre to the Egyptian centre (Knowles, 1969b). If Indian types are crossed with those of other areas, then sterile types appear in F_2. Interactions of recessive alleles at three loci are involved. This incipient isolation indicates that Indian populations have been separated from others for a long time (Carapetian, 1973).

8. Further Developments: Safflower was established in the 1940s on a small commercial scale in the western and northern Great Plains of the USA after a programme of testing at the University of Nebraska. However, production ceased in the 1960s, primarily because of competition from improved wheats and problems with weeds and disease. Safflower has been grown in California and Arizona since 1950. The whole area of production was grown under contract to companies interested in processing or selling the seed.

The development of the crop in countries of Asia, Europe and Africa has been severely handicapped by insects. Many of them adapted to survival on other Compositae including the weedy safflower species. Most serious among them is the safflower fly (*Acallthlophilus helianthi*). The safflower of the Deccan region of India suffers little damage from that pest, apparently since alternate hosts are not abundant or the crop is so early that it escapes damage.

Safflower is handled as a self-pollinated crop, though plants are bagged to prevent outcrossing by insects. Efforts to develop hybrid varieties, using a thin-hulled female parent which had a form of structural male-sterility, were unsuccessful, primarily because there were enough selfed female plants to affect yields adversely.

There have been many attempts to the development of disease resistance. Cultivars have been bred with resistance to wilt caused by both *Fusarium oxysporum f. carthami* and *Verticillium albo-atrum* and to rust caused by *Puccinia carthami*. A large germplasm reservoir is maintained by the US Department of Agriculture at the Plant Introduction Station at Washington State University, Pullman, Washington.

Belhaj *et al.* (2013) investigated salinity tolerance on a large panel of safflower composed of 265 entries selected among the international collection. All salinity levels, biological and grain yield and flower number averaged 6, 1.9 t ha-1 and 300 flowers per meter square respectively. Salinity reduced the latter variables by 50, 75 and 25% respectively. Some genotypes maintained their biological and growing yields at 7 and 2t ha-1 respectively. Relationships between yield components showed a threshold of 300 flowers and 15 dSm-1, which enabled many genotypes. Stability analysis using genotypic ecovalence parameter showed that 10 genotypes were adapted to high salinity and 20 genotypes were adapted to intermediate salinity. Based on yield building patterns and yield stability there were 13 genotypes that were selected out of the collection to constitute a nursery that is representative of the original collection over the whole range of variation for salinity response.

Oztuk *et al.* (2008) determined the responses of the hybrid and open-pollinated safflower genotypes to irrigated and non-irrigated conditions in a highland environment in Turkey. Safflower genotypes tested were well adapted to the cool and short-season conditions in this region. The response of the seed yield to genotype varied depending on the growing seasons. The non-irrigated plants produced nearly the same seed yield as irrigated ones. Average seed yields of safflower genotypes tested were 914.3 and 928.0 kg/ha in 2001, and 1143.6 and 1139.9kg/ha ini 2002 years for irrigated and non-irrigated experiments respectively. In general, the genotypes differed in all of the investigated traits. In both irrigated and non-irrigated experiments genotype X year interactions were very significant for all parameters. These experiments showed that in semi-arid and highland environments safflower has a big potential value as an oilseed crop under dry land conditions.

9. Epilogue: The main objective for all breeding programmes is disease resistance. A higher level of resistance to root rot caused by *Phytophthora drechsleri* is needed in all areas in which irrigation is practised. Greater tolerance to foliar diseases caused by *Alternaria* spp., *Puccinia* and *Botrytis cinerea* would permit more widespread cultivation in the USA and would facilitate establishment of the crop in damper areas elsewhere.

Higher increase in oil contents will be developed, with levels exceeding 50%, both by reducing the hull and by raising the level of oil at the expense of protein. A mutant type originated in the Ganges delta has an oil with high levels of oleic acid, making it chemically like olive oil (Knowles, 1969a). The characteristic has been bred into cultivars in the USA, so there are two types in commercial production: (i) nigh linoleic (poly-unsaturated) and (ii) high oleic (mono-unsaturated) types. Winter types are developed in Iran and the United States that tolerate temperatures of 15°C and which permit the culture of safflower as a winter crop.

9.8 SESAME

Sesamumorientale L. [(Syn.) *S.indicum* L.] Pedaliaceae

Resume: *It is thought originated, according to some, in Africa, while others believed its origin in India. De Candolle (1886) and Vavilov (1926) lent support to the latter standpoint. This is supported by the greater variability of S. indicum in India and the oldest archaeological record from Harappa dating 500-1700 BC. Thus, it appears polytopic in origin. The cultivated species and its forms are diploids having basic chromosome number x=13. It is usually used for edible purposes.*

1. Introduction: Occurrence of wild species of *Sesamum*, cultivation of *S. angustifolium* its use as vegetable and oil are suggestive of its origin in Africa (Bruce, 1953; Eckey, 1954). Hilderbrant (1932) lent support to this viewpoint. Ancient Sanskrit literature *e.g.* Bhavprakashnighantu (mentioning three varieties of seeds), Manusmruti (seeds offerinig to Bhramins in the Sraddha ceremony) and also in Atharveda and Ashtadhyayi by Panini, etc. contain information of uses of Sesame. Vavilov (1951) regarded India as the basic centre of origin. Marco Polo started his voyage in 1271 and reached India. In his travelogue, he mentioned production of sesame oil in the Malabar coast of India.

There is a dearth of archaeological, prehistorical and literary references to sesame from the Middle East, Egypt, Iran and India dating from about 4300 years BP. The evidences from Malaya and China are also ancient. All of them lend support to the high esteem in which both the seed and oil were held by the peoples of these regions and the many uses in cooking, medicine, toiletry and ritual. There are records of the crop in the Assyrian tablets, in the writings of Herodotus, Xenoplion, Theophrastus, Pliny, Marco Polo and of other ancient travellers and chroniclers. Archaeological evidence suggests that sesame was being cultivated in Palestine and Syna during the chalcolithic period, about 3000 BC.

Many earlier writers (for example, Hiltebrandt, Burkill, Dalziel and Porteres) considered Africa to be the ultimate source of the crop. This view was based on the presence of nearly two-thirds of *Sesamum* species in tropical Africa and the dominant position of the crop in the economies of (and the presence of variability in) several African countries. Even if sesame were known in the Middle East from prehistoric times, this region does not possess any wild *Sesamum* species. We can only speculate as to whence in Africa the Middle East could have received the crop.

The largest sesame acreage in the world is in India. It is also an important crop in Burma. The plant shows considerable variability in both these countries. Its antiquity in India is doubtless. Charred sesame seeds have been recorded from Harappa (ca. 3500-1700 BC.), the oldest archaeological site in the region. Etymology supports the idea of antiquity. The Sanskrit word for oil, 'taila' is almost synonymous with that for sesame, 'til'.

2. Plant Characteristics: An erect annual upto 2. 0 m in height, stem being furrowed, leaves lower broad and lobed and upper nearly lanceolate; flowers bell-shaped, white, pink or mauve borne simply or in groups of 2 to 3 in leaf axils; fruits bony capsules containing numerous seeds, seeds compressed, nearly pear-shaped, white, brown or black.

3. Parts Used: Seeds.

4. Chemical Constituents: Oil consist of the following: (i) Fatty acid: Oleic (37-50%), linoleic (37-47%), palmitic (7-9%), (ii) stearic Acid (4-5%), (iii) sesamin (0.5-1. 0%), (iv) sesamolin (0.3-0.15%).

5. Uses And Applications: Usually, sesame oil is used for edible purposes. It is used in the manufacture of cooking fats, margarine and salad oils. Its lower grade is used in the soap and paint industries. It is also as a lubricant and illuminant. It is highly prized in medicine. The oil-cake is fed to dairy cattle, poultry and pigs.

6. Karyotaxonomic Evidence: About 37 species belongs to the genus *Sesamum*, besides *S. indicum* (the predominant cultivated species), *S. angustifolium, S. radiatum* and *Ceratotheca sesamoides* are also cultivated to a small extent in Africa. Most species are found in three regions *viz.,* tropical Africa, India, including Sri Lanka and the far East. About 25 taxa occur in tropical Africa, nine in the Indian region, four in the East Indies; and one each is described from Crete and Brazil. A few taxa occur in more than one region and two, *capense* and *schenckii*, occur in all the three major regions of distribution.

The chromosome numbers of only three genera are known. Somatic chromosome numbers of nine *Sesamum* taxa have been determined, as 2n=2x=26 in 5 cases; 2n=4x=32 in 3 cases; and 2n=8x=64 in 2 cases. The basic numbers are thus x=8, 13. The predominant cultivated species, *indicum*, and the two widely distributed forms, *capense* and *schenckii*, are diploids sharing the basic x=13. Interspecific crosses in about 24 combinations involved *indicum, schenckii, grandiflorum, alatum, capense* (all 2n=26), *prostratum, laciniatum, angolense* (all 2n=32), *radiatum* and *occidentale* (both 2n=64). Of these, about two-thirds have been successful. The results from these studies and from distributional and morphological data suggest close affinities between *radiatum* and *occidentale*, between *prostratum* and *laciniatum* and between *indicum* and its ssp. *malabaricum*. They also indicate that *grandiflorum* may be synonymous with *schenckii*, and also *ekambaramii* and *capense* with *alatum*.

7. History: De Candolle (1886) suggested that India might have received sesame from the Far East. This is now known to be wrong. Both Watt (1890) and Vavilov (1926) considered that it had polytopic origins. The regions other than India in which they considered sesame to have originated were Ethiopia (including Somalia and Eritrea) and central Asia. It is difficult to assume that the crop could have originated in central Asia or the Middle East except that its use in these areas since prehistoric times has been well chronicled No wild taxa occur there, the variability present in the cultivated *indicum* is no greater than that found in several other regions. The crop today is of only local importance in the region.

Nayar and Mehra (1970) proposed that sesame could have originated in either the Ethiopian region or in peninsular India or even both independently. The two regions appear botanically contiguous as they share several species and genera, have long had cultural and commercial contacts over both land and sea. They have at least two wild species (*capense* and *schenckii*) in common. Moreover, a wild taxon *viz., S. indicum* ssp. *malabaricum*, which is completely interfertile with the cultivated *indicum*. It occurs on the Malabar coast. It may be either an escape from cultivation or a companion weed species of the cultivated *indicum*.

The progenitor species of the cultivated *indicum* is unknown as no wild sesame is known except the ssp. *malabaricum*, Among the truly wild forms, the two species *capense* (including *alatum* and *ekambaramii*) and *schenckii* (including *grandiflorum*) occur in Africa, India and the Far East and thus have the widest distributions in the genus. They both possess the same chromosome number as the cultivated *indicum* (2n=2x=26). Both produced viable F_1 hybrids with *indicum*. Both the F_1 hybrids are sterile but *schenckii* (as *grandiflorum*) – *indicum* shows some end-season fertility. Based on morphological differences between the seeds of *capense* and indicum, Nayar and Mehra (1970) advanced *schenckii* as the more likely ancestral species. Earlier, Porteres had suggested that sesame arose through hybridization between *alatum* (=capense) and *radiatum* and even that the related *Ceratotheca sesamoides* might have contributed to some of the cultivated races. Zukovzky (1950) however, felt that *radiatum* could not have been a progenitor of *indicum* because it possesses a narcotic odour which the cultivated species does not have. It has quite a different chromosome number.

8. Further Developments: A great range of variation exists in case of Sesame. It has been amply documented by the worldwide surveys of Vavilov (1926) and Hiltebrandt (1932). While tropical Africa, the Middle East, the Indian subcontinent and China have been growing sesame from ancient times. The crop has been introduced to East Africa (south of Ethiopia presumably, in fact, a reintroduction) and America only in the last 200 years.

The crop is predominantly self-fertilized, though some cross-fertilization by insects also takes place. High yield and, to a lesser extent, disease resistance, particularly to phyllody virus, have long been the major breeding objectives. The oil content of sesame is comparatively high but inter varietal variation is known to occur for this character. Attention has been paid in several countries to adaptation to mechanized harvesting which requires near-synchronous flowering and indehiscent capsules.

Pandey *et al.,* (2015) assessed morphological and genetic diversity of sesame (*S. indicum* L.). They characterized 37 quantitative and qualitative traits of 60 genotypes following IPGRI morphological descriptors. They detected altogether 27 alleles among the 60 genotypes. Cluster analysis based on morphological and molecular markers classified sesame genotypes into two major groups. Mantel test showed an insignificant correlation between phenotypic and molecular marker information. The genotypes belonging to the same geographical area did not always occupy the same cluster.

Adit *et al.,* (2016) studied genetic estimates and diversity of 33 genotypes of sesame representing different eco-geographical regions for number of morphological characters. Sesame genotypes showed considerable genetic variability and divergence. Additive genetic effect was more prevalent for capsule per plant. Correlation at phenotypic level showed that seed yield per plant was significant and positive associated with plant height, number of branches per plant, capsule length and number of capsule per plant. This will lead towards high yield. The cluster analysis helped in grouping the genotypes into different clusters having specific characteristic traits which may be helpful in selecting parents for future breeding programmes. Grouping of genotypes based on multivariate analysis was independent of origin of cultivars.

9. Epilogue: Weiss (1971) showed that world production is at best static, in contrast to the great increases in production being recerded by other sources of edible oils such as coconut, soybeans, oil palm and olive. Consequently, share of sesame of world trade has been declining. This is true despite the facts that sesame oil is one of the best qualitatively. The oil content of the seed is comparatively high. The crop is well adapted to a wide range of conditions. The growth period is relatively short. The present average oil yield is the lowest of any major oil crop.

The crop has better potential than is now realized. Attention should be paid to improvement and management. Because of its intrinsic qualities of good palatability, high nutritive value and stability of both oil and meal, sesame will always be more acceptable to the discriminating consumer than most of the other edible vegetable oils.

9.9 SOYBEAN

Glycine max (L.) Merr. (Leguminosae-Papilionatae)

> ***Resume:*** *Glycine first emerged as a domesticate around the 11th century BC in the eastern half of northern China. It has been cultivated in China from prehistoric times, as evident from Chinese writings dating back to 2800 BC. The cultigen G. max is diploid and G. soja is thought its probable ancestor. It is the first largest source of vegetable oil. It is also rich source of proteins.*

1. Introduction: It is called 'King of Legumes' or 'Cinderella crop'. It is one of the oldest crops grown by man. In China, the Emperor Chin-nung established a ceremony about 2800 BC. The reigning emperor has himself to sow the rice, while newly wed couples, in order to bring them good luck and assure them of many children, used to sow one of other sacred crop species, Soybean was one of them. Thus Soybean has been cultivated in China from pre-historic times. It was important food source in China, Manchuria, Korea and Japan since the ancient time. On the basis of a great diversity in the forms of soybean, Fakuda (1993) thought Manchuria as the centre of origin. It was carried by French missionaries in 1740, and also to the Royal Botanic Gardens, Kew in 1790. In USA, it was brought in 1804 but no recognition was received. Recently, however, it is recognized there also and ranks high among the legume crops. It was introduced in India in about 1880.

2. Plant Characteristics: An erect, much branched, pubescent annuals, reaching 1. 8 m high; large leaves alternate, stipulate, long-petioled and mostly 3-foliate, leaflets being ovate-lanceolate; flowers white to deep purple, small, borne in axillary racemes; pods 2-3 seeded, pods constricted between the seeds, seeds small, globose, creamy-white to grey, brown to black

3. Parts Used: Seeds.

4. Chemical Constituents: Protein (43%), carbohydrate (33.3%), oil (20%), minerals (calcium, phosphorous, iron, magnesium), vitamins (vitamin C, D and A) and fat (17.7%), thiamine, riboflavin, niacin, etc.

5. Uses And Applications: It is called a 'miracle bean' being the single largest source of vegetable oil. In China, Japan and East Asian countries, soybeans are mainly used as a food. Seeds can be eaten green, dry or sprouted. Unripe seeds are consumed as a vegetable. Roasted and salted seeds are used in candies and cakes. Soybean milk, by grinding boiled seeds with water, is used for infant feeding. This milk is also made into butter milk, curd, chees or paneer, and other milk products. Boiled seeds are fermented by mould and then made into cakes. Flour is important constitute of food stuffs and bakery products, *e.g.* biscuits, cakes, breads, etc. Soybean oil, apart from cookies, is widely used in making soaps, printing inks, greases, lubricants, glycerin, oil-cloth, putty, resins, etc. It is used in the paints, varnish and enamel industries, cosmetics and detergent. Soya meal is a good feed for cattle, pigs and poultry. Stalks and leaves constitute good fodder, either as hay or as silage, besides as a green manure.

6. Karyotaxonomic Evidence: Nearly 280 species, subspecies and taxonomic varieties have been included under the genus *Glycine*. Hermann (1962) and Verdcourt (1966, 1970) clarified the taxonomy of the genus *Glycine*. The nine species are placed under three subgenera *viz.*, *Glycine*, *Bracteata* and *Soja*. Cytogenetic work other than chromosome counting has been done on the species in the subgenus *Glycine*. All the species in this subgenus are perennials. None of them appeared to be useful in intensive agriculture. Reciprocal crosses between *G. tabacina* (2n=80) and *G. tomentella* (2n=ca. 80) produced F_1 hybrids. They were highly sterile with from 78 to 81 chromosomes. Crossing the species in the subgenus *Glycine* with the soybean have remained unsuccessful. *Glycine wightii* contains both diploid and tetraploid forms. Attempts to cross *G. wightii* with the soybean were unsuccessful. The chromosomes of *G. wightii* are much larger than those of the species in the subgenera *Glycine* and *Soja*. *Glycine soja* is an annual twining vine with small, hard, roundish, black to dark brown seed. It grows wild in the eastern half of the People's Republic of China, adjacent areas of the USSR, Korea, Japan and Taiwan. Hybrids between *G. soja* and the soybean have been produced by many research workers. There are few cytogenetic barriers to gene flow between *G. soja* and the soybean. *Glycine* *max* is an annual, erect, coarse herb that has never been found in the wild. On the basis of chromosome number, chromosome size, morphology, geographical distribution and electrophoretic banding patterns of seed proteins, *G. soja* is most thought probably the ancestor of the soybean. The differences between the domesticate, *G. max* and its putative ancestor, *viz.*, *G. soja* are the characteristics such as increased seed size, increased oil, decreased protein content in seed, erect habit, larger plant and reduced shattering of mature seeds (Hadley and Hymowitz, 1973).

7. History: The eastern half of northern China, is the area in which the soybean first emerged as a domesticate around the 11th century BC. Probably, the 11th century BC. date will be pushed back in time as additional archaeological evidence is unearthed in the People's Republic of China. The migration of the soybean from the eastern half of northern China to southern China, Korea, Japan and southeast Asia probably took place during the expansion of the Chou dynasty (ca. 11th to 3rd century BC.) (Hymowitz, 1970).

By 1712, Europeans had become aware of the potential of soybeans. Expansion of soybean production in Europe was not successful. The soybean was first mentioned in US literature in 1804. However, the expansion of soybean production in the USA did not come about until the third decade of the 20th century. The earliest known date for introduction into Brazil is 1882. Soybean production has undergone a rapid expansion in Brazil.

8. Further Developments: During the first three decades of the 20th century, production was confined to the Orient, China, Indonesia, Japan and Korea. Soybeans began their almost incredible rise to prominence since 1924 in the United States. Earlier, most US farmers considered the soybean as a hay crop. Between 1924 and 1973, the soybean area in the USA and total production increased from soybeans became the most important cash crop in the United States in 1973 overtaking wheat and maize (Koch, 1974).

The US Department of Agriculture initiated a programme, after 1940, to maintain germplasm and to screen the germplasm for certain economic traits *e.g.* chemical components of seed and sources for resistance to specific pathogens. The germplasm collection housed at the University of Illinois, Urbana, and at the Delta Branch Experiment Station, Stoneville, Mississippi reached about 4,500 entries. It includes plant introductions, cultivars released in the USA, genetic stocks and wild species.

In contemporary breeding programme, the major emphasis was on the development of high-yielding varieties with high in oil and protein content. Characteristics such as resistance to pathogens and pests, lodging, seed quality, shatter susceptibility of mature seed and height of first pods were important traits thought in breeding programmes. All soybeans, in commerce, must have the objective during various breeding programmes such as yellow seedcoats and cotyledons.

Soybeans are completely self-fertile, and the amount of outcrossing under natural conditions is about 1%. However, in certain genotypes outcrossing can occur up to 10%. Breeding methods were dominated by the self-fertilizing property of the soybean. Varieties with specific desirable traits such as pod-shattering, resistance, good seed longevity, suitable for mechanical harvesting, food uses, summer cultivation, high protein and oil content, have been introduced.

Brown-Guedira *et al.,* (2000) evaluated genetic diversity of soybean introductions and North American ancestors using RAPD and SSR markers. They assessed relationship of 18 major ancestors of North American soybean germlasm with 87 plant introductions that are potential new sources of genetic variation for soybean breeding programmes. The most stable grouping was among ancestors that corresponded with known relationships based on pedigree and maturity. Several groups of PIs are distinct from the majority of the ancestors. These genotypes may be useful to breeders to utilize genetically diverse introductions in soybean improvement. Zhang *et al.,* (2013) determined genetic diversity of soybean using EST-SSR markers. They studied 48 vegetable soybean accessions from China, Japan and USA. For the first time, the development of EST-SSRs in vegetable soybean is reported. Twenty-two EST-SSRs in vegetable soybean showed high polymorphism in vegtetable soybean and was successfully used to investigate the genetic diversity from countries mentioned. These result in high transferability to vegetable soybean from grain soybean. These new EST-SSR markers would be useful in vegetable soybean germplasm conservation, cultivar identification, molecular mapping and marker-assisted selection in the future.

Ngalamu *et al.* (2013) evaluated genetic variability of five soybean genotypes and assessed genotype X environment effect on seed yield and yield related traits. They used genotypes *viz.*, NA 50.9 RG; TGx1904-6F, TGx 1740-2F, TGx 1937-1F and Soja. The effect of genotype (G), environment (E) and GXE interactions on pod number per plant, plant height, first pod height, number of branches per plant, leaf, area, number of days to 50% flowering and seed yield found significant as P=0.05. The highest mean seed yield was obtained from TGx 1937-1F (0.98t/ha). Besides TGx 1740-2F, TGx1904-6F and Soja were significantly higher than NA 5009 RG in all environments for seed yield. TGx 1937-1F was an intermediate maturity and best in terms of number of pods per plant, number of branches per plant and leaf area. Correlation coefficient for seed yield showed significant association with days to 50% flowering and leaf area. The authors concluded that the best genotype for seed yield across the environments was TGx 1937-1F and TGx 1740-2F, TGTx 1904-6F and Soja were intermediate and NA 50.09 RG was the least. Thus, partitioning GXE into adaptability and phenotypic stability will positively address the information gap on association of trails to yield.

9. Epilogue: The continued need for soybean protein to meet the demand for more animal products and especially for use in human foods will certainly continue. In the past, much of the increase in production has come about from an absolute increase in total area planted to soybeans. However, there is a limited amount of land that can be used to grow the crop. There will always be a need of breakthrough in yield and this must come about from the development of new, higher-yielding cultivars.

The soybean industry is based entirely on germplasm introduced from the Orient. Unfortunately, the germplasm base needed for the improvement of the crop is rapidly being depleted. The dwindling of these genetic resources is obviously due to the impact of modern technology and the population explosion. The habitats of the wild and weedy relatives of the soybean are being covered by modern developments.

Soybean 'land races' have originated in traditional farming areas in a number of ways. Individual farm families have grown different lines for centuries. The two US Department of Agriculture seed explorers spent three years (1929 to 1931) collecting over 4,000 soybean seed samples from northeastern China, Japan and Korea. Only one-third of their collection still survived. A new effort to collect soybean seed samples from the Orient to provide new germplasm for contemporary plant breeders is needed.

In future, research priorities should be assigned for fundamental genetic research. Soybean breeding has been directed towards the needs of temperate countries. This trend should be directed toward breeding the crop for subtropical and tropical countries. Investigations have already begun in India, Brazil and Colombia, Nigeria and Uganda. The difficulties in growing soybeans in the lowland subtropics and tropics will be to overcome the day length responsiveness of the species.

9.10 SUNFLOWER

Helianthus Spp. (Asteraceae)

Resume: *Sunflowers are believed to be native of New World. It was domesticated possibly in the Central United State sometimes in the 1st millennium BC. Achenes of the domesticated sunflowers have been recovered in several archaeological sites in the central and eastern states. It was introduced in Europe in the 16th century. Dodoens gave the first account of the sunflowers in Europe in 1568. Sunflowers have 2n=34 chromosomes. Helianthus annuus is an important source of oil and has good demand being non-cholesterol and anti-cholesterol properties. It is used for edible purposes and in the manufacture of industrial products.*

1. Introduction: The genus *Helianthus* has provided man with two food plants *viz.,* (i) the sunflower (*H. annuus* L.) and (ii) Jerusalem artichoke (or topinambour) (*H. tuberosus* L.). The former is important for oil containing achenes, while the latter for tubers containing carbohydrates (inulin). Both are native to temperate North America. They were used as food plants by the Indians. The sunflowers, grown for its seed, in 1970 became the world's second most important supplier of vegetable oil. It is now widely grown throughout the world, with the USSR responsible for more than two-thirds of the world's production. Argentina and the Balkan countries are other important sunflower-growing areas.

2. Plant Characteristics:

(i) Sunflower: Erect, subshrubby 1-3 m tall annuals; lower, leaves opposite, upper ones alternate, broadly ovate to elliptic, 20-30 cm long, thick, chartaceous; flowering heads with yellow rays and disc florets brown, 7-15 across or even more; achenes obovate, black.

(ii) Jerusalem Artichoke: Perennial herbaceous plant, 1. 5-3 m height; tubers elongate and uneven leaves opposite, lower leaves alternate, hairy, broad, ovoid, acute; flowers yellow, crowded in capitutum.

3. Parts Used:

(i) Sunflower: Achenes

(ii) Jerusalem Artichoke: Tubers.

4. Chemical Constituents:

(i) Sunflower: Fatty acid (myristic, steric, palmitic, oleic, linoleic); glucerides (oleodilinolein, stearo-oleolinolein, palmito-oleolinotein), vitamin A, D and E phosphorides.

(ii) Jerusalem Artichoke (Tubers): Carbohydrate (inulin 8-18%), protein (0. 9-3. 25%), sodium, potassium iron, chlorine, iodine).

5. Uses And Applications:

(i) Sunflower: Sunflower oil is a very good cooking medium. It has good demand owing to non-cholesterol and anti-cholesterol properties. It is an excellent salad oil and used in margarines. The oil has semi-drying properties and hence used in making paints, varnish and soap.

(ii) Jerusalem Artichoke: The tubers are cooked, pickled or eaten raw. The carbohydrate food is in the form of inulin which is good for diabetic patients. It is also used as a source of levulose and industrial alcohol.

6. Karyotaxonomic Evidence: The genus *Helianthus* includes some 70 species and is divided into four sections (Heiser *et al.,* 1969). *Helianthus annuus* and 13 other species belong to the section Annui. This section comprises taprooted species, mostly annuals, confined to the western part of the United States. All of them are diploid (x=17, 2n=34). The species most closely related to *H. annuus* is *H. argophyllus*, native to Texas. Hybrids between the two show only moderate reduction in fertility. It is possible to secure hybrids of *H. annuus* with most of the other annual species. However, the fertility of these hybrids is greatly reduced. In addition to the domesticated, monocephalic types, *H. annuus* comprises both weedy and wild races. These are widespread in western and central North America, and occur also as ornamental forms. The wild and weedy forms are much branched plants. They bear much smaller heads and achenes as compared to the domesticated sunflowers.

Helianthus tuberosus is a part of the section Divaricati. The 30 species of this section are perennials and concentrated in the eastern and central part of the United States. They include diploid, tetraploid and hexaploid species. The Jerusalem artichoke is a hexaploid. It is fairly common in the eastern half of the United States as a wild plant. Some of the wild types differ little from the domesticated form except for smaller tubers. Nearly fertile hybrids have been derived between *H. tuberosus* and several of the other hexaploid species. Hybrids with low fertility have been secured in crosses with the tetraploid species. It has been impossible to obtain hybrids of any of the diploid perennials with *H. tuberosus*. Hybrids, however, have been secured between *H. annuus* and *H. tuberosus* which exhibit varying degrees of fertility.

Widespread self-incompatibility exists in the genus. Only *H. agrestis*, some races of *H. argophyllus* and some varieties of the domesticated sunflower, *H. annuus*, show self-compatibility. The fact that the Jerusalem artichoke often fails to set seed in cultivation probably is the result of a single clone being grown. It also fails to flower when grown in regions where the growing seasons are short. Considerable varietal diversity of the Jerusalem artichoke is still found in Europe. Apparently nearly all under cultivation in the United States are derived from the same clone.

7. History: Seeds of sunflower were probably an important wild food source to early man in the western United States. In time, the sunflower became a camp-following weed and was introduced from the western to the central part of the country. Somewhere in the latter area the sunflower appears to have been domesticated. As a domesticated plant, it was carried both eastward and to the southwest. Achenes of the domesticated sunflower have been found in several archaeological sites in the central and eastern states. So far only wild sunflowers have been reported from archaeological sites in the southwest. It seems probable that domestication occurred sometime in the first millenium BC. Possibly, the sunflower was domesticated in the central United States before the people there had acquired maize, beans and squash from Mexico. When the Europeans arrived, they found the sunflower cultivated in many places from southern Canada to Mexico. The seeds were consumed directly and used for their oil. The plant was also used medicinally as well as in other ways.

The sunflower went from the Americas to Europe in the 16th century. Several herbalists indicated that the source was Peru, but this is almost certainly an error. In all probability, the first introductions were from Mexico to Spain, followed by introductions from other parts of North America. Dodoens gave the first account of the sunflower in Europe in 1568. The plant spread through Europe, grown more as a novelty than as a food plant, until it reached Russia. The sunflower was readily adopted in Russia only shortly before its arrival there. The Church had laid down strict restrictions on the eating of oily foods on certain Holy days and the newly introduced sunflower was not on the list of prohibited foods. Its use for oil in Russia was suggested in 1779. Selection for high oil content began in 1860.

The Jerusalem artichoke was observed in cultivation in Samuel de Champlain. Massachusetts at Nausett Harbor, Massachusetts, in 1605 (*cf.* Shoemaker, 1927). It is clear that the American Indians used it as a food plant but to what extent it was cultivated is unknown. It may have been that the Indians depended primarily upon wild plants. The first tubers to reach Europe, however, appear to have been slightly larger than those of wild plants. So it is likely that it was cultivated to some extent and that some selection had taken place. In Europe, the Jerusalem artichoke had an enthusiastic reception. But it was shortly to fall into disfavour and come to be more food for swine than man. Champlain in his account of the plant mentioned that the 'roots' had a taste similar to the artichoke. So there is no difficulty in explaining the adoption of the name artichoke. It had usually been assumed that the 'Jerusalem' represented a corruption of 'girasole' (the Italian name for sunflower) until Salaman (1940) showed that the name 'girasole' was not established until after the Jerusalem artichoke had already received that name. He suggested that 'Jerusalem' was an English corruption of Ter Neusen, from which place the tubers had been introduced into England. The history of the equally ridiculous name, topinambour, has also been traced by Salaman. In 1613, six natives of the Topinambous tribe of Brazil took to France. Apparently some street hawker appropriated their name for the rather newly introduced tubers in order to increase their sales value. Some botanists thought the plant to be native of Brazil because of this name.

8. Further Developments: In Russia, the sunflower is the important oil crop. Subsequent breeding work resulted in greatly improved plants. Oil content has been increased from around 28% to 50%. Considerable work was done to control of fungal diseases and insect pests. Resistance to several major diseases was secured through hybridization with *H. tuberosus*. Dwarf varieties were developed that could more readily be mechanically harvested. The sunflower became an important oil crop in Argentina when that country was cut off from its olive oil source in Spain during the Spanish Civil War. During the Second World War, sunflower growing was attempted in several new countries. The effort was mostly unsuccessful in England. Later, the sunflower did achieve some success in Canada. The sunflower has never been more than a minor crop in its homeland. There has also been some interest in the sunflower in several tropical countries. While the sunflower has developed into a major crop, the Jerusalem artichoke continues to be of only very minor importance. There has been some renewal of interest in it as a food for humans in the United States. Rudorf (1958) reviewed breeding in Jerusalem artichoke.

Mandel *et al.,* (2011) studied genetic diversity and population structure in cultivated sunflower comparing its wild ancestor. Their results are based on 433 cultivated accessions from North America and Europe, as well as a rang-wide collection of 24 wild sunflower populations. They observed that cultivated sunflower harbors roughly two-thirds of the total genetic diversity presented in wild sunflower. Population structure analyses revealed that wild sunflower can be subdivided into four genetically distinct population clusters throughout its North American range. The cultivated sunflower gene poor could be split into two main clusters.

Shamshad Mohd (2014) assessed genetic diversity in sunflower germplasm consisting 31 lines. They analysed agronomic and yield traits *viz.,* plant height, head diameter, volume weight, 100 seed weight and seed yield per plant. The study indicated the presence of high level of genetic diversity which can be exploited in hybrids breedinig programmes.

9. Epilogue: The discovery of cytoplasmic male sterility in the sunflower derived from crosses of *H. annuus* with *H. petiolaris* is reported (Leclercq. 1969). Then, with the finding of fertility restorer genes; the practical production of hybrid sunflowers became a reality. In trials, the hybrid sunflowers have shown the same dramatic increases in yield. Some of the American Indian varieties of the sunflower have been preserved by the United States Department of Agriculture. Several varieties have already disappeared. A tremendous reservoir of genes exists among the wild and weed sunflowers which occupy a wide area in North America. Steps should be taken to preserve the older cultivated varieties as well as the wild and weedy sunflowers.

9.11 SWEDES AND RAPES

Brassica napus L. (Brassicaceae)

Resume: *Brassica napus is thought of recent origin, possibly from between B. oleracea and B. campestris by spontaneous hybridization. It has 2n=38 chromosomes and is an amphidiploid. The oil is edible when cold pressed and used in the manufacture of industrial products.*

1. Introduction: *Brassica napus* serves two purposes, one as forage and another as an oil-seed. The swede remains an important animal fodder in northern Europe, Russia and New Zealand and provides a minor vegetable for human consumption. Forage rape yields a leafy fodder for sheep in northern

Europe and New Zealand. Of the oil-seeds, there are both annual and biennial forms. The latter (winter oil-seed rape) has predominated in the recent spectacular development of oil-seed production. A broad parallelism exists between this species and *B. campestris*, which is, after all, a progenitor of *B. napus*, both provide biennial fodder bulb crops (which are also eaten as vegetables), leafy crops (fodder in one and vegetable in the other) and oil-seeds.

2. Plant Characteristics: Biennial grows to 100 cm high; leaves alternate, lower leaves glaucous and pinnatifid, leaf lobes toothed, upper leaves clasping the stem; flowers yellow, 4-merous; fruits siliqua, splits open when mature, seeds many, hard, round.

3. Parts Used: Seeds.

4. Chemical Constituents: Oil contains 50% acid and glucosinolates.

5. Uses And Applications: The oil is edible when cold pressed. It is used for greasing loaves of bread before baking seeds are used for animal feed. Oil is used as biodiesel.

6. Karyotaxonomic Evidence: *Brassica napus* (2n=38, AACC) has been demonstrated experimentally to be an amphidiploid of *B. campestris* (2n=20, AA) and *B. oleracea* (2n=18, CC) (U, 1935). The parental species are extremely difficult to cross artificially, due to endosperm deficiency leading to embryo abortion. Artificial forms of *B. napus* have been successfully raised, with or without embryo culture, by crossing diploid parents, followed by colchicine treatment of progeny, and by crossing autotetraploid forms of the parents. The spontaneous formation of *B. napus* is likely to have been an extremely rare event. This may be so because of post-fertilization barriers as well as differences in flower colour and form leading to discrimination by insect pollinators. These may likely to reduce the chances of inter-specific hybridization still further.

Brassica napus is generally self-fertile. Its parent species both have effective sporophytic incompatibility systems. It is also tolerant of inbreeding. *Brassica napus* is known to cross only with extreme difficulty with *B. oleracea* but readily with *B. campestris* (Yarnell, 1956). Natural introgression with the latter appears possible and has occurred in recent plant breeding. The oriental *B. napella* has been proved to be completely inter-fertile with *B. napus* and may be regarded as a variety of it (Olsson, 1954). A separate origin appears probable, presumably from local *B. campestris* and cultivated *B. oleracea* introduced from Europe.

7. History: It is doubtful whether or not *B. napus* grows in truly wild form. Linnaeus recorded it in sandy coastal areas of Sweden but the plants he saw may have been escapes. If wild napus exists, it must be a European Mediterranean species which originated in the area of overlap between *B. oleracea* and the much more widely distributed *B. campestris*.

Morphological forms, paralleling those found in *B. napus*, occur in *B. campestris*, the true turnips being equivalent to the swede. There are annual and biennial oil-seed forms of both species. On this basis, Olsson (1960) suggested that *B. napus* could have arisen several times by spontaneous hybridization of different forms of *B. campestris* and *B. oleracea*. Thus, the swede could have originated in mediaeval gardens where turnips and kale grew side by side. Another view that the swede could have arisen by selection from an oil-seed form. However, this is perhaps less likely. It seems plausible that forage rape originated from oil-seed forms.

Since the middle ages, rape has been recorded as an oil-seed crop in Europe. The species is unfortunately unknown (Appelqvist and Ohlson, 1972). By the early 19th century, British rape (*B. napus*) and Continental rape (*B. campestris*) were distinguishable on leaf characters. At that time, rape, almost certainly the biennial form of *B. napus*, was being used as an autumn forage for sheep. It was grazed lightly and left to yield an oil crop in the following summer. The Swiss botanist Casper Bauhin recorded Swedes first in Europe in 1620. Introduction into Britain appears to have been from Sweden around 1775-1780. Probably, it existed earlier than this (Boswell, 1949).

In brief, *B. napus* may or may not be a cultigen. Its domestication is recent. The swedes and rapes are probably only a few hundred years old. The oil seeds may or may not be somewhat older. A multiple origin from different parental combinations of *campestris* and *oleracea* is possible.

8. Further Developments: *Brassica napus* is a tolerant inbreeder. From the 19th century onwards, mass and line selection, with a good deal of inbreeding, was the standard method of improvement. Land races were replaced by named cultivars.

Artificial swede derived from crossing turnips with kale or kohlrabi gave useful breeding material (Namai and Hosoda, 1967). An artificial forage rape, in which, marrowstem kale was one parent, has been favourable to stock than a leading commercial cultivar. Some artificial oil-seed rapes are valuable with regard to winter hardiness and seed yield (Olsson, 1963).

Artificial 'heading' forms of *B. napus* have been produced in Japan from Chinese cabbage (*B. campestris*) and cabbage (*B. oleracea*). An entirely new morphological type is introduced into the group (Shinohara and Kanno, 1961). Artificial forms may best be regarded as a means whereby new variation may be introgressed into *B. napus* from the parental species. Considerable heterosis has been recorded in crosses between natural and artificial lines (Sarashirna, 1967).

It was possible, by individual plant selection, to breed cultivars with improved oil content from old land races. Heritabiliry studies showed that selection for oil content in winter rape can give better results than selection for seed yield (Olsson and Andersson, 1963). Advances in the breeding of oil-seed rape have taken place in recent years, largely due to improved analytical techniques and better chemical knowledge of the oil components and their genetic control. The most important feature has been the development of a technique for extraction and analysis of oil from single seeds without impairing viability. Improved crop management and advances in plant breeding have resulted in considerable improvements in seed yield of oil-seed rape (Downey, 1971). New characters have been introgressed into oilseed rape from turnip-rape. This method is used in Japan as a regular breeding technique (Shiga, 1970). The amphidiploid napocampestris (AAAACC, 2n=6x=58) has been explored by several breeders but without success. Oil-seed combinations were too infertile and forage types too low in dry matter content and/or yield.

9. Epilogue: The breeding of F_1 hybrid varieties and an interspecific hybridization are appearent trends. Thus swedes, with their exceptional potential for high-quality fodder production, are known to show yield heterosis (McNaughton and Munro, 1972) and could be bred as hybrids by the use of the self-incompatibility system. Suitable S-alleles can be introgrcssed from *campestris* relatively easily, from *oleracea* only with much more difficulty. In the oil-seed rapes, a cytoplasm-restorer male-sterility system has been identified (Shiga and Baba, 1973). AAC hybrids have been shown to have considerable potential as a rape-like crop if means of producing F_1 seed in commercial quantity could be developed (Mackay, 1973).

References:

Adil Iqbal, Akhtar Rumana, Tamina Begum and Tapesh Dasgupta (2016) Genetic estimates and diversity study in Sesame (Sesamum indicum L.) IOSR *Jour. Agri. Vet. Sci.* 9(8):01-05.

Almeida, F.J. de (1963) Acerca do melhorarnento da oliveira. Bat. Funta naco Azeite, 65-66: 1-18.

Aoba, T. (1972) Inheritance of seed colour and seed coat type in B. juncea. F. Yamagata *Agr. For. Soc.* 29:28-30.

Appelqvist, L.A. and Ohlson, R. (1972) Rapeseed. Elsevier Publishing Co. London, UK.

Arpna Kumara, Paul Satish and Vikas Sharma (2018) Genetic diversity analysis using RAPD and ISSR markers revealed discrete genetic makeup in relation to fibre and oil content in Linum usitatissimum L. genotypes. *The Nucleus* 61(1):45-53.

Bari, G. (1971) Effects of chronic and acute irradiation on morphological characters and seed yield in flax. Radiation Bot. 11:293-302.

Beccari, O. (1917) The origin and dispersal of cocos nucifera. *Philippines J. Sci.* 12:27-43.

Beinaert, A. and R. Vanderweijen (1941) Contribution a l'etude genetique et biometrique des varietes d' Elaeis guineensts. *Bull. INEAC., Serie Sci.* 27, pp. 101.

Belhaj, F.M. Dakheel, A.J., McCann, I., Abu Rumman, G., Shabbir, G.M. and A.A. Gailani (2013) Selection of high yielding and stable soft lower (Canthamus tinctorius L.) genotypes under salinity stress. *Agricultural science Research Journal* 3:273-283.

Benichow, A. (1962) Recherches critiques sur l'olivier de Laperrine. Mem. Inst. Reck Sahar. Alger 6, pp. 55.

Besnard, G., Baradat, P. and A. Berville (2001) Genetic relationships in the olive (Olea europaea L.) reflect multilocal selection of cultivars. *Theor. Appl. Genet.* 102:251-258.

Borkar, V.H. and V.M. Dharangutikar (2014) Evaluation of groundnut (Archis hypogaea L.) genotypes for physiological traits. *International Journal of Scientific And Research Publications* 4(1):1-8.

Boswell, V.R. (1949) Our vegetable travelers. *Naml geogr. Mag.* 96:145-217.

Bragalia, L., Manco, A. Giani, S., Hatzopoulos, P. and D. Breviaria (2017) *American Journal* of Plant Sciences 8:3475-3489.

Brown-Guediro, G. L., Thompson, J.A., Nelson, R.L. and M.L. Warburton (2000) Evaluation of genetic diversity of soybean introductions and North American Ancestors using RAPD and SSR markers. *Crop. Sci.* 40:815-823.

Bruce, E.A. (1953) Pedaliaceae. Flora of Tropical East Africa (Ed. W.B. Turrill and E. Milne-Redhead). Royal Botanical Garden, Kew, England.

Cantarelli, C. (1962) Cultivar e pigmentazione delle olive: essaine analitico dei prolifenoli delle drupe di diverse cultivar dell'Italia centrale. Atti I Conv. naz. Olivicolo Olearia, Spoleto, pp. 283-293.

Carapetian, J. (1973) Inheritance, cytology and histology of genic sterility in cultivated safflower (Carthamus tinctorills). Thesis, University of California, Davis, USA.

Chandravati, Maurya, R. Singh, P.K., Ranade, S.A. and H.K. Yadav (2014) Diversity analysis in Indian genotypes of linsed (*Linum usitatissimum* L.) using AFLP markers. *Gene* 549(1):171-178.

Chevalier, A. (1933) Histoire de l'arachide. Rev. Bot. appl. Agr. trop. 13:146-147, 722-752.

Ciferri R. (l950a) Dati ed ipotesi sull'origine e l'evoluzione dell'olivo. Olearia, pp. 115-122.

Ciferri, R. (I 950b) Elements pour I'erude de I'origine et de I'evolution de l'olivier cultivee, *Act. XIII Congr. internal. Oliicult* 1:189-194.

Cook, O., F. (1910) History of the coconut palm in America. Contrib. U. S. *Nat. Herb.* 14:271-342.

De Candolle, A. (1886) Origin of Cultivated Plants (Translated from 2nd Ed. in French, 1959), Hafner, New York, USA.

Downey, R.K. (1971) Agricultural and genetic potentials of Cruciferous oilseed crops. *J. Am. Oil Chew. Soc.* 48:718-722.

Eckey, E.W. (1954) Vegetable Fats And Oils. Amer. Chem. Soc. Monograph, 123. Reinhold Publishing Corp. New York, USA.

Fakuda, Y. (1933) Cytogenetical studies on the wild and cultivated Manchurian Soybeans (Glycine L.). *Jap. J. Bot.* 6:489-506.

Fior, H.H. (1955) Host-parasite interaction in flax rust-its genetics and other implications. Phytopath. 45:680-685.

Frandsen, K.J. (1943) The experimental formation of B. juncea. Dansk Bot. Archiu. 11:1-17.

Fremond, Y., Ziller, R. and de Nuee de Lamothe, M. (1966) Le cocotier. Paris.

Geethalakshmi, P., Parthasarathy, V.A. and V.Niral (2005) Genetic diversity among coconut (Cocos nucifera L.) genotypes using isozymes. *Asian Journal of Plant Sciences* 4(6):678-683.

Gibbons, R.W., Bunting, A. H. and J. Smartt (1972) The classification of varieties of groundnut (Arachis hypogaea) *Euphytica.* 21:78-85.

Gill, K.S. and D.M. Yermanos (1967) Cytogenetic studies on the genus Linum, I. Hybrids among taxa with 15 as the haploid chromosome number. *Crop. Sci.,* 7:623-627.

Gregory, W.C. and M.P. Gregory (1967) Induced mutation and species hybridization and de-speciation of Arachis. *Ciencia e Cultura* 19:166-174.

Gregory, W.C., Gregory, M.P., Krapovickas, A., Smith, B.W. and J.A. Yarbrough (1973) Structures and genetic resources of peanuts. *In:* Peanuts - Culture and Uses. Stillwater, Oklahoma, USA, Ch. 3.

Gregory, W.C., Smith, B.W. and J. Yarbrough (1951) Morphology, Genetics And Breeding. *In:* The Peanut: The Unpredictable Legume: A Symposium, National Fertilizer Association, Washington D. C., USA.

Hadfield, J.W. (1953) Linen flax fibre production in New Zealand. NZ Linen Flax Corp. and NZ Deps. Agric., Indust. and Comm., and Sci. indust. Res., New Zealand.

Hadley, H.H. and T. Hymowitz (1973) Speciation and cytogenetics. Soybeans, Production and Uses. Agronomy Series (Ed. B. E. Caldwell) American Society of Agronomy, Madison, Wisconsin, USA pp. 97-116. (Ed. Caldwell, Madison & Wisconsin) pp. 97 -116.

Hammons, R.O. (1973) Early history and origin of the peanut. *In:* Peanuts - Culture And Uses. Stillwater, Okla-homa, USA, Ch. 2.

Harberd, D.J. (1972) A contribution to the cytotaxonomy of Brassica and its allies. *Bot. J. Linn. Soc.* 65:1-23.

Hardon, J.J. and G.Y. Tan (1969) Interspecific hybrids in the genus Elaeis. I. Crossability, cytogenetics and fertility of F1 hybrids of E. guineensis × E. oleifera. *Euphytica* 18:372-379.

Harries, H.C. (1971) Coconut varieties in America. *Oleagineux Ami.* 26(4):235-242.

Hartley, C.W.S. (1967) The oil palm. Longmans, London, U.K.

Heiser, C.B. Smith, D.M., Clevenger, S.B. and W.C. Martin (1969) The North American sunflowers (Helianthus). *Mem. Torrey bot. Club* 22:1-218.

Hermann, F.J. (1962) A revision of the genus Glycine and its immediate allies. U. 5. D. A. *Agric. Res. Servo tech. Bull.* 1268, pp. 82.

Hiderbrandt, V.M. (1932) Sesame (Sesamum indicum L.) *Bull. Appl. Bot. Pl. Breed. Ser.* 9:1-114.

Higgins, B.B. (1951) Origin and early history of the peanut. *In:* The peanut the unpredictable legume. Washington, D.C., Ch. 2.

Husted, L. (1936) Cytological studies on the peanut, Arachis. II. Chromosome number, morphology and behavior and their application to the origin of cultivated forms. *Cytologia* 7:396-423.

Hymowitz, T. (1970) On the domestication of the soybean. *Econ. Bot.* 24:408-421.

Imrie, B.C. and P.F. Knowles (1970) Inheritance studies in interspecific hybrids between Carthamus flauescens and C. tinctorius. *Crop Sci.* 10:349-352.

Knowles, P.F. (1969a) Modification of quantity and quality and safflower oil through plant breeding. *J. Amr. Oil Chem. Soc.* 46:130-132.

Knowles, P.F. (1969b) Centers of plant diversity and conservation of crop germ plasm: Safflower. *Econ. Bot.* 23:324-329.

Koch, C. (1974) Soybean digest blue book, Hudson, Iowa, USA.

Krapovickas, A . (1968) Origen, variabilidad y difusion del mani (Arachis hypogaea). Aetasy Memorias XXXVII Congreso International Americanistas 2:517-534. [English translation by (Ed.) J. Smartt in P. J. Ucko and G. W. Dimbleby, The domestication and exploitation of plants and animals. London (1969)].

Krapovickas, A. (1973) Evolution of the genus Arachis. In Rom Moav (ed.), Agricultural genetics. New York, USA. 135-151.

Krapovickas, A. and V.A. Rigoni (1957) Nuevas especies de Arachis vinculadas al problema del origen del mani. *Darminiana* 11:431-455.

Leclercq, P. (1969) Une sterilite male cytoplasmique chez le Tournesol. *Ann. Amelior. Plantes* 19:99-106.

Mackay, G.R. (1973) Interspecific hybrids between forage rape (Brassica napus) and turnip (Brassica campestris ssp rapifera) as alternatives to forage rape. 1. An exploratory study with single pair crosses. *Euphytica* 22:495-499.

Mandel, J.R., Decaine, J.M., Marek, L.F. and J.M. Burke (2011) Genetic diversity and population structure in cultivated sunflower and a comparison to its wild progenitor, Helianthus annnus L. *Theoretical And Applied Genetics* 123(5):693-704.

McNaughton, I.H. and Munro, 1. K. (1972) Heterosis and its possible exploitation in swedes. Brassica napus ssp. rapifera). *Euphytica* 21:518-522.

Mehra, K.L. (1968) History and ethnobotany of mustard in India. *Adv. Front. Pl. Sci.* 19:51-59.

Meunier, J. and J.P. Gascon (1972) General scheme for oil palm improvement at the IRHO. *Oliaginacx* 27:1-12.

Namai, H. and Hosoda, T. (1967) On breeding of Brassica napus obtained (rom artificially induced amphidiploids. III. On the breeding of synthetic rutabaga (B. napus var, rapifera). *Jap. J. Breed.* 17:38-48.

Nayar, N.M. and K.L. Mehra (1970) Sesame: its uses, botany, cytogenetics and origin. *Econ. Bot.* 24:20-31.

Ngalamu, T., Ashraf MA and S. Meska (2013) Soybean (Glycine max L.) genotype and environment interaction effect on yield and other related traits. *American Journal of Experimental Agriculture* 3(4):977-987.

Ninan, K.N. (1989) Edible Oi Seeds: Growth, Area Responses And Prospects. Oxford & IBH Publishing Co. Pvt. Ltd. New Delhi, India.

Norden, A.J. (1973) Breeding of the cultivated peanut (Arachis hypagaea). In Peanuts - culture and uses. Stillwater, Oklahoma, USA, Ch. 5.

Olsson, G. (1960) Species crosses within the genus Brassica, II. Artificial Brassica napus. *Hereditas* 46:351-86.

Olsson, G. (1963) Induced polyploids in Brassica. *In:* Akerberg and Hagberg (Ed.), Recent Plant Breeding Research, Svaloj, 1946-1961. New York, USA and London, UK.

Ozturk, E., Ozer, H. and T. Polat (2008) Growth and yield of safflower genotypes grown under irrigated and non-irrigated conditions in a highland environment. *Plant Soil Environ.* 54(10):453-460.

Pandey, S.K., Das, A., Rai, P. and T. Dasgupta (2015) Morphological and genetic diversity assessment of sesame (Selamum indicum L.) accessions differing in origin. *Physiol. Mol. Biol. Plants* 21(4):519-529.

Pichel, R. (1956). L'arnelioration du palmier au Congo Bclge. *Bull. Agric. Congo Beige* 48 pp. 67.

Purseglovc, J.W. (1963) Some problems of the origin and distribution of tropical crops. *Genet. agraria* 17:104-122.

Rosengarten, F. Jr. (1969) Mustard seed. *In:* The Book of Spices. Jove Pblications INC New York, USA. pp. 295-305.

Rudorf, W. (1958) Topinambour, Helianthus tuberosus L. *Handbuch der Pflunzenziichtung* 3:327-341.

Salaman, R.N. (1940) Why 'Jerusalem' Artichoke? *F. Roy. Hort, Soc. NS.* 65:338-348, 376-383.

Sarashima, M. (1967) Studies on the breeding of artificially synthesised soiling forage rape. *Jap. J. Breed.* 17:26-31.

Shamsad Mohd, Dillon, S. K., Tyagi Vikrant and Akhtar Javed (2014) Assessment of genetic diversity in sunflower (Helianthus annus L.) germplasm. *Int. J. Agri. & Food Sci. Technol.* 5(4):267-272.

Shiga, T. (1970) Rape breeding by interspecific crossing between Brassica napus and Brassica campestris in Japan. *Fap. agrie. Res. Quarl.* 5(4):5-10.

Shiga, T. and S. Baba (1973) Cytoplasmic male sterility in oil seed rape, Brassica napus, and its utilization to breeding. *Jap. J. Breed.* 23:187-197.

Shinohara, S. and M. Kanno (1961) The species hybrid 'Hakuran' between common cabbage and Chinese cabbage. *Agric. Hart., Tokyo* 36:1189-1190.

Shoemaker, M.D. (1927) The Jerusalem artichoke as a crop plant. USDA. Tech. Bull. 33. pp. 32.

Smartt, J. (1964). Ph.D. Thesis, North Carolin State University. U.S.A.

Smartt, J. and W. C. Gregory (1967) Interspecific crosscompatibility between the cultivated peanut Arachis hypagaea and other members of the genus Arachis. *Oleagmeux* 22:455-459.

Souina, D., Sean B.N.T., Akain, D.K. and N.y Nicolas (2016) Genotype x environment interaction and kernel yield-stability of groundnut (*Arachis hypogaea* L.) in Northern Cameron. *J. App. Biol. Biotech*-4(1):1-7.

Tammes, P.L.M. and R.A. Whitehead (1969) Coconut. *In:* Outlinies of Perennial Crop Breeding In The Tropics (Ed. Ferwerda, F.P. and F. Wit) Wageningen, The Netherlands. pp. 175-188.

Tripura,U., Paramguru, P., Suresh, J., Kumaravadivel, N., Subramanian, A. and N Shoba (2018) Performance of indigenous and exotic coconut germplasm for yield and nut quality under Aliyarnagar Condition. *Int. Curr. Microbial. App. Sci.* 7(2):2611-2617.

U. (1935) Genome analysis in Brassica. *Jap. J. Bot.* 7:389-452.

U.N. (1935) Genome analysis in Brassica with special reference to the experimental formation of B. napus and peculiar mode of fertilization. *Jap. J. Bot.* 7:389-452.

Vaughan, J.G., Hemingway, J.S. and H.J. Schofield (1963) Contributions to a study of variation in Brassica juncea. *J. Linn. Soc. (Bot.)* 58:435-447.

Vavilov (1926) Studies in the origin of cultivated plants. *Bull Appl. Bot.* 16:1-245.

Vavilov, N.I. (1951) The origin, variation, immunity and breeding of cultivated plants. *Chronica Bot.* 13:1-366.

Verdcourt, B. (1966) A proposal concerning Glycine. *Taxon* 15:34-36.

Verdcourt, B. (1970) Studies in the Leguminosae-Papilionoideae for the 'Flora of Tropical East Africa': II. *Kew Bull.* 24:235-307.

Verdoorn, I.C. (1956) The Oleaceae of southern Africa. *Bothalia* 6:549-639.

Watt, G. (1890) Dictionary of The Economic Products of India. Cosmos Publication, New Delhi, India.

Weiss, E.A. (1971) Castor, sesame and safflower, Leonard Hill Books, London, UK.

Wester, P.J. (1920) The Coconut Palm: Its Culture and Uses. *Philipp. Agric. Bur. Bull.* 35:18-21.

Whitehead, R.A. (1965) Speed of germination, a characteristic of possible taxonomic significance in Cocos nucifera. *Trap. Agriculture Trin.* 42:369-372.

Whitehead, R.A. (1966a) Sample survey and collection of coconut germ plasm in the Pacific Islands. H.M.S.O., London.

Whitehead, R.A. (1966b) Some notes on dwarf coconut palms in Jamaica. *Trop. Agriculture Trin.* 43:277-294.

Whitehead, R.A. (1968a) Collection of coconut germ plasm from the Indian/ Malaysian region, Peru and the Seychelle Islands (1966-1967) and testing for resistance to Lethal. Yellotsing disease ill Jamaica. F. A. a., Rome.

Whitehead, R.A. (1968b) Selecting and breeding coconut palms (Cocos nucifera) resistant to lethal yellowing disease. A review of recent work in Jamaica. *Euphytica* 17:81-101.

Wynne, J.C. (1974) Breeding Potential of The Subspecies of Arachis hypogaea L: Inheritance of Branching Pattern, Estimates Combining Ability and Photoperiodic response. Ph.D. Thesis, North Carolina State University, U.S.A.

Yarnell, S.H. (1956) Cytogenetics of the vegetable crops. II. *Crucifers. Bot. Reo.,* 22:81-166.

Zeven, A.C. (1967) The semi-wild oil palm and its industry in Africa. Agricultural Research Report 698. Agricultural University, Wageningen, Netherlands.

Zeven, A.C. (1972) The partial and complete domestication of the oil palm (Elaeis guineensis). *Econ. Bot.* 26:274-279.

Zhang Gu-Wen, Sheng-Chun Xu, Wei-hua Mao, Qi-zan, Hu and Ya-Ming Gong (2013) Determination of the genetic diversity of vegetable soybean [Glycine max (L.) Merr.] using EST-SSR markers. *J. Zhejjang Univ. Sci.* B 14(4):279-288.

Zukozky, P.M. (1950) Cultivated plants and their wild relatives. (Abridged Translation By P. S. Hudson, Comm. Agric. Bureau, Farnham Royal, Bucks, UK. 1962, pp. 107.

10

Starch and Sugars

(I) STARCH

INTRODUCTION

Plant conduct the process of photosynthesis utilizing sun's energy and prepare their own reserve food. Starch is one of the most important and widely occurring products of the vegetable kingdom. It constitutes a chief type of reserve food for green plants. It is a complex carbohydrate. It is stored up in thin-walled cells in the form of grains. There are several types of starch differing in the size and shape of the grains and other physical and microscopic characteristics. The most important sources of starch are the cereal grains and underground tubers. Legumes, nuts and other plant organs may also contain appreciable amounts of starch. Starch serves as a staple food for animals and the greater portion of mankind. It has many industrial applications as well.

Starch is composed of two kinds of polysaccharides, amylose and amylopectin, exclusively composed of D-glucose residues with α-(1$\rightarrow$4) linkages in a linear amylose and α-(1$\rightarrow$4) linkages and ~50% α-(1$\rightarrow$6) branch linkages in amylopectin, both combined in a water-insoluble granule. It is partly crystalline and variable in size, shape and morphology. These are dependent on its biological source.

Starch is the common carbohydrate used in human diet. It generally occurs in many staple foods, especially hailed from cereals *e.g.* maize, wheat, sorghum and rice, etc. They are also available in root vegetables like sago, arrowroot, sweet potatoes, taro, yams, etc. Starch is the main way plant cells store energy in the form of glucose. This is the principal function of starch. Starch plays a big role in providing energy. The only function in the human diet is to convert into glucose for energy. Starch industry extract and refines starches from seeds, roots, tubers after grinding, washing, sieving and drying. Starch is used as sweetner in various food and drinks in the form of glucose syrups, dextrose, fructose syrup, maltodextrin, sugar alcohols, etc. They are used in food processing, pharmaceuticals, papermaking, clothing starch, corrugated board adhesives, adhesives or glues, etc.

10.1 CASSAVA

Manihot esculenta Crantz. (Euphorbiaceae)

Resume: *It is a native of South America. It is a staple food there of the native people. It is important tropical root crop and next to the sweet potato in importance. It has chromosome number n=18. Manihot species, in general, are thought to be segmental allotetraploid. The edible tubers are rich source of starch. They are eaten either raw or cooked and useful for the manufacture of industrial products.*

1. Introduction: The cassava, also called manioc, mandioc or yucca, is next to the sweet potato and most important of the tropical root crops and furnishes the basic food for millions of people. Among the tropical staples, it produces exceptional carbohydrate yields, much higher than those of maize or rice and second only to yams (de Vries *et al.,* 1967). It is mainly grown by peasant farmers, for many of whom it is the primary staple, but it is also a cash crop, being used to produce industrial starches, tapioca and livestock feeds.

It is native of South America. Over 150 varieties are known, most of which are used by the natives of food especially in south and central America at least twice a day throughout the year. Largest cassava is now grown in Africa, other five top cassava-producing countries are Brazil, Thailand, Nigeria, Zaire and Indonesia. It is usually propagated by stem cuttings and the plant matures in 6 to 12 months.

2. Plant Characteristics: A shrubby perennial attaining a height of 03 m or more, erect stem marked by leaf scars, bearing a cluster of 5 to 10 tubers; leaves crowded at top, spirally arranged, long-petioled, palmately compound, the lobes being 3 to 7, and obovate-lanceolate, green or variously pigmented; small flowers borne in axillary racemes with both sexes in the same inflorescence.

3. Parts Used: Root tubers.

4. Chemical Constituents: Tubers: Starch (30%), (protein or oil absent), minerals (rich in calcium), vitamin-C.

5. Uses And Applications: The tubers are eaten raw or cooked or boiled. A flour-like meal, called 'farminha' is baked into thin cakes (known as cassava bread). It is fermented into an intoxicating beverage. It is used in making tapiocea and cassava. Its flour is useful for sizing materials and laundering. Beer and other alcoholic drinks are also made from cassava.

6. Karyotaxonomic Evidence: The genus *Manihot* belong to the section Fruticosae, which contains shrubs adapted to savanna, grassland or desert. It is a cultigen, unknown in the wild state. The genus occurs naturally only in the western hemisphere between the southwest USA (33°N) and Argentina (33°S).

It shows most diversity in two areas: (i) one in north-eastern Brazil, extending towards Paraguay, and (ii) the other in western and southern Mexico. All species can be intercrossed, but they show evidence of being reproductively isolated in nature.

Species so far studied have 2n=36 chromosomes and show regular bivalent pairing. Studies of pachytene karyology have given evidence of polyploidy in both cassava and *M. glaziovii* (sect. Arboreae). There are three nucleolar chromosomes, which is high for true diploids. Duplication occurs for some of the chromosomes. *Manihot* species are segmental allotetraploids derived from a combination of two diploid taxa whose haploid complement had six chromosomes in common but differed in the other three (Magoon *et al.,* (1969).

7. History: The ancestors of cassava were suggested among the first plants to be used as food when man first migrated southwards into Central and South America. There seems good evidence that cassava flour was important in the 2nd and 3rd millennia BC. (Latharap, 1973; Reichel-Dolmatoff, 1965). The evidence is uncertain to decide whether present-day cassavas have descended from a single species or from several. *Manihot aesculifolia, M. rubricaulis* and *M. pringlei* are its closest relatives (Rogers and Appan, 1973). Each have a more or less erect growth habit and tuberous roots; *M. pringlei* is particularly interesting in having a low content of the poisonous cyanogenic glycosides. It seems likely that the variability of cultivated forms has been enhanced by hybridization with several wild forms. Most wild species have a propensity for colonizing disturbed areas and there would have been ample opportunity for gene exchange in areas adjacent to cultivation. Gene exchange produced hybrid swarms from which both new cultivated and wild forms were derived (Rogers, 1963, 1965). Selection in one direction by man and in the other by nature would have provided the kind of disruptive selection. It would have produced diversity both between and within the wild and cultivated forms and could have enhanced reproductive isolation.

Selection, during domestification, was for large roots, for more erect and less branched growth and for the ability to establish easily from stem cuttings. These pressures evoked correlated responses in several other characters. For example, branching occurs when inflorescences are formed, so the less branched plants selected were inevitably less floriferous. Selection for the ability to establish rapidly from cuttings probably favoured genotypes.

Forms completely devoid of cyanogenic glycosides are unknown. Forms with very low contents of glycoside have evolved locally but the bitter ones are still preferred by some tribes, particularly where predators are a problem. Processes for fermenting or heating the roots to remove the poison prior to consumption must have been developed at an early stage.

8. Further Developments: During post-Columbian period, Cassava spread rapidly from South America. It reached on the west coast of Africa, via the Gulf of Benin and the river Congo at the end of the 16th century; and on the east coast, via the islands of Reunion, Madagascar and Zanzibar at the end of the 18th century. Cultivation spread inland from both sides. The crop arrived in India about 1800. Its importance in the Old World increased slowly until the middle of the 19th century when it became widely grown as a famine reserve on account of its tolerance of drought locust attack and poor husbandry. Forms also established by cuttings. Self-sown seedlings frequently became established. The more successful of them were named and fixed by vegetative propagation. This gave diversity to the crop both in South America and in its new habitats. The process possibly proceeded faster in Africa, where virus degeneration of old stocks was more serious. The superiority of virus-free seedlings consequently became more conspicuous.

Until the 1920s, controlled breeding did not begin. Interspecific crossing and backcrossing was started in Java by that time and later in East Africa in 1937. In both places, the emphasis was on hybrids between cassava and the tree species *M. glaziovii* (Ceara rubber) and *M. dichotoma* (Jequie Manicoba rubber). These two-three species had already been introduced from South America as sources of rubber and were readily available to the breeders. Resistance to virus diseases was the main objective of breeding in Africa and *M. glaziovii* and *M. melanobasis* (a closely related shrub type, but misnamed) proved the best sources of resistance genes. But this early work also illustrated the value of interspecific crossing to widen the genetic base. Thus, *M. glaziovii* also contributed genes for improved vigour, drought resistance and resistance to bacterial disease. *Manihot melanobasis* contributed genes for modifying the leaf canopy and increasing the protein content of the roots. Breeding work elsewhere placed greater emphasis on local adaptation, low cyanogenic glycoside content and suitability for industrial starch.

Colombo *et al.,* (1998) investigated the genetic diversity of 31 Brazilian cassava clones using RAPD markers. They studied 31 distinct cassava landraces and cultivars originating in different Brazilian regions with different cultivation purposes. Both sets of variates revealed identical relationships among the cultivars. Multivariate analysis of genetic similarities placed genotypes destinated for consumption 'in nature' in one group, and cultivars useful for flour production in another. The power of RAPD markers is shown over botanical descriptors in studying genetic diversity, identifying duplicates, as well as validating or improving a core collection. The latter is particularly important in vegetatively propagated crop like cassava.

9. Epilogue: Cassava's many wild relatives contain reservoirs of untapped germplasm for crop improvement (Rogers and Appan, 1970), Species from the whole of the Fruticosae section of the genus can be crossed readily with cassava and, even with the tree species. The ecological range of cassava should be extended by crossing with species adapted to drought lower temperatures or acid soils. The roots of most of the wild species are high in cyanogenic glucosides. The emphasis should be on selection of clones with minimal levels. The objectives of breeding are changing only slightly, though some change may arise from the needs of mechanization.

10.2 SWEET POTATO

Ipomoea batatas (L.) Lam. (Convolvulaceae)

Resume: *Based on cross-disciplinary studies, it is concluded that sweet potato is American in origin. It's the closest wild relatives are American and Mexico is the centre of diversity. The related species has basic chromosome number x=15, as also of the cultivated I. batatas which is hexaploid. It is basically a starchy root tuber. The tubers are consumed raw or cooked and also used in the manufacture of industrial products.*

1. Introduction: Sweet potato is well known in the post-Columbian history. In Europe, it was introduced in 1492, nearly before 80 years before the Irish potato (*Solanum tuberosum* L.) reached Europe. The Spaniards introduced it to the Philippines, while the Portuguese introduced it early to their settlements in Africa and Asia. It reached China in 1594 and Japan in 1698.

Cross-disciplinary evidence suggests its origin in American regions. It is unknown in the wild state. Basically, the sweet potato is a starchy vegetable, confined mainly to tropical and sub-tropical regions. The tubers gained varying popularity and importance. The value of tubers and leafage as stock food was early recognized in Australia. They have been utilized in Japan as a source of industrial starch. It is in the emergent countries of the tropics, however, that the sweet potato makes its most valuable contribution as a crop, grown in mixed clonal plantings in the fields of subsistence farmers.

2. Plant Characteristics: A trailing vine with elongated root tubers; leaves simple, variable in size and shape, mostly ovate, entire or deeply dissected, often with purple pigmentation; flowers large, funnel-shaped, reddish-purple borne either singly or in cymose inflorescence; fruits capsules containing four, small black, flat-sided seeds.

3. Parts Used: Root tubers.

4. Chemical Constituents: Tubers: Carbohydrate (27%, including sugar and starch, fats 0.2%), protein (1.5-2.0%), yellow-fleshed tubers rich in vitamin-A.

5. Uses And Applications: The tubers are consumed fresh, boiled, baked or as canned or in dehydrated form. After fermentation, alcohol is produced. Sweet potato is utilized in the manufacture of adhesives in textile and paper sizing.

6. Karyotaxonomic Evidence: Sweet potato, *Ipomoea batatas* is spaced in the section Batatas with the wild species *I. gracilis, I. tiliacea, I. trifida, I. triloba* (Van Ooststrom and Hoogland, 1953). House (1908) had earlier grouped *I. batatas* with 25 wild species under section Batatas, sub-section Aequisepalae to include *I. tiliacea, I. trifida, I. triloba* and had added *I. trichocarpa* and *I. lacunosa*. Cytotaxonomic study by Nishiyama (1971) suggested a new classification of *I. batatas* and its wild relatives on the basis of chromosome number and genome analysis. These results are presented in Table-1 below:

Table 1:

	Taxon	Synonyms	2n (x = 15)	Constitution
(i)	*I. batatas*			
	var. *batatas*	—	90	BBBBBB
	f. trifida	*I. trifida*	90	BBBBBB
	var. *littoralis*	*J. littoralis*	60	BBBB
	var. *leucantha*	*I. leucantha*	30	BB
(ii)	*I. triloba*			
	var. *triloba*	—	30	AA
	f. lacunosa	*I. lacunosa*	30	AA
	f. trichocarpa	*I. trichocarpa*	30	AA
	f. ramoni	*I. ramoni*	30	AA
(iii)	*I. gracilis*	—	60	--
(iv)	*I. tiliacea*	—	60	--

Nishiyama's revision is open to some doubts: (i) on cytological grounds, the range of irregularities such as multivalents and univalents at meiosis in inter-specific hybrids between wild *Ipomoea* forms may be paralleled within *I. batatas* itself (Ting and Kehr, 1953; Yen, 1974), (ii) on morphological grounds, no wild species examined by Martin and Jones (1972) (especially *I. trifida* but also including *I. lacunosa, I. gracilis, I. trichocarpa* and *I. triloba*) having sufficient similarities with *I. batatas* to suggest direct evolutionary relationships.

7. History: A general agreement based on cross-disciplinary evidence reached to a conclusion that sweet potato is of American origin. The closest wild relatives are American, and Mexico has been inferred as the centre of diversity for the Batatas section of *Ipomoea* by Martin and Jones (1972) and Nishiyama (1971). The comparative study of variation in cultivated populations from America, southeast Asia and Oceania shows a Vavilovian effect in that the American, specifically western South American, population shows the widest range of variability (Yen, 1974). Archaeological remains of the tuber from Peru, including representations in pottery, antedate any direct or indirect evidence for antiquity of the plant from any other area in the world (Towle, 1961; Yen, 1974). Engel (1970) reported tuber remains from a cave in Chilca Canyon, Peru, carbon-dated at between 8,000 and 10,000 BC. It is not certain whether these remains of sweet potato and other tubers are representative of very early agriculture or whether they are the product of plant gathering.

There is uncertainty about the evolutionary steps by which the crop originated. Ting and Kehr (1953) suggested hybridization between two unknown wild species followed by allohexaploidy (2n= 2x15+ 4x15 = 90). Nishiyama (1971), however, regards the sweet potato as an autohexaploid originated directly from *I. trifida*. He places *I. leucantha* (2x), *I. littoralis* (4x) and *I. trifida* (3x and 6x) in a scheme resembling that of Ting and Kehr. Jones (1967) and Martin and Jones (1972) have suggested that the tuberless *I. trifida* evolved in parallel with *I. batatas* from ancient species, while the self-fertile diploid *Ipomoea* forms evolved more recently from earlier self-incompatible ancestors. Self-iricornpatibility is a characteristic of most sweet potatoes (Martin, 1965, 1968)

Martin (*loc. cit.)* defined its genetic basis in terms of S-type genes with complex epistatic relationships depending on the alleles present. Subsequently, Martin has addressed himself to the general problem of infertility as the result of self-pollination and hybridization. He suggested an over-riding physiological sterility caused by chromosomal aberrations and weak gamete production superimposed on the incompatibility system.

The role of breeding system has been a conservative agent in the evolution of the crop. Vegetative reproduction must have had a quasi-apomictic effect. The out-crossing habit also tended to produce variation. The absence of conscious breeding by seed and the biological barriers of sterility and incompatibility were further impeding factors.

Tuber remains from archaeological sites have been also recovered from Hawaii, New Zealand and Easter Island, Structural remains from New Zealand early in prehistory indicate that storage practices were developed to convert the tropical perennial plant into an annual in response to seasonal temperate climates. This adaptation was the forerunner of modern practices. Yen (*cf.* Riley *et al.,* 1971) thought that the introduction of 8th plant from America to Polynesia occurred before the 8th century AD.; but Brand (*cf.* Riley *et al.,* 1971) attributed it to Spanish travellers after AD. 1500.

8. Further Developments: The widespread adoption of the plant in America by the 15th century, are demonstrated by records of sweet potato in the chronicles of Columbus's first expedition to Hispaniola. The Caribbean region was to be the source of the earliest sweet potato stocks in Europe. By Shakespearean times, the plant was well known in England.

Portuguese voyagers following the track of Da Gama, carried the plant eastwards from the Caribbean, Brazil and Europe to Africa, India, southeast Asia and Indonesia. From their colonies in western South America and Mexico, the Spaniards took the plant westward to Guam and the Philippines. The confluence of these lines of introduction, together with the earlier American-Polynesian diffusion, is indicated by studies of variability in clonal populations collected largely from subsistence cultivators (Yen, 1974).

Geographically separated populations exhibited somewhat different ranges of variation affecting some 40 morphological and physiological (*e.g.* cold and disease reaction) characters. There have been some effect of human selection on two characters *viz.,* tolerance of cold and a preference for a long-vined growth habit in hilly South America and New Guinea. The movement of the plant into new environments has not resulted in speciation or the formation of recognizably differentiated races or chromosome number.

The cataloguing of regional collections aroused scientific interest in sweet potato late in the 19th century in America. These catalogues are invaluable, for they record varieties now extremely rare or extinct. Programmes on conscious improvement by clonal selection and the investigation of hybridization and inheritance in the species were started in America. Similar programmes were soon instituted in other parts of the United States and in Japan, the Philippines and India. Selection for quality and yield in clonal and hybrid populations resistance to fungal and viral disease were their concerns. In the United States, cultivation was concerned on variety, the 'yam' (unfortunately the name is coined so), whose tuber flesh is very sweet, soft-textured, bright orange in colour and with generally high carotene content. It is different from the 'Jersey' type which is yellow or white fleshed. At one time, the southern states produced mainly the 'yam' type, while California favoured the 'Jersey'. After 1960, the 'yam' predominated in all sweet potato-producing areas.

Zang *et al.,* (2000) assessed genetic diversity of sweet potato. They evaluated 69 sweet potato cultivars from four geographical regions including 13 countries of Latin America. These were randomly sampled and fingerprinted using ALFP markers. They revealed highest genetic diversity found in Central America. However, they reported lowest genetic diversity in Peru-Ecuador. The within-region variation was the major source of molecular variance. The between regions variation is statistically significant. Based on the results obtained, these authors opined that Central America is the primary centre of diversity and most the centre of origin of sweet potato. They considered Peru-Ecuador as a secondary centre of sweet potato diversity.

9. Epilogue: The objective for disease resistance may continue in future breeding programmes. Other objectives, involving machine harvesting and the use of tubers in processed foods have considerable impact. In developing countries, practical objectives are fairly well defined, *viz.,* disease and pest resistance, and nutritive value. In areas in which sweet potato forms a dominant part of the diet and protein intake is low, Oomen *et al.,* (1961) suggested the upgrading of amino-acid content in the sweet potato by breeding, as varieties show considerable variability in this respect.

Some innovations in the approaches to breeding are apparent. Jones (1965) proposed a modification of the polycross method of open pollination with the aid of insects, as an alternative to the generally applied pedigree methods. This method also designed to maximize variability and to promote the expression of desirable epistatic effects. Moreover, the conservation of variability is inherent in the procedure and this must be a major concern in breeding for subsistence agriculture. Tai *et al.,* (1967) proposed the utilization of native selection in the selection stages of breeding programmes.

10.3 YAMS

Dioscorea Spp. (Dioscoreaceae)

Resume: *Yams belong to a large genus Dioscorea which contains more than 600 species in the world. Of these, only eleven species are grown for their tubers. They are domesticated independently in the Old and New World. All others, except D. trifida L. f. (called cush-cush yam) are found in wet tropics of the Old World. Only D. trifida was domesticated in the New World. The evolutionary separation of the Old and New World species is observed in their cytology, chemistry and vascular anatomy. Old World Dioscoreas contain alkaloids. High degree of polyploidy occurs within Old World yams with chromosomes counts of 50 to 140. The tubers are consumed after cooking. They are a cheap source of carbohydrate and protein.*

1. Introduction: In many tropical countries, yams are a staple food crop. Yams are also important in parts of southeast Asia and Oceania especially Melanesia, and in the Caribbean and neighbouring parts of tropical America (Coursey, 1967). They are essentially crops of subsistence agriculture. There has been some semicommercial production for local markets in West Africa and the Caribbean. Yams are primarily a starchy staple. Nevertheless, their protein content is appreciable and contributes significantly to dietary value. Apart from food yams, interest has recently developed in several other *Dioscorea* spp. as sources of steroidal sapogenins for pharmaceutical purposes (Anonymous, 1972).

2. Plant Characteristics: Dioecious perennial climbers, deeply rooted and bearing root tubers of variable size and shape, stem-axis armed with spines or spineless; leaves simple, reticulately veined, opposite or alternate, cordate or deeped lobed at base, and long-petioled; flowers small, unisexual in axillary racemes, panicles or cymes on male or female plants; fruits generally winged capsules.

3. Parts Used: Tubers.

4. Chemical Constituents: Tubers: Carbohydrate (15-25%), protein (4-8%), vitamin-C.

5. Uses And Applications: Tubers are edible, they are baked, boiled or ground into flour. They are also used as vegetable.

6. Karyotaxonomic Evidence: The genus *Dioscorea* contains about 600 species. It is by far the large genus in the family Dioscoreaceae. Burkill (1960) indicated the emergence of the ancestral proto-Dioscoreaceae in southeast Asia. Old and New World separated by the formation of the Atlantic Ocean at the Cretaceous. The species of *Dioscorea* followed divergent evolution. There are consistent differences in vascular anatomy (Ayensu in Anonymous, 1972) while only Old World *Dioscoreas* contain alkaloidal toxins (Coursey, 1967). Desiccation of the Middle East in the Miocene separated African from Asiatic species. The evolutionary separation of Old and New World groups is reflected in their cytology as well. The Old World species have chromosome numbers based on x=10 (except for the section Borderea and the genus *Tamus*, both based on x=12). The primitive section Stenophora has 2n=2x=20, but in many tropical species related to the food yams, 2n=4x=40 (Burkill, 1960; Coursey, 1967). The New World Dioscoreas and the related *Rajania*, have x=9, most being tetraploids or hexaploids (2n=36, 54) (Coursey, 1967; Ayensu and Coursey, 1972).

High degrees of polyploidy occur within Old World yams. Chromosome counts of 50, 60, 70, 80, 90, 100, 120 and 140 are noted (Coursey, 1967). Most of these are from cultivated forms, but some of the highest counts are from wild material, indicating that the relationship between cultivation and polyploidy stressed by Burkill (1960) is by no means certain. Among cultivated yams aneuploids (2n=38, 52, 55, 66 and 81) are reported and, rarely (Coursey, 1967) variation in chromosome number has even been recorded within a single individual.

Food yams are derived mainly from the twelve species. Some 50 or 60 other species are gathered or cultivated to a very limited degree (Coursey, 1967; Alexander and Coursey, 1969). The species exploited for pharmaceutical purposes are quite distinct. Interest has concentrated largely on six, although others have been utilized (Coursey and Martin, 1970; Martin in Anonymous, 1972; Coursey, 1972).

7. History: The beginnings of the evolution of yams as crop plants lie earlier in history, within the aboriginal cultures. The domestication of the yams took place differently in Asia, Africa and tropical America involving different species. Yam based agricultural systems evolved only through external cultural influences, subsequent to the initiation of grain-crop agriculture (Coursey, 1967). Culture contacts between grain-using intrusive and non-grain-using aboriginal populations probably took place in Asia, Africa and northern South America, around 5000 BP., contributing to the present form of yam-based agriculture (Alexander and Coursey, 1969).

The cultigen *D. alata* appears to have arisen in Asia under human selection from wild forms related to *D. hamiltonii* and *D. persimilis* in the northcentral parts of the southeast Asian peninsula. *Dioscorea esculenta* is indigenous to the same area and cultivated forms seem to have been derived by selection from the wild (Alexander and Coursey, 1969; Coursey, 1975). The major centre of diversity of cultivars of both species is now in Papua-New Guinea. The lesser diversity in other areas may have resulted from erosion of a former greater variability.

Their ancient domestication evidenced by vege-cultural civilizations in southeast Asia, at radiocarbon dates of 10000 BP. or more, associated with mesolithic Hoabinhian ceramics. Yams had evolved as domestic plants and have been taken across the Pacific in the Polynesian migrations. They have been associated, even earlier, with sophisticated premetallic cultures in Melanesia. Both species were taken to Madagascar and, probably, to the East African littoral by later Malaysian migrations (ca. 1500 B. P.) without influencing African yam domestication (Coursey, 1972, 1975). The domestication of the warm-temperate species, *D. opposita* and *D. japonica*, may be associated with the Jomon culture in Japan.

The evolution of African yams as crop plants was initially independent of external influences. Of the two major species, *D. cayenensis* is indigenous to much of forest zone Africa, while *D. rotundata*, a cultigen, like its Asiatic counterpart *D. alata*, is so close to *D. cayenensis* that it has been regarded as a sub-species (*cf.* Ayensu and Coursey, 1972). These African species belong, like *D. alata*, to section Enantiophyllum of *Dioscorea. Dioscorea rotundata*, the most important of the African yams, appears to be of hybrid origin. One parent is undoubtedly *D. cayenensis*, a forest species, adapted to a short dry season and having limited tuber dormancy. The other must be a savanna species; *D. praehensilis* have the greatest resemblance to some *D. rotundata* forms.

Hybrids may often have arisen spontaneously in the ecotonal forest/savanna mosaic. However, they were not suited to the local ecology. They did not persist until the advent of Mesolithic man who, with his concepts of 'protection' of food plant resources, created new ecological niches which favoured survival of the archetypal *D. rotundata*. His recognition of this superior food resource led to the many, eventually clonal, cultivars known today (Ayensu and Coursey, 1972; Coursey, 1975).

In the New World, yams were of secondary importance in pre-Columbian times due to availability of cassava. *Dioscorea trifida* is clearly an Amerindian domesticate. Distribution of cultivars suggests domestication on the borders of

Brazil and Guyana, which is followed by spread through the Caribbean (Ayensu and Coursey, 1972).

Yams are primarily Old World domesticates. A close parallel exists between those of Asia and Africa, botanically and in cultural importance as well (Coursey, 1972). They are domesticates of the tropical/ equatorial 'non-centres'. (Harlan, 1971). They diffuse in both space and time, where crops developed by a slow, widely spread process of evolution towards symbiosis with man. These contrast sharply with the subtropical Vavilovian centres where grain crops were domesticated in closely defined geographical areas during the so-called 'Neolithic Revolution'. Within these 'non-centres' yams of tropical regions, rather than those of equatorial ones regions are inherently best adapted to exploitation by man.

At very early times, a continuous ring of gardening cultures once linked Oceania to Africa in the tropical latitudes (Lomax and Berkowitz, 1972). These cultures have survived only at the West African and Melanesian extremities and among relict peoples of mainland Asia and the Philippines. The massive 'incursion of higher cultures', subsequent to the 'Neolithic Revolution', have virtually eliminated older cultural patterns in the areas between. Within these gardening cultures that yam domestication began, with a ritualization of man's relationship with his originally wild food plants. This permitted the emergence of forms suited to man's requirements but possibly ill-adapted to survival in the wild (*cf.* Ayensu and Coursey, 1972). Criteria of domestication are hard to define. Loss of defence mechanisms such as toxicity, deep-burying tuber habit, spininess, reduction of sexual fertility and high polyploidy are indicative (Coursey, 1975).

8. Further Developments: Evolution of yams as crop plants began in recent times when the paleo-colonial Lusitano-Iberian expansion into the tropics around 500 BP. brought the Indo-Pacific area, Africa and America into closer contact. Asiatic and, later, African yams were widely used for the victualling of ships. *Dioscorea alata* was taken to West Africa through this agency, while the Atlantic slave trade resulted in both this species and the African *D. rotundata* and *D. cayenensis* being taken to the Caribbean. Early historic records of *D. alata* in West Africa and of African yams in the Americas exist (*cf.* Coursey, 1967). These transfers have resulted in relatively few cultivars being taken outside their continents of origin. The Asiatic *D. esculenta* was only taken to Africa and America within the 19th century. Jamaican and Puerto Rican forms of *D. cayenensis* are superior to those in Africa. This suggested that transdomestication has taken place.

Spontaneous hybridization must have contributed to the ancestry of some yams. Propagation has been entirely vegetative. Each cultivar is a single clone. Improvement must have been mostly by selection of somatic mutants rather than of sexual seedlings. Vegetative propagation has led to reduced sexual fertility, resulting from limited production and incomplete opening of flowers and failure of pollen release.

After development in biochemical industry, the drug Dioscoreas emerged as crop plants. These created a demand for steroidal saponins for manufacture of pharmaceuticals (Coursey, 1967). Major demand has been met from wild yams from arid regions of South Africa and Mexico. Gathering has been disciplined by legal sanctions into more systematic collecting, with precautions to ensure regeneration. Attempt are also made to cultivate some species (*cf.* Coursey and Martin, 1970; Anonymous, 1972). This progress towards domestication constitutes a recapitulation over decades of what took place over millenia with the food yams (Coursey, 1972). The drug yams are still essentially wild species, and seed freely.

Sebsese *et al.,* (2013) investigated genetic diversity and species delimitation of cultivated and wild Guidnea yams from southwest Ethiopia determined by AFLP markers. They used AFLP genetic fingerprinting to evaluate and characterize 43 individual plants belonging to different populations of wild and cultivated Guinea yams. Ordination and cluster analyses of AFLP data failed to produce any clear species boundary between either the Guinea yam accessions or between them and their wild relatives. AFLP analyses also revealed a higher genetic divergence among cultivated Guinea yam accessions of the sheko cultivars. Ordination and cluster analysis using UPGMA revealed no clear boundaries between members of the complex. These authors opined that the taxonomy of these species needs to be revised using other markers.

Chair *et al.,* (2016) studied plastid phylogenetics of Oceania yams (*Dioscorea* spp.) using plastid trnL-F and rpl32-trnL(UAG) sequences and nine nuclear co-dominant microsatellites. The dischotomy *D. nummularia* in semi-perennial and annual types was reflected by molecular markers, but their genetic difference is more complex. Nuclear markers revealed that some annul yams shared alleles with *D. alata* and semi-perennial *D. nummularia*, suggesting a hybrid origin. This explains their intermediate morphotypes and the difficulty met in classifying them.

9. Epilogue: Yam production has been a decline over recent decades in comparison to cassava and rice. Critical factors responsible are high labour requirements and consequent costs of production. Yams are often the preferred staple and because of population increase, there will not be an absolute decline. Demand will probably continue.

Developmental plans for food yams as crops should be: (i) towards improved clones for the small farmer for growing under virtually horticultural conditions at relatively high cost, (ii) towards attempts to modify the crop better to adapt for mechanized agriculture. The variation in yam cultivars is vast. We can experiment simply by selection and dissemination of the best existing clones. For more radical improvement, breeding will be required, necessitating study of the reduced sexuality of many clones to overcome low fertility. Few species are reported in danger of extinction. There is an urgent need to collect such forms before further erosion occurs. Even apparently inferior clones may have value for particular breeding characteristics or for different ecosystems.

(II) SUGARS

INTRODUCTION:

Green plant, in general, manufacture sugars and so are to be found in all individuals. So much of the manufactured products, however, is used directly in the metabolism of the plant that comparatively little is accumulated. Storage sugars are to be found in various parts: (i) roots, as in the case of beets, carrots, parsnips etc. (ii) stems, as in sugar cane, ,maize, sorghum, and the sugar maple, (iii) in flowers, such as the palm; (v) in bulbs like the onion: and (v) in many fruits. Several types of sugars are to be found, chief among which are sucrose (cane sugar), glucose (grape sugar) and fructose (fruit sugar). These sugars serve as a reserve food supply for the plant itself.

One of the most necessary foods for mankind is sugar. The day has passed when it was considered merely as a luxury to be used for flavoring purposes. It constitutes a perfect food as it is in a form that can be readily assimilated by the human body. It is mainly valued as an energy producer, and it is particularly well adapted for use after any type of muscular exertion. Although primarily a food, such a vast industry has been built up in connection with the extraction of the sugar from plant tissues, its purification, and refining, that it seems best to consider sugar as an industrial plant product. Moreover, sugar has become an extremely important industrial chemical with numerous different derivatives.

One of the most valuable products of the plant world is sugar. It is surpassed in importance only by wheat, corn, rice, and potatoes. Considering its importance, it is surprising that the sources from which it is obtained are few in number. Only the sugar cane, sugar beet, sugar maple, maize, sorghum, and a few palms are of commercial interest. In all of these, sucrose, a disaccharide of glucose and fructose; is the type of sugar stored. To humans, sucrose tastes sweeter than either of its constituent monosaccharide.

Apart from sweetening, sugar perform a variety of functions in food products. It is used as a preservative in baked foods, to hold moisture and prevent staleness. It enhances texture and colours. It is useful in preserves, jam, jellies, canning, freezing and beverages. It is involved in fermentation, textile industry and detergents.

10.4 SUGAR BEET

Beta vulgaris L. (Chenopodiaceae)

Resume: *It is a second important source of sugar after sugarcane sugar. The cultivated varieties of Beta vulgaris are said to be descended from a common ancestor B. martima L. called 'Sea Beat' or 'Wild Beat' of the coastal regions of Europe, North Africa, the Azores and from Asia minor to the East Indies. B. vulgaris has 2n=18 chromosomes as diploid species. Its related species are also autopolyploid. Romans used it as a cattle feed. It featured also in Roman recipes of 2nd and 3rd centuries. Apart from sugar extraction, it is widely consumed as salad or boiled.*

1. Introduction: Probably, the use of beet dates from prehistoric times when the leaves were almost used as potherbs. Aristotle mentioned red chards and Theophrastus mentioned red light green and dark green chards used in the 4th century BC. The Romans used beet as feed for animals and man. It was called Beta by the Romans. It was taken from Italy to northern Europe by the barbarian invaders. By the 16th century, it was widely used for feeding animals, particularly during the winter. Red beet or beetroot featured in Roman recipes of the 2nd and 3rd centuries. It was recorded in English recipes of the 14th century. In Germany, it was first described in 1557 when it was referred to as Roman beet.

Crystalline sugar was a scarce luxury in the western world before the 17th century. Originally, all sugar derived from sugar cane grown in the tropics but, at the present time, beet sugar accounts for nearly half the total world production of the refined product. Production of sugar beet is mainly in temperate climates, especially Europe, the USSR and North America, but it is also grown in the sub-tropics under irrigation.

2. Plant Characteristics: A true biennial, with a thickened root and short stem; leaves simple, arranged in a closed spiral; flowers small, green in cymes, arranged on a branched spike; fruits joined together by the swollen perianth, containing small, kidney-shaped, brown seeds.

3. Parts Used: Swollen root.

4. Chemical Constituents: Mainly sugar (20%).

5. Uses And Applications: Beetroot is consumed as salad or boiled. It is extracted for sugar. Byproducts of sugar industry, especially molasses are used for alcohol manufacture

6. Karyotaxonomic Evidence: The Old World genus *Beta* is confined to Europe. It is divided into three subgeneric taxa *viz.*, (i) sections Beta (Vulgares), (ii) Patellares and (iii) Corallinae. The section Beta has the extensive distribution and may be the most primitive. The Corallinae occupy the eastern part of the range while the Patellares are restricted to the southwest, chiefly in the Atlantic Islands and the western Mediterranean. Cultivated beets are all originated from the section Beta, other sections have species of limited culinary value. Useful characters occur throughout the genus and attempts have been made to incorporate them into sugar beet. Interspecific hybrids involving members of different sections are more difficult to make than those derived from the same section and are less easily exploited. Systems of classification have been proposed. (Coons, 1954; Krasochkin, 1959). They cris-cross specific and intraspecific boundaries. Broadly, however, the section Beta comprises: (i) *Beta vulgaris* ssp. vulgaris (contains many cultivars, including sugar beet, beetroot, mangolds and fodder beet; ssp. cicla embraces the chards; ssp. maritima, sea beet, is thought to be ancestral to most if not all, cultivars), (ii) *B. atriplicifofia*, (iii) *B. macrocarpa* (of little agricultural significance) and (iv) *B. patula*.

Beta maritima L. (Sea Beet) grows in coastal Europe, Africa, Azores, Asia Minor and East Indies. Diploid chromosome numbers within the genus are: 2n=2x=18 (*e.g. B. vulgaris*), 2n=4x=36 (*e.g. B. corollifora*) and 2n=6x=54 (*e.g. B. trigyna*). Spontaneous chromosome doubling results in the occasional occurrence of autopolyploid individuals.

7. History: Maregraf noted the presence of sugar in sap from the roots of fodder beet in 1947. His pupil *viz.,* Achard, laid the foundations of the improvement of the crop plant and the development of an extraction process. Achard received a subsidy from the King of Prussia and secured a ten-year monopoly for the manufacture of beet sugar. The first factory was built at Kunern, Silesia, in 1801. Silesian sweet fodder beet, in which sugar contents up to 6. 2% were recorded, was introduced into France in 1775. In 1811, Napoleon published the first of a series of decrees requiring beet to be grown, He established schools for its study. This action was prompted by disruption to the supply of cane sugar from the West Indies to France caused by the British naval blockade of continental ports. Later, the import of sugar to France was restricted in order to prevent the beet industry from succumbing to competition from cane sugar produced more cheaply in the colonies.

8. Further Developments: During the 19th century, conscious selection was started primarily in Germany and France. Rapid advances were made in breeding for greater sugar percentages, facilitated by the introduction of polarimetric methods of estimation. This trend was encouraged in Germany by the imposition of excise duty. It was levied on the weight of beet. Government support proved to be essential for the development of beet industries, both in Europe and in the USA. The first successful factory was established in 1879 in California.

Outbreeding and largely anemophily is noted sugar beet (Dark, 1971). Varieties grown in the first half of the 20th century were multilines based on 20-30 parental stocks. The parents were maintained by sib-mating, with mass selection at each generation for root size and sucrose content. Pollination in the final generation was random, with all plants producing pollen and seed. The commercial root crop consisted of 95-97% interline hybrids.

Experiments on polyploids began in the late 1930s, It soon became evident that, although tetraploids were slightly inferior to the parental diploids, some triploid combinations were very promising. Anisoploid varieties, produced from random pollination of parental mixtures of diploid and tetraploid lines, were marketed in some European countries after the late war. They were also replaced by monogerm varieties, many of which are triploid, and thus also derived from the work on polyploidy (Savitsky, 1962; Sedlmayr, 1964; Hornsey, 1975). Wholly triploid varieties are produced with the aid of cytoplasmic male sterility (Owen, 1945).

It is important that the monogerm character allows the possibility of total mechanization of crop production. It was first exploited in the USSR and the USA. It has been also adopted by breeders in most other countries. A monogerm plant bears flowers singly at most nodes of the inflorescence. The monogerm character is an inflorescence character and is determined by the genotype of the seed-parent. Hence, it was possible to use a multigerm pollinator on a male-sterile monogerm in commercial seed production (Savitsky, 1954).

Want and Goldmen (1999) investigated genetic distance and diversity in table beat and sugar best accessions using DNA RAPD. Results from these investigations demonstrated that table beet are genetically distinct from standard table beet germplasm, likely because of their origin from an introgression program designed to incorporate sugar beet genes. Characterization of sugar beet-specific genomic regions retained in inbred lines of table beet may shed additional light on the value of sugar beet genes in table beet improvement.

In parts of the USA the curly-top virus, which is transmitted by the leafhopper (*Eutettix tenetlus*), made beet growing unprofitable until resistant varieties were produced. Yellowing viruses transmitted by aphids occur more widely but are generally less damaging. Selection for tolerance to yellowing has met with some success. Varieties resistant to the leaf-spot caused by *Cercospora beticola* were developed in the USA and parts of Europe, while resistance to root infecting fungi such as *Aphanomyces cochlioides* and *Sclerotium rolfsii* has also been developed in countries where they were important.

It was possible to select for resistance to bolting by applying artificial conditions of vernalization to plants at the four-leaf stage (Campbell, 1953). This work provided the stimulus for intensive screening of breeding material for commercial use. The bolting-resistant varieties that were produced enabled beet to be grown further north than had hitherto been possible.

9. Epilogue: Emphasis in breeding was on the production of improved varieties with the monogerm character because of constant rise in costs of production. The main objective of breeding programmes are to produce monogerm varieties with improved field emergence and the potential for heavier yields of extractable sugar from roots that can be harvested efficiently by machine. The trend to earlier drilling in temperate climates calls for still greater resistance to bolting. The increased cost of chemicals used for controlling insect pests calls for increased effort to produce varieties more resistant both to the aphids and to the virus diseases they transmit. The breeder must also know the development of improved herbicides for use with sugar beet to ensure their selectivity with new varieties.

With genetic improvements to monogerm diploids, any advantage that the triploid may have had in the past may well not apply in the future. The popularity of triploid varieties in Europe has not been shared in the USA. Tetraploid, anisoploid and triploid varieties suffer from the disadvantage of producing a propagation of aneuploids. It, therefore, it appears that the success of a new variety, in future, is more likely to be related to the amount and type of effort that has gone into selecting its components, instead of any inherent varietal advantages.

10.5 SUGARCANE

Saccharum officinarum L. (Poaceae)

Resume: *It is thought originated in the South Pacific, probably New Guinea. It has been cultivated in India since prehistoric times and was an important crop by the end of 4th century BC. It was then introduced in various parts of the Old World and was introduced in the New World by Columbus. It is polyploid crop. Saccharum robustum with n=80 is its probable primary ancestor. It is primary source of sugar (sucrose). Its use is indispensible for human consumption. The various uses are obvious. It also finds a place in various industrial products.*

1. Introduction: It is supposed that sugarcane have originated in the South Pacific (probably New Guinea). It spread throughout most of southeast Asia at an early date. It is well known to have been cultivated in India since pre-historic times and was an important crop there by the end of the 4th century BC. During the invasion of India (327 BC.), Alexander's army found the local people obtaining 'honey' from reeds without the aid of bees. The methods of growing cane and making sugar reached east to Indochina and west to Arabian countries and Europe. Sugarcane was being cultivated in China before the 1st century BC. It reached Persia (Iran) in the beginning of the 6th century AD. from where it was carried to Egypt (AD. 641) and Spain (AD. 755) by the Arabs. Shortly after AD. 1400, sugarcane was introduced to Madeira and the Azores in the Atlantic ocean. Columbus is credited with having introduced it into the New World from the Canary Islands. In the 15th and 16th centuries, the Portuguese and Spanish explorers, colonisers and merchants introduced sugarcane throughout the world, reaching Santo Domingo and Hispaniola (now the Dominican Republic) by 1509, Mexico by 1520, Brazil by 1532 and Peru by 1533. Sugarcane was introduced into Louisiana and Florida about the middle of the 18th century and was brought from the West Indies.

In the history of sugarcane, major event was the importation of thick-stemmed varieties of *S. officinarum* from Tahiti to Jamaica in 1791 by Captain Bligh, known for the famous mutiny on the Bounty. After a few years of their introduction, these new forms 'Bourbon' or 'Lahaina' replaced the thin varieties belonging to two closely related species *S. sinense* Roxb. and *S. barberi* Jeswiet. The newly introduced varieties spread throughout the world widely.

When the British Navy blockaded supplies of raw cane sugar from the West Indies to France, Napoleon issued a decree on 25th March, 1811, subsidizing the establishment of a sugar beet industry. He declared that as far as France was concerned the British Government could dump its sugar from the West Indies into river Thames and that they did not need it. He was subjected to much ridicule him squeezing sugarbeet into his cup of coffee; another picture showed the nurse offering the root to Napoleon's little son, the king of Rome, with the caption, 'Suck deer, suck, your father says, "It is sugar".

Charaka wrote the oldest treatise extant on Hindu medicine. The term 'Sarkara' appears to have given origin to the names other foreign countries. The name 'Sugar' is derived from the Sanskrit word 'Sarkara', meaning gravel or sand. The earliest crude sugar made from the juice of the sugarcane was like sand. This original name changed, during its journey, the 'sukkar' in Arabic, 'Sakharon' in Greek, 'Sucre' in French, and finally to 'Sugar' in English.

However, there are several sources of information that carry the Hindu knowledge in cane sugar considerably further back than can perhaps be shown for any other nation. The sugar cane plant and the cane-sugar are frequently mentioned in the Institutes of Manu. These Institutes were begun two or three centuries before the advent of Christ. It may thus convinced that the reference to a field of sugarcane, in Manusmriti, is more than 2000 years old. It may thus be accepted that sugarcane was very generally cultivated in India, during the time the Institutes of Manu were written and so sugar-cane has been cultivated in India for at least 3,000 years. In the 'Ain-o-Akbari' (Abul Fazl, 1590 AD.) description of the methods of cultivation, manufacture of all forms of sugar; and the distillation of spirits from it are given in detail.

The Spaniards carried the cultivation and manufacture of sugarcane to the Canary Islands in the 15th century but prior to that in 1420 the Portuguese had conveyed it from Sicily to Madeira and to St. Thomas Island. In 1506 it was taken from the Canary Islands to San Domingo. The Dutch first established sugar works in Brazil in 1580, and then they carried the art of sugar manufacture to the West Indies in 1655. Sugar was manufactured by the English in Barbadoes in 1643 and in Jamaica in 1664. Sugarcane was first introduced into the United States of America in Louisiana in 1741.

2. Plant Characteristics: A rapid growing perennial grass, reaching a height of 8 to 12 ft. and 2 in. in diameter; grows in clumps with bamboo like, noded stems, stem solid with tough rind and numerous fibrous strands, containing sugary juice, colour of stem ranging from green to purple-red or violet; leaves distichous, lamina long 1 to 1. 5 m in length, 2. 5 to 10 cm wide, flowers in silvery-white terminal feathery panicles.

3. Parts Used: Culms (Stem).

4. Chemical Constituents: Sucrose.

5. Uses And Applications: Sugar is an essential food. It is used variously *e.g.* in the manufacture of alcoholic beverages, soft drinks, chocolates, ice-cream and confectionery. It is used in canned fruits, jam and jellies. Baking industry use it as a sweetening agent. It is also employed in medicines, tanning of leather, silvering of mirrors, making adhesives, explosives, etc.

6. Karyotaxonomic Evidence: Members of the genus *Saccharum* themselves are high polyploids (2n=2x=20) and diploids are unknown. Great intraspecific variability of chromosome number and much aneuploidy are characteristic (Price, 1963, 1965) as dilated below:

(i) *S. robustum*: Chromosomes 2n=60, 80 are frequent but putative interspecific hybrids with numbers in the range 63 to about 200 occur. Wild species are distributed from Borneo to the New Hebrides with much variability in New Guinea. This is a 'hybrid species' derived from natural crossing between *S. spontaneum* and *Miscanthus floridulus* in the New Guinea area (Daniels, 1973).

(ii) *S. spontaneum:* Chromosomes range 2n=40 to 128. Nodes occur at 2n=64, 80, 96, 112, 120, with the first centred in India, the last in Africa and the middle three nodes in southeast Asia from New Guinea to Taiwan. A very variable wild species distributed from Africa to the Solomon Islands and Japan, with a centre of diversity in India.

(iii) *S. officinarum:* Chromosomes are 2n=80 (with a few, probably hybrid, exceptions). These are the cultivated 'noble' canes distributed throughout the tropics.

(iv) *S. sinens:* Chromosomes are 2n=82-124. These are old cultivated canes and are natural hybrids of *officinarum* and *spontaneum* chromosome numbers broadly correspond with expectations based on putative parentage *Saccharum barberi.*

Wild spontaneum *X robustum* hybrids have been recognized with 2n=80-101. Also intergeneric *Saccharum X Miscanthus* hybrids are recognized with very high numbers in the range 2n=114-about 205.

7. History: The essential early steps of evolution in sugarcane took place in or near New Guinea. On grounds of morphology and chromosome number, *S. robustum* (2n=80) is the probable primary ancestor. Human selection would have been for characteristics such as chewing plant and in favour of clones having a sweet juice and low fibre content. The New Guinea area is still the primary centre of diversity of 80-chromosome *officinarum* clones. Subsequent evolution followed migration of *officinarum* ('noble') canes northwestwards to continental Asia where they hybridized with local *S. spontaneum* to produce the hybrid clones grouped under *S. sinense*. These, the 'thin' canes, grow in the seasonal monsoon climates of the north-east India-southern China area. They became the foundation of local syrup production The noble canes also migrated eastwards across the Pacific. Somatic mutation has contributed materially to diversity near the centre of origin. The crop was known to Europeans in classical times only by repute. It reached the Mediterranean after the Moorish conquest of Spain. Early European-controlled cultivations of cane in the tropics used the Creole cane, a clone primary available then.

8. Further Developments: At the end of 19th century, cane breeding began in Java and Barbados simultaneously. The need for new varieties was the sugar-producing potential of the noble canes and the hardiness, disease resistance and ratooning capacity of the hybrid *sinense* types. The first interspecific crosses were made as early as 1893 in Java but success with this approach did not arrive until the 1920. The key event in cane breeding was the production of POJ 2878 in 1921 in Java. It was the first of the so-called 'nobilized' canes, These canes are all essentially derivatives of *S. spontaneum* back crossed to noble types.

Cytological studies by Bremer (Bremer, 1961-63) revealed a very curious situation which was not fully understood. The cross noble (female, 2n=80) X wild *spontaneum* (male) yielded progeny having the somatic complement of the female parent (80 chromosomes) plus the gametic number of the male. The reciprocal cross, gave the expected number. This behaviour implied some sort of restitution on the female side and also intense selection for the products of restitution. Selfs or crosses among nobles yielded only 80-chromosome progeny. Cytology of the restitution process was uncertain. The fact of female restitution was confirmed. Its effect was to increase chromosome numbers and to hasten phenotypic convergence on the recurrent parent. Restitution occurs for two generations but not for the third. So, there was a limit to increase of chromosome numbers. Actual pedigrees (*e.g.* Bremer, 1962; Price, 1963) are complicated by the presence of some natural hybrid parents, by uncertainties and by some aneuploidy but the broad pattern was clear. The proportions of wild chromosomes in the backcross generations were somewhat over half what would be expected in the absence of restitution. The canes with chromosome numbers in the range 100-125 and a bout 5-10% wild contribution were what was wanted. Modern canes were derived from extensive recombination among materials of this kind throughout the world.

Economic characters usually have a fairly high additive component in reasonable justification of the usual policy of exploiting proven parents in a fairly wide range of crosses. This was indicated by biometrical genetic studies (Brown, Daniels and Latter, 1968, 1969). Criteria of selection were such as sugar yield qualified by various disease resistances, rationing capacity, freedom from flowering and good plant habit in relation to harvesting. Many intergeneric hybrids have been recorded, between *Saccharum* and *Erianthus, Miscanihus, Miscanthidium, Narenga, Imperata, Sorghum, Sclerostachya* and *Zea*. But they have had no practical significance.

Sugar canes are day-length sensitive and flower in response to light periods somewhat greater than 12 hours. In Cultivation, flowering is undesirable (Coleman, 1969). It hinders growth and may lead to 'pithiness' and late-tillering; at middle latitudes intense selection against flowering is therefore practised. For the cane breeder, flowering is, of course, essential. It is promoted by careful choice of site for breeding station supplemented by photoperiod control for refractory clones.

9. Epilogue: Selection of new clones from large populations of nobilized seedlings with chromosome numbers in the range 100-125 appear well established and most unlikely to change. Biometrical sophistication, however, allied to computerized data handling, is surely yielding a steady improvement in efficiency (Brown *et al.,* 1968-71). Non-flowering mutants of otherwise desirable canes that flower too much have already been induced. Criteria of selection also appeared to be essentially stable, though there is an increasing emphasis on juice quality, time of maturity and aptitude for mechanical harvesting. The most profitable cane is not the one with highest sugar yield but rather that which, at high yield, has a sweet juice, thus minimizing costs of harvesting non-sugars.

It has been the general realization that the genetic base is very narrow (Arcenaux, 1967). A few clones recur in nearly every pedigree. The modern canes are founded on perhaps 20 nobles and fewer than 10 *spontaneum* or *spontaneum* derivatives. Some 30 *spontaneum* lines have been introduced to breeding programmes. A mass selection approach to noble improvement was initiated in Barbados.

References:

Alexander, J. and D.G. Coursey (1969) The origins of yam cultivation. *In:* Ucko and Dimbleby (eds.). The Domestication And Exploitation Of Plants And Animals. (Ed. P.J. Ucko and G.W. Dimbleby) Aldine Publishing Co. Chicago, USA.

Anonymous (1972) Primer simposio internacional sobre Dioscoreas. Publ. Esp. Inst, Nal. Invest. For., Mexico, 8, pp. 130.

Arcenaux, G. (1967) Cultivated sugarcanes of the world and their botanical derivation. Proc. Int. Soc. Sugar Cane Tech. Congr., Puerto Rico (1965), 12: 844-854.

Ayensu, E.S. and D.G. Coursey (1972) Guinea yams. *Econ. Bot.* 26:301-318.

Bremer, G. (1961-63) Problems in breeding and cytology of sugarcane. *Euphytica* 10: 59-78, 121-133, 229-243, 352-342; 11:65-80; 12:178-188.

Brown, A.H.D., Daniels, J. and B.H.D. Latter (1968, 1969) Quantitative genetics of sugarcane, I-III. *Theor. appl. Genet.* 38:361-369; 39:1-10, 79-87.

Brown, A.H.D., Daniels, J. and N.D. Stevenson (1971) The mass selection reservoir and sugarcane selection. *Theor. appl. Genet.* 41:174-180.

Burkill, I.H. (1960) The organography and the evolution of the Dioscoreaceae, the family of the yams. *J. Linn. Soc. L (Bot.)* 56:319-420.

Campbell, G.K.G. (1953) Selection of sugar beet for resistance to bolting. Publ. Inst, internl Rech. better. XVI Winter Congr. pp. 5.

Chair Hana, Julie Sardos, Anthea Supply, Pierre Mournet, Roger Malapa and Vincent Lebot (2016) Plastid phylogenetics of Occania yam (*Dioscorea* spp., Dioscoreaceae) reveals natural interspecific hybridization of the greater yam (*D. alata*). *J. Linn. Bot. Soc.* 180(3):319-333.

Coleman, R.E. (1969) Physiology of flowering in sugarcane. Proc. Int. Soc. Sugar Cane Tech. Congr. *Taiwan* (1968) 13:92-100.

Colombi Carlos, Gerand Second, Tereza Losada Valle and Andre Charrier (1998) Genet. Mol. Biol. 21(1):S1415-4557...Google Chrome.

Coons, G.H. (1954) The wild species of Beta. *Proc. Amer. Soc. Sugar Beet. Technol.* 8:142-147.

Coursey, D.G. (1967) Yams. Longmans, London, U. K.

Coursey, D.G. (1972) The civilizations of the yam: interrelationships of man and yams in Africa and the Indo-Pacific region. *Archaeol, Phys. Anthropol. Oceania* 7:215-231.

Coursey, D.G. (1976) The origins and domestication of yams in Africa. (Ed. J. R. Harlan *et al.,*). *In:* Origins of African Plant Domestication. The Hague, Netherlands, pp. 383-408.

Coursey, D.G. and F.W. Martin (1970) The past and future of yams as crop plants. Proc, 2nd intnl Symp. trap, Root Crops, Hawaii 1:87-90; 99-101.

Daniels, J. (1973) Origin of S. officinarum and S. robustum. ISSCT Sugarcane *Breeders Netosl.* 32:4-13.

Dark, S.O.S. (1971) Experiments on the cross-pollination of sugar beet in the field. *J. natnl Inst, agric. Bot.* 12:242-266.

De Vries, C.A., Ferwerda, J.D. and M. Flach (1967) Choice of food crops in relation to actual and potential production in the tropics. *Neth. J. Agric. Sci.* 15:241-248.

Engel, F. (1970) Exploration of the Chilca Canyon, Peru. Current *Anthropol.* 11:55-58.

Harlan, J.R. (1971) Agricultural origins: centres and noncentres. *Science,* N.Y. 154:468-474.

Hornsey, K.G. (1975) The exploitation of polyploidy in sugar beet breeding. *J. Agric. Sci. Camt.* 84:543-557.

House, H.D. (1908) The North American species of the genus Ipomoea. *Ann. N.Y. Acad. Sci.* 18:181-263.

Jones, A . (1965) A proposed breeding procedure for sweet potato. *Crop Sci.* 5:191-192.

Jones, A. (1967) Should Nishiyama's K123 (Ipomoea trifida) be designated I. Batatas? *Econ. Bot.* 21:163-166.

Krasochkin, V.T. (1959) Review of the genus Beta. B*ull. appl. Bot. Genet. Pl. Breeding,* Leningrad 32:3-36.

Lathrap, D.W. (1973) The antiquity and importance of long-distance trade relationships in the moist tropics of Pre-Columbian South America. *World Archaeology* 5:170-186.

Lomax, A. and N. Berkowitz (1972) The evolutionary taxonomy of culture. *Science,* N.Y. 177:228-239.

Magoon, M.L., Krishnan, R. and K.V. Bai (1969) Morphology of the pachytene chromosomes and meiosis in Manihot esculenta. *Cytologia* 34:612-624.

Martin, F.W. (1965) Incompatibility in the sweet potato. *A review. Econ. Bot.* 19:406-415.

Martin, F.W. (1968) The system of self-incompatibility in Ipomoea. *J. Hered.* 59: 262-267.

Martin, F.W. and A. Jones (1972) The species of Ipomoea closely related to the sweet potato. *Econ. Bot.* 26:201-15.

Nishiyama, I. (1971) Evolution and domestication of the sweet potato. *Bot. Mag., Tokyo* 84:377-287.

Oomen, H.A.P.C., Spoon, W., Heesterrnan, J.E., Ruinard, J., Luyken, R. and P. Slump (1961) The sweet potato as the staff of life of the highland Papuan. *Trop. Geogr. Medicine* 13:55-66.

Owen, F.V. (1945) Cytoplasmically inherited male sterility in sugar beets. *F. agric. Res.* 71:423-440.

Price, S. (1963) Cytogenetics of modern sugar cane. *Econ. Bot.* 17:97-106.

Price, S. (1965) Cytology of Saccharum robust urn and related sympatric species and natural hybrids. USDA *agric. Res. Servo tech. Bull.* 1337:47.

Reiche1-Dolmantoff, G. (1965) Colombia: Ancient peoples and places. Vol. 44. Frederick A. Praeger Publishers, London, UK.

Riley, C.L., Rielley, J.C., Pennington, C.W. and R.L. Rands (Ed.). (1971) Man Across The Sea: Problems of Pre-Columbian Contact. University of Texas, Austin and London.

Rogers, D.J. (1963) Studies of Manihot esculenta and related species. Bull. Torrey *Bot. Club* 90:43-54.

Rogers, D.J. (1965) Some botanical and ethnological considerations of Manihot esculenta. *Econ. Bot.* 19:369-377.

Rogers, D.J. and S.G. Appan (1970) Untapped genetic resources for cassava improvement. Proc. Znd intnl. Symp. Tropical Root and Tuber Crops, Hawaii, pp. 72-75

Rogers, D.J. and S.G. Appan (1973) Manihot, Manillotaides (Euphorbiaceae), Flora Neotropica. Monograph 13, Hafner Press, New York, USA.

Savitsky, V.F. (1954) Inheritance of the number of flowers in flower clusters of Beta vulgaris. *Proc. Amer. Soc. Sugar Beet Technol.* 8:3-15.

Savitsky, V.F. (1962). Sucrose and weight of root in tetraploid monogerm and multigerm sugar beet populations under different mating systems. *F. Amer. Soc. Sugar Beet Technol.* 11:676-711.

Sebsebe Demissew, Fay, M.F., Smith, R.J., Nordal P and P Willkin (2013) Genetic diversity and species delimitation in the cultivated and wild yams ALFP (Dioscorea spp.) from Southwest Ethiopea as determined by markers. *Genetic Resources And Crop Evolution* 60(4):1365-1375.

Sedlmayr, K. (1964) Monogerme Zuckerriiben, ihre Genetik, Zuchtung und Bedeutung fur den Zuckerriibenbau, *Ziichter* 34:45-51.

Ting, Y. C. and A.E. Kehr (1953) Meiotic studies in the sweet potato. *J. Hered.* 44: 207-211.

Towle, M. (1961) The ethnobotany of pre-Columbian Peru. Viking Fund Publications In Anthropology No. 30, New York, USA.

Van Ooststroom, S.J. and R.D. Hoogland (1953) Convolvulaceae, *In:* Flora Malesiana Ser. I, 388-512 Nordhoff-Kolff N.V. Djakarta.

Wang Min and I.R. Goldman (1999) Genetic distance and diversity in table beet and sugar beet accessions measured by randomly amplified polymorphic amplified polymorphic DNA. *J. Amer. Soc. Hort. Sci.* 124(6):630-635.

Yen, D.E. (1974) The sweet potato and Oceania. B. P. Bishop Museum Bulletin, 236, Honolulu.

Zhang Dapeng, Jim Cervantes, Zosimo Hauman, Edward Carey and Marc Ghislain (2000) Assessing genetic diversity of sweet potato [Ipomoeas batatas (L.) Lam.] cultivars from tropical America using AFLP. *Genetic Resources And Crop Evolution* 47(6):659-665.